Lecture Notes in Mathematics

Edited by A. Dold and B. Eckmann

1242

Functional Analysis II

with Contributions by J. Hoffmann-Jørgensen et al.

Edited by S. Kurepa, H. Kraljević and D. Butković

Springer-Verlag
Berlin Heidelberg New York London Paris Tokyo

Editors

Davor Butković
Department of Applied Mathematics
Electroengineering Faculty
Unska 3
41000 Zagreb
Yugoslavia

Svetozar Kurepa
Hrvoje Kraljević
Department of Mathematics
P.O.Box 187
41001 Zagreb
Yugoslavia

Mathematics Subject Classification (1980): 10C05, 28A35, 28C20, 39B50, 40C05, 46C05, 46D05, 47A20, 47B37, 47B50, 60A10, 60J10, 60J25, 60J45

ISBN 3-540-17833-3 Springer-Verlag Berlin Heidelberg New York
ISBN 0-387-17833-3 Springer-Verlag New York Berlin Heidelberg

Printed in Germany

Printing and binding: Druckhaus Beltz, Hemsbach/Bergstr.
2146/3140-543210

FOREWORD

This volume contains lecture notes given at postgraduate school and conference on Functional Analysis held from November 3 to November 17, 1985, at the Inter-University Center of Postgraduate Studies, Dubrovnik, Yugoslavia.

The lectures and communications were devoted to several parts of functional analysis but centered mainly on functional-analytic aspects of probability theory and operator theory on Hilbert and Banach spaces. There were six series of lectures:

1. Subspaces and operators in Krein spaces, characteristic functions and related topics, given by A.Dijksma (University of Groningen), H.Langer (Technische Universität Dresden) and H.S.V. de Snoo (University of Groningen);

2. Quadratic and sesquilinear forms, given by S.Kurepa (University of Zagreb) and J.Vukman (University of Maribor);

3. Measure-theoretical topics in probability; the general marginal problem, given by J.Hoffmann-Jørgensen (Aarhus Universitet);

4. Markov processes and potential theory, given by Z.R.Pop-Stojanović and Murali Rao (University of Florida, Gainesville);

5. Unitary representations of the groups GL(n), given by H.Kraljević and M.Tadić (University of Zagreb);

6. On representations of the Heisenberg groups, given by M.Duflo (Universitè Paris VII).

This volume contains the somewhat enlarged lecture notes of the first four series of lectures. Besides these the volume contains three papers connected with some one-hour lectures.

We use this opportunity to express our thanks to the institutions whose financial support made the conference possible. These are: Samoupravna interesna zajednica za znanstveni rad SRH - SIZ I; Savez

R/P SIZ-ova za naučnu djelatnost SFRJ; Department of Mathematics, University of Zagreb.

We are also grateful to the Inter-University center of Postgraduate Studies in Dubrovnik, where the lectures and the conference were held.

Finally, the authors are grateful to Springer-Verlag for its prompt publication of these proceedings.

S.Kurepa
H.Kraljević
D.Butković

CONTENTS

ADDRESSES OF THE AUTHORS

Davor BUTKOVIĆ, Department of Applied Mathematics, Electroengineering Faculty, Unska 3, 41000 Zagreb, Yugoslavia

Aad DIJKSMA, Department of Mathematics, University of Groningen, 9700 AV Groningen, The Netherlands

Neven ELEZOVIĆ, Department of Applied Mathematics, Electroengineering Faculty, Unska 3, 41000 Zagreb, Yugoslavia

Jørgen HOFFMANN-JØRGENSEN, Aarhus Universitet, Mathematisk Institut Ny Munkegade, DK-8000, Aarhus C, Denmark

Hrvoje KRALJEVIĆ, Department of Mathematics, University of Zagreb, Marulićev trg 19, 41000 Zagreb, Yugoslavia

Svetozar KUREPA, Department of Mathematics, University of Zagreb, Marulićev trg 19, 41000 Zagreb, Yugoslavia

Heinz LANGER, Technische Universität Dresden, Sektion Mathematik, Mommsenstrasse 13, 8027 Dresden, DDR

Zoran R.POP-STOJANOVIĆ, Department of Mathematics, University of Florida, Gainesville, Fl 32611, USA

Nikola SARAPA, Department of mathematics, University of Zagreb, Marulićev trg 19, 41000 Zagreb, Yugoslavia

Hendrik S.V.de SNOO, Department of Mathematics, University of Groningen, 9700 AV Groningen, The Netherlands

Salih SULJAGIĆ, Department of Mathematics, Civil Engineering Institute, J.Rakuše 1, 41000 Zagreb, Yugoslavia

UNITARY COLLIGATIONS IN KREIN SPACES AND THEIR ROLE IN THE EXTENSION THEORY OF ISOMETRIES AND SYMMETRIC LINEAR RELATIONS IN HILBERT SPACES.

Aad Dijksma, Heinz Langer, Henk de Snoo*)

INTRODUCTION

Let H be a Hilbert space and let S be a closed linear relation in H, i.e., S is a subspace of H^2 such that $S \subset S^* \subset H^2$ (for the definition of S^* and other definitions see Section 1). Furthermore, let K be a Krein space such that $H \subset K$ and the Krein space inner product on K coincides on H with the Hilbert space inner product on H; we denote this situation by $H \underset{s}{\subset} K$. In this case H is an orthocomplemented subspace of K (see [4]). The corresponding orthogonal projection from K onto H is denoted by P_H. We consider a selfadjoint relation A in K with a nonempty resolvent set $\rho(A)$, such that $S \subset A$, that is, A is a selfadjoint extension of S in K. We define $P_H^{(2)}A = \{\{P_H f, P_H g\} \mid \{f,g\} \in A\}$; it is clear that $S \subset P^{(2)}A$. With A we associate the socalled Štraus extension T of S in H, which is by definition $T = (T(\ell))_{\ell \in ¢ \cup \{\infty\}}$, where $T(\ell)$ is given by

$$(0.1)\qquad \begin{cases} T(\ell) = \{\{P_H f, P_H g\} \mid \{f,g\} \in A,\ g - \ell f \in H\}, & \ell \in ¢, \\ T(\infty) = \{\{\ f, P_H g\} \mid \{f,g\} \in A,\ f \in H\}. & \end{cases}$$

It is clear that $T(\ell) \subset T(\bar{\ell})^*$, $\ell \in ¢$, and $T(\infty) \subset T(\infty)^*$. In particular, $T(\ell)$, for real values of ℓ, and $T(\infty)$ are symmetric. Moreover, the following relations are easy to verify

$$(0.2)\qquad S \subset A \cap H^2 = (P_H^{(2)}A)^* \subset T(\ell) \subset P_H^{(2)}A \subset S^*, \quad \ell \in ¢ \cup \{\infty\},$$

and $(T(\ell) - \ell)^{-1} = \{\{g - \ell f, P_H f\} \mid \{f,g\} \in A, g - \ell f \in H\}$, $\ell \in ¢$. Since we suppose that $\rho(A) \neq \emptyset$, for $\ell \in \rho(A)$ the resolvent $R_A(\ell) = (A - \ell)^{-1} = \{\{g - \ell f, f\} \mid \{f,g\} \in A\}$ belongs to $L(K)$ and hence we obtain that $(T(\ell) - \ell)^{-1} = P_H R_A(\ell)|_H$, $\ell \in \rho(A)$. Here we denote by $L(K)$ the set of bounded linear operators in K; $L(K_1, K_2)$ is the set of all bounded linear operators from the Krein space K_1 into the Krein space K_2. The function $\ell \to P_H R_A(\ell)|_H$, $\ell \in \rho(A)$, with values in $L(H)$ is the socalled generalized resolvent of S associated with the extension A.

In this paper we describe for a given symmetric linear relation S in a Hilbert space H the selfadjoint extensions A (with nonempty resolvent sets) in Krein spaces

---

*) This work was supported by the Netherlands Organization for the Advancement of Pure Research (Z.W.O.).

$K \underset{s}{\supset} H$ in terms of the corresponding Štraus extensions of S in H. This description is given by means of characteristic functions which are holomorphic mappings between the defect subspaces of S. The basic result needed here is the fact that every function θ holomorphic at 0 and with values in $L(F,G)$, where F and G are Hilbert spaces, is the characteristic function of a unitary colligation with some Krein space as inner space. This result was first proved (in a more general form) by T.Ya Azizov [2], [3]. It leads in an easy way to the description of all unitary extensions U in Krein spaces of a given isometric operator V in a Hilbert space. (See also Azizov [3].) It turns out that with each such pair V,U, there can be associated three colligations. Their characteristic functions play an essential role in our description of unitary extensions. The extension problem for symmetric subspaces formulated above can then be solved in an easy way by means of the Cayley transformation. For the case that the larger space K is a π_κ - space instead of a general Krein space we have considered this problem in [9], [10] and [11] (compare also [16], [19]). The holomorphic functions, which are involved there, are characterized by the fact that a certain kernel associated with them has a finite number κ of negative squares. It turns out that for the more general extension problem with K being a Krein space, the holomorphy is already sufficient.

As an illustration of our results we consider in the Hilbert space $H = L^2(0,\infty)$ the maximal differential operator S^* associated with the differential expression $-(\frac{d}{dx})^2 + q(x)$, where q is a real valued locally integrable potential. We suppose that at ∞ the limit point case holds, i.e., that we only have to require a boundary condition at 0. Then each Štraus extension is described by two holomorphic functions α and β defined on an open set $U \subset \mathbb{C} \setminus \mathbb{R}$, which is symmetric with respect to the real axis $\mathbb{R}$, and satisfying $|\alpha(\ell)| + |\beta(\ell)| \neq 0$, $\overline{\alpha(\bar\ell)}\beta(\ell) = \alpha(\ell)\overline{\beta(\bar\ell)}$, $\ell \in U : T(\ell) = \{\{f, -f'' + qf\} | f \in D(S^*)$, $\alpha(\ell)f(0) + \beta(\ell)f'(0) = 0\}$. The corresponding extension space $K \underset{s}{\supset} H$ is a π_κ - space if and only if the kernel $(\ell - \bar\lambda)^{-1}(\alpha(\ell)\overline{\beta(\lambda)} - \beta(\ell)\overline{\alpha(\lambda)}$, $\ell,\lambda \in U$ and $\ell \neq \bar\lambda$, has κ negative squares. In particular, the extension space is a Hilbert space if and only if $(\mathrm{Im}\ \ell)^{-1}\mathrm{Im}\ \alpha(\ell)\overline{\beta(\ell)} \geq 0$, $\ell \in U$.

We briefly outline the contents of this paper. In Section 1 we collect the preliminary facts about linear relations. In Section 2 we consider unitary colligations and their characteristic functions. We give a proof of Azizov's theorem (see Theorem 2.2) mentioned above which is more direct than Azizov's proof. We give a reproducing kernel type of construction and apply a fundamental statement of M.G. Krein about operators in spaces with two norms, see Lemma 1.1 in Section 1. Unitary extensions in Krein spaces of an isometry V in a Hilbert space H are considered in Section 3. The main result there is the description of the generalized coresolvents of V corresponding to unitary extensions in terms of a characteristic function of a unitary colligation, see Theorem 3.2. A proof of this theorem can be given by means of Theorem 1 of [12], but here we base the proof on Theorem 2.2. In Section 4 we discuss symmetric linear relations, their characteristic functions defined on $\mathbb{C} \setminus \mathbb{R}$ and the boundary behaviour of these functions on the real line. Štraus extensions and generalized resolvents of a symmetric linear relation in a Hilbert space are characterized in Section 5 in the way described above. These characterizations are deduced from the results in Sections 2 and 3 by means of the Cayley transformation. As this transformation is defined by means of a fixed point $\mu \in \mathbb{C} \setminus \mathbb{R}$, they depend on this parameter. All selfadjoint extensions considered here have the property that μ belongs to their resolvent sets. In Section 6 we calculate the characteristic function of the minimal symmetric linear relation associated with a canonical differential system on a compact interval and the minimal symmetric operator associated with a Sturm-Liouville differential expression and their Štraus extensions. Finally, in Section 7 we show how A.V. Štraus' formal definition of a characteristic function, see [29], is related to the characteristic function of a unitary colligation. In each section we give additional references to papers containing results related to the ones treated in that section.

At the Conference on Functional Analysis in Dubrovnik, November 3-17, 1985, we gave six lectures on "Subspaces and operators in Krein spaces, characteristic functions and related topics", in which, besides the results of this paper, we presented also results from [9], [10], [11] and [12]. These lectures were of an expository character. The present notes contain a somewhat extended version of the material which we discussed at the conference and include new results and their proofs. The authors wish to express their gratitude to the organizers of the conference for the opportunity given to deliver these lectures and for their hospitality.

1. PRELIMINARY RESULTS

Let K be a Banach space over $\mathbb{C}$ and let K^2 be the product space consisting of all pairs $\{f,g\}$, $f,g \in K$, with the linear structure defined by $\alpha\{f,g\} + \beta\{h,k\} = \{\alpha f + \beta h, \alpha g + \beta k\}$, $f,g,h,k \in K$, $\alpha,\beta \in \mathbb{C}$, and provide K^2 with the usual topology. A linear relation A in K is a linear manifold in K^2, $A \subset K^2$. For linear relations A and B in K we define

$$D(A) = \{f \in K \mid \{f,g\} \in A \text{ for some } g \in K\}, \text{ the domain of } A,$$
$$R(A) = \{g \in K \mid \{f,g\} \in A \text{ for some } f \in K\}, \text{ the range of } A,$$
$$\nu(A) = \{f \in K \mid \{f,0\} \in A\}, \text{ the null space of } A,$$
$$A(0) = \{g \in K \mid \{0,g\} \in A\}, \text{ the multivalued part of } A,$$
$$A^{-1} = \{\{g,f\} \mid \{f,g\} \in A\}, \text{ the inverse of } A,$$
$$AB = \{\{f,g\} \mid \{f,h\} \in B, \{h,g\} \in A \text{ for some } h \in K\},$$
$$A + B = \{\{f,g+h\} \mid \{f,g\} \in A, \{f,h\} \in B\},$$
$$A \dotplus B = \{\{f+h,g+k\} \mid \{f,g\} \in A, \{h,k\} \in B\}.$$

The sum $A \dotplus B$ is called direct if $A \cap B = \{\{0,0\}\}$. A linear relation A in K is (the graph of) an operator if and only if $A(0) = \{0\}$. If $A(0) = \{0\}$ we use the notation $g = Af$ to denote $\{f,g\} \in A$. The identity operator in K is denoted by I and we write $\lambda A, A + \lambda, \lambda \in \mathbb{C}$, instead of $(\lambda I)A$ and $A+\lambda I$.

For a linear relation A in K we define the set $\gamma(A)$ of points of regular type by $\gamma(A)=\{\lambda\in\mathbb{C}\,|\,(A-\lambda)^{-1}$ is a bounded operator$\}$ and the resolvent set $\rho(A)$ by $\rho(A)=\{\lambda\in\gamma(A)\,|\,R(A-\lambda)$ is dense in $K\}$. The sets $\gamma(A)$ and $\rho(A)$ are open. A linear relation A in K is called a subspace if it is closed in K^2. Any linear relation A has a closure which will be denoted by A^c and we have $\gamma(A)=\gamma(A^c)$. We note that A is closed if and only if $R(A-\lambda)$ is closed for some (and hence for all) $\lambda\in\gamma(A)$. For a subspace $A\subset K^2$ with nonempty resolvent set we define the resolvent operator $R_A:\rho(A)\to L(K)$ by $R_A(\lambda)=(A-\lambda)^{-1}$, $\lambda\in\rho(A)$. It satisfies $R_A(\lambda)-R_A(\mu)=(\lambda-\mu)R_A(\lambda)R_A(\mu),\lambda,\mu\in\rho(A)$, and hence R_A is holomorphic on $\rho(A)$.

For a linear relation A in K and $\mu\in\mathbb{C}$ we define the Cayley transform $C_\mu(A)$ and the inverse Cayley transform $F_\mu(A)$ by $C_\mu(A)=\{\{g-\mu f,g-\bar{\mu}f\}\,|\,\{f,g\}\in A\}$ and $F_\mu(A)=\{\{g-f,\mu g-\bar{\mu}f\}\,|\,\{f,g\}\in A\}$. We have for $\mu\in\mathbb{C}\setminus\mathbb{R}$ and linear relations A and B

$$
\begin{aligned}
&A=C_\mu(F_\mu(A))=F_\mu(C_\mu(A)),\\
&A\subset B\Leftrightarrow C_\mu(A)\subset C_\mu(B)\Leftrightarrow F_\mu(A)\subset F_\mu(B),\\
&C_{\bar{\mu}}(A)=C_\mu(A)^{-1},\ F_{\bar{\mu}}(A)=F_\mu(A)^{-1},\\
&C_\mu(A\dot{+}B)=C_\mu(A)\dot{+}C_\mu(B),\ F_\mu(A\dot{+}B)=F_\mu(A)\dot{+}F_\mu(B)\\
&A \text{ closed}\Leftrightarrow C_\mu(A)\text{ closed}\Leftrightarrow F_\mu(A)\text{ closed.}
\end{aligned}
$$

If A is closed and if $\mu\in\rho(A)$, then $C_\mu(A)\in L(K)$ and $C_\mu(A)=I+(\mu-\bar{\mu})R_A(\mu)$. For any relation A in K we have the formal identity

$$(1.1)\qquad (C_\mu(A)-\frac{\lambda-\bar{\mu}}{\lambda-\mu})^{-1}=\frac{\mu-\lambda}{\mu-\bar{\mu}}(I+(\lambda-\mu)(A-\lambda)^{-1}),\ \mu\in\mathbb{C}\setminus\mathbb{R},\ \lambda\neq\mu.$$

Hence if A is closed and $\mu\in\rho(A)\setminus\mathbb{R}$, then $\frac{\lambda-\bar{\mu}}{\lambda-\mu}\in\rho(C_\mu(A))$ if and only if $\lambda\in\rho(A)$.

We recall the definition of a Krein space. Let K be a linear space over $\mathbb{C}$ and suppose that $[\ ,\ ]:K\times K\to\mathbb{C}$ is an inner product (i.e., a sesquilinear form) on K, which is non-degenerated. Then $(K,[\ ,\ ])$ is a Krein space if

$$(1.2)\qquad K=K_+ + K_-,$$

where $K_\pm\subset K_-$ are linear manifolds such that $(K_\pm,\pm[\ ,\ ]|_{K_\pm\times K_\pm})$ are Hilbert spaces and $[K_+,K_-]=\{0\}$. In a Krein space K the decomposition (1.2) is a direct sum; it is called a fundamental decomposition. Let $P_\pm:K\to K_\pm$ be the corresponding projections and put $J=P_+-P_-$. Then $(\ ,\ ):K\times K\to\mathbb{C}$ defined by $(f,g)=[Jf,g]$ is a positive definite inner product on K and $(K,(\ ,\ ))$ is a Hilbert space. The operator J is called the fundamental symmetry corresponding to the fundamental decomposition (1.2). In general, there are many fundamental decompositions. If K is decomposed as in (1.2) and $K=K'_+ + K'_-$ is another fundamental decomposition with corresponding projections $P'_\pm$ and symmetry J', then dim $K_\pm$ = dim $K'_\pm$ and the norms corresponding to (,) and (,)', defined by (f,g)' = [J'f,g], are equivalent and hence generate the same topology. All topological notions on a Krein space are defined with respect to this norm topology. The Krein space is called a Pontryagin space of index κ or π_κ - space if $\kappa = \dim K_- < \infty$. We shall denote orthogonal complements with respect to [,] by [⊥] and orthogonal sums and differences by [+] and [-] (or ⊕ and ⊖ in the Hilbert space case).

Now suppose K is a Krein space and provide K^2 with the inner product defined by $[\{f,g\},\{h,k\}] = [f,h] + [g,k], \{f,g\},\{h,k\} \in K^2$. For a set $A \subset K^2$ we define its adjoint A^+ by $A^+ = \{\{f,g\} \in K^2 \mid [g,h] = [f,k] \text{ for all } \{h,k\} \in A\}$. If K is a Hilbert space we shall denote the adjoint of A by A^*. Note that A^+ is a subspace and $A^+ = JA^*J$, where J is a fundamental symmetry and A^* is the adjoint of A in the Hilbert space $(K,[J.,.])$. For linear relations A and B in K we recall that

$$A \subset B \Rightarrow B^+ \subset A^+,$$
$$A^{++} = A^c, \ (A^c)^+ = A^+,$$
$$(A+\lambda)^+ = A^+ + \bar{\lambda}, \ (\lambda A)^+ = \bar{\lambda} A^+, \qquad \lambda \in \mathbb{C},$$
$$(A^+)^{-1} = (A^{-1})^+,$$
$$R(A)^{[\perp]} = \nu(A^+), \ D(A)^{[\perp]} = A^+(0),$$
$$R(A^+)^{[\perp]} = \nu(A^c), \ D(A^+)^{[\perp]} = A^c(0).$$

If A is a subspace, $R(A)$ is closed if and only if $R(A^+)$ is closed. In terms of Cayley transforms we have for a linear relation A in K, $(C_\mu(A))^+ = C_{\bar{\mu}}(A^+)$, $(F_\mu(A))^+ = F_{\bar{\mu}}(A^+)$, $\mu \in \mathbb{C} \setminus \mathbb{R}$.

A linear relation A in K is called

dissipative if $\operatorname{Im}[g,f] \geq 0$, $\{f,g\} \in A$,
symmetric if $\operatorname{Im}[g,f] = 0$, $\{f,g\} \in A$, or, equivalently, if $A \subset A^+$,
selfadjoint if $A = A^+$,
contractive if $[g,g] \leq [f,f]$, $\{f,g\} \in A$,
isometric if $[g,g] = [f,f]$, $\{f,g\} \in A$, or, equivalently, if $A^{-1} \subset A^+$,
unitary if $A^{-1} = A^+$.

In terms of the Cayley transform we have the following correspondences for a linear relation A in K

±A is dissipative if and only if $C_\mu(A)$ is contractive, $\mu \in \mathbb{C}_\mp$,
A is symmetric if and only if $C_\mu(A)$ is isometric, $\mu \in \mathbb{C} \setminus \mathbb{R}$,
A is selfadjoint if and only if $C_\mu(A)$ is unitary, $\mu \in \mathbb{C} \setminus \mathbb{R}$.

They follow easily from the equality $[g-\bar{\mu}f, g-\bar{\mu}f] - [g-\mu f, g-\mu f] = 4(\operatorname{Im}\mu)\operatorname{Im}[g,f]$, where $\{f,g\} \in K^2$, $\mu \in \mathbb{C} \setminus \mathbb{R}$.

In the sequel we shall be interested in selfadjoint relations A in a Krein space K with nonempty resolvent set $\rho(A)$. As an example of an A with $\rho(A) = \emptyset$ we consider a Hilbert space H and a densely defined selfadjoint operator T in H. We provide the space $K = H \times H$ with the inner product $[f,g] = (Jf,g), f,g \in K$, where $J = \begin{pmatrix} 0 & I \\ I & 0 \end{pmatrix}$. Thus K is a Krein space. In K we consider the operator $A = \begin{pmatrix} 0 & I \\ I & 0 \end{pmatrix}\begin{pmatrix} T & 0 \\ 0 & 0 \end{pmatrix} = \begin{pmatrix} 0 & 0 \\ T & 0 \end{pmatrix}$. Then $A^+ = JA^*J = A$, so that A is a densely defined selfadjoint operator in K. It is easy to see that $R(A - \ell) = D(T) \times H$, $\nu(A - \ell) = \{0\}$, $\ell \in \mathbb{C} \setminus \{0\}$, and that $R(A) = \{0\} \times R(T)$, $\nu(A) = \nu(T) \times H$. Therefore $\rho(A) = \emptyset$. Note that for $\mu \in \mathbb{C} \setminus \mathbb{R}$ the Cayley transform is a densely defined unitary operator with dense range, but it does not belong to $L(K)$. Similar remarks may be made for the selfadjoint relation A^{-1}. We shall not have an occasion to use unitary relations. In the following when we speak of a unitary operator U in K, we mean an operator U such that $D(U) = R(U) = K$ and U leaves the inner product of K invariant. If A is a selfadjoint relation in K and $\rho(A) \neq \emptyset$ then with $\mu \in \rho(A) \setminus \mathbb{R}$, $W = C_\mu(A)$ is a unitary operator. Conversely, if W is a unitary operator in K then for any $\mu \in \mathbb{C} \setminus \mathbb{R}$, $A = F_\mu(W)$ is a selfadjoint relation in K and $\mu \in \rho(A)$. For the results so far in this section and for related topics we refer to [4], [13], [14] and [18] and the references cited in these papers.

Let $(H, (\ ,\ ))$ be a Hilbert space and suppose we are given an inner product $[\ ,\ ]$ on H which is bounded, i.e., there exists a constant C such that $|[x,y]| \leq C\|x\| \|y\|$, $x,y \in H$, where $\|x\| = (x,x)^{\frac{1}{2}}$. Then there exists a bounded linear operator G on H such that $[x,y] = (Gx,y)$, $x,y \in H$, and the space H admits a decomposition $H = H_+ \oplus H_- \oplus H_0$, orthogonal sum, where $H_0 = \nu(G)$ is the isotropic subspace of H with respect to the inner product $[\ ,\ ]$ and H_+, H_- are subspaces of H invariant under G such that $[x,x] > 0 (< 0)$, $0 \neq x \in H_+ (H_-$, respectively). E.g., $H_+ = E_+ H$, the range of the spectral projection E_+ of G corresponding to the interval $(0,\infty)$. Let $(K_+, [\ ,\ ])((K_-, -[\ ,\ ]))$ be the Hilbert space completion of $(H_+, [\ ,\ ])((H_-, -[\ ,\ ]$, respectively), form the space $K = K_+ + K_-$ and provide it with the inner product $[\ ,\ ]$ defined by $[x,y] = [x_+,y_+] + [x_-,y_-]$, $x = x_+ + x_-$, $y = y_+ + y_-$, $x_\pm, y_\pm \in K_\pm$. Then $(K, [\ ,\ ])$ is a Krein space containing the space $H_+ \oplus H_-$, which can be considered as the factor space $\hat{H} = H/H_0$, and is dense in K. We shall call K the Krein space associated with H (with respect to the given bounded inner product $[\ ,\ ]$).

The following lemma is a slight generalization of a result of M.G. Krein, which was later also proved in a similar form independently by W.T. Reid [27], P. Lax [24] and J. Dieudonné [8]. For convenience of the reader we give a complete proof.

LEMMA 1.1. Let H_j be Hilbert spaces and suppose that on each H_j there is given a bounded inner product $[\ ,\ ]_j$, $j = 1,2$. Furthermore, assume that we are given linear operators $U_0 \in L(H_1, H_2)$ and $V_0 \in L(H_2, H_1)$ such that

$$[U_0 x, y]_2 = [x, V_0 y]_1, \quad x \in H_1, \ y \in H_2. \tag{1.3}$$

Then U_0, V_0 *generate linear operators* $\hat{U}_0 \in L(\hat{H}_1, \hat{H}_2), \hat{V}_0 \in L(\hat{H}_2, \hat{H}_1)$ *and these operators can be extended by continuity to operators* U,V *between the associated Krein spaces* K_1 *and* $K_2 : U \in L(K_1, K_2), V \in L(K_2, K_1)$ *and* $[Ux,y]_2 = [x,Vy]_1$, $x \in K_1, y \in K_2$.

PROOF. The elements of $\hat{H}_j$ are equivalence classes of elements of H_j; by $\hat{x} \in \hat{H}_j$ we denote the equivalence class containing $x \in H_j$, $j = 1,2$. Let $\hat{U}_0 : \hat{H}_1 \to \hat{H}_2$ and $\hat{V}_0 : \hat{H}_2 \to \hat{H}_1$ be defined by $\hat{U}_0 \hat{x} = (U_0 x)\hat{}$, $\hat{V}_0 \hat{y} = (V_0 y)\hat{}$, $x \in H_1$, $y \in H_2$. Relation (1.3) implies that $U_0 H_1^0 \subset H_2^0$ and $V_0 H_2^0 \subset H_1^0$ and it follows from these inclusions that $\hat{U}_0$ and $\hat{V}_0$ are well defined. One can easily verify that $\hat{U}_0 \in L(\hat{H}_1, \hat{H}_2)$, $\hat{V}_0 \in L(\hat{H}_2, \hat{H}_1)$ and

$$[\hat{U}_0 \hat{x}, \hat{y}]_2 = [\hat{x}, \hat{V}_0 \hat{y}]_1, \quad \hat{x} \in \hat{H}_1, \ \hat{y} \in \hat{H}_2,$$

where $[\hat{u},\hat{v}]_j = [u,v]_j, u,v \in H_j$, $j = 1,2$. From these considerations it follows that without loss of generality we may and shall assume that the isotropic subspaces H_1^0 and H_2^0 consist only of the zero element, so that now we only have to show that U_0 and V_0 can be extended by continuity to operators U and V having the properties stated in the lemma. To that end we introduce on K_j the positive definite inner product $\{x,y\}_j = [x_+, y_+]_j - [x_-, y_-]_j$, where $x = x_+ + x_-$, $y = y_+ + y_-$, $x_\pm \in K_j^\pm$, $y \in K_j^\pm$ and $K_j = K_j^+ \oplus K_j^-$ is the fundamental decomposition of K_j that corresponds to the decomposition $H_j = H_j^+ \oplus H_j^-$, $j = 1,2$. Now, for $j = 1,2$, $(K_j, \{\ ,\ \}_j)$ is a Hilbert space and it is easy to check that (1.3) implies that

$$\{U_0 x, y\}_2 = \{x, \tilde{V}_0 y\}_1, \quad x \in H_1, \ y \in H_2, \tag{1.4}$$

where $\tilde{V}_0 \in L(H_2, H_1)$ is given by the matrix representation

$$\tilde{V}_0 = \begin{pmatrix} V_{11} & -V_{12} \\ -V_{21} & V_{22} \end{pmatrix} : \begin{pmatrix} H_2^+ \\ H_2^- \end{pmatrix} \to \begin{pmatrix} H_1^+ \\ H_1^- \end{pmatrix}$$

if V_0 has the representation

$$V_0 = \begin{pmatrix} V_{11} & V_{12} \\ V_{21} & V_{22} \end{pmatrix} : \begin{pmatrix} H_2^+ \\ H_2^- \end{pmatrix} \to \begin{pmatrix} H_1^+ \\ H_1^- \end{pmatrix},$$

where $\begin{pmatrix} H_j^+ \\ H_j^- \end{pmatrix}$ denotes the $[\ ,\ ]_j$ - orthogonal direct sum $H_j^+ \oplus H_j^-$, $j = 1,2$. Using the Cauchy Schwarz' inequality several times we find that (1.4) implies, for $x \in H_1$,

$$\begin{aligned}\{U_0 x, U_0 x\}_2 &= \{\tilde{V}_0 U_0 x, x\}_1 \le \{\tilde{V}_0 U_0 x, \tilde{V}_0 U_0 x\}_1^{\frac{1}{2}} \{x,x\}_1^{\frac{1}{2}} \\ &= \{(\tilde{V}_0 U_0)^2 x, x\}_1^{\frac{1}{2}} \{x,x\}_1^{\frac{1}{2}} \le \ldots \le \{(\tilde{V}_0 U_0)^{2^n} x, x\}_1^{2^{-n}} \{x,x\}_1^{1-2^{-n}}, \quad n \in \mathbb{N}.\end{aligned}$$

As the inner product $\{\ ,\ \}_1$ is bounded with respect to the norm $\|\ \ \|_1$ of the Hilbert space H_1 : for some $C > 0$ $|\{x,y\}_1| \le C\|x\|_1 \|y\|_1$, $x,y \in H_1$ and also the operators U_0 and $\tilde{V}_0$ are bounded with norms less than γ, say, it follows that

$$\{U_0x,U_0x\}_2 \le (C\|(\tilde{V}_0U_0)^{2^n}x\|_1 \|x\|_1)^{2^{-n}} \{x,x\}_1^{1-2^{-n}}$$

$$\le C^{2^{-n}} \gamma^2 \|x\|_1^{2^{-n+1}} \{x,x\}_1^{1-2^{-n}}$$

and hence, letting $n \to \infty$, we obtain $\{U_0x,U_0x\}_2 \le \gamma^2\{x,x\}_1$, $x \in H_1$. It follows that U_0 is bounded on the dense subset H_1 of the Krein space K_1, and hence that it can be extended by continuity to a bounded operator $U \in L(K_1,K_2)$. The conditions of the lemma are symmetric with respect to U_0 and V_0 and therefore also V_0 can be extended by continuity to an operator $V \in L(K_2,K_1)$. The last equality in the statement is an easy consequence of (1.3). This completes the proof of the lemma.

COROLLARY 1.2. Suppose that in addition to the conditions of Lemma 1.1 we have that $H_1 = H_2 =: H$ and $[\ ,\ ]_1 = [\ ,\ ]_2 =: [\ ,\ ]$. Then $\rho(U_0) \cap \rho(V_0)^* \subset \rho(U) \cap \rho(V)^*$. Moreover, if σ is a spectral set of U_0 such that σ^* is a spectral set of V_0, then $\sigma \cap \sigma(U) \ne \emptyset$.

PROOF. Let $\lambda \in \rho(U_0) \cap \rho(V_0)^*$. Then $[(U_0 - \lambda)^{-1}x,y] = [x,(V_0 - \bar{\lambda})^{-1}y]$, $x,y \in H$ and hence, by Lemma 1.1, $(U_0 - \lambda)^{-1}$ has a continuous extension to all of K, which coincides with $(U - \lambda)^{-1}$ as can easily be seen. If σ is a spectral set of U_0 we consider the Riesz projection

$$-\frac{1}{2\pi i} \oint_{C_\sigma} (U_0 - \lambda)^{-1} d\lambda \ne 0.$$

According to the conditions about σ and the first statement of the corollary the contour C_σ can be chosen such that $C_\sigma \subset \rho(U)$ and hence

$$-\frac{1}{2\pi i} \oint_{C_\sigma} (U - \lambda)^{-1} d\lambda \ne 0,$$

which implies the last statement of the corollary.

2. CHARACTERISTIC FUNCTIONS OF UNITARY COLLIGATIONS IN KREIN SPACES

Let K, F and G be Krein spaces. The inner product of these and other Krein spaces will be denoted by $[\ ,\ ]$. Only when necessary we shall be more specific by making use of subscripts. Let $U : \binom{K}{F} \to \binom{K}{G}$ be a unitary operator. Here we denote by $\binom{K}{F}$, for example, the Krein space which is the orthogonal direct sum of K and F. The quadruple $\Delta = (K,F,G;U)$ is called a unitary colligation, K is called the inner space, F and G the left and right outer spaces, respectively, and U is called the connecting operator. In the matrix representation $U = \begin{pmatrix} T & F \\ G & H \end{pmatrix} : \begin{pmatrix} K \\ F \end{pmatrix} \to \begin{pmatrix} K \\ G \end{pmatrix}$, where $T \in L(K)$, $F \in L(F,K)$, $G \in L(K,G)$ and $H \in L(F,G)$, the operator T is called the basic operator. We sometimes write $\Delta = (K,F,G;T,F,G,H)$. The fact that U is unitary can be expressed by the relations

$$T^+T + G^+G = I, \quad F^+F + H^+H = I, \quad F^+T + H^+G = 0$$

and

$$TT^+ + FF^+ = I, \quad GG^+ + HH^+ = I, \quad TG^+ + FH^+ = 0,$$

which follows from the relations $U^+U = I$ and $UU^+ = I$, respectively. Here we use I and 0 to denote the identity and zero operator. It should be clear from the context in which space they act. The adjoint Δ^+ of Δ is the unitary colligation defined by $\Delta^+ = (K,G,F;T^+,G^+,F^+,H^+)$.

A unitary colligation $\Delta = (K,F,G;T,F,G,H)$ is called closely connected if the linear span of all elements of the form T^nFf or $T^{+(m)}G^+g, n,m = 0,1,2,..,$ $f \in F$ and $g \in G$, is dense in K. Following Arov [1] we say that two unitary colligations $\Delta = (K,F,G;T,F,G,H)$ and $\Delta_1 = (K_1,F,G;T_1,F_1,G_1,H_1)$ with the same right and left outer spaces, are weakly isomorphic if $H = H_1$ and if there exist dense subspaces $L \subset K$ and $L_1 \subset K_1$ and a linear bijection $Z : L \to L_1$ with the properties : $TL \subset L$, $T_1L_1 \subset L_1$, $FF \subset L$, $ZT = T_1Z$ on L, $F_1 = ZF$, $G_1Z = G$ on L and $[Zh,Z\tilde{h}] = [h,\tilde{h}], h,\tilde{h} \in L$. If L and L_1 can be chosen so that $L = K$ and $L_1 = K_1$ then we say that Δ and Δ_1 are unitarily equivalent. Clearly, these concepts define equivalence relations; they coincide if K and K_1 are Pontryagin spaces, see [4].

The characteristic function θ_Δ of the unitary colligation $\Delta = (K,F,G;T,F,G,H)$ is defined by $\theta_\Delta(z) = H + zG(I - zT)^{-1}F$, $z \in \{z \mid z^{-1} \in \rho(T)\} \cup \{0\}$. It satisfies the following relations:

$$(2.1) \quad \begin{cases} \theta_\Delta(z) - \theta_\Delta(w) = (z-w)G(I-zT)^{-1}(I-wT)^{-1}F, \\ I - \theta_\Delta(w)^+\theta_\Delta(z) = (1-z\bar{w})F^+(I-\bar{w}T^+)^{-1}(I-zT)^{-1}F, \\ I - \theta_\Delta(z)\theta_\Delta(w)^+ = (1-z\bar{w})G(I-zT)^{-1}(I-\bar{w}T^+)^{-1}G^+, \end{cases}$$

see M.S. Brodskii [5]. Moreover, $\theta_\Delta(z)^+ = \theta_{\Delta^+}(\bar{z})$.

Let $\Delta = (K,F,G;U)$ be a unitary colligation in which the left and right outer spaces coincide, i.e., $F = G$. We shall call the inner space K in Δ minimal if $K = \text{c.l.s.}\{(U-z)^{-1}f \mid f \in F, z \in \rho(U)\}$.

PROPOSITION 2.1. Let $\Delta = (K,F,F;U)$ be a unitary colligation. Then Δ is closely connected if and only if K is minimal.

PROOF. We shall use the fact that a matrix with operator entries

$$\begin{pmatrix} A & B \\ C & D \end{pmatrix} : \begin{pmatrix} K \\ F \end{pmatrix} \to \begin{pmatrix} K \\ F \end{pmatrix}$$

is invertible, if A and $R = D - CA^{-1}B$ are invertible, in which case its inverse is given by

$$\begin{pmatrix} A^{-1} + A^{-1}BR^{-1}CA^{-1} & -A^{-1}BR^{-1} \\ -R^{-1}CA^{-1} & R^{-1} \end{pmatrix}.$$

It implies that if U has the matrix representation introduced above, then for large values of $|z|$

$$P_K(U - z)^{-1}|_F = -(T - z)^{-1}F(\theta(\tfrac{1}{z}) - z)^{-1},$$

where P_K is the orthogonal projection from $\binom{K}{F}$ onto K and $\theta = \theta_\Delta$. On the other hand making use of the relation $(U - z)^{-1} = (U^+ - \frac{1}{z})^{-1}(-\frac{1}{z}U^+)$ we find that for small values of $|z|$

$$P_K(U - z)^{-1}|_F = (zT^+ - I)^{-1}G^+(z\theta(\bar{z})^+ - I)^{-1}.$$

From these two equalities and the fact that $\lim_{z \to \infty} z(U - z)^{-1} = -I$, the proposition follows.

We denote by $S(F,G)$ the class of all $L(F,G)$ valued functions which are defined and holomorphic on a neighbourhood of 0 in the unit disc $\mathbb{D} = \{z \in \mathbb{C} \mid |z| < 1\}$. The domain of holomorphy of $\theta \in S(F,G)$ inside the unit disc will be denoted by D_θ. Clearly, we have that $\theta_\Delta \in S(F,G)$ for all unitary colligations Δ with left and right outer spaces F and G, respectively. The converse is also true, namely:

THEOREM 2.2. Let $\theta \in S(F,G)$ and D be a simply connected open set with a smooth boundary ∂D such that $0 \in D$ and $D \cup \partial D \subset D_\theta$. Then there exists a unitary colligation $\Delta = (K,F,G;U)$ such that $D_{\theta_\Delta} \supset D$ and $\theta = \theta_\Delta$ on D. The colligation Δ can be constructed so that it is closely connected, in which case it is uniquely determined up to weak isomorphisms.

This theorem (in less detailed form) was stated and proved by T. Ya Azizov in [2], [3]. Azizov's proof starts from a result of D.Z. Arov [1] that an arbitrary operator valued function, holomorphic at $z = 0$, is the characteristic function of a (not necessarily unitary) colligation and applies C. Davis' statement [7] that any bounded operator has a unitary dilation in a Krein space. Here we shall give a longer but more explicit proof making use of Lemma 1.1.

PROOF. We divide the proof of the theorem in seven parts. In parts (i)-(vi) we consider the case that D is an open disc around 0 and in part (vii) we shall briefly outline the proof of the general case in which D is an arbitrary set satisfying the conditions of the theorem. Let $0<r<1$ be such that the closed disc$\{z\in\mathbb{C}\,|\,|z|\leq r\}\subset D_\theta$ and put $R=r^{-1}$.

(i) Construction of Hilbert spaces L, H_1 and H_2. We choose and fix two fundamental decompositions of G and F: $G=G_+[+]G_-$ and $F=F_+[+]F_-$, and denote the corresponding positive definite inner products and norms by $(\ ,\ )$ and $\|\ \ \|$, respectively. Now we define the Hilbert spaces $H_+(G)$ and $H_-(F)$ by

$$H_+(G)=\{g\,|\,g(t)=\sum_{n=0}^{\infty} t^n g_n,\ |t|<R,\ g_n\in G,\ \sum_{n=0}^{\infty} R^{2n}\|g_n\|^2<\infty\},$$

$$H_-(F)=\{f\,|\,f(s)=\sum_{n=1}^{\infty} s^{-n} f_{-n},\ |s|>r,\ f_{-n}\in F,\ \sum_{n=1}^{\infty} R^{2n}\|f_{-n}\|^2<\infty\},$$

with inner products $(g,\tilde g)=\sum_{n=0}^{\infty} R^{2n}(g_n,\tilde g_n)$, $g,\tilde g\in H_+(G)$ and $(f,\tilde f)=\sum_{n=1}^{\infty} R^{2n}(f_{-n},\tilde f_{-n})$, $f,\tilde f$ $H_-(F)$, respectively. The series $\sum_{n=0}^{\infty} t^n g_n$ defining $g\in H_+(G)$ converges in G for $|t|<R$ and if $t\to Re^{i\varphi}$ (nontangentially) then $g(t)$ tends for almost all $\varphi\in[0,2\pi)$ and also in the mean to $g(Re^{i\varphi})=\sum_{n=0}^{\infty} R^n e^{in\varphi} g_n$ and this series converges in the mean. Hence, for $g,\tilde g\in H_+(G)$,

$$(g,\tilde g)=\frac{1}{2\pi}\int_0^{2\pi}(g(Re^{i\varphi}),\tilde g(Re^{i\varphi}))d\varphi=\frac{1}{2\pi i}\oint_{|t|=R}\frac{1}{t}(g(t),\tilde g(t))dt.$$

Furthermore, for $g\in H_+(G)$ we have

$$\frac{1}{2\pi i}\oint_{|t|=R} t^n g(t)dt=\frac{R^{n+1}}{2\pi}\int_0^{2\pi} e^{i(n+1)\varphi} g(Re^{i\varphi})d\varphi=0,n=0,1,\ldots, \tag{2.2}$$

where the integrals are to be understood in the weak sense. Similarly, the series $\sum_{n=1}^{\infty} s^{-n} f_{-n}$ defining $f\in H_-(F)$ converges in F for $|s|>r$ and has for almost all $\varphi\in[0,2\pi)$ nontangential boundary values $f(re^{i\varphi})=\sum_{n=1}^{\infty} re^{-in\varphi} f_{-n}$, which converges in the mean (comp. [34]). Hence for $f,\tilde f\in H_-(F)$

$$(f,\tilde f)=\frac{1}{2\pi}\int_0^{2\pi}(f(re^{i\varphi}),\tilde f(re^{i\varphi}))d\varphi=\frac{1}{2\pi i}\oint_{|s|=r}\frac{1}{t}(f(t),\tilde f(t))dt.$$

Moreover, we have that for $f\in H_-(F)$

$$\frac{1}{2\pi i}\oint_{|s|=r} s^{-n} f(s)ds=\frac{R^{n-1}}{2\pi}\int_0^{2\pi} e^{-i(n-1)\varphi} f(re^{i\varphi})d\varphi=0,n=1,2,\ldots,$$

where the integrals are to be understood in the weak sense.

Finally, we define the Hilbert spaces L, H_1 and H_2 by

$$L=\begin{pmatrix}H_+(G)\\ H_-(F)\end{pmatrix},\ H_1=\binom{L}{F},\ H_2=\binom{L}{G},$$

where on the right hand sides of these equalities are direct sums orthogonal with respect to the Hilbert space inner products. If $F=G$, then the elements of L can be identified with functions defined and holomorphic in the annulus $r<|t|<R$ and with values in $F=G$.
(ii) Definition of the Krein spaces K, K_1 and K_2. We extend θ to the domain $|z|\geq R$ by $\theta(z)=\theta(\frac{1}{\bar{z}})^*$, $|z|\geq R$. Note that now $\theta(z)\in L(F,G)$ if $z\in D_\theta$ and $\theta(z)\in L(G,F)$ if $|z|\geq R$. On L we define the following inner product

$$[\begin{pmatrix} g\\ f\end{pmatrix},\begin{pmatrix} \tilde{g}\\ \tilde{f}\end{pmatrix}]_L =$$

$$-\frac{1}{4\pi^2}\oint_{|t|=R}\oint_{|s|=R}[\frac{\theta(s)^+\theta(t)-I}{t\bar{s}-1}g(t),\tilde{g}(s)]dt\overline{ds}+\frac{1}{4\pi^2}\oint_{|t|=r}\oint_{|s|=R}[\frac{\theta(t)-\theta(s)^+}{1-t\bar{s}}f(t),\tilde{g}(s)]dt\overline{ds}$$

$$+\frac{1}{4\pi^2}\oint_{|t|=R}\oint_{|s|=R}[\frac{\theta(s)^+-\theta(t)}{1-t\bar{s}}g(t),\tilde{f}(s)]dt\overline{ds}+\frac{1}{4\pi^2}\oint_{|t|=r}\oint_{|s|=r}[\frac{I-\theta(s)^+\theta(t)}{1-t\bar{s}}f(t),\tilde{f}(s)]dt\overline{ds},$$

where $g,\tilde{g}\in H_+(G)$, $f,\tilde{f}\in H_-(F)$. Using simple estimates on the integrals and the fact that the norms of $\theta(z)$ on the circles $|z|=r$ and $|z|=R$ are uniformly bounded, one can easily convince oneself that this inner product is bounded on the Hilbert space L. Finally, we define the inner products $[\ ,\ ]_j$ on H_j, $j=1,2$ by

$$[\begin{pmatrix} h\\ f_0\end{pmatrix},\begin{pmatrix} \tilde{h}\\ \tilde{f}_0\end{pmatrix}]_1=[h,\tilde{h}]_L+[f_0,\tilde{f}_0],\quad h,\tilde{h}\in L,\ f_0,\tilde{f}_0\in F,$$

$$[\begin{pmatrix} h\\ g_0\end{pmatrix},\begin{pmatrix} \tilde{h}\\ \tilde{g}_0\end{pmatrix}]_2=[h,\tilde{h}]_L+[g_0,\tilde{g}_0],\quad h,\tilde{h}\in L,\ g_0,\tilde{g}_0\in G.$$

Clearly, these inner products are bounded on the Hilbert spaces H_1 and H_2, respectively. Let K, K_1 and K_2 be the Krein spaces associated with L, H_1 and H_2 with respect to the bounded inner products $[\ ,\ ]_L$, $[\ ,\ ]_1$ and $[\ ,\ ]_2$, respectively. If we denote the isotropic parts of L, H_1 and H_2 with respect to these inner products by L^0, H_1^0 and H_2^0 then we have $H_j^0=\begin{pmatrix} L^0\\ \{0\}\end{pmatrix}$, $j=1,2$, and from this it can easily be seen that

$$\text{(2.3)}\qquad K_1=\begin{pmatrix} K\\ F\end{pmatrix},\quad K_2=\begin{pmatrix} K\\ G\end{pmatrix},$$

direct sums, orthogonal with respect to the Krein space inner products of K, F and G.
(iii) Construction of the unitary colligation Δ. Let

$$U_0=\begin{pmatrix} T_{11} & 0 & 0\\ T_{21} & T & F\\ G_1 & G & H\end{pmatrix}:\begin{pmatrix} H_+(G)\\ H_-(F)\\ F\end{pmatrix}\to\begin{pmatrix} H_+(G)\\ H_-(F)\\ G\end{pmatrix}$$

be defined by :

$$(T_{11}g)(t) = \frac{1}{t}(g(t)-g(0)),\ g\in H_+(G),$$

$$(T_{21}g)(s) = \frac{1}{s} I_R(g),\ I_R(g) = \frac{1}{2\pi i}\oint_{|t|=R} \frac{1}{t}\,\theta(t)g(t)dt,\ g\in H_+(G),$$

$$(Tf)(s) = \frac{1}{s} f(s),\ f\in H_-(F),$$

$$(Ff_0)(s) = \frac{1}{s} f_0,\quad f_0\in F,$$

$$G_1 g = \theta(0)I_R(g)-g_0, g_0 = g(0),\ g\in H_+(G),$$

$$Gf = \frac{1}{2\pi i}\oint_{|s|=r} \frac{1}{s}\,\theta(s)f(s)ds,\ f\in H_-(F),$$

$$Hf_0 = \theta(0)f_0,\ f_0\in F.$$

Note that the Cauchy formula implies that for all $f_0\in F$ and all $z\in ¢$ with $|z|<r$

(2.4)
$$(H+zG(I-zT)^{-1}F)f_0 = \theta(0)f_0 + z\frac{1}{2\pi i}\oint_{|s|=r}\frac{1}{s}\theta(s)\frac{1}{1-zs^{-1}}\frac{1}{s}f_0ds =$$
$$= \theta(0)f_0 + z\frac{1}{2\pi i}\oint_{|s|=r}\frac{\theta(s)}{s(s-z)}f_0ds$$
$$= \theta(0)f_0 + z\left(\frac{\theta(z)}{z}f_0 - \frac{\theta(0)}{z}f_0\right) = \theta(z)f_0.$$

It is not difficult to verify that U_0 is a bijection with inverse

$$U_0^{-1} = V_0 = \begin{pmatrix} W_{11} & W_{12} & W_{13} \\ 0 & W_{22} & 0 \\ W_{31} & W_{32} & W_{33}\end{pmatrix} : \begin{pmatrix} H_+(G) \\ H_-(F) \\ G\end{pmatrix} \to \begin{pmatrix} H_+(G) \\ H_-(F) \\ F\end{pmatrix}$$

where

$$(W_{11}g)(t) = tg(t),\ g\in H_+(G),$$

$$(W_{12}f)(t) = \frac{1}{2\pi i}\oint_{|s|=r}\theta(s)f(s)ds,\ f\in H_-(F),$$

$$(W_{13}g_0)(t) = -g_0,\ g_0\in G,$$

$$(W_{22}f)(s) = sf(s) - \frac{1}{2\pi i}\oint_{|s|=r} f(s)ds,\ f\in H_-(F),$$

$$W_{31}g = -\frac{1}{2\pi i}\oint_{|t|=R}\theta(t)g(t)dt,\ g\in H_+(G),$$

$$W_{32}f = \frac{1}{2\pi i}\oint_{|s|=r}(I-\theta(0)^*\theta(s))f(s)ds,\ f\in H_-(F),$$

$$W_{33}g_0 = \theta(0)^*g_0,\ g_0\in G.$$

As before, simple estimates show that U_0 and U_0^{-1} are bounded, i.e., $U_0 \in L(H_1,H_2)$ and $U_0^{-1} \in L(H_2,H_1)$. Furthermore,

$$(2.5) \qquad \left[U_0 \begin{pmatrix} g \\ f \\ f_0 \end{pmatrix}, U_0 \begin{pmatrix} \tilde{g} \\ \tilde{f} \\ \tilde{f}_0 \end{pmatrix} \right]_2 = \left[\begin{pmatrix} g \\ f \\ f_0 \end{pmatrix}, \begin{pmatrix} \tilde{g} \\ \tilde{f} \\ \tilde{f}_0 \end{pmatrix} \right]_1$$

for all $g,\tilde{g} \in H_+(G)$, $f,\tilde{f} \in H_-(F)$ and $f_0,\tilde{f}_0 \in F$. The verification of this relation is tedious but straightforward. For example, using (2.2) we have that for $g,\tilde{g} \in H_+(G)$

$$\left[U_0 \begin{pmatrix} g \\ 0 \\ 0 \end{pmatrix}, U_0 \begin{pmatrix} \tilde{g} \\ 0 \\ 0 \end{pmatrix} \right]_2$$

is the sum of the following five terms:

(1) $$-\frac{1}{4\pi^2} \oint_{|t|=R} \oint_{|s|=R} \left[\frac{\theta(s)^+\theta(t) - I}{t\bar{s} - 1} (T_{11}g)(t), (T_{11}\tilde{g})(s)\right] dt\overline{ds}$$

$$= \frac{-1}{4\pi^2} \oint_{|t|=R} \oint_{|s|=R} \left[\frac{\theta(s)^+\theta(t) - I}{t\bar{s} - 1} \frac{g(t) - g(0)}{t}, \frac{\tilde{g}(s) - \tilde{g}(0)}{s}\right] dt\overline{ds}$$

$$= -\frac{1}{4\pi^2} \oint_{|t|=R} \oint_{|s|=R} \left[\frac{\theta(s)^+\theta(t) - I}{t\bar{s} - 1} \frac{g(t)}{t}, \frac{\tilde{g}(s)}{s}\right] dt\overline{ds}$$

(as the other three terms are equal to 0 : $\oint_{|z|=R} \frac{k(z)}{z} dz = 2\pi i\, k(\infty) = 0$ for the corresponding integrands k)

$$= -\frac{1}{4\pi^2} \oint_{|t|=R} \oint_{|s|=R} \left\{\left[\frac{\theta(s)^+\theta(t) - I}{t\bar{s} - 1} g(t), \tilde{g}(s)\right] - \right.$$

$$\left. - \left[\frac{g(t)}{t}, \frac{\tilde{g}(s)}{s}\right] + \left[\frac{\theta(t)}{t} g(t), \frac{\theta(s)}{s} \tilde{g}(s)\right]\right\} dt\overline{ds}$$

$$= \left[\begin{pmatrix} g \\ 0 \\ 0 \end{pmatrix}, \begin{pmatrix} \tilde{g} \\ 0 \\ 0 \end{pmatrix} \right]_1 - [g(0), \tilde{g}(0)] + [I_R(g), I_R(\tilde{g})],$$

(2) $$\frac{1}{4\pi^2} \oint_{|t|=r} \oint_{|s|=R} \left[\frac{\theta(t) - \theta(s)^+}{1 - t\bar{s}} (T_{21}g)(t), (T_{11}\tilde{g})(s)\right] dt\overline{ds}$$

$$= \frac{1}{4\pi^2} \oint_{|t|=r} \oint_{|s|=R} \left[\frac{\theta(t) - \theta(s)^+}{1 - t\bar{s}} \frac{1}{t} I_R(g), \frac{\tilde{g}(s) - \tilde{g}(0)}{s}\right] dt\overline{ds}$$

$$= -\frac{1}{2\pi i} \oint_{|s|=R} \left[(\theta(0) - \theta(s)^+) I_R(g), \frac{\tilde{g}(s) - \tilde{g}(0)}{s}\right] \overline{ds}$$

(because of Cauchy's formula $\oint_{|t|=r} \frac{k(z)}{z}\,dz = 2\pi i\; k(0)$)

$$= \frac{1}{2\pi i} \oint_{|s|=R} [I_R(g), \tfrac{1}{s}\,\theta(s)(\tilde{g}(s) - \tilde{g}(0))]\overline{ds}$$

$$= -[I_R(g), I_R(\tilde{g})] + [I_R(g), \theta^+(0)\tilde{g}(0)],$$

$$(3)\quad \frac{1}{4\pi^2} \oint_{|t|=R} \oint_{|s|=r} [\frac{\theta(s)^+ - \theta(t)}{1 - t\bar{s}}\,(T_{11}g)(t), (T_{21}\tilde{g})(s)]dt\overline{ds}$$

$$= \ldots = -[I_R(g), I_R(\tilde{g})] + [\theta(0)^+ g(0), I_R(\tilde{g})],$$

$$(4)\quad \frac{1}{4\pi^2} \oint_{|t|=r} \oint_{|s|=r} [\frac{I - \theta(s)^+\theta(t)}{1 - t\bar{s}}\,(T_{21}g)(t), (T_{21}\tilde{g})(s)]dt\overline{ds}$$

$$= [I_R(g), I_R(\tilde{g})] - [\theta(0)I_R(g), \theta(0)I_R(\tilde{g})],$$

and

$$(5)\quad [G_1 g, G_1\tilde{g}] = [g(0), \tilde{g}(0)] - [\theta(0)I_R(g), \tilde{g}(0)] - [g(0), \theta(0)I_R(\tilde{g})] + [\theta(0)I_R(g), \theta(0)I_R(\tilde{g})].$$

Hence

$$\left[U_0\begin{pmatrix} g \\ 0 \\ 0 \end{pmatrix}, U_0\begin{pmatrix} \tilde{g} \\ 0 \\ 0 \end{pmatrix}\right]_2 = \left[\begin{pmatrix} g \\ 0 \\ 0 \end{pmatrix}, \begin{pmatrix} \tilde{g} \\ 0 \\ 0 \end{pmatrix}\right]_1.$$

In a similar way one can prove that

$$\left[U_0\begin{pmatrix} g \\ 0 \\ 0 \end{pmatrix}, U_0\begin{pmatrix} 0 \\ \tilde{f} \\ \tilde{f}_0 \end{pmatrix}\right]_2 = \left[\begin{pmatrix} g \\ 0 \\ 0 \end{pmatrix}, \begin{pmatrix} 0 \\ \tilde{f} \\ \tilde{f}_0 \end{pmatrix}\right]_1,$$

where $g \in H_+(\mathcal{G})$, $\tilde{f} \in H_-(\mathcal{F})$ and $\tilde{f}_0 \in \mathcal{F}$. Finally, one can check that, with a slight deviation of our notation,

$$[\begin{pmatrix} T & F \\ G & H \end{pmatrix}\begin{pmatrix} f \\ f_0 \end{pmatrix}, \begin{pmatrix} T & F \\ G & H \end{pmatrix}\begin{pmatrix} \tilde{f} \\ \tilde{f}_0 \end{pmatrix}]_2 = [\begin{pmatrix} f \\ f_0 \end{pmatrix}, \begin{pmatrix} \tilde{f} \\ \tilde{f}_0 \end{pmatrix}]_1.$$

This isometric property accounts for the fact that the first row of the matrix representation of U_0 has two zero entries. From (2.5) it follows that

$$(2.6)\qquad \left[U_0\begin{pmatrix} g\\ f\\ f_0\end{pmatrix},\begin{pmatrix}\tilde g\\ \tilde f\\ \tilde g_0\end{pmatrix}\right]_2=\left[\begin{pmatrix} g\\ f\\ f_0\end{pmatrix},V_0\begin{pmatrix}\tilde g\\ \tilde f\\ \tilde g_0\end{pmatrix}\right]_1,$$

$$(2.7)\qquad \left[\begin{pmatrix} g\\ f\\ g_0\end{pmatrix},\begin{pmatrix}\tilde g\\ \tilde f\\ \tilde g_0\end{pmatrix}\right]_2=\left[V_0\begin{pmatrix} g\\ f\\ g_0\end{pmatrix},V_0\begin{pmatrix}\tilde g\\ \tilde f\\ \tilde g_0\end{pmatrix}\right]_2$$

for all $g,\tilde g\in H_+(G)$, $f,\tilde f\in H_-(F)$, $f_0\in F$ and g_0, $\tilde g_0\in G$.

We now apply Lemma 1.1 with H_j, $[\ ,\ ]_j$, K_j, $j=1,2$, as in parts (i) and (ii) and U_0, V_0 as above. Let U and V be as in the conclusion of Lemma 1.1. Then it follows from (2.5), (2.6) and (2.7) that $[Uk_1,U\tilde k_1]_2=[k_1,\tilde k_1]_1$, $[Uk_1,k_2]_2=[k_1,Vk_2]_1$, and $[k_2,\tilde k_2]_2=[Vk_2,V\tilde k_2]_1$ for all $k_j,\tilde k_j\in K_j$, $j=1,2$. Hence $U^+U=I$, $U=V^+$ and $V^+V=1$, which implies that U is unitary. From (2.3) it follows that $\Delta=(K,F,G;U)$ is a unitary colligation.

(iv) $\theta=\theta_\Delta$ <u>on</u> $|z|<r$. The notation for equivalence classes in the factor spaces which we are going to use is the same as in the beginning of the proof of Lemma 1.1. The operator $U\in L(K_1,K_2)$ is the continuous extension of the operator

$$\hat U_0=\begin{pmatrix}\hat U_{11} & \hat U_{12}\\ \hat U_{21} & \hat U_{22}\end{pmatrix}:\begin{pmatrix}\hat L\\ F\end{pmatrix}\to\begin{pmatrix}\hat L\\ G\end{pmatrix},$$

where

$$\hat L=L/L^0,$$

$$\hat U_{11}\binom{g}{f}^{\wedge}=\begin{pmatrix}T_{11}g\\ T_{21}g+Tf\end{pmatrix}^{\wedge}\in\hat L,\ \hat U_{21}\binom{g}{f}^{\wedge}=G_1g+Gf\in G,\ \binom{g}{f}^{\wedge}\in\hat L,\ g\in H_+(G),\ f\in H_-(F),$$

$$\hat U_{12}f_0=\begin{pmatrix}0\\ Ff_0\end{pmatrix}^{\wedge}\in\hat L,\ \hat U_{22}f_0=Hf_0\in G,\ f_0\in F.$$

Hence $U=\begin{pmatrix}U_{11} & U_{12}\\ U_{21} & U_{22}\end{pmatrix}:\binom{K}{F}\to\binom{K}{G}$, where U_{11} and U_{21} are the continuous extension of $\hat U_{11}$ and $\hat U_{21}$ to all of K, $U_{12}=\hat U_{12}$ and $U_{22}=\hat U_{22}=H$. It now follows that for all $f_0\in F$ and $z\in\mathbb{C}$ with $|z|<r$

$$\theta_\Delta(z)f_0=U_{22}f_0+zU_{21}(I-zU_{11})^{-1}U_{12}f_0$$

$$=U_{22}f_0+zU_{21}\left(\sum_{n=0}^{\infty}z^nU_{11}^n\right)U_{12}f_0=Hf_0+z\hat U_{21}\left(\sum_{n=0}^{\infty}z^n\hat U_{11}^n\right)\binom{0}{Ff_0}^{\wedge}$$

$$=Hf_0+z\hat U_{21}\begin{pmatrix}0\\ \sum_{n=0}^{\infty}z^nT^nFf_0\end{pmatrix}^{\wedge}=Hf_0+zG\sum_{n=0}^{\infty}z^nT^nFf_0$$

$$=H+zG(I-zT)^{-1}F)f_0=\theta(z)f_0,$$

see (2.4), which implies $\theta = \theta_\Delta$ on $|z| < r$.

(v) Δ is closely connected. We have

$$(2.8)\qquad U_{11}^n U_{12} f_0 = \begin{pmatrix} 0 \\ T^n F f_0 \end{pmatrix}^\wedge = \begin{pmatrix} 0 \\ s^{-n-1} f_0 \end{pmatrix}, \qquad n = 0,1,\ldots,\ f_0 \in F.$$

In order to determine $U_{11}^{+(n)} U_{21}^+ g_0$, we recall that $V = U^+$ is the continuous extension of the operator

$$\hat{V}_0 = \begin{pmatrix} \hat{V}_{11} & \hat{V}_{12} \\ \hat{V}_{21} & \hat{V}_{22} \end{pmatrix} : \begin{pmatrix} \hat{L} \\ G \end{pmatrix} \to \begin{pmatrix} \hat{L} \\ F \end{pmatrix}, \text{where } \hat{V}_{11} \binom{g}{f}^\wedge = \begin{pmatrix} W_{11} g + W_{12} f \\ W_{22} f \end{pmatrix} \in \hat{L},$$

$\hat{V}_{21} \binom{g}{f}^\wedge = W_{31} g + W_{32} f \in F$, $\binom{g}{f}^\wedge \in \hat{L}$, $g \in H_+(G)$, $f \in H_-(F)$ and

$$\hat{V}_{12} g_0 = \begin{pmatrix} W_{13} g_0 \\ 0 \end{pmatrix}^\wedge,\ \hat{V}_{22} g_0 = W_{33} g_0,\ g_0 \in G.$$

Consequently, $V = \begin{pmatrix} U_{11}^+ & U_{21}^+ \\ U_{12}^+ & U_{22}^+ \end{pmatrix}$, where U_{11}^+ and U_{12}^+ coincide with the continuous extensions of $\hat{V}_{11}$ and $\hat{V}_{21}$ to all of K, $U_{21}^+ = \hat{V}_{12}$ and $U_{22}^+ = H^+ = \hat{V}_{22}$. We obtain for $g_0 \in G$

$$(2.9)\qquad U_{11}^{+(n)} U_{21}^+ g_0 = \hat{V}_{11}^n \hat{V}_{12} g_0 = \begin{pmatrix} W_{11}^n W_{13} g_0 \\ 0 \end{pmatrix}^\wedge = \begin{pmatrix} -t^n g_0 \\ 0 \end{pmatrix}^\wedge, \quad n = 0,1,2,\ldots .$$

As the linear span of all elements of the form $f(s) = s^{-n-1} f_0$, $n = 0,1,2,\ldots$, $f_0 \in F$ ($g(t) = t^n g_0$, $n = 0,1,2,\ldots$, $g_0 \in G$) is dense in $H_-(F)$ ($H_+(G)$, respectively), it follows from (2.8) and (2.9) that Δ is closely connected.

(vi) Δ is unique up to weak isomorphisms. Let $\Delta = (K,F,G;T,F,G,H)$ and $\Delta_1 = (K_1,F,G;T_1,F_1,G_1,H_1)$ be two closely connected unitary colligations such that $\theta_\Delta(z) = \theta_{\Delta_1}(z)$, $z \in D_{\theta_\Delta} \cap D_{\theta_{\Delta_1}}$. Then, putting $z = 0$, we find that $H = H_1$ and $G(I - zT)^{-1} F = G_1 (I - zT_1)^{-1} F_1$. From this equality and the last two equalities in (2.1) we obtain $GT^m F = G_1 T_1^m F_1$, $F^+ T^{+(m)} T^n F = F_1^+ T_1^{+(m)} T_1^n F_1$, and $GT^m T^{+(n)} G^+ = G_1 T_1^m T_1^{+(n)} G_1^+$, $m,n = 0,1,2,\ldots$. Let $L_0(L_1)$ be the linear span of all elements of the form $T^m Ff$ and $T^{+(n)} G^+ g$, ($T_1^m F_1 f$ and $T_1^{+(n)} G_1^+ g$, respectively), $m,n = 0,1,2,\ldots$, $f \in F$, $g \in G$. Then $L_0(L_1)$ is dense in $K(K_1)$ as $\Delta(\Delta_1)$ is closely connected. Define the linear map $Z : L_0 \to L_1$ such that $Z(T^m Ff) = T_1^m F_1 f$, $Z(T^{+(n)} G^+ g) = T_1^{+(n)} G_1^+ g$, $m,n = 0,1,2,\ldots$, $f \in F$ and $g \in G$. It can be shown that Z is well defined and that Z has all the properties required to prove that Δ and Δ_1 are weakly isomorphic (see M.S. Brodskii [5]). We leave the details to the reader.

(vii) The general case. Let D satisfy the conditions stated in the theorem and put $\tilde{D} = \{\bar{z}^{-1} \mid z \in D\}$. Then we can repeat the construction given above by first replacing $H_+(G)(H_-(F))$ by the completion of the linear manifolds $P_+(G)(P_-(F))$ of all polynomials $p(t)(q(s)$ with $q(\infty) = 0)$ in $t(s^{-1})$ and values in $G(F)$ with respect to the norms

$\oint_{\partial D} \frac{1}{t} \| p(t) \|^2 dt$ ($\oint_{\partial \tilde{D}} \frac{1}{s} \| q(s) \|^2 ds$, respectively), and by extending θ to $\tilde{D}$ by putting $\theta(z) = \theta(\frac{1}{\bar{z}})^*$, $z \in \tilde{D}$. Then we continue as in parts (ii) - (v) using the same relations with the elements in $H_+(G)$ and $H_-(F)$ replaced by polynomials in $P_+(G)$ and $P_-(F)$ and with each contour integral over the circle with radius r or R replaced by the contour integral over ∂D or $\partial \tilde{D}$, respectively. These relations can then be extended by continuity to the completions of $P_+(G)$ and $P_-(F)$. We leave further details to the reader.

REMARK. The proof of Theorem 2.2 given above is modelled after the proof of a similar theorem in [11]. There we constructed the space L and the operator U_0 in a different way: The space $\tilde{L}$ corresponding to L was defined as the linear span of all formal finite sums

$$\sum_z \varepsilon_z x_z, \quad x_z \in F \text{ if } |z| < r, \quad x_z \in G \text{ if } |z| > R,$$

in which ε_z is a symbol associated with each $z \in \{z \in \mathbb{C} \,|\, |z| < r \text{ or } |z| > R\}$, provided with the inner product $[\sum_z \varepsilon_z x_z, \sum_z \varepsilon_w y_w]_{\tilde{L}} = \sum_{z,w} [\tilde{S}_\theta(z,w) x_z, y_w]$, where $\tilde{S}_\theta$ is the kernel associated with $\theta \in S(F,G)$, (extended as in part (ii) of the proof given above) defined by

$$\tilde{S}_\theta(z,w) = \begin{cases} \dfrac{I - \theta(w)^+ \theta(z)}{1 - z\bar{w}}, & |z|, |w| < r, \\[2ex] -\bar{w}\,\dfrac{\theta(z) - \theta(w)^+}{1 - z\bar{w}}, & |z| < r, |w| > R, \\[2ex] z\,\dfrac{\theta(z) - \theta(w)^+}{1 - z\bar{w}}, & |z| > R, |w| < r, \\[2ex] -z\bar{w}\,\dfrac{I - \theta(w)^+ \theta(z)}{1 - z\bar{w}}, & |z|, |w| > R. \end{cases} \tag{2.10}$$

The operator $\tilde{U}_0 = \begin{pmatrix} T_0 & F_0 \\ G_0 & H_0 \end{pmatrix} : \begin{pmatrix} \tilde{L} \\ F \end{pmatrix} \to \begin{pmatrix} \tilde{L} \\ G \end{pmatrix}$ corresponding to U_0 was defined by

$$T_0 \varepsilon_z x = \begin{cases} \dfrac{1}{z}(\varepsilon_z x - \varepsilon_0 x), & |z| < r, \ x \in F, \\[2ex] \dfrac{1}{z}\varepsilon_z x - \varepsilon_0 \theta(z) x, & |z| > R, \ x \in G, \end{cases}$$

$$F_0 x = \varepsilon_0 x, \quad x \in F,$$

$$G_0 \varepsilon_z x = \begin{cases} \dfrac{1}{z}(\theta(z) - \theta(0))x, & |z| < r, \ x \in F, \\[2ex] (I - \theta(0)\theta(z))x & |z| > R, \ x \in G, \end{cases}$$

$$H_0 x = \theta(0) x, \quad x \in F.$$

There is a close relation between the two constructions, i.e., between $(\tilde{L}, [\ ,\ ]_{\tilde{L}})$ and $\tilde{U}_0$ defined above and $(L, [\ ,\ ]_L)$ and U_0 defined in parts (ii) and (iii) of the proof of Theorem 2.2. It can be described as follows: Let $\tilde{Z} : \tilde{L} \to L$ be the linear map such that

$$\tilde{Z}\varepsilon_z x = \begin{cases} \binom{g}{0} \in L \text{ with } g(t) = \dfrac{zx}{t-z}, & \text{if } |z| > R, \\ \binom{0}{f} \in L \text{ with } f(s) = \dfrac{x}{s-z}, & \text{if } |z| < r. \end{cases}$$

Then

(i) $\tilde{Z}$ is injective and $\tilde{Z}\tilde{L}$ is dense in L,

(ii) $[\tilde{Z} \sum_z \varepsilon_z x_z, \tilde{Z} \sum_w \varepsilon_w y_w]_L = [\sum_z \varepsilon_z x_z, \sum_w \varepsilon_w y_w]_{\tilde{L}}$

and

(iii) $\begin{pmatrix} \tilde{Z} & 0 \\ 0 & I \end{pmatrix} \tilde{U}_0 = U_0 \begin{pmatrix} \tilde{Z} & 0 \\ 0 & I \end{pmatrix}$,

which implies that the two constructions are weakly isomorphic. We leave the proof of (i) - (iii) to the reader.

Let K be a Krein space and let $K = K_+[+]K_-$ be a fundamental decomposition of K. The numbers $n_\pm(K) = \dim K_\pm$ depend only on K, not on the chosen fundamental decomposition of K. Let $\theta \in S(F,G)$ and let $\Delta = (K,F,G;U)$ be a unitary colligation such that $\theta = \theta_\Delta$ on some neighbourhood O in the disc $\mathbb{D}$, see Theorem 2.2. In the next theorem we relate the numbers $n_\pm(K)$ of the inner space of Δ to a property of the kernel $\tilde{S}_\theta$ (see (2.10) which we describe now. As $\tilde{S}_\theta(z,w)^+ = \tilde{S}_\theta(w,z)$, for each choice of $n \in \mathbb{N}$, $z_i \in \mathbb{C}$ and $x_i \in F \cup G$, $i = 1,2,\dots,n$, satisfying

$$\text{(2.11)} \qquad \begin{cases} |z_i| < r \text{ or } |z_i| > R, \\ x_i \in F \text{ if } |z_i| < r,\ x_i \in G \text{ if } |z_i| > R,\ i = 1,2,\dots,n, \end{cases}$$

the $n \times n$ matrix

$$\text{(2.12)} \qquad ([\tilde{S}_\theta(z_i,z_j)x_i,x_j])_{i,j=1,2,\dots,n}$$

is Hermitian and, consequently, has only real eigenvalues. We say $\tilde{S}_\theta$ has κ positive (negative) squares if for each choice this matrix has at most κ and for at least one choice exactly κ positive (negative) eigenvalues. Furthermore, we say that $\tilde{S}_\theta$ has infinitely many positive (negative) squares if for each $m \in \mathbb{N}$ a matrix of the form (2.12) can be found which has at least m positive (negative) eigenvalues. We denote the number of positive (negative) squares of $\tilde{S}_\theta$ by $n_+(\theta)$ ($n_-(\theta)$, respectively).

THEOREM 2.3. Let $\theta \in S(F,G)$ and let $\Delta = (K,F,G;U)$ be a unitary colligation such that

(*) $$\theta = \theta_\Delta \text{ on } O,$$

where O is a neighbourhood of 0 in $\mathbb{D}$. Then

(i) $n_+(\theta) \leq n_+(K)$, $n_-(\theta) \leq n_-(K)$,

(ii) if Δ is closely connected, $n_+(\theta) = n_+(K)$ and $n_-(\theta) = n_-(K)$,

(iii) if $n_+(\theta)$ or $n_-(\theta)$ is finite and Δ is closely connected, Δ is, up to unitary equivalence, uniquely determined by (*).

The proof of this result is quite simple and therefore omitted here.

When F,G are Hilbert spaces, we denote by $S_\kappa(F,G)$, $\kappa \in \mathbb{N} \cup \{0\}$ the set of all $\theta \in S(F,G)$ for which $n_-(\theta) = \kappa$, i.e., for which the kernel $\tilde{S}_\theta$ has κ negative squares. By Theorems 2.2 and 2.3 each $\theta \in S_\kappa(F,G)$ is the characteristic function of a closely connected unitary colligation in which the inner space is a π_κ - space and which is uniquely determined up to unitary equivalence. For $\kappa = 0$ the class $S_\kappa(F,G)$ coincides with the class of all contraction valued mappings defined and holomorphic on $\mathbb{D}$, whereas for $\kappa \geq 1$ the elements of $S_\kappa(F,G)$ are holomorphic on $\mathbb{D}$ with the exception of κ poles (including their multiplicities). It can be shown that each $\theta \in S_\kappa(F,G)$ can be factored as $\theta(z) = B(z)^{-1}\tilde{\theta}(z)$ where $\tilde{\theta} \in S_\kappa(F,G)$ and B is a finite Blaschke-Potapov product, see [11], [16], [17]. This factorization and Theorems 2.2 and 2.3 can be used to prove a result concerning the boundary behaviour of $\theta \in S_\kappa(F,G)$: For $\xi \in \partial\mathbb{D}$ we define the operator $\theta(\xi)$ acting on elements ψ from

$$D_0(\theta(\xi)) = \{\psi \in F \mid \lim_{z \hat{\to} \xi} \frac{\|\psi\| - \|\theta(z)\psi\|}{1-|z|} \text{ exists}\}$$

by $\theta(\xi)\psi = \lim_{z \hat{\to} \xi} \theta(z)\psi$, strongly, where by $\lim_{z \hat{\to} \xi}$ we denote the nontangential limit as $z \in \mathbb{D}$ tends to $\xi \in \partial\mathbb{D}$. Evidently, for $\xi \in \partial\mathbb{D}$, $\theta(\xi)$ is an isometry on $D_0(\theta(\xi))$. If $\Delta = (K,F,G;T,F,G,H)$ is a closely connected unitary colligation such that $\theta = \theta_\Delta$, then, for $\xi \in \partial\mathbb{D}$, $D_0(\theta(\xi)) = \{\psi \in F \mid F\psi \in R(I - \xi T)\}$ and $\theta(\xi)\psi = (H + \xi G(I - \xi T)^{-1}F)\psi$, $\psi \in D_0(\theta(\xi))$, see [11]. This result will be applied in later sections. We refer to [11] for a list of references about S_κ functions.

3. ISOMETRIES, UNITARY EXTENSIONS AND CHARACTERISTIC FUNCTIONS

Let H be a Hilbert space and let V be a closed isometric operator in H with domain $D(V)$ and range $R(V)$. The characteristic function X of the isometric operator V is defined by $X(z) = zP_{D(V)^\perp}(I - z\tilde{V})^{-1}|_{R(V)^\perp}$, $z \in \mathbb{D}$, where $\tilde{V}$ denotes the trivial extension of V to all of H, i.e., $\tilde{V}x = Vx$ if $x \in D(V)$, $\tilde{V}x = 0$ if $x \in D(V)^\perp$ and $\tilde{V}$ is linear. The operator $P_{D(V)^\perp}(I - z\tilde{V})^{-1}$ appearing in the definition of X is the projection from H onto $D(V)^\perp$ parallel to $R(I - zV)$; $H = D(V)^\perp + R(I - zV)$, direct sum, $z \in \mathbb{D}$. The Hilbert space H can be decomposed as $H = H_u \oplus H_s$ and V can be written as $V = V_u \oplus V_s$, where V_u is a unitary operator in the subspace H_u and V_s is an isometry in H_s which does not have a nontrivial unitary part. V_s is called the simple (or completely nonunitary) part of V and V is called simple if it is equal to its simple part. The characteristic function X of the isometry V in H coincides with the characteristic function of its simple part V_s in H_s. Obviously, $X \in S_0(R(V)^\perp, D(V)^\perp)$, $X(0) = 0$ and X is the characteristic function of the unitary colligation $(H, R(V)^\perp, D(V)^\perp; \tilde{V}, I|_{R(V)^\perp}, P_{D(V)^\perp}, 0)$. This colligation is closely connected if and only if V is simple. Conversely, if F and G are Hilbert spaces and $X \in S_0(F,G)$ satisfies $X(0) = 0$, then applying Theorems 2.2 and 2.3 we find that $X = \theta_\Delta$ where $\Delta = (H,F,G;T,F,G,0)$ is a closely connected unitary colligation in which H is a Hilbert space, T is a partial isometry with simple isometric part V, say, and $F(G^*)$ is an isometry mapping $F(G)$ onto $R(V)^\perp = \nu(T^*)$ ($D(V)^\perp = \nu(T)$, respectively). Identifying the spaces F and G with $R(V)^\perp$ and $D(V)^\perp$ respectively, we obtain the well-known result that X is the characteristic function of a simple isometry V in a Hilbert space H.

Concerning the boundary behaviour of the characteristic function X of the isometry V in the Hilbert space H we have that for $\xi \in \partial\mathbb{D}$ $D_0(X(\xi))$ is the set of all $\psi \in R(V)^\perp$ for which there exists a unique element $\varphi \in D(V)^\perp$ such that $\psi - \varphi \in R(I - \xi V)$ and then $X(\xi)\psi = \xi\varphi$, $\psi \in D_0(X(\xi))$.

With the isometry V we associate two other unitary colligations corresponding to a unitary extension of V in a Krein space: Let W be a unitary extension of V in the Krein space K, i.e., $H \subseteq_s K$ and W is a unitary operator in K such that $V \subset W$. Then W has the following matrix decomposition:

$$(3.11) \qquad W = \begin{pmatrix} T & F & 0 \\ G & H & 0 \\ 0 & 0 & V \end{pmatrix} : \begin{pmatrix} \hat{H} \\ D(V)^\perp \\ D(V) \end{pmatrix} \to \begin{pmatrix} \hat{H} \\ R(V)^\perp \\ R(V) \end{pmatrix},$$

where $\hat{H} = K[-]H$, the orthogonal complement of H in K. This representation gives rise to the unitary colligations

$$(3.2) \qquad (\hat{H}, D(V)^\perp, R(V)^\perp; U), \quad U = \begin{pmatrix} T & F \\ G & H \end{pmatrix}$$

and

$$(3.3) \qquad (\hat{H}, H, H; T, (F \ 0), \begin{pmatrix} G \\ 0 \end{pmatrix}, \begin{pmatrix} H & 0 \\ 0 & V \end{pmatrix})$$

with characteristic functions $\theta \in S(D(V)^{\perp}, R(V)^{\perp})$ and $\tilde{\theta} \in S(H,H)$, respectively. It is easy to verify that

$$(3.4) \qquad \tilde{\theta}(z) = \begin{pmatrix} \theta(z) & 0 \\ 0 & V \end{pmatrix} : \begin{pmatrix} D(V)^{\perp} \\ D(V) \end{pmatrix} \to \begin{pmatrix} R(V)^{\perp} \\ R(V) \end{pmatrix},$$

$$P_{R(V)^{\perp}}\tilde{\theta}(z) = P_{R(V)^{\perp}}\tilde{\theta}(z)P_{D(V)^{\perp}} = \tilde{\theta}(z)P_{D(V)^{\perp}},$$

$$\theta(z) = P_{R(V)^{\perp}}\tilde{\theta}(z)|_{D(V)^{\perp}},$$

$$\theta(z)P_{D(V)^{\perp}} = P_{R(V)^{\perp}}\tilde{\theta}(z)P_{D(V)^{\perp}} = \tilde{\theta}(z) - \tilde{V}.$$

Moreover, if we put $Q(z) = \{\{P_H h, P_H Wh\} | h \in K, (I - zW)h \in H\}$, $z \in \mathbb{C}$, then $Q(z)$ coincides with the graph of $\tilde{\theta}(z)$ in H^2 for $z \in D_{\tilde{\theta}}$. The following proposition gives some relations between the characteristic functions X, θ and $\tilde{\theta}$.

PROPOSITION 3.1. For z in some neighbourhood $O \subset \mathbb{D}$ of 0 the following equalities are valid:

(i) $P_{D(V)^{\perp}}(I - z\tilde{\theta}(z))^{-1}|_{D(V)^{\perp}} = (I - X(z)\theta(z))^{-1}$,

(ii) $P_{R(V)^{\perp}}(I - z\tilde{\theta}(z))^{-1}|_{R(V)^{\perp}} = (I - \theta(z)X(z))^{-1}$,

(iii) $P_{D(V)^{\perp}}(I - z\tilde{\theta}(z))^{-1}|_{R(V)^{\perp}} = \frac{1}{z}(I - X(z)\theta(z))^{-1}X(z)$,

(iv) $P_{R(V)^{\perp}}(I - z\tilde{\theta}(z))^{-1}\tilde{\theta}(z))|_{D(V)^{\perp}} = (I - \theta(z)X(z))^{-1}\theta(z)$.

PROOF. From the identity $I - z\tilde{\theta}(z) = I - z\tilde{V} - z\theta(z)P_{D(V)^{\perp}}$, it follows that $(I - z\tilde{\theta}(z))^{-1} = (I - z(I - z\tilde{V})^{-1}\theta(z)P_{D(V)^{\perp}})^{-1}(I - z\tilde{V})^{-1}$ and $(I - z\tilde{\theta}(z))^{-1} = (I - z\tilde{V})^{-1}$ $(I - z\theta(z)P_{D(V)^{\perp}}(I - z\tilde{V}))^{-1}$. Hence $P_{D(V)^{\perp}}(I - z\tilde{\theta}(z))^{-1} = P_{D(V)^{\perp}}(I - X(z)\theta(z)P_{D(V)^{\perp}})^{-1}(I - z\tilde{V})^{-1}$ and $(I - z\tilde{\theta}(z))^{-1}|_{R(V)^{\perp}} = (I - z\tilde{V})^{-1}(I - \theta(z)X(z))^{-1}$. Of the last two identities the first one leads to (i) and the second one to (ii). Item (iii) is obtained by using either one of these identities, and item (iv) follows directly from (ii).

The unitary extension W of V in K also admits the matrix representation

$$W = \begin{pmatrix} T & FP_{D(V)^{\perp}} \\ G & HP_{D(V)^{\perp}} + VP_{D(V)} \end{pmatrix} : \begin{pmatrix} \hat{H} \\ H \end{pmatrix} \to \begin{pmatrix} \hat{H} \\ H \end{pmatrix}.$$

Thus we obtain that $P_H(I - zW)^{-1}|_H = (I - z\tilde{\theta}(z))^{-1}$ and, on account of the equality,

$(I+zW)(I-zW)^{-1} = 2(I-zW)^{-1} - I$, that $P_H(I+zW)(I+zW)^{-1}|_H = 2(I-z\tilde{\theta}(z))^{-1} - I =$ $(I+z\tilde{\theta}(z))(I-z\tilde{\theta}(z))^{-1}$. This proves the first part of the following theorem, which establishes a one to one correspondence between the generalized resolvents (or coresolvents, see also [19]) of V and the functions $\theta \in S(R(V)^{\perp}, D(V)^{\perp})$, which are connected with $\tilde{\theta}$ via (3.4).

THEOREM 3.2. Let V be a closed isometric operator in a Hilbert space H. (i) If W is a unitary extension of V in a Krein space K with matrix representation (3.1) and $\tilde{\theta}$ is the characteristic function of the unitary colligation (3.3), then

$$(3.6) \qquad P_H(I+zW)(I-zW)^{-1}|_H = (I+z\tilde{\theta}(z))(I-z\tilde{\theta}(z))^{-1}, \; z \in O,$$

where $O \subset \mathbb{D}$ is some neighbourhood of 0. (ii) Conversely, if $\theta \in S(R(V)^{\perp}, D(V)^{\perp})$ and $\tilde{\theta}$ is given by (3.4), then there exists a unitary extension W of V in a Krein space K such that (3.6) holds for some neighbourhood $O \subset \mathbb{D}$ of 0. The Krein space K can be chosen minimal, that is

$$K = \text{c.l.s. } \{(W-z)^{-1}f \mid f \in H, \; z \in \rho(W)\},$$

in which case W is uniquely determined up to weak isomorphisms.

The last statement means that if for $j = 1,2$, W_j is a unitary extension of V in the Krein space K_j such that (3.6) with $W = W_j$ holds and K_j is minimal, then W_1 and W_2 are weakly isomorphic, i.e., there exist dense subsets $L_j \subset K_j$ such that $W_j L_j \subset L_j$ and a bijection Z from L_1 onto L_2, which preserves the indefinite inner product: $[Zx,Zy] = [x,y]$, $x,y \in L_1$, and has the property $ZW_1x = W_2Zx$, $x \in L_1$.

PROOF OF PART (ii). It follows from Theorem 2.2 that θ is the characteristic function of a closely connected unitary colligation of the form (3.2). Define W by (3.1), then W admits the representation (3.6) and the unitary colligation Δ given by $\hat{\Delta} = (\hat{H}, H, H; T, FP_{D(V)^{\perp}}, G, HP_{D(V)^{\perp}} + VP_{D(V)})$ is closely connected. It follows from Proposition 2.1 that K is minimal. The uniqueness now follows in a standard way from the relation

$$[(W-z)^{-1}f, (W-w)^{-1}g] = [\frac{F(z)+F(w)^*}{2(1-z\bar{w})} f, g], \; f,g \in H,$$

where F(z) is defined by the left hand side of (3.6).

Part (ii) of Theorem 3.2 can also be proved by making use of Theorem 1 in [12] which states that a function defined and holomorphic in a neighbourhood of 0 and with values in $L(F,G)$ can be written as the coresolvent of a unitary operator in a Krein space.

COROLLARY 3.3. *Let* V *be a closed isometric operator in a Hilbert space* H. *Let* W *be a unitary extension of* V *in a Krein space* K *and let* $\theta \in S(R(V)^{\perp}, D(V)^{\perp})$ *be associated with* W *as in Theorem 3.2. Then for* z *in some neighbourhood* $O \subset \mathbb{D}$ *of* 0, *we have*

$$P_{D(V)^{\perp}}(I+zW)(I-zW)^{-1}|_{D(V)^{\perp}} = (I+X(z)\theta(z))(I-X(z)\theta(z))^{-1},$$

$$P_{R(V)^{\perp}}(I+zW)(I-zW)^{-1}|_{R(V)^{\perp}} = (I+\theta(z)X(z))(I-\theta(z)X(z))^{-1}.$$

The proof of the corollary follows easily from the formula (see (3.6)) $P_H(I-zW)(I-zW)^{-1}|_H = 2(I-z\tilde{\theta}(z))^{-1} - I$ and items (i) and (ii) of Proposition 3.1

Of course all results in this section can be extended by analytic continuation, if the space K is a Pontryagin space. The formulas in this section go back at least to [16] and [19]. In our treatment we make systematic use of $P_H(I-zW)^{-1}|_H = (I-z\tilde{\theta}(z))^{-1}$, a formula that appears in a somewhat different context in [22], [23]. Let V be a closed isometric operator in a Hilbert space H, let $\theta \in S(D(V)^{\perp}, R(V)^{\perp})$ and let $\tilde{\theta}$ be given by (3.4). Put $F(z) = (I+z\tilde{\theta}(z))(I-z\tilde{\theta}(z))^{-1}$. Then $\tilde{\theta}(z) = \frac{1}{z}(F(z)-I)(F(z)+I)^{-1}$ and

$$\frac{F(z)+F(w)^*}{1-z\bar{w}} = 2(I-\bar{w}\tilde{\theta}(w)^*)^{-1}\,\frac{I-z\bar{w}\tilde{\theta}(w)^*\tilde{\theta}(z)}{1-z\bar{w}}\,(I-z\theta(z))^{-1}, \tag{3.7}$$

$$\frac{I-\tilde{\theta}(w)^*\tilde{\theta}(z)}{1-z\bar{w}} = \begin{pmatrix} \dfrac{I-\theta(w)^*\theta(z)}{1-z\bar{w}} & 0 \\ 0 & 0 \end{pmatrix}. \tag{3.8}$$

It follows that the kernel on the left hand side of (3.7), the kernel in the middle of the right hand side of (3.7) and, by [11], the kernel on the left hand side of (3.8) and therefore also the kernel in the upper left corner of the matrix on the right hand side, all have the same number of negative squares. Hence F and θ as members of the Carathéodory and Schur classes, respectively, have the same number of negative squares. The kernel on the left hand side of (3.7) has been considered in [12].

4. SYMMETRIC LINEAR RELATIONS IN A HILBERT SPACE

Let S be a symmetric linear relation in a Hilbert space H, i.e., $S \subset S^* \subset H^2$. It is straightforward to obtain $\|g-\ell f\| \geq |\mathrm{Im}\,\ell|\,\|f\|$, $\{f,g\} \in S$, $\ell \in \mathbb{C}\setminus\mathbb{R}$. This shows that any point in $\mathbb{C}\setminus\mathbb{R}$ is a point of regular type. Hence if S is a subspace (closed linear relation), then $R(S-\ell)$ is closed for any $\ell \in \mathbb{C}\setminus\mathbb{R}$. Thus we obtain the orthogonal decomposition $H = R(S-\bar{\ell}) \oplus \nu(S^*-\ell)$, $\ell \in \mathbb{C}\setminus\mathbb{R}$, with corresponding orthogonal projection $P_{\nu(S^*-\ell)}$ from H onto $\nu(S^*-\ell)$. In the following we use the notation $M_\ell(S^*) = \{\{\chi,\ell\chi\} \mid \chi \in \nu(S^*-\ell)\}$, $\ell \in \mathbb{C}$, and $M_\infty(S^*) = \{\{0,\chi\} \mid \chi \in S^*(0)\}$.

PROPOSITION 4.1. *(Von Neumann's formula) Let* S *be a symmetric subspace in the Hilbert space* $\mathcal{H}^2$. *Then* $S^* = S \dotplus M_\mu(S^*) \dotplus M_{\bar\mu}(S^*)$, $\mu \in \mathbb{C}\setminus\mathbb{R}$, *direct sums.*

PROOF. Evidently, $S \dotplus M_\mu(S^*) \dotplus M_{\bar\mu}(S^*)$ is a direct sum, which is contained in S^*. In order to prove the converse inclusion, we let $\{f,g\} \in S^*$. Then $g - \bar\mu f \in R(S^* - \bar\mu) \subset \mathcal{H} = R(S - \bar\mu) \oplus \nu(S^* - \mu)$, $\mu \in \mathbb{C}\setminus\mathbb{R}$. Hence there exist elements $\{h,k\} \in S$ and $\varphi \in \nu(S^* - \mu)$, such that $g - \bar\mu f = k - \bar\mu h + (\mu - \bar\mu)\varphi$. If we introduce $\psi = f - h - \varphi$, then $\{f,g\} = \{h,k\} + \{\varphi,\mu\varphi\} + \{\psi,\bar\mu\psi\}$, which shows that $\{\psi,\bar\mu\psi\} \in S^*$. This completes the proof.

Our interest will be in describing (families of) subspaces H, which are between S and S^*. The next proposition deals with the intermediate dissipative subspaces.

PROPOSITION 4.2. *Let* S *be a symmetric subspace in the Hilbert space* $\mathcal{H}^2$. *The following statements are equivalent:*

(i) $-H(H)$ *is a dissipative relation with* $S \subset H \subset S^*$,

(ii) *for* $\mu \in \mathbb{C}^+(\mathbb{C}^-)$ *there exist linear manifolds* $N_{\bar\mu} \subset M_{\bar\mu}(S^*)$, $N_\mu \subset M_\mu(S^*)$ *and a contraction* V_μ *from* $N_{\bar\mu}$ *onto* N_μ, *such that* $H = S \dotplus (I - V_\mu)N_{\bar\mu}$, *direct sum.*

Putting $N_{\bar\mu} = N_{\bar\mu}(H)$, $N_\mu = N_\mu(H)$, $V_\mu = V_\mu(H)$, *to indicate the correspondence between* (i) *and and* (ii) *we have*

$$D(N_{\bar\mu}(H)) = R(H - \mu) \cap \nu(S^* - \bar\mu),$$

$$D(N_\mu(H)) = R(H - \bar\mu) \cap \nu(S^* - \mu),$$

$$V_\mu(H)\{\psi,\bar\mu\psi\} = \{C_\mu(H)\psi,\ \mu C_\mu(H)\psi\}, \quad \{\psi,\bar\mu\psi\} \in N_{\bar\mu}(H).$$

The relation H *is closed if and only if* $N_{\bar\mu}(H)$ *is closed, which is the case if and only if* $V_\mu(H)$ *is a closed operator. The relation* $-H(H)$ *is maximal dissipative if and only if for* $\mu \in \mathbb{C}^+(\mathbb{C}^-)$ $N_{\bar\mu} = M_{\bar\mu}(S^*)$.

PROOF. We know (see Section 1) that $-H(H)$ is dissipative if and only if for $\mu \in \mathbb{C}^+(\mathbb{C}^-)$ $C_\mu(H)$ is contractive. If $-H(H)$ is dissipative and $\mu \in \mathbb{C}^+(\mathbb{C}^-)$, then $C_\mu(H)$ maps $R(H - \mu) \cap \nu(S^* - \bar\mu)$ onto $R(H - \bar\mu) \cap \nu(S^* - \mu)$ as $C_\mu(S)$ maps $R(S - \mu)$ onto $R(S - \bar\mu)$, compare [33]. Hence $C_\mu(H) = C_\mu(S) \oplus W$, where W, the restriction of $C_\mu(H)$ to $R(H - \mu) \cap \nu(S^* - \bar\mu)$, maps $R(H - \mu) \cap \nu(S^* - \bar\mu)$ onto $R(H - \bar\mu) \cap \nu(S^* - \mu)$. This implies that $H = S \dotplus \{\{\psi - W\psi, \bar\mu\psi - \mu W\psi\} \mid \psi \in R(H - \mu) \cap \nu(S^* - \bar\mu)\}$ and thus (ii) follows from (i). Conversely, if H is given as in (ii), then it is straightforward to show that $-H(H)$ is dissipative if $\mu \in \mathbb{C}^+(\mathbb{C}^-)$. The last part of the proposition is also easy to prove and is omitted.

COROLLARY 4.3. Let S be a symmetric subspace in the Hilbert space H^2. The following statements are equivalent:

(i) H is a symmetric linear relation with $S \subset H \subset S^*$,

(ii) for $\mu \in ¢ \setminus \mathbb{R}$ there exist linear manifolds $N_{\bar\mu} \subset N_{\bar\mu}(S^*)$, $N_\mu \subset N_\mu(S^*)$ and an isometry V_μ from $N_{\bar\mu}$ onto N_μ, such that $H = S \dotplus (I - V_\mu)N_{\bar\mu}$, direct sum.

With the notation as in Proposition 4.2 we have $V_\mu(H)^{-1} = V_{\bar\mu}(H)$ and $M_\mu(S^*) = M_\mu(H^*) \oplus (N_\mu(H))^c$, $\mu \in ¢ \setminus \mathbb{R}$.

PROOF. We just have to observe that H is symmetric if and only if H and $-H$ are dissipative. So the representation $H = S \dotplus (I - V_\mu(H))N_{\bar\mu}(H)$ is valid for any $\mu \in ¢ \setminus \mathbb{R}$, which yields $V_\mu(H)^{-1} = V_{\bar\mu}(H)$. The decomposition $H = R(H - \bar\mu)^c \oplus \nu(H^* - \mu)$, $\mu \in ¢ \setminus \mathbb{R}$, and the inclusion $S \subset H$ lead to

$$\nu(S^* - \mu) = (\nu(S^* - \mu) \cap R(H - \bar\mu)^c) \oplus \nu(H^* - \mu) = (\nu(S^* - \mu) \cap R(H - \bar\mu))^c \oplus \nu(H^* - \mu).$$

By Proposition 4.2 we obtain $\nu(S^* - \mu) = (D(N_\mu(H)))^c \oplus \nu(H^* - \mu)$, which is equivalent to $M_\mu(S^*) = M_\mu(H^*) \oplus (N_\mu(H))^c$.

For a symmetric subspace S in a Hilbert space H^2 we define $S(\ell) = S \dotplus M_\ell(S^*)$, $\ell \in ¢ \cup \{\infty\}$. Then we have $S \subset S(\ell) \subset S^*$, $\ell \in ¢ \cup \{\infty\}$, and $\pm S(\ell)$ is maximal dissipative for $\ell \in ¢^\pm$, and $S(\ell)^* = S(\bar\ell)$, $\ell \in ¢ \setminus \mathbb{R}$. Also $S(\ell)$ is symmetric for $\ell \in \mathbb{R} \cup \{\infty\}$, but note that $S(\ell)$ need not be closed for $\ell \in \mathbb{R} \cup \{\infty\}$. For $\mu \in ¢ \setminus \mathbb{R}$ we define the half plane $¢_\mu$ by $¢_\mu = \{\ell \in ¢ \mid (\mathrm{Im}\ell)(\mathrm{Im}\mu) > 0\}$. It is clear that $C_\mu(S(\ell))$, $\ell \in ¢_{\bar\mu}$, is a contraction, defined on all of H, and that $C_\mu(S(\ell))$ is an isometric operator for $\ell \in \mathbb{R} \cup \{\infty\}$, but not necessarily defined everywhere. Furthermore, for any $\mu \in ¢ \setminus \mathbb{R}$, $C_{\bar\mu}(S(\mu))$ is a partial isometry on H: it is isometric on $R(S - \bar\mu)$ with range $R(S - \mu)$ and null space $\nu(S^* - \mu)$.

The characteristic function Y_μ of the symmetric subspace S relative to $\mu \in ¢ \setminus \mathbb{R}$ is defined as the characteristic function of the isometric operator $C_{\bar\mu}(S)$. Hence (see Section 3) $Y_\mu(z) = zP_{\nu(S^* - \mu)}(I - zC_{\bar\mu}(S(\mu)))^{-1}|_{\nu(S^* - \bar\mu)}$, where we have used the relations $D(C_{\bar\mu}(S))^\perp = \nu(S^* - \mu)$, $R(C_{\bar\mu}(S))^\perp = \nu(S^* - \bar\mu)$. It is the characteristic function of the unitary colligation $(H, \nu(S^* - \bar\mu), \nu(S^* - \mu); C_{\bar\mu}(S(\mu)), I|_{\nu(S^* - \bar\mu)}, P_{\nu(S^* - \mu)}, 0)$ and with $z = \frac{\ell - \bar\mu}{\ell - \mu}$, $\ell \in ¢_{\bar\mu}$, the operator $P_{\nu(S^* - \mu)}(I - zC_{\bar\mu}(S(\mu)))^{-1}$ is the projection of H onto $\nu(S^* - \mu)$ parallel to $R(S - \ell)$; $H = \nu(S^* - \mu) + R(S - \ell)$, direct sum, $\ell \in ¢_{\bar\mu}$. Hence it is clear that $Y_{\bar\mu}(z) = Y_\mu(\bar z)^*$, or, in other words, that $Y_{\bar\mu}$ is the characteristic function of the isometric operator $C_\mu(S)$. As to the boundary behaviour of Y_μ on $\mathbb{R}$ we note that with $\xi = \frac{\lambda - \bar\mu}{\lambda - \mu}$, $\lambda \in \mathbb{R}$, $D_0(Y_\mu(\xi))$ is the set of all $\psi \in \nu(S^* - \bar\mu)$ for which there exists a unique $\varphi \in \nu(S^* - \mu)$ such that $\psi - \varphi \in R(S - \lambda)$ and then $Y_\mu(\xi)\psi = \xi\varphi$, $\psi \in D_0(Y_\mu(\xi))$. Furthermore, $D_0(Y_\mu(1))$ is the set of all $\psi \in \nu(S^* - \bar\mu)$ for which there exists a unique

$\varphi \in \nu(S^* - \mu)$ such that $\psi - \varphi \in D(S)$ and then $Y_\mu(1)\psi = \varphi$, $\psi \in D_0(Y_\mu(1))$.

A symmetric subspace S in a Hilbert space H^2 is called simple if it satisfies one of the following three equivalent conditions: (i) there is no non-trivial orthogonal decomposition $H = H_1 \oplus H_2$ $S = S_1 \oplus S_2$, where $S_1 \subset H_1^2$ is symmetric and $S_2 \subset H_2^2$ is selfadjoint, (ii) for some (and hence for all) $\mu \in ¢ \setminus \mathbb{R}$, the isometry $C_\mu(S)$ is simple and (iii) $\cap \{\nu(S^* - \ell) \mid \ell \in ¢ \setminus \mathbb{R}\} = \{0\}$. Clearly, a simple symmetric subspace is an operator, cf. [20]. As for isometric operators, the characteristic function of a symmetrix subspace coincides with its simple part defined in the obvious way, and conversely, if $Y \in S_0(F,G)$ with $Y(0) = 0$, where F and G are Hilbert spaces then Y can be identified with the characteristic function of a symmetric subspace in a Hilbert space H^2. The following result shows that the characteristic function Y_μ of a symmetric sucspace $S \subset H^2$ can be described in another way.

PROPOSITION 4.4. <u>We have for</u> $\mu \in ¢ \setminus \mathbb{R}$,

$$Y_\mu\left(\frac{\ell - \bar{\mu}}{\ell - \mu}\right) = C_\mu(S(\ell))\big|_{\nu(S^* - \bar{\mu})}, \quad \ell \in ¢_{\bar{\mu}}.$$

PROOF. By proposition 4.2 we obtain $S(\ell) = S \dotplus (I - V_\mu(S(\ell)))M_{\bar{\mu}}(S^*)$, $\ell \in ¢_{\bar{\mu}}$. Hence for all $\psi \in \nu(S^* - \bar{\mu})$ we have that $C_\mu(S(\ell))\psi \in \nu(S^* - \mu)$. The previous formula also implies $(I - V_\mu(S(\ell)))M_{\bar{\mu}}(S^*) \subset S \dotplus M_\ell(S^*)$, which shows that for all $\psi \in \nu(S^* - \bar{\mu})$ $(\ell - \bar{\mu})\psi - (\ell - \mu)C_\mu(S(\ell))\psi \in R(S - \ell)$, $\ell \in ¢_{\bar{\mu}}$. Hence $\frac{\ell - \mu}{\ell - \bar{\mu}} C_\mu(S(\ell))\psi$ is the parallel projection of $\psi \in \nu(S^* - \bar{\mu})$ onto $\nu(S^* - \mu)$, parallel to $R(S - \ell)$, which implies the equality in the proposition.

The next result concerns the boundary behaviour of Y_μ. Together with Proposition 4.4 it shows that, in a way, $C_\mu(S(\lambda))$, $\lambda \in \mathbb{R} \cup \{\infty\}$ can be considered as the limiting value of $C_\mu(S(\ell))$, $\ell \in ¢_\mu$ with $\ell \to \lambda$.

PROPOSITION 4.5. <u>For</u> $\lambda \in \mathbb{R} \cup \{\infty\}$ <u>we have</u>

$$D_0\left(Y_\mu\left(\frac{\lambda - \bar{\mu}}{\lambda - \mu}\right)\right) = D(N_{\bar{\mu}}(S(\lambda))), \quad Y_\mu\left(\frac{\lambda - \bar{\mu}}{\lambda - \mu}\right)\psi = C_\mu(S(\lambda))\psi, \quad \psi \in D_0\left(Y_\mu\left(\frac{\lambda - \bar{\mu}}{\lambda - \mu}\right)\right),$$

<u>where, if</u> $\lambda = \infty$, $\frac{\lambda - \bar{\mu}}{\lambda - \mu}$ <u>has to be replaced by</u> 1.

PROOF. By proposition 4.2 we have $S(\lambda) = S \dotplus (I - V_\mu(S(\lambda)))N_{\bar{\mu}}(S(\lambda))$, $\lambda \in \mathbb{R} \cup \{\infty\}$. Hence for all $\psi \in D(N_{\bar{\mu}}(S(\lambda)))$ we have that $C_\mu(S(\lambda))\psi \in \nu(S^* - \mu)$. The previous formula also implies $(I - V_\mu(S(\lambda)))N_{\bar{\mu}}(S(\lambda)) \subset S \dotplus M_\lambda(S^*)$, $\lambda \in \mathbb{R} \cup \{\infty\}$. This shows that for all $\psi \in D(N_{\bar{\mu}}(S(\lambda)))$, $(\lambda - \bar{\mu})\psi - (\lambda - \mu)C_\mu(S(\lambda))\psi \in R(S - \lambda)$ if $\lambda \in \mathbb{R}$, and $\psi - C_\mu(S(\infty))\psi \in D(S)$ if $\lambda = \infty$. Conversely, if $\psi \in \nu(S^* - \bar{\mu})$ and $\varphi \in \nu(S^* - \mu)$ are such that $(\lambda - \bar{\mu})\psi - (\lambda - \mu)\varphi \in R(S - \lambda)$

if $\lambda \in \mathbb{R}$, or $\psi - \varphi \in D(S)$ if $\lambda = \infty$, then $\psi \in D(N_{\bar{\mu}}(S(\lambda)))$ and $\varphi = C_\mu(S(\lambda))\psi$ for $\lambda \in \mathbb{R} \cup \{\infty\}$. We show this first for $\lambda \in \mathbb{R}$. In this case $(\lambda - \bar{\mu})\psi - (\lambda - \mu)\varphi = k - \lambda h$ for some $\{h,k\} \in S$. If we define $\chi = \psi - \varphi + h$, then it is clear that $\{\chi, \lambda\chi\} \in S^*$ and $\{\psi, \bar{\mu}\psi\} - \{\varphi, \mu\varphi\} = \{-h, -k\} + \{\chi, \lambda\chi\} \in S \dot{+} M_\lambda(S^*) = S(\lambda)$. If $\lambda = \infty$, then $\psi - \varphi \in D(S)$ implies that $\psi - \varphi = h$ for some $\{h,k\} \in S$. If we define $\chi = \bar{\mu}\psi - \mu\varphi - k$, then it is clear that $\{0,\chi\} \in S^*$ and $\{\psi, \bar{\mu}\psi\} - \{\varphi, \mu\varphi\} = \{h,k\} + \{0,\chi\} \in S \dot{+} M_\infty(S^*) = S(\infty)$. Hence $D(N_{\bar{\mu}}(S(\lambda)))$, $\lambda \in \mathbb{R}$, is the set of all $\psi \in \nu(S^* - \bar{\mu})$ for which there exists a uniquely determined element $\varphi \in \nu(S^* - \mu)$ such that $(\lambda - \bar{\mu})\psi - (\lambda - \mu)\varphi \in R(S - \lambda)$, and $D(N_{\bar{\mu}}(S(\infty)))$ is the set of all $\psi \in \nu(S^* - \bar{\mu})$ for which there exists a uniquely determined element $\varphi \in \nu(S^* - \mu)$ such that $\psi - \varphi \in D(S)$. The rest of the argument now follows from the characterization of $D_0(Y_\mu(\xi))$, $\xi \in \partial\mathbb{D}$, which we stated just before the statement of Proposition 4.4.

Many results in this section go back to Štraus [29], [33] for the case of (not necessarily densely defined) symmetric operators. Here we present such results from the point of view of unitary colligations.

5. ŠTRAUS EXTENSIONS AND GENERALIZED RESOLVENTS OF SYMMETRIC SUBSPACES

Let S be a symmetric subspace in a Hilbert space H^2 and let $\mu \in \mathbb{C} \setminus \mathbb{R}$ be fixed. We shall now show that all Štraus extensions $T(\ell)$ of S corresponding to all selfadjoint extensions A of S in Krein spaces $K \underset{s}{\supseteq} H$ with $\mu \in \rho(A)$ can be characterized by means of elements $\theta \in S(\nu(S^* - \bar{\mu}), \nu(S^* - \mu))$. First, we shall prove that $T(\ell)$ corresponding to a selfadjoint extension A of S in the Krein space K with $\mu \in \rho(A)$ can be written as

(5.1) $$T(\ell) = S \dot{+} \{\{\psi - \theta(z(\ell))\psi,\ \bar{\mu}\psi - \mu\theta(z(\ell))\psi\} | \psi \in \nu(S^* - \bar{\mu})\},$$
direct sum, $\ell \in \mathbb{C}_\mu$ with $z(\ell) \in D_\theta$,

where $z(\ell) = (\ell - \mu)(\ell - \bar{\mu})^{-1}$. As $T(\bar{\ell}) = T(\ell)^*$, $\ell \in \rho(A)$, formula (5.1) and Proposition 4.1 then imply that

(5.2) $$T(\bar{\ell}) = S \dot{+} \{\{\varphi - \theta(z(\ell))^*\varphi, \mu\varphi - \bar{\mu}\theta(z(\ell))^*\varphi\} | \varphi \in \nu(S^* - \mu)\},$$
direct sum, $\ell \in C_\mu$ with $z(\ell) \in D_\theta$.

In case K is a π_κ-space formula (5.1) leads to a description of $T(\lambda)$ for real values of λ and of $T(\infty)$ in terms of boundary values of θ on $\partial\mathbb{D}$.

To see that (5.1) holds, put $W = C_\mu(A)$. Then W is a unitary operator in K extending the isometry $V = C_\mu(S)$ in H. Such extensions were considered in Section 3 and in the following we shall use the notations introduced there. In particular, the matrix decomposition (3.1) of W with $D(V)^\perp = \nu(S^* - \bar{\mu})$ and $R(V)^\perp = \nu(S^* - \mu)$ and (3.2) imply that $C_\mu(A) = W = V \dotplus U = C_\mu(S) \dotplus U$, direct sum, and hence

$$A = S \dotplus F_\mu(U),\ \text{direct sum}, \tag{5.3}$$

where U is identified with its graph. Writing (5.3) in full detail and putting $\nu(S^* - \bar{\mu}) = F$, we obtain

$$A = S \dotplus \{\{\begin{pmatrix}\hat{\psi} - (T\hat{\psi} + F\psi) \\ \psi - (G\hat{\psi} + H\psi)\end{pmatrix}, \begin{pmatrix}\bar{\mu}\hat{\psi} - \mu(T\hat{\psi} + F\psi) \\ \bar{\mu}\psi - \mu(G\hat{\psi} + H\psi)\end{pmatrix}\} \mid \psi \in F,\ \hat{\psi} \in \hat{H}\},\ \text{direct sum}. \tag{5.4}$$

A direct consequence of this formula is a description of $A \cap H^2$:
$A \cap H^2 = S \dotplus \{\{\psi - H\psi, \bar{\mu}\psi - \mu H\psi\} \mid \psi \in \nu(F)\}$, direct sum. Using definition (0.1) we obtain from formula (5.4)

$$(5.5)\quad \begin{cases} T(\ell) = S \dotplus \{\{\psi - (G\hat{\psi} + H\psi), \bar{\mu}\psi - \mu(G\hat{\psi} + H\psi)\} \mid \psi \in F, \hat{\psi} \in \hat{H}, (\ell - \mu)\hat{\psi} = (\ell - \mu)(T\hat{\psi} + F\psi)\}, \\ \text{direct sum},\ \ell \in ¢, \\ T(\infty) = S \dotplus \{\{\psi - (G\hat{\psi} + H\psi), \bar{\mu}\psi - \mu(G\hat{\psi} + H\psi)\} \mid \psi \in F, \hat{\psi} \in \hat{H}, \hat{\psi} = T\hat{\psi} + F\psi\},\ \text{direct sum}. \end{cases}$$

Now (5.5) implies that the Štraus extension $T(\ell)$ of S associated with A via (0.1) can be written as in (5.1), where θ is the characteristic function of the unitary colligation $(\hat{H}, \nu(S^* - \bar{\mu}), \nu(S^* - \bar{\mu}); U)$. This proves the first part of the following theorem.

THEOREM 5.1. Let S be a symmetric subspace in H^2, where H is a Hilbert space and let $\mu \in ¢ \setminus \mathbb{R}$ be fixed. (i) If $T(\ell)$ is a Štraus extension of S corresponding to a selfadjoint extension A of S in K^2 with $\mu \in \rho(A)$, where K is a Krein space, then there exists a $\theta \in S(\nu(S^* - \bar{\mu}), \nu(S^* - \mu))$ such that $T(\ell)$ is given by (5.1). (ii) Conversely, if for some $\theta \in S(\nu(S^* - \bar{\mu}), \nu(S^* - \mu))$, $T(\ell)$ is given by (5.1), then there exists a selfadjoint extension A of S in a Krein space K with $\mu \in \rho(A)$, such that the Štraus extension of S corresponding to A coincides with $T(\ell)$ for all ℓ in some open set $O \subset ¢_\mu$ containing μ. The Krein space K can be chosen minimal in the sense that $K = \text{c.l.s.}\{(A - \ell)^{-1} f \mid f \in H\} \cup H$ in which case A is uniquely determined up to weak isomorphisms.

Here the last statement means that if for $j = 1,2$, A_j is a selfadjoint extension of S in the Krein space K_j, which is minimal in the above sense, and $\mu \in \rho(A_j)$, then the unitary operators $C_\mu(A_1)$ and $C_\mu(A_2)$ are weakly isomorphic. As to the proof of the second part of the theorem, Theorem 2.2 implies that θ is the characteristic function of some closely connected unitary colligation $\Delta = (K, \nu(S^* - \bar{\mu}), \nu(S^* - \mu); U)$, i.e., $\theta = \theta_\Delta$ on some open set in $\mathbb{D}$ containing 0 and one can easily verify that the subspace $A \subset K^2$ defined by (5.3) has the desired properties. Lemma 2.1 and the equality (1.1) imply the last statement of the theorem. We note that $C_\mu(T(\ell)) = V + \theta(z(\ell)) = \tilde{\theta}(z(\ell))$. It is possible to express the entries in the matrix representation of the unitary operator U, given by (3.2) in terms of the resolvent operator R_A of A:

PROPOSITION 5.2. For the entries in the matrix representation of U we have

$$T = \hat{P}(I + (\mu - \bar{\mu})R_A(\mu))|_{\hat{H}} \quad , \; F = (\mu - \bar{\mu})\hat{P}R_A(\mu)|_{\nu(S^* - \bar{\mu})},$$

$$G = (\mu - \bar{\mu})PR_A(\mu)|_{\hat{H}} \quad , \; H = P(I + (\mu - \bar{\mu})R_A(\mu))|_{\nu(S^* - \bar{\mu})},$$

and

$$T^+ = \hat{P}(I + (\bar{\mu} - \mu)R_A(\bar{\mu}))|_{\hat{H}} \quad , \; F^+ = (\bar{\mu} - \mu)PR_A(\bar{\mu})|_{\hat{H}},$$

$$G^+ = (\bar{\mu} - \mu)\hat{P}R_A(\bar{\mu})|_{\nu(S^* - \mu)}, \; H^+ = P(I + (\bar{\mu} - \mu)R_A(\bar{\mu}))|_{\nu(S^* - \mu)}$$

where $\hat{P} = P_{\hat{H}}$ and $P = P_H$.

PROOF. We have

$$T = \hat{P}C_\mu(A)|_{\hat{H}} = \{\{g - \mu f, \hat{P}(g - \bar{\mu}f)\} | \{f,g\} \in A, \; g - \mu f \in \hat{H}\} = \{\{h, \hat{P}(I + (\mu - \bar{\mu})R_A(\mu))h\} | h \in \hat{H}\},$$

where we have used $\{f,g\} \in A$ if and only if $R_A(\mu)(g - \mu f) = f$. Similarly we have

$$G = PC_\mu(A)|_{\hat{H}} = \{\{h, P(I + (\mu - \bar{\mu})R_A(\mu))h\} | h \in \hat{H}\} = \{\{h, (\mu - \bar{\mu})PR_A(\mu)h\} | h \in \hat{H}\},$$

$$F = \hat{P}C_\mu(A)|_{\nu(S^* - \bar{\mu})} = \{\{h, \hat{P}(I + (\mu - \bar{\mu})R_A(\mu))h\} | h \in \nu(S^* - \bar{\mu})\} =$$

$$= \{\{h, (\mu - \bar{\mu})\hat{P}R_A(\mu)h\} | h \in \nu(S^* - \bar{\mu})\},$$

$$H = PC_\mu(A)|_{\nu(S^* - \bar{\mu})} = \{\{h, P(I + (\mu - \bar{\mu})R_A(\mu))h\} | h \in \nu(S^* - \bar{\mu})\}.$$

The remaining identities easily follow from the relations $C_\mu(A)^+ = C_{\bar{\mu}}(A^+) = C_{\bar{\mu}}(A)$.

In the next theorem the selfadjoint extensions A of S in Krein spaces with $\mu \in \rho(A)$ are characterized by means of the generalized resolvents of S without making use of a fixed selfadjoint extension of S. We leave the proof to the reader.

THEOREM 5.3. Let S be a symmetric subspace in H^2, where H is a Hilbert space and let $\mu \in \mathbb{C} \setminus \mathbb{R}$ be fixed. (i) If A is a selfadjoint extension of S in a Krein space K with $\mu \in \rho(A)$ and with $W = C_\mu(A)$, $V = C_\mu(S)$, $\tilde{\theta}$ is the characteristic function of the unitary collligation (3.3), then

$$(5.6)\qquad P_H(A-\ell)^{-1}|_H = \frac{1}{\ell-\bar\mu}\left(\tilde\theta\left(\frac{\ell-\mu}{\ell-\bar\mu}\right)-I\right)\left(I-\frac{\ell-\mu}{\ell-\bar\mu}\,\tilde\theta\left(\frac{\ell-\mu}{\ell-\bar\mu}\right)\right)^{-1},\ \ell\in O,$$

where O is an open set in $\mathbb{C}_\mu$ containing μ. (ii) Conversely, if $\theta\in S(\nu(S^*-\bar\mu),\nu(S^*-\mu))$ and $\tilde\theta$ is given by (3.4) with $V=C_\mu(S)$ then there exists a selfadjoint extension A of S in a Krein space K with $\mu\in\rho(A)$ such that (5.6) holds for some open set O in C_μ containing μ. The Krein space K can be chosen minimal, i.e., $K=\text{c.l.s.}\{(A-\ell)^{-1}f\,|\,f\in H,\ \ell\in\rho(A)\}\cup H$, in which case A is uniquely determined up to weak isomorphisms.

COROLLARY 5.4. Let S be a symmetric subspace in H^2 and let $\mu\in\mathbb{C}\setminus\mathbb{R}$. Let A be a selfadjoint extension of S in a Krein space K with $\mu\in\rho(A)$ and let $\theta\in S(\nu(S^*-\bar\mu),\nu(S^*-\mu))$ be associated with A as in Theorem 5.2. Then for ℓ in an open set O in $\mathbb{C}_\mu$ containing μ, we have

$$P_{\nu(S^*-\bar\mu)}(A-\ell)^{-1}|_{\nu(S^*-\bar\mu)} = \left(\frac{1}{\ell-\mu}X\left(\frac{\ell-\mu}{\ell-\bar\mu}\right)\theta\left(\frac{\ell-\mu}{\ell-\bar\mu}\right)-\frac{1}{\ell-\bar\mu}\right)\left(I-X\left(\frac{\ell-\mu}{\ell-\bar\mu}\right)\theta\left(\frac{\ell-\mu}{\ell-\bar\mu}\right)\right)^{-1},$$

$$P_{\nu(S^*-\mu)}(A-\ell)^{-1}|_{\nu(S^*-\mu)} = \left(\frac{1}{\ell-\mu}\theta\left(\frac{\ell-\mu}{\ell-\bar\mu}\right)X\left(\frac{\ell-\mu}{\ell-\bar\mu}\right)-\frac{1}{\ell-\bar\mu}\right)\left(I-\theta\left(\frac{\ell-\mu}{\ell-\bar\mu}\right)X\left(\frac{\ell-\mu}{\ell-\bar\mu}\right)\right)^{-1},$$

where X is the characteristic function of the isometry $C_\mu(S)$.

In Theorems 5.1 and 5.3 the extension space K is a Pontryagin space if and only if $\theta\in S_\kappa(\nu(S^*-\bar\mu),\nu(S^*-\mu))$ for some $\kappa\in\mathbb{N}\cup\{0\}$. In that case the minimality of K in the last statements of Theorems 5.1 and 5.3 implies that A in fact is uniquely determined up to unitary equivalence. Moreover, in that case, if the Štraus extension $T(\ell)$ and $\theta\in S_\kappa(\nu(S^*-\bar\mu),\nu(S^*-\mu))$ are connected by (5.1) then with $\xi=\frac{\lambda-\mu}{\lambda-\bar\mu}$ if $\lambda\in\mathbb{R}$, and $\xi=1$ if $\lambda=\infty$, we have

$$(5.7)\qquad T(\lambda)=S\dotplus\{\{\psi-\theta(\xi)\psi,\ \bar\mu\psi-\mu\theta(\xi)\psi\}\,|\,\psi\in D_0(\theta(\xi))\},$$

where

$$D_0(\theta(\xi))=\{\psi\in\nu(S^*-\bar\mu)\,|\,\lim_{z\hat\to\xi}\frac{\|\psi\|-\|\theta(z)\psi\|}{1-|z|}\ \text{exists}\},$$

and

$$\theta(\xi)\psi=\lim_{z\hat\to\xi}\theta(z)\psi,\quad \psi\in D_0(\theta(\xi)).$$

In case $\kappa=0$ the descriptions (5.1) and (5.6) go back to Straus [30], [31] and [33] for the situation of not necessarily densely defined symmetric operators and certain admissable selfadjoint extensions. The description (5.1) for Hilbert space extensions can be found in [13] and was extended to the case of Pontryagin space extensions in [9]. The description (5.7) can be found in [10].

6. CANONICAL DIFFERENTIAL RELATIONS

We consider the first order linear differential equation $Jx'(t) = \lambda\Delta(t)x(t)$ for a $\mathbb{C}^n$ valued function x on a closed bounded interval [a,b], under the following conditions

(i) J is a constant $n \times n$ matrix such that $J^* = J^{-1} = -J$,

(ii) Δ is an integrable $n \times n$ matrix valued function on [a,b], such that $\Delta(t) = \Delta(t)^*$ and $\Delta(t) \geq 0$ for almost all $t \in [a,b]$,

(iii) $\int_a^b e^*\Delta(t)e\,dt > 0$ for all $e \in \mathbb{C}^n$, $e \neq 0$.

Let $H = L^2(\Delta)$ be the Hilbert space consisting of all equivalence classes f of $\mathbb{C}^n$ valued functions $\hat{f}$ on [a,b] such that $\int_a^b \hat{f}(t)^*\Delta(t)\hat{f}(t)dt < \infty$, provided with the inner product $[f,g] = \int_a^b \hat{g}(t)^*\Delta(t)\hat{f}(t)dt$, $f,g \in H$. By S_{max} we denote the subset of H^2, consisting of all $\{f,g\}$ such that there is a pair of representative functions $\hat{f}$ and $\hat{g}$ for f and g respectively, satisfying $\hat{f}$ is absolutely continuous on [a,b] and $J\hat{f}'(t) = \Delta\hat{g}(t)$, for almost all $t \in [a,b]$. Evidently, S_{max} is a linear relation in H. With each pair $\{f,g\} \in S_{max}$ there is associated a unique absolutely continuous function $\hat{f}$, which is an element of the equivalence class f. For $\{f,g\},\{h,k\} \in S_{max}$ and $a \leq p < q \leq b$ we have

$$(6.1)\qquad \int_p^q \hat{h}(t)^*\Delta(t)\hat{g}(t)dt - \int_p^q \hat{k}(t)^*\Delta(t)\hat{f}(t)dt = \int_p^q \hat{h}(t)^*J\hat{f}'(t)dt + \int_p^q \hat{h}'(t)^*J\hat{f}(t)dt$$
$$= \int_p^q \frac{d}{dt}(\hat{h}(t)^*J\hat{f}(t))dt = \hat{h}(q)^*J\hat{f}(q) - \hat{h}(p)^*J\hat{f}(p),$$

which is Lagrange's identity or Green's formula. In particular, we have with $\{f,g\},\{h,k\} \in S_{max}$

$$(6.2)\qquad [g,h] - [f,k] = (\hat{h}(a)^* \quad \hat{h}(b)^*)\begin{pmatrix} -J & 0 \\ 0 & J \end{pmatrix}\begin{pmatrix} \hat{f}(a) \\ \hat{f}(b) \end{pmatrix}.$$

It can be shown that

$$(6.3)\qquad S = \{\{f,g\} \in S_{max} \mid \hat{f}(a) = \hat{f}(b) = 0\}$$

is a closed symmetric linear relation in H, and that $S_{max} = S^*$. Hence we have $S_{max} = S \dotplus M_{\bar{\mu}}(S_{max}) \dotplus M_{\mu}(S_{max})$, $\mu \in \mathbb{C} \setminus \mathbb{R}$. It is clear in our situation that $\dim \nu(S_{max} - \bar{\mu}) = \dim \nu(S_{max} - \mu) = n$. Hence S has selfadjoint extensions in H and they are given by $\{\{f,g\} \in S_{max} \mid A\hat{f}(a) + B\hat{f}(b) = 0\}$, where A, B are $n \times n$ matrices such that $AJA^* = BJB^*$, $\operatorname{rank}(A : B) = n$. Here $(A : B)$ denotes the $n \times 2n$ matrix by putting the matrix B to the right of the matrix A in the indicated manner.

By $M(.,\lambda)$ we denote the $n \times n$ matrix valued function, which solves the initial value problem

$$\begin{cases} J \frac{d}{dt} M(t,\lambda) = \lambda\Delta(t)M(t,\lambda), \text{ for almost all } t \in [a,b], \\ M(a,\lambda) = I, \end{cases}$$

for all $\lambda \in ¢$. Then by (6.1) we obtain for $t \in [a,b]$ and $\lambda, \xi \in ¢$, that $M(t,\xi)^* JM(t,\lambda) - J = (\lambda - \overline{\xi}) \int_a^t M(s,\xi)^* \Delta(s)M(s,\lambda)ds$, and, in particular,

(6.4) $\qquad M(t,\overline{\lambda})^* JM(t,\lambda) = J, \; t \in [a,b], \; \lambda \in ¢.$

Introducing the entire $n \times n$ matrix valued function $M(\lambda)$, $\lambda \in ¢$, by $M(\lambda) = M(b,\lambda)$, the monodromy matrix, we obtain

(6.5) $\qquad M(\xi)^* JM(\lambda) - J = (\lambda - \overline{\xi})[M(.,\lambda),M(.,\xi)], \; \lambda,\xi \in ¢.$

In particular, we have $M(\lambda)^* JM(\lambda) - J = (\lambda - \overline{\lambda})[M(.,\lambda),M(.,\lambda)]$, $\lambda \in ¢$, so that

(6.6) $$\frac{M(\lambda)^* JM(\lambda) - J}{\lambda - \overline{\lambda}} \geq 0, \; \lambda \in ¢ \setminus \mathbb{R}.$$

In fact, we have by our assumption (iii), that

(6.7) $$\frac{M(\lambda)^* JM(\lambda) - J}{\lambda - \overline{\lambda}} > 0, \; \lambda \in ¢ \setminus \mathbb{R}.$$

Next we determine the characteristic function Y_μ of the subspace S, i.e., the characteristic function of the unitary colligation

$$(H, \nu(S^* - \overline{\mu}), \; \nu(S^* - \mu); \; C_{\overline{\mu}}(S(\mu)), \; I|_{\nu(S^* - \overline{\mu})}, \; P_{\nu(S^* - \mu)}, 0),$$

where S is the closed minimal linear relation (6.3). Note that bases for $\nu(S^* - \overline{\mu})$ and $\nu(S^* - \mu)$ are given by the columns of $M(.,\overline{\mu})$ and $M(.,\mu)$ respectively.

PROPOSITION 6.1. <u>For</u> $\ell \in ¢_{\overline{\mu}}$ <u>we have</u>

$$Y_\mu\left(\frac{\ell - \overline{\mu}}{\ell - \mu}\right)M(.,\overline{\mu}) = M(.,\mu)(M(\ell) - M(\mu))^{-1}(M(\ell) - M(\overline{\mu})).$$

PROOF. By Proposition 4.5 we have $Y_\mu\left(\frac{\ell - \overline{\mu}}{\ell - \mu}\right) = C_\mu(S(\ell))|_{\nu(S^* - \overline{\mu})}$, $\ell \in ¢_{\overline{\mu}}$ and for all $\psi \in \nu(S^* - \overline{\mu})$ we know that $C_\mu(S(\ell))\psi \in \nu(S^* - \mu)$ and $(\ell - \overline{\mu})\psi - (\ell - \mu)C_\mu(S(\ell))\psi \in R(S - \ell)$, $\ell \in ¢_{\overline{\mu}}$, cf. the proof of Proposition 4.5. This implies $C_\mu(S(\ell))M(.,\overline{\mu}) = M(.,\mu)A(\ell)$, $\ell \in ¢_{\overline{\mu}}$, for some holomorphic $n \times n$ matrix $A(\ell)$, and

$$(\ell-\bar{\mu})M(.,\bar{\mu}) - (\ell-\mu)M(.,\mu)A(\ell) \in R(S-\ell) = \nu(S^*-\bar{\ell})^{\perp}, \quad \ell \in ¢_{\bar{\mu}}.$$

This leads to $(\ell-\bar{\mu})[M(.,\bar{\mu}),M(.,\bar{\ell})] - (\ell-\mu)[M(.,\mu),M(.,\bar{\ell})]A(\ell) = 0$, so that by (6.5) we obtain $M(\bar{\ell})^*JM(\bar{\mu}) - J = (M(\bar{\ell})^*JM(\mu) - J)A(\ell)$. Multiplying this from the left by $M(\ell)J$ and using (6.4) we get $M(\ell) - M(\bar{\mu}) = (M(\ell) - M(\mu))A(\ell)$, $\ell \in ¢_{\bar{\mu}}$. It remains to show that $M(\ell) - M(\mu)$ is invertible for all $\ell \in ¢_{\bar{\mu}}$. If this were not true, then for some $\ell \in ¢_{\bar{\mu}}$ and some non-trivial $x \in ¢^n$ we have $M(\ell)x = M(\mu)x$. But this leads to the contradiction

$$0 < x^* \frac{M(\ell)^*JM(\ell) - J}{\ell - \bar{\ell}} x = \frac{\mu - \bar{\mu}}{\ell - \bar{\ell}} x^* \frac{M(\mu)^*JM(\mu) - J}{\mu - \bar{\mu}} x < 0.$$

This completes the proof.

The following proposition gives a relation between boundary values and the characteristic function, which determines a selfadjoint extension of S.

PROPOSITION 6.2. (i) <u>Let $\tilde{A}$ be a selfadjoint extension of</u> S <u>in a Krein space</u> K^2 <u>with a non real number</u> $\mu \in \rho(\tilde{A})$, <u>so that there exists a function</u> $\theta \in S(\nu(S^*-\bar{\mu}),\nu(S^*-\bar{\mu}))$ <u>such that</u> (5.1) <u>holds for all</u> $\ell \in ¢_\mu$ <u>with</u> $z(\ell) \in D_\theta$. <u>Then there exist holomorphic</u> $n \times n$ <u>matrix valued functions</u> A <u>and</u> B, <u>defined for all</u> $\ell \in ¢_\mu$ <u>with</u> $z(\ell) \in D_\theta$, <u>such that</u>

(a) $\operatorname{rank}(A(\ell) : B(\ell)) = n$,

(b) $A(\mu) + B(\mu)M(\mu)$ <u>is invertible,</u>

(c) $T(\ell) = \{\{f,g\} \in S_{max} \mid A(\ell)\hat{f}(a) + B(\ell)\hat{f}(b) = 0\}$,

(d) $\theta(z(\ell))M(.,\bar{\mu}) = M(.,\mu)(A(\ell) + B(\ell)M(\mu))^{-1}(A(\ell) + B(\ell)M(\bar{\mu}))$, <u>whenever the inverse exists.</u>

(ii) <u>Conversely let</u> $U \subset ¢_\mu$ <u>be a neighbourhood of</u> $\mu \in ¢ \setminus \mathbb{R}$, <u>on which functions</u> A <u>and</u> B <u>are given which satisfy</u> (a) <u>and</u> (b), <u>and which define</u> $T(\ell)$ <u>by</u> (c). <u>Then there exists a minimal selfadjoint extension</u> $\tilde{A}$ <u>of</u> S <u>in a Krein space</u> K^2 <u>with</u> $\mu \in \rho(\tilde{A})$, <u>such that on a possibly smaller neighbourhood</u> $\tilde{U} \subset U$ <u>of</u> μ, <u>the corresponding Štraus extension coincides with</u> $T(\ell)$ <u>given by</u> (c), <u>and the corresponding</u> $\theta(z(\ell))$ <u>is given by</u> (d).

REMARK. Locally around μ condition (a) follows from condition (b).

PROOF. By (5.2) we know that for all $\ell \in ¢_\mu$ with $z(\ell) \in D_\theta$

$$T(\ell) = T(\bar{\ell})^* = \{\{f,g\} \in S_{max} \mid [g,\varphi - \theta(z(\ell))^*\varphi] = [f,\mu\varphi - \bar{\mu}\theta(z(\ell))^*\varphi],\ \varphi \in \nu(S^* - \mu)\}.$$

Now we can write $\theta(z(\ell))^*M(.,\mu) = M(.,\bar{\mu})\Omega(\ell)^*$, with Ω a holomorphic $n \times n$ matrix valued function. By Green's formula (6.2) we obtain

$$T(\ell) = \{\{f,g\} \in S_{max} \mid (I - \Omega(\ell) \quad M(\mu)^* - \Omega(\ell)M(\overline{\mu})^*)\begin{pmatrix} -J & 0 \\ 0 & J \end{pmatrix}\begin{pmatrix} \hat{f}(a) \\ \hat{f}(b) \end{pmatrix} = 0\}.$$

If we introduce $A(\ell)$ and $B(\ell)$ by

$$(A(\ell) : B(\ell)) = (I - \Omega(\ell) \quad M(\mu)^* - \Omega(\ell)M(\overline{\mu})^*)\begin{pmatrix} -J & 0 \\ 0 & J \end{pmatrix},$$

then we obtain $\operatorname{rank}(A(\ell) : B(\ell)) = n$ (by a similar argument as given in the proof of Proposition 6.1) and $T(\ell) = \{\{f,g\} \in S_{max} | A(\ell)\hat{f}(a) + B(\ell)\hat{f}(b) = 0\}$. Since $\nu(T(\ell) - \ell) = P_H\nu(\tilde{A} - \ell)$ and $\mu \in \rho(\tilde{A})$, we obtain $\nu(T(\mu) - \mu) = \{0\}$, which implies that $A(\mu) + B(\mu)M(\mu)$ is invertible. In order to write $\theta(z(\ell))$ in terms of a fractional linear transformation we define $\theta(z(\ell))M(.,\overline{\mu}) = M(.,\mu)\Gamma(\ell)$. By (5.1) $M(.,\overline{\mu}) - M(.,\mu)\Gamma(\ell)$ must satisfy the boundary conditions, so that $A(\ell)(I - \Gamma(\ell)) + B(\ell)(M(\overline{\mu}) - M(\mu)\Gamma(\ell)) = 0$ for all $\ell \in \mathbb{C}_\mu$ with $z(\ell) \in D_\theta$, and thus $(A(\ell) + B(\ell)M(\mu))\Gamma(\ell) = A(\ell) + B(\ell)M(\overline{\mu})$. As to the converse, let A and B be given on U by (a) and (b). Then we define $\theta(z(\ell))$ by (d) on a neighbourhood of μ, as an element of $S(\nu(S^* - \overline{\mu}),\nu(S^* - \mu))$. By Theorem 5.1 θ defines a minimal selfadjoint extension $\tilde{A}$ such that (5.1) holds. By definition it is clear that around μ we have $S \dot{+} \{\{(I - \theta(z(\ell)))\psi, \overline{\mu}\psi - \mu\theta(z(\ell))\psi\} | \psi \in \nu(S^* - \overline{\mu})\} = \{\{f,g\} \in S_{max} | A(\ell)\hat{f}(a) + B(\ell)\hat{f}(b) = 0\}$.

We close this section with a brief description of the corresponding characteristic functions for a singular Sturm-Liouville operator, as mentioned in the Introduction. Let H be $L^2(0,\infty)$ and let S be the minimal operator associated with $L = -D^2 + q$ where q is a locally integrable real function on $[0,\infty)$. We assume the limit point case, i.e., $\dim \nu(S^* - \overline{\mu}) = \dim \nu(S^* - \mu) = 1$. For $\ell \in \mathbb{C}$ let $u_1(.,\ell)$ and $u_2(.,\ell)$ be solutions of $Ly = \ell y$, which satisfy $u_1(0,\ell) = 1$, $u_1'(0,\ell) = 0$, $u_2(0,\ell) = 0$, $u_2'(0,\ell) = -1$. For any $\ell \in \mathbb{C} \setminus \mathbb{R}$ there is a uniquely determined function $m(\ell)$ such that $\psi(.,\ell) = u_2(.,\ell) + m(\ell)u_1(.,\ell) \in H$. It is well-known that m is holomorphic in $\mathbb{C} \setminus \mathbb{R}$ and $\frac{\operatorname{Im} m(\ell)}{\operatorname{Im} \ell} > 0$, $\overline{m(\ell)} = m(\overline{\ell})$, $\ell \in \mathbb{C} \setminus \mathbb{R}$. We choose $\psi(.,\overline{\mu})$ and $\psi(.,\mu)$ as bases of $\nu(S^* - \overline{\mu})$ and $\nu(S^* - \mu)$ respectively, and note that $\|\psi(.,\overline{\mu})\| = \|\psi(.,\mu)\|$. For $\ell \in \mathbb{C}_{\overline{\mu}}$ we find

$$Y_\mu\left(\frac{\ell - \overline{\mu}}{\ell - \mu}\right)\psi(.,\overline{\mu}) = \frac{m(\ell) - m(\overline{\mu})}{m(\ell) - m(\mu)}\,\psi(.,\mu).$$

In a similar way we obtain for a Štraus extension, as given in the Introduction, that

$$\theta\left(\frac{\ell - \mu}{\ell - \overline{\mu}}\right)\psi(.,\overline{\mu}) = \frac{\beta(\ell) - \alpha(\ell)m(\overline{\mu})}{\beta(\ell) - \alpha(\ell)m(\mu)}\,\psi(.,\mu),$$

for ℓ in a neighbourhood of μ.

In the case of nth order differential operators in $H = L^2(a,b)$ the formula in Proposition 6.2 appears in Štraus' dissertation [28]. The characteristic function Y_μ for the singular differential operator $-D^2 + q$ can be found in [32].

For a treatment of canonical differential relations we refer to [25], while for further information concerning (6.1) we refer to [21]. For Proposition 6.2 in the case of Pontryagin space extensions we refer to [9].

Any entire $n \times n$ matrix valued function M is called J-inner, if it satisfies (6.6) for some J with $J^* = J^{-1} = -J$. It was shown by Potapov [26] that any J-inner function M with $M(0) = I$, is the monodromy matrix of a canonical system, see also [15]. If the J-inner function is strict, i.e., if it satisfies (6.7) then the transform $Z(\ell) = (M(\ell) - M(\mu))^{-1}(M(\ell) - M(\bar{\mu}))$ is defined for all $\ell \in \mathbb{C}_{\bar{\mu}}$. In this case it follows directly from (6.7) that for all $\ell \in \mathbb{C}_{\bar{\mu}}$

$$Z(\ell)^* \frac{M(\mu)^* JM(\mu) - J}{\mu - \bar{\mu}} Z(\ell) \leq \frac{M(\bar{\mu})^* JM(\bar{\mu}) - J}{\bar{\mu} - \mu}.$$

7. CHARACTERISTIC FUNCTIONS ACCORDING TO A.V. ŠTRAUS

In this section we shall show how the definition of the characteristic function of A.V. Štraus is related to the notions discussed here. Let K be a Krein space and let E be a closed linear relation in K with nonempty resolvent set. For $\mu \in \rho(E) \setminus \mathbb{R}$ we provide the subspaces E and E^+ by the following inner products

$$[\{f,g\},\{h,k\}]_E = -\frac{2\mathrm{Im}\mu}{i}([g,h] - [f,k]), \quad \{f,g\},\{h,k\} \in E,$$

and

$$[\{f,g\},\{h,k\}]_{E^+} = \frac{2\mathrm{Im}\mu}{i}([g,h] - [f,k]), \quad \{f,g\},\{h,k\} \in E^+.$$

We assume that there exist Krein spaces L and L_+ and isometries Γ from E into L and Γ_+ from E^+ into L_+, such that $R(\Gamma)$ is dense in L and $R(\Gamma_+)$ is dense in L_+.

For $\ell \in \rho(E)$ we define a mapping $S_{E^+}(\ell)$ from E^+ into E by

$$S_{E^+}(\ell)\{f,g\} = \{R_E(\ell)(g - \ell f), \ (I + \ell R_E(\ell))(g - \ell f)\}, \quad \{f,g\} \in E^+.$$

Let $\{f,g\} \in E^+$, $\{h,k\} \in E$, then obviously we have $S_{E^+}(\ell)\{f,g\} = \{h,k\}$ if and only if $g - \ell f = k - \ell h$. We define the characteristic function $\chi_{E^+}(\ell)$, $\ell \in \rho(E)$ as a mapping from $R(\Gamma_+) \subset L_+$ into $R(\Gamma) \subset L$ by the following diagram

$$\begin{array}{ccc} E^+ & \xrightarrow{\;S_{E^+}(\ell)\;} & E \\ \Gamma_+ \downarrow & & \downarrow \Gamma \\ L_+ & \xrightarrow[\;\chi_{E^+}(\ell)\;]{} & L \end{array}$$

It is clear that this mapping is well defined.

As an application we consider a canonical differential relation as in Section 6. We define $E = \{\{h,k\} \in S_{max} | \hat{h}(b) = 0\}$. Then it is not difficult to show that $\rho(E) = ¢$, and $E^+ = \{\{f,g\} \in S_{max} | \hat{f}(a) = 0\}$. From (6.2) it follows with $\mu = \frac{1}{2}i$, that Γ and Γ^+ with $\Gamma\{h,k\} = \hat{h}(a)$, $\Gamma^+\{f,g\} = \hat{f}(b)$, define isometries from E and E^+ onto $(¢^n, J)$, respectively. Now for $\{f,g\} \in E^+$, $\{h,k\} \in E$ we have $S_{E^+}(\ell)\{f,g\} = \{h,k\}$, $\ell \in ¢$ if and only if $k - \ell h = g - \ell f$ or $g - k = \ell(f - h)$. Since E^+, $E \subset S_{max}$, we have $\{f - h, g - k\} \in S_{max}$ so that $\hat{f} - \hat{h} = M(.,\ell)c$ for some $c \in ¢^n$. In particular, this shows $c = -\hat{h}(a)$ and hence $\hat{f}(b) = -M(b,\ell)\hat{h}(a)$. From this we obtain $\chi_{E^+}(\ell) = -M(\ell)$.

Next we shall show how such characteristic functions are related to characteristic functions of unitary colligations. For this we consider a unitary colligation $\Delta = (K,F,G;T,F,G,H)$, where K, F and G are Krein spaces. To that end we introduce the following notations. If $E \subset K^2$ is a subspace with $\mu \in \rho(E) \setminus \mathbb{R}$ we define the mappings Φ_μ and Ψ_μ by

$$\Phi_\mu\{h,k\} = k - \mu h, \quad \{h,k\} \in E,$$

and

$$\Psi_\mu h = \frac{1}{\mu - \bar{\mu}} \{(C_\mu(E) - I)h, \ (\mu C_\mu(E) - \bar{\mu})h\}, \quad h \in K.$$

Note that $\Psi_\mu \circ \Phi_\mu = I_E$ and $\Phi_\mu \circ \Psi_\mu = I_K$. We shall use a similar notation for E^+ and $\bar{\mu}$.

PROPOSITION 7.1. Let K be a Krein space and let F,G be Krein spaces. Let $\Delta = (K,F,G;T,F,G,H)$ be a unitary colligation, for which $R(F^+)$ is dense in F and $R(G)$ is dense in G. We define

$$E = F_\mu(T), \quad \Gamma = G \circ \Phi_\mu, \quad \Gamma_+ = F^+ \circ \Phi_{\bar{\mu}}, \quad \mu \in ¢ \setminus \mathbb{R}.$$

Then E is a subspace in K^2 with $\mu \in \rho(E)$, Γ maps E isometrically onto $R(G)$, and Γ_+ maps E^+ isometrically onto $R(F^+)$, and we have

$$\theta_\Delta\left(\frac{\ell - \mu}{\ell - \bar{\mu}}\right)\Big|_{R(F^+)} = -\chi_{E^+}(\ell), \quad \frac{\ell - \bar{\mu}}{\ell - \mu} \in \rho(T).$$

PROOF. From the definition it follows that E is a subspace with $\mu \in \rho(E)$. It is straightforward to show that Γ maps E isometrically onto $R(G)$ and Γ_+ maps E^+ isometrically onto $R(F^+)$. As $T = C_\mu(E)$, $\mu \in \rho(E)$, we recall that $\ell \in \rho(E)$ if and only if $\frac{\ell - \bar{\mu}}{\ell - \mu} \in \rho(T)$. It is clear that the following diagram commutes for $\ell \in \rho(E)$:

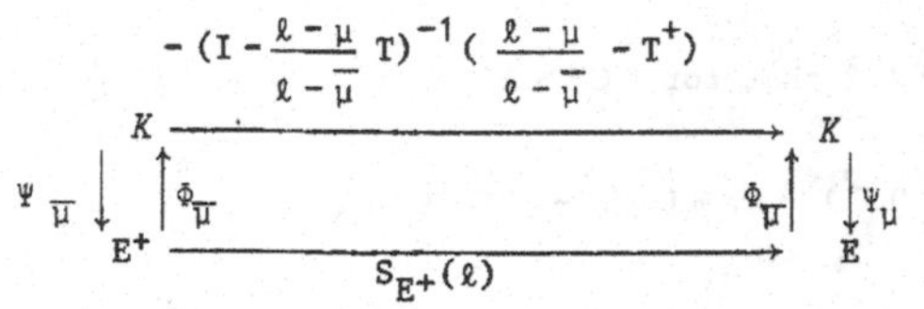

This shows that

$$\chi_E(\ell)\Gamma_+\Psi_{\overline{\mu}} = -\Gamma\Psi_\mu\left(I - \frac{\ell-\mu}{\ell-\overline{\mu}}T\right)^{-1}\left(\frac{\ell-\mu}{\ell-\overline{\mu}} - T^+\right),$$

or

$$\chi_E(\ell)F^+ = -G\left(I - \frac{\ell-\mu}{\ell-\overline{\mu}}T\right)^{-1}\left(\frac{\ell-\mu}{\ell-\overline{\mu}} - T^+\right), \qquad \ell \in \rho(E).$$

It is not difficult to check that

$$\theta_\Delta(z)F^+ = -G(I - zT)^{-1}(T^+ - z), \quad \frac{1}{z} \in \rho(T).$$

Hence we obtain

$$\theta_\Delta\left(\frac{\ell-\mu}{\ell-\overline{\mu}}\right)F^+ = -\chi_E(\ell)F^+, \quad \ell \in \rho(E).$$

REMARK. Consider a unitary colligation $\Delta = (K,F,G;T,F,G,H)$, where K is a Krein space, and F and G are Hilbert spaces. It is clear that H maps $\nu(F)$ isometrically onto $\nu(G^*)$. Thus we can reduce the unitary colligation by deleting the isometric part of H. Hence we obtain a unitary colligation of the form $\Delta' = (K,F',G';T,F',G,H')$ where $R(F'^+)$ is dense in F' and $R(G)$ is dense in G'. Note that

$$\theta_\Delta(z) = \theta_{\Delta'}(z) \oplus H|_{\nu(F)}.$$

We consider the situation of Section 5. Let S be a symmetric subspace in H^2, where H is a Hilbert space, and let A be a selfadjoint extension in K^2 of S, with $H \subseteq_s K$ and $\rho(A) \neq \emptyset$. Reducing the colligation corresponding to the previous Remark is equivalent to replacing S by $A \cap H^2$ in the extension problem, compare (0.2). Applying Proposition 5.2 we obtain

$$E = \{\{\hat{P}R_A(\mu)h,\ (I + \mu\hat{P}R_A(\mu))h\} \mid h \in \hat{H}\},$$

$$E^+ = \{\{\hat{P}R_A(\overline{\mu})h,\ (I + \overline{\mu}\hat{P}R_A(\overline{\mu}))h\} \mid h \in \hat{H}\},$$

$$\Gamma(\{\hat{P}R_A(\mu)h,\ (I + \mu\hat{P}R_A(\mu))h\}) = (\mu - \overline{\mu})PR_A(\mu)h, \quad h \in \hat{H},$$

$$\Gamma_+(\{\hat{P}R_A(\overline{\mu})h,\ (I + \overline{\mu}\hat{P}R_A(\overline{\mu}))h\} = (\overline{\mu} - \mu)PR_A(\overline{\mu})h, \quad h \in \hat{H},$$

where $\hat{P} = P_{\hat{H}}$ and $P = P_H$. We have

$$R(\Gamma)^c = \nu((A \cap H^2)^* - \mu), \quad R(\Gamma_+)^c = \nu((A \cap H^2)^* - \overline{\mu}).$$

We recall from Corollary 4.3 that for $\xi \in \mathbb{C} \setminus \mathbb{R}$ we have

$$\nu(S^* - \xi) = \nu((A \cap H^2)^* - \xi) \oplus (\nu(S^* - \xi) \cap R(A \cap H^2 - \overline{\xi}))^c.$$

It is with these subspaces E and E^+, and with these mappings Γ and Γ^+ that A.V. Štraus [29] showed that the characteristic function in Theorem 5.1 is a characteristic function in his sense.

Finally, we show that the characteristic functions of A.V. Štraus are all described by Proposition 7.1. We omit the proof, which is straightforward.

PROPOSITION 7.2. Let K be a Krein space and let $E \subset K^2$ be a subspace with $\rho(E) \neq \emptyset$. Let L and L_+ be Krein spaces and let Γ and Γ^+ be isometries from E into L, and E^+ into L_+, respectively, such that $R(\Gamma)$ is dense in L and $R(\Gamma^+)$ is dense in L_+. We define for $\mu \in \rho(E) \setminus \mathbb{R}$

$$T = C_\mu(E),\quad F^+ = \Gamma_+ \circ \Psi_{\bar{\mu}},\quad G = \Gamma \circ \Psi_\mu,\quad H = -(\chi_{E^+}(\mu))^c .$$

Then $\Delta = (K, L_+, L; T, F, G, H)$ is a unitary colligation and

$$\theta_\Delta\left(\frac{\ell - \mu}{\ell - \bar{\mu}}\right)\Big|_{R(F^+)} = -\chi_{E^+}(\ell),\quad \ell \in \rho(E).$$

The original definition of A.V. Štraus [29] was concerned with the characteristic function of a densely defined operator in a Hilbert space, whereas in his case the spaces L and L_+ were allowed to be inner product spaces. This definition of Štraus was extended to relations in inner product spaces by Čunaeva and Vernik [6].

REFERENCES

[1] D.Z. Arov, Passive linear stationary dynamical systems, Sibirskii Matem. Ž. 20 :2(114) (1979), 211-228.

[2] T. Ya Azizov, On the theory of extensions of isometric and symmetric operators in spaces with an indefinite metric, Preprint Voronesh University, 1982; Deposited paper n°. 3420-82.

[3] T. Ya Azizov, Extensions of J - isometric and J - unitary operators, Functional Anal. i Pr. ložen 18 (1984), 57-58 (English translation: Functional Anal. Appl. (1984), 46-48).

[4] J. Bognar, Indefinite Inner Product Spaces, Springer-Verlag, Berlin-Heidelberg-New York, 1974.

[5] M.S. Brodskii, Unitary operator colligations and their characteristic functions, Uspekhi Mat. Nauk 33:4 (1978), 141-168 (English translation: Russian Math. Surveys 33:4 (1978), 159-191).

[6] M.S. Cunaeva and A.N. Vernik, The characteristic function of a linear relation in a space with an indefinite metric, Functional Analysis, Ul'yanovsk. Gos. Ped. Inst., Ul'yanovsk, 16 (1981) 42-52.

[7] C. Davis, J - unitary dilation of a general operator, Acta Sci. Math. (Szeged) 31 (1970), 75-86.

[8] J. Dieudonné, Quasi-hermitian operators. Proc. Internat. Symposium Linear Spaces, Jerusalem 1961.

[9] A. Dijksma, H. Langer and H.S.V. de Snoo, Selfadjoint π_κ - extensions of symmetric subspaces: an abstract approach to boundary problems with spectral parameter in the boundary conditions, Integral equations and operator theory 7 (1984), 459-515.

[10] A. Dijksma, H. Langer and H.S.V. de Snoo, Unitary colligations in π_κ - spaces, characteristic functions and Štraus extensions, to appear in Pacific J. Math.

[11] A. Dijksma, H. Langer and H.S.V. de Snoo, Characteristic functions of unitary operator colligations in π_κ - spaces, to be published.

[12] A. Dijksma, H. Langer and H.S.V. de Snoo, Representations of holomorphic functions by means of resolvents of unitary or selfadjoint operators in Krein spaces, to be published.

[13] A. Dijksma and H.S.V. de Snoo, Selfadjoint extensions of symmetric subspaces, Pacific J. Math. 54 (1974), 71-100.

[14] A. Dijksma and H.S.V. de Snoo, Symmetric and selfadjoint relations in Krein spaces I, to be published.

[15] I.C. Gohberg and M.G. Krein, Theory of Volterra operators in Hilbert space and its applications (Russian), Moscow 1967; English transl.: Amer. Math. Soc. Transl. of Math. Monographs, 24, 1970.

[16] M.G. Krein and H. Langer, Über die verallgemeinerten Resolventen und die charakteristische Funktion eines isometrischen Operators im Raume Π_κ, Hilbert Space Operators and Operator Algebras (Proc. Int. Conf., Tihany, 1970) Colloqu. Math. Soc. János Bolyai, no. 5, North-Holland, Amsterdam (1972), 353-399.

[17] M.G. Krein and H. Langer, Some propositions on analytic matrix functions related to the theory of operators in the space Π_κ, Acta Sci. Math. (Szeged) 43 (1981), 181-205.

[18] H. Langer, Spectral functions of definitizable operators in Krein spaces in "Functional Analysis, Proceedings Dubrovnik 1981", Lecture Notes in Mathematics 948, Springer Verlag, Berlin-Heidelberg-New York, 1982.

[19] H. Langer and P. Sorjonen, Verallgemeinerte Resolventen hermitescher und isometrischer Operatoren im Pontryaginraum, Ann. Acad. Sci. Fennicae A.I. Math. 561 (1974), 1-45.

[20] H. Langer and B. Textorius, On generalized resolvents and Q - functions of symmetric linear relations (subspaces) in Hilbert spaces, Pacific J. Math. 72 (1977), 135-165.

[21] H. Langer and B. Textorius, L - resolvent matrices of symmetric linear relations with equal defect numbers; applications to canonical differential relations, Integral equations and operator theory 5 (1982), 208-243.

[22] H. Langer and B. Textorius, Generalized resolvents of contractions, Acta. Sci. Math. (Szeged) 44 (1982), 125-131.

[23] H. Langer and B. Textorius, Generalized resolvents of dual pairs of contractions, Invariant subspaces and other topics (Proc. 6-th Int. Conf. on Operator Theory, Romania, 1981) Birkhauser Verlag, Basel (1982), 103-118.

[24] P.D. Lax, Symmetrizable linear transformations, Comm. Pure Appl. Math. VII (1954), 633-647.

[25] B. Orcutt, Canonical differential equations, University of Virginia Ph. D. Thesis, 1969.

[26] V.P. Potapov, The multiplicative structure of J - contractive matrix functions, Trudy Moskov. Mat. Obšč. 4 (1955). 125-236 (Russian).

[27] W.T. Reid, Symmetrizable completely continuous linear transformations in Hilbert space, Duke Math. J. 18 (1951), 41-56.

[28] A.V. Štraus, On some questions in the theory of symmetric operators, Moscow State University, Dissertation, 1960.

[29] A.V. Štraus, Characteristic functions of linear operators, Izv. Akad. Nauk SSSR Ser. Mat. 24 (1960) 43-74 (English translation: Amer. Math. Soc. Transl. (2) 40 (1964) 1-37).

[30] A.V. Štraus, On the extensions of symmetric operators depending on a parameter, Izv. Akad. Nauk SSSR Ser. Mat. 29 (1965) 1389-1416 (English translation: Amer. Math. Soc. Transl. (2) 61 (1967) 113-141).

[31] A.V. Štraus, On one-parameter families of extensions of a symmetric operator, Izv. Akad. Nauk SSSR Ser. Mat. 30 (1966) 1325-1352 (English translation: Amer. Math. Soc. Transl. (2) 90 (1970) 135-164).

[32] A.V. Štraus, On the extensions and the characteristic function of a symmetric operator, Izv. Akad. Nauk SSSR Ser. Mat. 32 (1968) 186-207 (English translation: Math. USSR Izv. 2 (1968) 181-204).

[33] A.V. Štraus, Extensions and generalized resolvents of a symmetric operator which is not densely defined, Izv. Akad. Nauk SSSR Ser. Mat. 34 (1970) 175-202 (English translation: Math. USSR Izv. 4 (1970) 179-208.

[34] B. Sz.-Nagy and C. Foias, Harmonic Analysis of Operators on Hilbert space, North-Holland Publishing Company, Amsterdam-London, 1970.

A. Dijksma and H.S.V. de Snoo
Mathematisch Instituut
Rijksuniversiteit Groningen
Postbus 800
9700 AV GRONINGEN
Nederland

H. Langer
Technische Universität
Sektion Mathematik
DDR-8027 Dresden
Mommsenstrasse 13
D.D.R.

QUADRATIC AND SESQUILINEAR FORMS. CONTRIBUTIONS TO CHARACTERIZATIONS OF INNER PRODUCT SPACES

Svetozar Kurepa

INTRODUCTION

It was M.Fréchet who in 1935 raised a question of characterization of inner product spaces in the category of linear metric spaces. Roughly speaking his result says the following: A normed space X is an inner product space if each three dimensional subspace of X looks like euclidean three dimensional space. His paper was followed by the paper of P.Jordan and J. von Neumann in which the authors simplified Fréchet's result: A normed space X is an inner product space if and only if each two dimensional subspace of X looks like euclidean plane, i.e. the norm $x \to |x|$ satisfies the parallelogram law:

$$|x+y|^2 + |x-y|^2 = 2|x|^2 + 2|y|^2 \quad (x,y \in X) .$$

Since that time many characterizations of the inner product spaces have been obtained.

If X is a vector space over a field Φ and $L: X \times X \to \Phi$ is a sesquilinear form on X, then the function $q(x) = L(x,x)$ satisfies the parallelogram law

$$q(x+y)+q(x-y) = 2q(x)+2q(y) \quad (x,y \in X)$$

and the homogeneity property

$$q(\lambda x) = |\lambda|^2 q(x) \quad (\lambda \in \Phi,\ x \in X).$$

Professor Israel Halperin in 1963 (The New Scottish Book) raised the following question: Suppose that a function $q: X \to \Phi$ satisfies the parallelogram law and the above said homogeneity condition. Does there exist a sesquilinear form $L: X \times X \to \Phi$ such that $q(x)=L(x,x)$ holds for all $x \in X$? In 1964 we answered this question. If $\Phi = C$, then the answer is YES and if $\Phi = \mathbb{R}$ the answer is NO. In the case of a real space we gave a complete description of such functionals q in terms of derivations on $\mathbb{R}$ and an algebraic basic set of X. The result for a complex space can be considered as a generalization of the above mentioned wellknown Jordan-von Neumann characterization of the inner product spaces. In 1973 P.Vrbová proved a simple Lemma, by which our proof was simplified in the case $\Phi = C$. Recently J.Vukman extended the Halperin problem and our results to functions $q: X \to A$, where X

is a complex vector space, A is a complex hermitian Banach *-algebra with an identity and X is a unitary left A-module. On the other hand in his forthcoming paper C.T.Ng extended our results to vector spaces over an arbitrary field of characteristic different than 2 .

In the present paper we treat the above mentioned problems and some related topics. As for the References we did not strive for completeness and have listed primarily papers directly connected with the subject and methods we are using here. The paper is divided in sections as follows:

1. Some preliminary notions
2. Jordan-von Neumann Theorem
3. Quadratic forms on normed spaces
4. Solution of Halperin's problem for complex spaces
5. Halperin's problem for additive functions
6. Some regularity properties of quadratic forms
7. Solution of Halperin's problem for real vector spaces
8. Quadratic forms conditioned on an algebraic basic set
9. Quadratic forms on abelian groups
10. Quadratic forms on groups

1. SOME PRELIMINARY NOTIONS

In what follows Φ will denote the field of real numbers $\mathbb{R}$ or the field of complex numbers $\mathbb{C}$. By $X, Y, \ldots$ we denote a vector space over Φ.

A <u>function</u> $A: X \to Y$ is

a) additive if $A(x+y) = Ax+Ay \quad (x,y \in X)$;

b) linear (linear operator) if it is additive and

$$A(\lambda x) = \lambda Ax \quad (\lambda \in \Phi,\ x \in X) .$$

A function $B: X_1 \times X_2 \to Y$ is a bilinear operator if

$$x_1 \to B(x_1,x_2) \quad \text{and} \quad x_2 \to B(x_1,x_2)$$

are linear operators.

A function $B: X \times X \to \Phi$ is a sesquilinear functional (form) if $x \to B(x,y)$ is linear and $y \to B(x,y)$ antilinear, i.e. $B(x,\mu_1 y_1+\mu_2 y_2) = \bar{\mu}_1 B(x,y_1)+ \bar{\mu}_2 B(x,y_2)$. A sesquilinear form B is hermitian if

$$B(x,y) = \overline{B(y,x)} \quad (x,y \in X).$$

A hermitian form B is positive semidefinite if $B(x,x) \geq 0$, $x \in X$. A positive semidefinite hermitian form B is positive definite if $B(x,x) = 0 \Longrightarrow x=0$. An ordered pair of a vector space X and a positive definite hermitian form B is an inner product space (a unitary space). In this case we write

$$(x|y) = B(x,y).$$

A function $x \to |x|$ from a vector space X into $\mathbb{R}$ is a norm on X if

1. $|x| \geq 0 \quad (x \in X)$, 2. $|x| = 0 \Rightarrow x = 0$,
3. $|\lambda x| = |\lambda|\,|x| \quad (\lambda \in \Phi, x \in X)$, 4. $|x+y| \leq |x| + |y| \quad (x,y \in X)$.

An ordered pair of a vector space X and a norm on X is a normed space.

2. JORDAN-VON NEUMANN THEOREM

If X is an inner product space then the function $x \to |x| = (x|x)^{1/2}$ is a norm on X and:

$$|x+y|^2 = (x+y|x+y) = (x|x) + (x|y) + (y|x) + (y|y),$$

$$|x-y|^2 = (x-y|x-y) = (x|x) - (x|y) - (y|x) + (y|y).$$

If we add these two equalities we get the parallelogram identity:

(1) $$|x+y|^2 + |x-y|^2 = 2|x|^2 + 2|y|^2 \qquad (x,y \in X).$$

Here the norm

(2) $$|x| = (x|x)^{1/2} \qquad (x \in X)$$

on X is deduced from the inner product in X. Identity (1) is characteristic for inner product spaces.

THEOREM 1. (P.Jordan-J.von Neumann [10], 1935). If a norm $x \to |x|$ on a vector space X satisfies the parallelogram law (1) then

(3) $$(x|y) = \frac{|x+y|^2 - |x-y|^2}{4}$$

is an inner product in X if $\Phi = \mathbb{R}$, resp.

(4) $$(x|y) = \frac{|x+y|^2 - |x-y|^2}{4} + i\,\frac{|x+iy|^2 - |x-iy|^2}{4}$$

if $\Phi = C$. In any case

(5) $$(x|x) = |x|^2 \qquad (x \in X).$$

First we prove

LEMMA 1. Let X be a vector space over Φ. If a function $q: X \to \Phi$ satisfies the parallelogram law

(6) $$q(x+y) + q(x-y) = 2q(x) + 2q(y) \qquad (x,y \in X)$$

then the function

$$(7) \qquad S(x,y) = q(x+y) - q(x-y) \qquad (x,y \in X)$$

is symmetric and biadditive, i.e. additive in each argument.

Proof of Lemma 1. From (6) for $x = y = 0$ we get $q(0) = 0$. For $x = 0$ from (6) we get $q(-y) = q(y)$, i.e. q is an even function and therefore S is a symmetric function:

$$S(x,y) = S(y,x) \qquad (x,y \in X).$$

For $x,y,u \in X$ we have:

$$\begin{aligned} S(x+y,2u) &= q(x+y+2u)-q(x+y-2u) \\ &= q((x+u)+(y+u))+q((x+u)-(y+u)) \\ &- q((x-u)+(y-u))-q((x-u)-(y-u)) \\ &= 2(q(x+u)+ q(y+u))-2(q(x-u)+q(y-u)) \\ &= 2S(x,u) + 2S(y,u). \end{aligned}$$

Hence

$$S(x+y,2u) = 2S(x,u) + 2S(y,u) .$$

From this for $y=0$ and $x=z$ we get $S(z,2u) = 2S(z,u)$ which for $z = x+y$ implies

$$S(x+y,u) = S(x,u) + S(y,u). \square$$

Proof of Theorem 1. Since the function $q(x) = \frac{1}{4}|x|^2$ satisfies the conditions of Lemma 1, the function

$$(8) \qquad S(x,y) = \frac{|x+y|^2-|x-y|^2}{4}$$

is biadditive and symmetric. But then

$$S(rx,y) = rS(x,y)$$

holds for any rational number r. From

$$|tx+y| = |(t-s)x+sx+y| \leq |t-s|\ |x|+|sx+y| \Longrightarrow$$

$$||tx+y| - |sx+y|| \leq |t-s|\ |x|$$

the function

$$t \to |tx+y|$$

is continuous on $\mathbb{R}$ for any $x,y \in X$. This implies that the function

$$t \to S(tx,y)$$

is continuous on $\mathbb{R}$ .Since $S(rx,y) = rS(x,y)$ for rational r we find

$$(9) \qquad S(tx,y) = tS(x,y) \quad (t \in \mathbb{R}\ ;\ x,y \in X).$$

From (9) one easily verifies that by (3) in the case of a real space and by (4) in the case of a complex space,the inner product is defined on X such that $(x|x) = |x|^2$ holds for all $x \in X$.$\square$

Remark 1. J.von Neumann and P.Jordan have proved that for any normed space X, there exists a constant C, $1 \leq C \leq 2$ such that

$$\frac{1}{C} \leq \frac{|x+y|^2+|x-y|^2}{2|x|^2+2|y|^2} \leq C \qquad (x,y \in X) \tag{10}$$

provided that $|x| + |y| > 0$ [10] . J.A.Clarkson [6] has found that

$$C = 2^{\frac{2-p}{p}} \quad (1 \leq p \leq 2) \ , \ C = 2^{\frac{p-2}{p}} \quad (2 \leq p)$$

for spaces $X = L_p$ and $X = l_p$.

Proof of (10). If $x,y \in X$ then

$$|x+y|^2 \leq (|x|+|y|)^2 \leq |x|^2+2|x||y|+|y|^2 \Rightarrow$$

$$|x+y|^2 \leq 2|x|^2 + 2|y|^2 , \ |x-y|^2 \leq 2|x|^2 + 2|y|^2 \tag{11}$$

Hence

$$\frac{|x+y|^2+|x-y|^2}{2|x|^2+2|y|^2} \leq 2 . \tag{12}$$

The left side of (12) is equal to 1 if $x \neq 0$ and $y = 0$. Thus

$$C = \sup \frac{|x+y|^2+|x-y|^2}{2|x|^2+2|y|^2} , \quad C \in [1,2] . \tag{13}$$

In (13) the sup is taken over all $x,y \in X$ for which $|x|+|y| > 0$.

If $|x|+|y| > 0$ then for $x' = x+y$, $y' = x-y$ we have $|x'|+|y'| > 0$ so that

$$\frac{2|x|^2+2|y|^2}{|x+y|^2+|x-y|^2} = \frac{|x'+y'|^2+|x'-y'|^2}{2|x'|^2+2|y'|^2} \leq C ,$$

i.e.

$$\frac{1}{C} \leq \frac{|x+y|^2+|x-y|^2}{2|x|^2 + 2|y|^2} \leq C . \square$$

Remark 2. Suppose that on a vector space X a distance d is defined and that the metric d is translation invariant, i.e. $d(x,y) = d(x-y,0) (x,y \in X)$. Then the function $|x| = d(x,0)$ $(x \in X)$ has properties:

$$|x| \geq 0 \ , |-x| = |x| \ , \ |x+y| \leq |x| + |y| \qquad (x,y \in X)$$

and $|x| = 0$ if and only if $x = 0$. Conversely, if on X a function $x \to |x|$ is defined with the above four properties, then by $d(x,y) = |x-y|$ a translation invariant metric on X is defined. Observe that we do not assume $x \to |x|$ to be a norm, i.e. the homogeneity $|\lambda x| = |\lambda| \, |x|$ of the function $x \to |x|$ is not assumed.

P.Jordan and J.von Neumann [10] have proved the following theorem (See: Corollary 10). Let X be a complex vector space with distance defined in terms of a function $x \to |x|$ from X into $\mathbb{R}$

so that

$$|x| > 0 \ \text{if} \ x \neq 0 \ , \ |x+y| \leq |x|+|y| \ , \ |ix| = |x|$$

and

$$\lim_{t \to 0} |tx| = 0 \ .$$

Then the identity

(14) $$|x+y|^2 + |x-y|^2 = 2|x|^2 + 2|y|^2$$

is characteristic for the existence of an inner product $(x|y)$ connected with the function $x \to |x|$ by the relation

(15) $$(x|y) = \frac{1}{4}(|x+y|^2 - |x-y|^2) - \frac{i}{4}(|ix+y|^2 - |ix-y|^2).$$

On p. 721 in [10] we read that the relation (14) occurred in a paper of E.Wigner and the authors, Annals of Math. 35 (1934), p. 32 and that the importance of this relation in Hilbert spaces has been pointed by F.Riesz, Acta Szeged, 7 (1934), p.36. The above quoted theorem of P.Jordan and von Neumann is an improvement of the following result of M.Fréchet [7]:

Suppose that on a complex vector space X a translation invariant metric is defined in terms of a function $x \to |x|$ such that

$$|x| > 0 \ \text{if} \ x \neq 0 \ , \ |x| = 0 \iff x = 0 \ , \ |x+y| \leq |x|+|y| \ , |ix| = |x| .$$

If the function $x \to |x|$ has the property that for all $\alpha, \beta, \gamma \in \mathbb{R}$ and all $x, y, z, w \in X$:

(16) $$|\alpha(x-w)+\beta(y-w)+\gamma(z-w)|^2 = \alpha^2|x-w|^2+\beta^2|y-w|^2+\gamma^2|z-w|^2$$

$$+ \beta\gamma(|y-w|^2+|z-w|^2-|y-z|^2) + \gamma\alpha(|z-w|^2+|x-w|^2-|x-z|^2)$$

$$+ \alpha\beta(|x-w|^2+|z-w|^2-|x-y|^2)$$

holds, then X is an inner product space with an inner product defined by

$$(x|y) = \frac{1}{2}(|x|^2+|y|^2-|x-y|^2)+\frac{i}{2}(|x|^2+|y|^2-|x-iy|^2) \quad (x,y \in X).$$

The geometric meaning of (16) is that any n-dimensional $(n=1,2,3)$ affine real subspace Y of X is the ordinary n-dimensional euclidean space. If for example $y_1, y_2, y_3, y_0 \in Y$ are such that y_1-y_0, y_2-y_0 and y_3-y_0 are independent and that any $y \in Y$ is of the form

$$y = y_0+s_1(y_1-y_0)+s_2(y_2-y_0)+s_3(y_3-y_0) \quad (s_1,s_2,s_3 \in \mathbb{R}),$$

then points T_0, T_1, T_2, T_3 in $\mathbb{R}^3$ can be found such that the function $y \to T = T_0+s_1(T_1-T_0)+s_2(T_2-T_0)+s_3(T_3-T_0)$ is an isometric bijection of Y onto $\mathbb{R}^3$.

The geometric meaning of (14) is that any affine real n-dimensional (n=1,2) subspace of X is the ordinary n-dimensional euclidean space. For details see [7].

3. QUADRATIC FORMS ON NORMED SPACES

DEFINITION. Let X be a vector space over a field Φ. A functional $q: X \to \Phi$ is a <u>quadratic form</u> or a <u>quadratic functional</u> on X if

(1) $$q(x+y) + q(x-y) = 2q(x) + 2q(y) \quad (x,y \in X)$$

holds true.

By Lemma 1, the function

(2) $$S(x,y) = q(x+y) - q(x-y) \quad (x,y \in X)$$

is symmetric, biadditive, $q(0) = 0$, $q(-y) = q(y)$. Hence

$$S(r_1x, r_2y) = r_1r_2\, S(x,y)$$

holds for all rational numbers r_1 and r_2. From (1) and this for $x=y$ and $r_1=r_2=2r$ we get $q(2x) = 4q(x)$ and

(3) $$q(rx) = r^2q(x) \quad (x \in X;\ r \in Q) .$$

THEOREM 2 ([17]). Let X be a normed space over a field Φ and $q: X \to \Phi$ a quadratic functional. The following properties of q are equivalent:

1. q is continuous at a point $x_0 \in X$;
2. q is bounded on a ball;
3. q is bounded on any ball;
4. $S(x,y) = q(x+y)-q(x-y)$ is continuous on $X \times X$;
5. q is continuous on X.

<u>Proof</u>. If $x_n \to 0$, then $x_0+x_n \to x_0$ and $x_0-x_n \to x_0$ so that $q(x_0+x_n) \to q(x_0)$ and $q(x_0-x_n) \to q(x_0)$. This and

$$2q(x_n) = q(x_0+x_n) + q(x_0-x_n) - 2q(x_0) \Longrightarrow q(x_n) \to 0.$$

Hence q is continuous at $x = 0$. We claim that q is bounded on the unit ball $|x| \le 1$. Otherwise for each n we find $x_n, |x_n| \le 1$ such that $|q(x_n)| \ge n^2$. But then

$$\left| q\left(\frac{x_n}{n}\right) \right| \ge 1 \quad \text{and} \quad \frac{x_n}{n} \to 0$$

contradicts the continuity of q in x=0. Thus

$$\sup_{|x|\le 1} |q(x)| < +\infty .$$

This and $q(nx) = n^2 q(x)$, $n \in \mathbb{N}$ implies that q is bounded on any ball.

Suppose that

$$M = 2 \sup\{|q(z)| : |z| \le 2\} < +\infty .$$

Take x,y: $|x| \le 1$, $|y| \le 1$. Then $|x+y|, |x-y| \le 2$ so that $|S(x,y)| \le M$. Hence

$$(4) \qquad |x| \le 1 , \ |y| \le 1 \Rightarrow |S(x,y)| \le M .$$

Since $t \to S(tx,y)$ is additive and bounded on the interval $[0,1]$ we find [1] $S(tx,y) = tS(x,y)$. Thus

$$(5) \qquad S(tx,sy) = ts\, S(x,y) .$$

If $x \ne 0$ and $y \ne 0$ are arbitrary vectors, then by using (4) and (5)

$$S(x,y) = S\left(|x| \cdot \frac{x}{|x|}, |y| \cdot \frac{y}{|y|}\right) = |x| \cdot |y| \cdot S\left(\frac{x}{|x|}, \frac{y}{|y|}\right)$$

we find

$$(6) \qquad |S(x,y)| \le M|x|\,|y| \qquad (x,y \in X).$$

If $u,v \in X$, then by using (6) we have

$$|S(x,y)-S(u,v)| = |S(x-u,y-v)+S(x-u,v)+S(u,y-v)|$$

$$\le M|x-u| \cdot |y-v| + M|x-u| \cdot |v| + M|u| \cdot |y-v|$$

which implies that S is continuous on $X \times X$. From $q(x) = \frac{1}{4} S(x,x)$ we find that q is continuous on X. □

COROLLARY 1. Let X be a normed space and $q: X \to \Phi$ a quadratic form on X. Suppose that q is continuous on X and that $q(ix) = q(x)$, $x \in X$ in the case $\Phi = \mathbb{C}$. Then by

$$(7) \qquad L(x,y) = \frac{q(x+y)-q(x-y)}{4} \qquad \text{if} \quad \Phi = \mathbb{R}$$

and by

$$(8) \qquad L(x,y) = \frac{q(x+y)-q(x-y)}{4} + i\,\frac{q(x+iy)-q(x-iy)}{4}$$

if $\Phi = \mathbb{C}$ a continuous sesquilinear functional on X is defined. Furthermore

$$(9) \qquad q(x) = L(x,x) \qquad (x \in X) .$$

Proof. Since $S(tx,sy) = tsS(x,y)$ $(t,s \in \mathbb{R};\ x,y \in X)$ one easily finds that L defined by (7) resp. (8) is sesquilinear and that

(9) holds. Continuity of q implies also the continuity of L. □

COROLLARY 2. Let X be a Hilbert space over a field Φ and $q: X \to \mathbb{R}$ a continuous quadratic form such that $q(ix) = q(x)$, $x \in X$ holds if $\Phi = C$.

Then, there exists a unique bounded selfadjoint operator $A: X \to X$ such that

(10) $$q(x) = (Ax|x) \qquad (x \in X).$$

Proof. The functional L defined by (7) or (8) is continuous and sesquilinear. By the Riesz representation theorem there exists a bounded linear operator $A: X \to X$ such that

$$L(x,y) = (Ax|y) \qquad (x,y \in X).$$

Since $q(x) = L(x,x)$ is real one finds $L(x,y) = \overline{L(y,x)}$, i.e.

$$(Ax|y) = (x|Ay) \qquad (x,y \in X). \ \square$$

4. SOLUTION OF HALPERIN'S PROBLEM FOR COMPLEX SPACES

If $L: X \times X \to \Phi$ is a sesquilinear form on a vector space X, then the functional $q(x) = L(x,x)$ $(x \in X)$ has the following two properties:

(1) $$q(x+y)+q(x-y) = 2q(x)+2q(y) \qquad (x,y \in X)$$

(2) $$q(\lambda x) = |\lambda|^2 q(x) \qquad (\lambda \in \Phi;\ x \in X).$$

On the other hand if $q: X \to \Phi$ satisfies conditions (1), (2) and if X is a normed space and q continuous, then by

(3) $$L(x,y) = \frac{1}{4}(q(x+y)-q(x-y)) \quad \text{if } \Phi = \mathbb{R} \text{ resp.}$$

(4) $$L(x,y) = \frac{1}{4}(q(x+y)-q(x-y)) + \frac{i}{4}(q(x+iy)-q(x-iy))$$

in the case $\Phi = C$ a sesquilinear form is defined such that $q(x) = L(x,x)$ $(x \in X)$ (Corollary 1). This shows that the theory of continuous quadratic forms on normed spaces is reduced to the theory of continuous sesquilinear forms. It is natural to ask whether the analogous situation holds in general spaces. In other words, if X is a vector space over Φ and $q: X \to \Phi$ satisfies (1) and (2) is there any sesquilinear form L such that $q(x)=L(x,x)$ $(x \in X)$. This problem has been raised by Prof. Israel Halperin in 1963 in the New Scottish Book and communicated to us by prof. Janos Aczél. The following theorem is an extension of the Jordan-von Neumann characterization of inner product spaces.

THEOREM 3 (S.Kurepa [21], 1964). Let X ba a complex vector space and $q: X \to C$ a functional which satisfies conditions (1) and (2).

Then by (4) a sesquilinear form L on X is defined such that $q(x) = L(x,x)$ $(x \in X)$.

LEMMA 2. Let $f: C \to C$ be an additive function. If

$$f(\lambda) = |\lambda|^2 f(\tfrac{1}{\lambda}) \quad (\lambda \in C\ ,\ \lambda \neq 0)\ ,$$

then

$$f(\lambda) = f(1)\ \mathrm{Re}\ \lambda \quad (\lambda \in C)\ .$$

Proof. If in addition $f(1) = 0$, then for $\lambda \neq 0$ and $\lambda \neq -1$ we have

$$f(\lambda) = f(1+\lambda) = |1+\lambda|^2 f(\tfrac{1}{1+\lambda}) = |1+\lambda|^2 f(1-\tfrac{\lambda}{1+\lambda})$$

$$= -|1+\lambda|^2 f(\tfrac{\lambda}{1+\lambda}) = -|1+\lambda|^2 \left|\tfrac{\lambda}{1+\lambda}\right|^2 f(\tfrac{1+\lambda}{\lambda}) = -|\lambda|^2 f(\tfrac{1}{\lambda}) = -f(\lambda)\ .$$

Thus $f(\lambda) = 0$ for each $\lambda \in C$.

If $f(1) \neq 0$, then the function $f_1(\lambda) = f(\lambda) - f(1)\mathrm{Re}(\lambda), \lambda \in C$ is additive, $f_1(1) = 0$ and $f_1(\lambda) = |\lambda|^2 f_1(\tfrac{1}{\lambda})$ holds for $\lambda \in C$, $\lambda \neq 0$. Thus $f_1 = 0$, i.e. $f(\lambda) = f(1)\ \mathrm{Re}\ \lambda$. □

LEMMA 2'. (P.Vrbová [31], 1973). If $g: C \to C$ is additive and

$$g(\lambda) = -|\lambda|^2 g(\tfrac{1}{\lambda}) \quad (\lambda \in C,\ \lambda \neq 0)$$

then $g(\lambda) = g(i)\ \mathrm{Im}\ \lambda$ $(\lambda \in C)$.

Proof. The function $f(\lambda) = g(i\lambda)$ satisfies the conditions of Lemma 2. Thus $g(i\lambda) = g(i)\mathrm{Re}\ \lambda$, i.e. $g(\lambda) = g(i)\mathrm{Im}\ \lambda$. □

REMARK. Formulation of Vrbová's Lemma in the form of Lemma 2 is essentially due to J.Vukman [33].

Proof of Theorem 3. According to Lemma 1 the functional

$$S(x,y) = q(x+y) - q(x-y) \quad (x,y \in X) \tag{5}$$

is biadditive. Hence the functional L defined by (4) is also biadditive. For $\lambda \in C$, $\lambda \neq 0$ by using (2) we have

$$S(\lambda x,y) = q(\lambda x+y) - q(\lambda x-y) = |\lambda|^2(q(x+\tfrac{y}{\lambda}) - q(x-\tfrac{y}{\lambda})) = |\lambda|^2 S(x,\tfrac{y}{\lambda}) \Rightarrow$$

$$L(\lambda x,y) = |\lambda|^2 L(x,\tfrac{y}{\lambda}),\ L(x,\lambda y) = |\lambda|^2 L(\tfrac{x}{\lambda},y)\ . \tag{6}$$

Now

$$4L(ix,y) = S(ix,y) + iS(ix,iy) = S(x,-iy) + iS(x,y)$$
$$= i(S(x,y) - iS(x,-iy)) = i(S(x,y) + iS(x,iy))$$
$$= 4iL(x,y) \Longrightarrow$$

(7) $$L(ix,y) = iL(x,y) .$$

Furthermore:

$$4L(x,iy) = S(x,iy) + iS(x,-y) = S(x,iy) - iS(x,y)$$
$$= -i(S(x,y) + iS(x,iy)) = -4iL(x,y) \Longrightarrow$$

(8) $$L(x,iy) = -iL(x,y) .$$

For $x,y \in X$ the function

$$\lambda \to f(\lambda) = L(\lambda x,y) + L(x,\lambda y)$$
$$= |\lambda|^2 L(x,\tfrac{y}{\lambda}) + |\lambda|^2 L(\tfrac{x}{\lambda},y) = |\lambda|^2 f(\tfrac{1}{\lambda})$$

from $\mathbb{C}$ into $\mathbb{C}$ satisfies the conditions of Lemma 2. Hence $f(\lambda) = f(1)\,\mathrm{Re}\,\lambda$. In particular $f(i\tau) = 0$ for $\tau \in \mathbb{R}$ so that (7) and (8) imply

(9) $$L(\tau x,y) = L(x,\tau y) .$$

If $\lambda = \sigma$, $\sigma \in \mathbb{R}$, then

$$L(\sigma x,y) + L(x,\sigma y) = f(\sigma) = \sigma f(1) = \sigma(L(x,y)+L(x,y))$$

together with (9) implies

$$L(\sigma x,y) = L(x,\sigma y) = \sigma L(x,y)$$

from which it follows that L is a sesquilinear form. □

J. Vukman has generalized Lemma 2 and Theorem 3 to some Banach *-algebras.

LEMMA 3 ([33]). Let A be a complex hermitian Banach *-algebra with an identity e. If $f: A \to A$ is an additive function such that

$$f(a) = af(a^{-1})\,a^*$$

holds for all normal invertible elements $a \in A$, then

$$f(a) = \frac{1}{2}\,(af(e) + f(e)a^*) \qquad (a \in A).$$

THEOREM 4 ([33]). Let X be a complex vector space and A as in Lemma 3. Suppose that X is a unitary left A-module. If $q: X \to A$ is such that

$$q(x+y)+q(x-y) = 2q(x)+2q(y) \qquad (x,y \in X)$$

$$q(ax) = aq(x)a^* \qquad (a \in A;\ x \in X)$$

then the function $L: X \times X \to A$ defined by

$$L(x,y) = \frac{1}{4}\,(q(x+y)-q(x-y)) + \frac{i}{4}(q(x+iy)-q(x-iy))$$

is biadditive and

$$L(ax,y) = aL(x,y),\ L(x,ay) = L(x,y)a^*\ ,\ q(x) = L(x,x)$$

holds for all $a \in A$ and all $x,y \in X$.

5. HALPERIN'S PROBLEM FOR ADDITIVE FUNCTIONS

In 1963 prof. I.Halperin raised and prof. J.Aczél communicated to us the following question:

Suppose that a function $f: \mathbb{R} \to \mathbb{R}$ is additive and that

$$f(t) = t^2\, f(\tfrac{1}{t}) \qquad (t \in \mathbb{R}\ ,\ t \neq 0)\ .$$

Is f continuous?

In connection with this and some related questions derivations on $\mathbb{R}$ are to be considered.

DEFINITION. An additive function $a: \mathbb{R} \to \mathbb{R}$ is a <u>derivation</u> on $\mathbb{R}$ if

$$a(ts) = ta(s) + sa(t) \qquad (t,s \in \mathbb{R}).$$

The existence and abundance of nontrivial derivations is proved in [34].

THEOREM 5 (S.Kurepa [20], 1964). Let $f,g: \mathbb{R} \to \mathbb{R}$, $g \neq 0$ be additive functions. If

$$g(t) = P(t)f(\tfrac{1}{t}) \qquad (t \in R\ ,\ t \neq 0)\ ,$$

where P is a continuous function such that $P(1) = 1$, then: $P(t)=t^2$, the function

$$a(t) = f(t) - tf(1) \qquad (t \in \mathbb{R})$$

is a derivation on $\mathbb{R}$ and

$$f(t) = \alpha t + a(t)\ ,\ g(t) = \alpha t - a(t)$$

where $\alpha = f(1) = g(1)$.

<u>Proof</u>. If we take $t \neq 0$ and a rational number $r \neq 0$, then

$$g(rt) = P(rt)f(\tfrac{1}{rt}) \Longrightarrow rg(t) = P(rt)\ \frac{1}{r}\ f(\tfrac{1}{t}) \Longrightarrow$$

$$(1) \qquad \left[\ \frac{P(rt)}{r^2} - P(t)\right] f(\tfrac{1}{t}) = 0\ .$$

Using continuity of P from (1) we find

$$[\frac{P(st)}{s^2} - P(t)]\, f(\frac{1}{t}) = 0 \qquad (s \in \mathbb{R},\ s \neq 0)\ .$$

Since $f \neq 0$ we get $P(st) = s^2 P(t)$ for all $s \neq 0$ and for at least one $t \neq 0$. If we take $s = \frac{1}{t}$ we get $P(t) = t^2 P(1) = t^2$. Hence $P(st) = s^2 t^2$, i.e. $P(s) = s^2$ for any $s \neq 0$. By continuity of P we find $P(s) = s^2$ for any $s \in \mathbb{R}$. Thus

(2) $$g(t) = t^2 f(\frac{1}{t}) \qquad (t \in R\ ,\ t \neq 0).$$

From (2) we find $g(1) = f(1)$. Set

(3) $$a(t) = f(t) - tf(1), \quad b(t) = g(t) - tg(1) \qquad (t \in \mathbb{R})\ .$$

Using (2) we find that additive functions a and b are related by the equation

(4) $$b(t) = t^2 a(\frac{1}{t}) \qquad (t \in R\ ,\ t \neq 0)\ .$$

Obviously $a(r) = b(r) = 0$ for any $r \in Q$. We have therefore

$$b(t) = b(1+t) = (1+t)^2 a(\frac{1}{1+t}) = (1+t)^2 a(1 - \frac{t}{1+t})$$

$$= -\,(1+t)^2 a(\frac{t}{1+t}) = -(1+t)^2 (\frac{t}{1+t})^2 b(\frac{1+t}{t}) = -t^2 b(\frac{1}{t}) = -a(t).$$

Thus

(5) $$b(t) = -a(t) \qquad (t \in \mathbb{R})$$

which together with (4) leads to

(6) $$a(t) = -t^2 a(\frac{1}{t}) \qquad (t \in R,\ t \neq 0)\ .$$

From (5) and (3) we find $f(t) + g(t) = 2tf(1)$. It remains to be proven that a is a derivation on $\mathbb{R}$. For a given $t \in \mathbb{R}$, $t^2 \neq 1$, $t \neq 0$ we apply a to the identity

$$\frac{2}{t^2-1} = \frac{1}{t-1} - \frac{1}{t+1}$$

and use (6) to get

(7) $$a(t^2) = 2ta(t) \qquad (t \in R).$$

Replacing in (7) t by $t+s$ and using (7) we get $a(ts)=ta(s)+sa(t)$ for all $t,s \in \mathbb{R}$. □

COROLLARY 3. If $f: \mathbb{R} \to \mathbb{R}$ is an additive function and if

(8) $$f(t) = t^2 f(\frac{1}{t}) \qquad (t \in R\ ,\ t \neq 0)\ ,$$

then $f(t) = tf(1)$ $(t \in \mathbb{R})$.

<u>Proof</u>. The relation (8) is nothing but (2) for $f = g$. Hence $2f(t) = f(t)+g(t) = 2tf(1)$, i.e. $f(t) = tf(1)$. □

Recently C.T.Ng [27] has generalized Theorem 5 by replacing the field $\mathbb{R}$ by any field k of characteristic $\neq 2$. The following is his result.

THEOREM 6 (C.T.Ng [27], 1985). Let additive $F,G: k \to k$ and multiplicative $M: k \to k$ be nonzero maps satisfying

$$F(x) + M(x)G(\tfrac{1}{x}) = 0 \quad (x \in k,\ x \neq 0) .$$

Then they are of the following three exclusive representations.

(i) For some (field) morphism $\varphi: k \to k$, $\varphi \neq 0$, some φ-derivation D which is an additive map satisfying $D(xy) = \varphi(x)D(y) + \varphi(y)D(x)$ for all x,y in k, and a constant $b \in k$ with $(D,b) \neq (0,0)$,

$$F = D + b\varphi, \quad G = D - b\varphi, \quad M = \varphi^2 .$$

(ii) For some morphisms $\varphi,\psi: k \to k$, $\varphi,\psi \neq 0$ and $\varphi \neq \psi$ and some constants $c_1, c_2 \in k$, $(c_1,c_2) \neq (0,0)$,

$$F = c_1\varphi - c_2\psi, \quad G = c_2\varphi - c_1\psi, \quad M = \varphi\psi .$$

(iii) For some constant $c \in k$ with $a = \sqrt{c} \notin k$, some embedding $\varphi: k \to k(a)$ with $\varphi \neq \overline{\varphi}$, and some nonzero constant $\lambda \in k(a)$,

$$F = \lambda\varphi + \overline{\lambda\varphi}, \quad G = -\overline{\lambda}\varphi - \lambda\overline{\varphi}, \quad M = \varphi\overline{\varphi} .$$

Here conjugacy in $k(a)$ is defined by $\overline{x+ya} = x - ya$, for all $x,y \in k$.

The converse is also true.

REMARK. Corollary 3 was independently obtained by S.Kurepa [20] and by W.B.Jurkat [11]. For some other generalizations of Theorem 5 see papers [12], [13] and [26].

6. SOME REGULARITY PROPERTIES OF A QUADRATIC FORM

In this section we consider $\mathbb{R}^n$ as an inner product space with a scalar product given by

$$(x|y) = \sum_{i=1}^{n} \xi_i \eta_i, \quad x = \sum_{i=1}^{n} \xi_i e_i, \quad y = \sum_{j=1}^{n} \eta_j e_j$$

where $e_1 = (1,0,\dots,0)$, $e_2 = (0,1,0,\dots)$ etc.

THEOREM 7 ([17]) If a quadratic form $q: \mathbb{R}^n \to \mathbb{R}$ is bounded on a set of positive Lebesgue's measure, then

$$q(x) = (Ax|x) \qquad (x \in \mathbb{R}^2) \tag{1}$$

holds with a symmetric linear operator $A: \mathbb{R}^n \to \mathbb{R}^n$.

The proof of Theorem 7 is based on the following lemma.

LEMMA 4 ([16],[19]). Let $K \subset \mathbb{R}^n$ be Lebesgue measurable set such that $0 < mK < +\infty$. Then, there exists a number $\rho > 0$ with the property that for every x from the ball $K(0,\rho) = \{y \in \mathbb{R}^n : |y| < \rho\}$ there are elements $s_1(x), s_2(x), s_3(x) \in K$ such that

(2) $$s_1(x) = s_2(x) - \frac{1}{2}x = s_3(x) - x .$$

Proof. Let u be a function defined on $\mathbb{R}^n$ by the equation

(3) $$u(x) = m(K - (K - \frac{x}{2}) \cap (K-x)).$$

If χ denotes the characteristic function of the set K, then

$$|u(x)-u(0)| = |\int\chi(t)(\chi(t+\frac{x}{2})\chi(t+x)-\chi(t)\chi(t+x) + \chi(t)\chi(t+x)-\chi(t))dt|$$
$$\leq \int|\chi(t+\frac{x}{2})-\chi(t)|dt + \int|\chi(t+x)-\chi(t)|\, dt ,$$

where integrals are taken over $\mathbb{R}^n$. Since the right side tends to zero as $x \to 0$ we find the function u continuous in $x = 0$. Since

$$u(0) = mK > 0 ,$$

there exists a number $\rho > 0$ such that $u(x) \neq 0$ for each $x \in K(0,\rho)$. But $u(x) \neq 0$ implies

$$K \cap (K - \frac{x}{2}) \cap K \neq \emptyset .$$

Hence for each $x \in K(0,\rho)$ there are $s_1(x), s_2(x), s_3(x) \in K$ such that $s_1(x) = s_2(x) - \frac{x}{2} = s_3(x) - x$. □

Proof of Theorem 7. By the assumption there is a measurable set $K \subset \mathbb{R}^n$, $0 < mK < +\infty$ on which the restriction of q is bounded. Let

(4) $$M = \sup\{|q(x)| : x \in K\}.$$

Replacing x by $x+y$ in $q(x+y)+q(x-y) = 2q(x) + 2q(y)$ we deduce

(5) $$\frac{1}{2}|q(2y)| = 2|q(y)| \leq 2|q(x+y)|+|q(x+2y)|+|q(x)| .$$

According to Lemma 4 there exists a number $\rho > 0$ such that for every $2y \in K(0,\rho)$ there is a corresponding element x with the property that

$$x, x + \frac{2y}{2}, x+2y \in K .$$

If for $2y \in K(0,\rho)$ we take such an x, then (5) implies

$$|q(y)| \leq 2M ,$$

i.e. the function q is bounded on a ball $K(0, \frac{\rho}{2})$. By Theorem 7 q is continuous and by Corollary 2 q is of the form

$$q(x) = (Ax|x) \qquad (x \in \mathbb{R}^n)$$

with a selfadjoint operator A. □

COROLLARY 4. Let $q\colon \mathbb{R}^n \to \mathbb{R}$ be a quadratic form. If there exists a measurable set $K \subset \mathbb{R}^n$ of positive Lebesgue's measure such that the restriction of q to K is measurable, then q is continuous.

Proof. Since K is the union of disjoint and measurable sets

$$K_i = \{x \in K\colon i-1 \le q(x) < i\} \quad , \ i \in Z$$

there is an integer j such that $mK_j > 0$. This and boundedness of q on K_j together with Theorem 7 imply $q(x) = (Ax|x)$ for any $x \in \mathbb{R}^n$ with a selfadjoint operator A. □

COROLLARY 5 ([17]). If a Lebesgue measurable function $f\colon \mathbb{R}\to\mathbb{R}$ satisfies the functional equation

(6) $$f(x+y) + f(x-y) = 2f(x)+2f(y) \qquad (x,y \in \mathbb{R}),$$

then

$$f(x) = x^2 f(1) \qquad (x \in \mathbb{R}) .$$

COROLLARY 6. An additive functional $g\colon \mathbb{R}^n \to \mathbb{R}$ which is bounded on a set of positive Lebesgue measure is of the form

$$g(x) = (x|a) \qquad (x \in \mathbb{R}^n) .$$

Proof. The functional $q(x) = (g(x))^2$, $x \in \mathbb{R}^n$ satisfies conditions of Theorem 7. Therefore $q(x) = (Ax|x)$, $x \in \mathbb{R}^n$ with a selfadjoint operator A. Now

$$q(x+y)-q(x-y) = (g(x+y))^2 - (g(x-y))^2 \Longrightarrow$$

$$(Ax|y) = g(x)\cdot g(y) .$$

Take y_0 such that $g(y_0) \neq 0$ and set $a = Ay_0/g(y_0)$ to get

$$g(x) = (x|a) \qquad (x \in \mathbb{R}^n) . \ \square$$

COROLLARY 7. Let X be a real vector space and $q\colon X \to \mathbb{R}$ a quadratic functional. If for each $x \in X$, there are two positive numbers $A_x > 0$ and $B_x > 0$ such that

$$|t| \le A_x \Rightarrow |q(tx)| \le B_x$$

then

$$q(tx) = t^2 q(x) \qquad (t \in \mathbb{R} : x \in X) .$$

Proof. For $z \in X$ and $t \in \mathbb{R}$ define

$$f(t) = q(tz).$$

Since q is quadratic, f satisfies the functional equation (6). Since $|t| \le A_z \Rightarrow |f(t)| \le B_z$, by Theorem 7 f is continuous, i.e.

$q(tz) = t^2 q(z)$. □

THEOREM 8 ([20]). Let X be a real vector space and $q: X \to \mathbb{R}$ a quadratic form on X. If

$$\sup\{|q(x)|: x \in \Delta\} < +\infty$$

holds for any segment Δ in X, then the functional

$$S(x,y) = q(x+y) - q(x-y)$$

is bilinear and

$$q(x) = \frac{1}{4} S(x,x) \quad (x \in X).$$

Proof. For $x,y \in X$ consider the following segments:

$$\Delta_1 = [y,y+x] = \{y+tx: t \in [0,1]\}, \quad \Delta_2 = [y,y-x].$$

Let $M_1, M_2 > 0$ be such that $|q(t)| \le M_k (z \in \Delta_k)$, $k=1,2$. For $t \in [0,1]$ we have

$$|S(tx,y)| = |q(tx+y) - q(tx-y)|$$
$$\le |q(y+tx)| + |q(y-tx)| \le M_1 + M_2.$$

Since $t \to S(tx,y)$ is additive and bounded on $[0,1]$ we find [1]

$$S(tx,y) = tS(x,y) \quad (t \in \mathbb{R};\ x,y \in X).$$ □

In [17] the following theorem is proved.

THEOREM 9. Let X be a real Hilbert space, a and b real numbers and $L,F,G: X \to \mathbb{R}$ functions such that

$$L(x+y) + aL(x-y) = 2F(x) + 2G(y) + 2b \quad (x,y \in X).$$

If for some $\varepsilon > 0$ $\sup\{|L(x)| : |x| \le \varepsilon\} < +\infty$, then a bounded self-adjoint operator $A: X \to X$ and a vector $x_0 \in X$ exist such that

$$L(x) = \frac{2}{1+a}(Ax|x) + (x|x_0) + L(0),$$
$$G(x) = (Ax|x) + \frac{1-a}{2}(x|x_0) + G(0),$$
$$F(x) = (Ax|x) + \frac{1+a}{2}(x|x_0) + \frac{1+a}{2}L(0) - G(0) - b$$

hold for any $x \in X$.

In the case $a \ne 1$ in all formulae above we have to put $A=0$.

In [18] the following theorem is proved:

THEOREM 10. Let to every real square matrix x of order n a real number $q(x)$ be attached in such a way that

$$q(x+y) + q(x-y) = 2q(x) + 2q(y)$$

holds for all matrices x,y and let

$$q(s^{-1}xs) = q(s)$$

hold for every matrix x and for each regular matrix s. If q is a continuous function then:

$$q(x) = a\left(\sum_{i=1}^{n} x_{ii}\right)^2 + b \sum_{1\le i<j\le n} \begin{vmatrix} x_{ii} & x_{ij} \\ y_{ji} & x_{jj} \end{vmatrix}$$

holds for all x with real constants a and b.

7. SOLUTION OF HALPERIN'S PROBLEM FOR REAL VECTOR SPACES

An additive function $a: \mathbb{R} \to \mathbb{R}$ is a derivation on $\mathbb{R}$ if

(1) $$a(ts) = ta(s) + sa(t) \qquad (t,s \in \mathbb{R}).$$

If e_1, e_2 is a basic set in a real two-dimensional vector space X and $a: \mathbb{R} \to \mathbb{R}$, $a \neq 0$ a derivation on $\mathbb{R}$ then the functional

(2) $$q(te_1+se_2) = \begin{vmatrix} a(t) & a(s) \\ t & s \end{vmatrix} \qquad (t,s \in \mathbb{R})$$

is quadratic. Furthermore, for $\lambda \in \mathbb{R}$ and $x = te_1+se_2$ we have

$$q(\lambda x) = \begin{vmatrix} a(\lambda t) & a(\lambda s) \\ \lambda t & \lambda s \end{vmatrix}$$

$$= \lambda(\lambda a(t)+ta(\lambda))s - \lambda(\lambda a(s)+sa(\lambda))t$$

$$= \lambda^2(a(t)s-ta(s)) = \lambda^2 \begin{vmatrix} a(t) & a(s) \\ t & s \end{vmatrix}.$$

Hence

(3) $$q(\lambda x) = \lambda^2 q(x) \qquad (\lambda \in \mathbb{R};\ x \in X).$$

Moreover

$$S(te_1,e_2) = q(te_1+e_2)-q(te_1-e_2)$$

$$= \begin{vmatrix} a(t) & a(1) \\ t & 1 \end{vmatrix} - \begin{vmatrix} a(t) & a(-1) \\ t & -1 \end{vmatrix} \Rightarrow$$

$$S(te_1,e_2) = 2a(t), \qquad (t \in \mathbb{R})$$

from which we see that $t \to S(te_1,e_2)$ is not continuous, because $t \to a(t)$ is not continuous. The following theorem shows that by (2) a characteristic quadratic form on a real vector space is X, with homogeneity (3) is given.

THEOREM 11 (S.Kurepa [20], 1964). Let X be a real vector space and $q: X \to \mathbb{R}$ a functional such that

(4) $$q(x+y) + q(x-y) = 2q(x) + 2q(y) \qquad (x,y \in X),$$

(5) $$q(tx) = t^2 q(x) \qquad (t \in \mathbb{R};\ x \in X).$$

If $(e_\alpha,\ 1 \leq \alpha < \Omega)$ is an algebraic basic set in X, then

(6) $$q(\sum_\alpha t_\alpha e_\alpha) = \sum_{1\leq\alpha,\beta<\Omega} b_{\alpha\beta} t_\alpha t_\beta + \sum_{1\leq\alpha<\beta<\Omega} \begin{vmatrix} a_{\alpha\beta}(t_\alpha) & a_{\alpha\beta}(t_\beta) \\ t_\alpha & t_\beta \end{vmatrix}$$

holds true for all $t_\alpha \in \mathbb{R}$, where in the sums only finite number of terms may be different from zero ; $b_{\alpha\beta} = b_{\beta\alpha}$ are real constants and $t \to a_{\alpha\beta}(t)$ are derivations on $\mathbb{R}$.

Conversely, if $(e_\alpha,\ 1 \leq \alpha < \Omega)$ is an algebraic basic set in X, $b_{\alpha\beta} = b_{\beta\alpha}$ real constants and $t \to a_{\alpha\beta}(t)$ derivations on $\mathbb{R}$, then by (6) a functional $q: X \to \mathbb{R}$ is defined and it satisfies (4) and (5).

Proof. It is easy to verify that a functional q defined by (6) satisfies (4) and (5) provided that $t \to a_{\alpha\beta}(t)$ are derivations on $\mathbb{R}$.

The proof that a functional q, which satisfies (4) and (5) is of the form (6) is carried in several steps and it will be done by the use of the following functions:

(7) $$S(t;\ x,y) = \frac{1}{4}\,(q(tx+y)-q(tx-y)) = \frac{1}{4}\,S(tx,y),$$

(8) $$a(t;x,y) = \frac{1}{2}\,(S(t;x,y) - S(t;y,x)),$$

(9) $$b(t;x,y) = \frac{1}{2}\,(S(t;x,y) + S(t;y,x)).$$

By Lemma 1 each of these functions is additive in each of its variables. Furthermore

$$a(t;y,x) = -a(t;x,y),\ b(t;y,x) = b(t;x,y).$$

From the additivity of a in the last argument one obtains

$$a(t;x,-y) = -a(t;x,y).$$

STEP 1. $t \to a(t;x,y)$ is a derivation on $\mathbb{R}$ and

$$b(t;x,y) = tb(1;x,y) = t\,\frac{q(x+y)-q(x-y)}{4}.$$

Proof. Using (5) we have:

$$S(t;x,y) = \frac{1}{4}\,(q(tx+y)-q(tx-y)) = \frac{t^2}{4}\,(q(x+\frac{y}{t})-q(x-\frac{y}{t}))$$
$$= t^2 S(\frac{1}{t};y,x).$$

Hence

$$a(t;x,y) = \frac{t^2}{2}(S(\tfrac{1}{t};y,x)-S(\tfrac{1}{t};x,y)) = -t^2a(\tfrac{1}{t};x,y)$$

which by Theorem 5 implies that the function $t \to a(t;x,y)$ is a derivation on $\mathbb{R}$. Next

$$b(t;x,y) = \frac{t^2}{2}(S(\tfrac{1}{t};y,x) + S(\tfrac{1}{t};x,y)) = t^2b(\tfrac{1}{t};x,y)$$

which by Corollary 3 implies

$$b(t;x,y) = tb(1;x,y) .$$

STEP 2.

(10) $$q(tx+y) = t^2q(x) + \frac{t}{2}(q(x+y)-q(x-y))+q(y)+2a(t;x,y) ,$$

(11) $$S(sx,y) = sS(x,y) + 4a(s;x,y) .$$

Proof. From (8) and (9) we get:

$$S(t;x,y) = b(t;x,y) + a(t;x,y) , \text{ i.e.}$$

(12) $$\frac{q(tx+y)-q(tx-y)}{4} = t\,\frac{q(x+y)-q(x-y)}{4} + a(t;x,y) .$$

From (4) we have

(13) $$\frac{q(tx+y)+q(tx-y)}{4} = \frac{q(tx)+q(y)}{2} .$$

If we add (12) and (13) and if we use $q(tx) = t^2q(x)$ we get (10). To get (11) we use (10):

$$q(sx+y) = s^2q(x)+ \frac{s}{2}(q(x+y)-q(x-y))+q(y)+2a(s;x,y) ,$$

$$q(sx-y) = s^2q(x)+ \frac{s}{2}(q(x-y)-q(x+y))+q(y)+2a(s;x,-y)$$

from which follows

(14) $$q(sx+y)-q(sx-y) = s(q(x+y)-q(x-y)) + 4a(s;x,y)$$

and this implies (11).

STEP 3.

(15) $$a(t;sx,y) = sa(t;x,y) = a(t;x,sy) .$$

Proof. By using (10) we have

$$q(ts\cdot x+y) = (ts)^2q(x)+ts\,\frac{q(x+y)-q(x-y)}{2} +q(y)+2a(ts;x,y) ,$$

$$q(t\cdot sx+y) = t^2q(sx)+t\,\frac{q(sx+y)-q(sx-y)}{2} + q(y)+2a(t;sx,y) .$$

This implies:

$$ts(q(x+y)-q(x-y))+4a(ts;x,y) = t(q(sx+y)-q(sx-y))+4a(t;sx,y)$$

(by use of (14)) $$= t\cdot s(q(x+y)-q(x-y))+4ta(s;x,y)+4a(t;sx,y) \Longrightarrow$$

(16) $$a(ts;x,y) = ta(s;x,y)+a(t;sx,y) .$$

Since $t \to a(t;x,y)$ is a derivation,(16) implies $a(t;sx,y) = = sa(t;x,y)$. Since $a(t;y,x) = -a(t;x,y)$ (15) holds too.

Thus, $t \to a(t;x,y)$ is a derivation on $\mathbb{R}$ and $(x,y) \to a(t;x,y)$ is a bilinear skew symmetric form.

STEP 4. If $e_1,\ldots,e_k$ $(k > 1)$ is a basic set in X, then

$$(17) \qquad q\left(\sum_{i=1}^{k} t_i e_i\right) = \sum_{i,j=1}^{k} b_{ij} t_i t_j + \sum_{1\le i<j\le k} \begin{vmatrix} a_{ij}(t_i) & a_{ij}(t_j) \\ t_i & t_j \end{vmatrix}$$

where

$$a_{ij}(t) = 2a(t;e_i,e_j), \quad b_{ij} = \frac{1}{4}(q(e_i+e_j)-q(e_i-e_j))$$
$$t \in \mathbb{R}, \ i,j = 1,\ldots,k .$$

<u>Proof.</u> Applying (10) for $x=e_1$, $y = \sum_{i=2}^{k} t_i e_i$, $t=t_1$ we have

$$q\left(\sum_{i=1}^{k} t_i e_i\right) = t_1^2 q(e_1)+q\left(\sum_{i=2}^{k} t_i e_i\right)+2t_1\frac{1}{4} S\left(e_1, \sum_{i=2}^{k} t_i e_i\right)$$
$$+ 2a\left(t_1;e_1,\sum_{i=2}^{k} t_i e_i\right) \quad \text{(use (15))}$$
$$=t_1^2 q(e_1)+q\left(\sum_{k=2}^{k} t_i e_i\right)+\frac{1}{2} t_1 \sum_{i=2}^{k} S(e_1,t_i e_i) + 2\sum_{i=2}^{k} t_i a(t_1;e_1,e_i)$$
$$\text{(use (11))} = t_1^2 q(e_1) + q\left(\sum_{i=2}^{k} t_i e_i\right) + 2\sum_{i=2}^{k} t_1 t_i \ \frac{1}{4} S(e_1,e_i)$$
$$+ 2 \sum_{i=2}^{k} (t_i a(t_1;e_1,e_i) - t_1 a(t_i;e_1,e_i))$$
$$= b_{11}t_1^2+2\sum_{i=2}^{k} b_{1i}t_1t_i+ \sum_{i=2}^{k} \begin{vmatrix} a_{1i}(t_1) & a_{1i}(t_i) \\ t_1 & t_i \end{vmatrix}$$
$$+ q\left(\sum_{i=2}^{k} t_i e_i\right) .$$

By the induction (17) follows.

STEP 5. (Proof of Theorem 11). Let $(e_\alpha, 1\le\alpha<\Omega)$ be an algebraic basic set in X. Set:

$$b_{\alpha\beta} = \frac{1}{4} (q(e_\alpha+e_\beta)-q(e_\alpha-e_\beta)) ,$$
$$a_{\alpha\beta}(t) = 2a(t;e_\alpha,e_\beta) \quad (\alpha < \beta;\ t \in \mathbb{R}) .$$

If $x = \sum t_\alpha e_\alpha = \sum_{i=1}^{m} t_{k_i} e_{k_i}$, then

$$q(x) = q\left(\sum_{i=1}^{m} t_{k_i} e_{k_i}\right) \quad \text{(by the STEP 4)}$$

$$= \sum_{i,j=1}^{m} b_{k_ik_j} t_{k_i} t_{k_j} + \sum_{k_i<k_j} \begin{vmatrix} a_{k_ik_j}(t_{k_i}) & a_{k_ik_j}(t_{k_j}) \\ t_{k_i} & t_{k_j} \end{vmatrix}$$

$$= \sum_{\alpha,\beta} b_{\alpha\beta} t_\alpha t_\beta + \sum_{1\le\alpha<\beta<\Omega} \begin{vmatrix} a_{\alpha\beta}(t_\alpha) & a_{\alpha\beta}(t_\beta) \\ t_\alpha & t_\beta \end{vmatrix} . \qquad \square$$

Extending Theorem 11 to a vector space over an arbitrary field of characteristic $\neq 2$, C.T.Ng has obtained the following theorems.

THEOREM 12. ([27]). Let V be a vector space over k . Let $T: V \to k$ be a nonzero M-homogeneous biadditive form for some $M: k\to k$, i.e.

$$T(\lambda x,\lambda y) = M(\lambda)T(x,y) \quad \text{for all } \lambda \in k,\ x,y \in V ,$$

then M is multiplicative, and is of one of the following three exclusive types:

$M = \varphi^2$ where $\varphi: k \to k$ is a nonzero (field) morphism

$M = \varphi\psi$ where $\varphi,\psi: k \to k$ are distinct nonzero morphisms

$M = \varphi\overline{\varphi}$ where $\varphi: k \to k(\sqrt{c})$ is a nontrivial embedding of k into an extension of k obtained by attaching a square root which is not in k .

Furthermore, such representations of M are essentially unique.

THEOREM 13. ([27]). Let nonzero $\varphi: k \to k$ be a morphism. The following statements concerning biadditive functions T on a vector space V over k are equivalent.

(1) T is φ^2-homogeneous.

(2) T satisfies the following laws

$$T(\lambda x,y) = D(\lambda,x,y)+\varphi(\lambda)T(x,y),\ T(x,\lambda y) = -D(\lambda,x,y)+\varphi(\lambda)T(x,y) ,$$

where $D: k \times V \times V \to k$ satisfies the following description:

$D(\cdot,x,y)$ is a φ-derivation on k for each fixed $x,y \in V$

$D(\lambda,\cdot,\cdot)$ is a (φ,φ)-bihomogeneous biadditive form on V for each fixed $\lambda \in k$.

(3) T can be constructed through the following steps:

(i) Choose an arbitrary basis $\Omega = (e_i)_{i\in I}$ for V over k.

(ii) Initiate T on $\Omega \times \Omega$ (into k) arbitrarily.

(iii) Initiate for each $(e,e')\in\Omega\times\Omega$ a φ-derivation on k, labelled by $D(\cdot,e,e')$, arbitrarily.

(iv) Extend T and $D(\lambda,\cdot,\cdot)$ above to $V \times V$ by the definitions

$$T(x,y) = \Sigma_{ij}\varphi(\mu_j)D(\lambda_i,e_i,e_j)-\varphi(\lambda_i)D(\mu_j,e_i,e_j)+ +\varphi(\lambda_i)\varphi(\mu_j)T(e_i,e_j) ,$$

$$D(\lambda,x,y) = \Sigma_{ij}\varphi(\lambda_i)\varphi(\mu_j)D(\lambda,e_i,e_j) ,$$

where $x = \Sigma_i\lambda_ie_i$ and $y = \Sigma_j\mu_je_j, \lambda_i, \mu_j \in k$.

THEOREM 14 ([27]). Let nonzero φ,ϕ: $k \to k$ be distinct morphisms. The following two statements concerning biadditive forms T on a vector space V over k are equivalent.

(1) T is $\varphi\phi$-homogeneous .

(2) T has the decomposition

$$T(x,y) = C_1(x,y) - C_2(x,y) ,$$

for some unique biadditive C_1, C_2: $V \times V \to k$, with (φ,ϕ)-bihomogeneous C_1, and (ϕ,φ)-bihomogeneous C_2 .

THEOREM 15. ([27]). Let $c \in k$ be such that $\sqrt{c} \notin k$, and let φ: $k \to k(\sqrt{c})$ be a nontrivial embedding. The following two statements concerning biadditive forms T on a vector space V over k are equivalent.

(1) T is $\varphi\overline{\varphi}$ -homogeneous.

(2) T has the unique decomposition

$$T(x,y) = \Lambda(x,y) + \overline{\Lambda}(x,y)$$

where Λ: $V \times V \to k(\sqrt{c})$ is $(\varphi,\overline{\varphi})$-bihomogeneous biadditive.

8. QUADRATIC FUNCTIONALS CONDITIONED ON AN ALGEBRAIC BASIC SET

In this section we raise (and solve a particular case of) the following problem:

<u>Problem Q</u>. Find all quadratic functionals q on a vector space X for which there is at least one algebraic basic set $(e_i, i\in I)$ of X such that the function

(1) $$\lambda \to q(\lambda e_i)$$

is continuous on Φ for each $i \in I$.

The above proposed problem is a generalization of the I.Halperin problem in which (1) was replaced by

(2) $$q(\lambda x) = |\lambda|^2 q(x) \quad (\lambda \in \Phi;\ x \in X).$$

We will prove that in the case of a complex vector space the Problem Q is reduced to two problems corresponding to the replacement of (1) by one of these conditions:

$$q(\lambda e_i) = \lambda^2 q(e_i) \quad (\lambda \in C,\ i \in I) \tag{1'}$$

$$q(\lambda e_i) = |\lambda|^2 q(e_i) \quad (\lambda \in C,\ i \in I). \tag{1''}$$

This follows from the following theorem.

THEOREM 16. ([23]). Let $q: X \to C$ be a quadratic functional on a complex vector space X. Then,

$$q = q_1 + q_2 + q_3 \tag{3}$$

where q_i $(i=1,2,3)$ is a quadratic functional on X and

$$q_1(rx) = r^2 q_1(x)\ ,\ q_2(rx) = |r|^2 q_2(x),\ q_3(rx) = \bar{r}^2 q_3(x) \tag{4}$$

holds for all $x \in X$ and all complex rational numbers r.

Proof. The function $S(x,y) = q(x+y)-q(x-y)$ is biadditive so that for a complex rational number $r = r_1+ir_2 (r_1, r_2 \in Q)$, we have

$$q(rx) = q(r_1x+r_2x) = q(r_1x) + \tfrac{1}{2}S(r_1x, ir_2x) + q(ir_2x) \Rightarrow$$

$$q(rx) = r_1^2 q(x) + r_1 r_2 \tfrac{1}{2} S(x,ix) + r^2 q(ix) \Longrightarrow \tag{5}$$

$$q(rx) = r^2 q_1(x) + |r|^2 q_2(x) + \bar{r}^2 q_3(x) \tag{6}$$

with

$$q_1(x) = \tfrac{1}{4}[\ q(x) + \tfrac{1}{2i}\ q((1+i)x) - \tfrac{1}{2i}\ q((1-i)x) - q(ix)],$$

$$q_2(x) = \tfrac{1}{2}\ [q(x)+q(ix)],$$

$$q_3(x) = \tfrac{1}{4}\ [q(x) - \tfrac{1}{2i} q((1+i)x) + \tfrac{1}{2i} q((1-i)x) - q(ix)\ ]$$

which is obtained from (5) by expressing r_1 and r_2 by the use of r and $\bar{r}$. Obviously q_i $(i=1,2,3)$ are quadratic functionals. If we replace x by ρx in (6), where ρ is a complex rational number, then by the use of $q(\rho r \cdot x) = q(\rho \cdot rx)$ we find

$$\rho^2(r^2 q_1(x) - q_1(rx)) + |\rho|^2(|r|^2 q_2(x) - q_2(rx)) + \bar{\rho}^2(\bar{r}^2 q_3(x) - q_3(rx)) = 0$$

for all rational and therefore for all complex numbers ρ. Hence (6) follows. From the definition of functionals q_i $(i=1,2,3)$ it follows that the decomposition (3), (4) is unique. □

COROLLARY 8 ([4]). If $q: X \to C$ is a quadratic functional and if for each $x \in X$ the functional $\lambda \to q(\lambda x)$ is continuous on C, then

(7) $q(\lambda x) = \lambda^2 q_1(x) + |\lambda|^2 q_2(x) + \bar{\lambda}^2 q_3(x) \quad (\lambda \in C;\ x \in X)$

holds true with quadratic functionals q_i $(i=1,2,3)$.

THEOREM 17 ([4]). If $q: X \to C$ is a quadratic functional and if

(8) $$|q(rx)| = |r^2 q(x)|$$

holds for each $x \in X$ and each complex rational number r, then

$$q(rx) = r^2 q(x) \quad (x \in X;\ r \text{ complex rational}),$$

or $$q(rx) = |r|^2 q(x) \quad (x \in X;\ r \text{ complex rational}),$$

or $$q(rx) = \bar{r}^2 q(x) \quad (x \in X;\ r \text{ complex rational}).$$

Since the proof is rather long it will not be given here.

COROLLARY 9 ([4]). If $q: X \to C$ is a quadratic functional and if

(9) $$|q(\lambda x)| = |\lambda^2 q(x)| \quad (x \in X;\ \lambda \in C)$$

then

$$q(\lambda x) = \lambda^2 q(x) \quad (\lambda \in C;\ x \in X),$$

or $$q(\lambda x) = |\lambda|^2 q(x) \quad (\lambda \in C;\ x \in X),$$

or $$q(\lambda x) = \bar{\lambda}^2 q(x) \quad (\lambda \in C;\ x \in X).$$

THEOREM 18 ([23]). Let X be a real vector space, $q: X \to \mathbb{R}$ a quadratic functional. Suppose that:

1. There is an algebraic basic set $(e_i, i \in I)$ of X such that

(10) $$q(te_i) = t^2 q(e_i) \quad (t \in R,\ i \in I)$$

holds true;

2. For any ordered pair $(i,j) \in I \times I$, $i \neq j$ there exists a set $A_{ij} \subset \mathbb{R} \times \mathbb{R}$ of strictly positive two-dimensional Lebesgue measure such that

(11) $$\inf\{q(te_i + se_j) : (t,s) \in A_{ij}\} > -\infty .$$

Then, the function

(12) $$S(x,y) = q(x+y) - q(x-y)$$

is bilinear.

<u>Proof</u>. By adding $2q(x)+2q(y) = q(x+y)+q(x-y)$ and (12) we get

$$\frac{1}{2} S(x,y) = q(x+y) - q(x) - q(y)$$

Take $x = te_i$, $y = se_j$ to get

(13) $$\frac{1}{2} S(te_i,se_j) = q(te_i+se_j)-t^2q(e_i)-s^2q(e_j) \quad .$$

Without loss of generality we assume that A_{ij} is a bounded set. From (11) and (13) we find

(14) $$\inf\{S(te_i,se_j): t \in A_{ij}(s)\} > -\infty$$

where for $s \in \mathbb{R}$ we define the section

(15) $$A_{ij}(s) = \{ t \in \mathbb{R}: (t,s) \in A_{ij} \} .$$

Now if $s \in \mathbb{R}$ is such that the linear Lebesgue measure of the section $A_{ij}(s)$ is strictly positive, then (14) implies

(16) $$S(te_i,se_j) = tS(e_i,se_j) \qquad (t \in \mathbb{R})$$

due to the fact that the function $t \to S(te_i,se_j)$ is additive. On the other hand it is well known that the set

$$B_{ij} = \{s \in R: mA_{ij}(s) > 0\}$$

has strictly positive Lebesgue measure. Thus (16) holds for any $s \in B_{ij}$. Set $u = s_1-s_2$ with $s_1,s_2 \in B_{ij}$. From (16) and the additivity of S in the second variable we have:

$$\begin{aligned} S(te_i,ue_j) &= S(te_i,s_1e_j) - S(te_i,s_2e_j) \\ &= tS(e_i,s_1e_j) - tS(e_i,s_2e_j) \\ &= tS(e_i,ue_j) , \end{aligned}$$

i.e. (16) holds for any s from the difference set

$$D_{ij} = B_{ij} - B_{ij} = \{u-v: u,v \in B_{ij}\} .$$

Since the set D_{ij} containes an interval $[-\varepsilon,\varepsilon], \varepsilon > 0$ it follows that (16) holds for any $s \in \mathbb{R}$. Thus

(17) $$S(te_i,se_j) = ts\, S(e_i,e_j) \qquad (t,s \in \mathbb{R} ;\ i,j \in I).$$

If $t \in \mathbb{R}$ and $x,y \in X$ are given, then

$$x = \sum_{i\in I} t_ie_i \quad , \quad y = \sum_{j\in I} s_je_j \Longrightarrow$$

$$S(tx,y) = \sum_{i,j\in I} S(tt_ie_i,s_je_j) = \sum_{i,j\in I} tt_is_jS(e_i,e_j) = tS(x,y) \ . \square$$

9. QUADRATIC FORMS ON ABELIAN GROUPS

In this section X denotes an abelian group in which division by two is defined , i.e. ($a \in A$ and $2a = 0 \Longrightarrow a = 0$). A function $q: X \to \mathbb{R}$ is a quadratic form on X if

(1) $$q(x+y)+q(x-y) = 2q(x) + 2q(y) \qquad (x,y \in X).$$

It is easy to see that Lemma 1 holds true, i.e. the function

(2) $$S(x,y) = q(x+y)-q(x-y) \qquad (x,y \in X)$$

is biadditive. Hence also the function

(3) $$(x|y) = \frac{1}{4}(q(x+y)-q(x-y)) \qquad (x,y \in X)$$

is biadditive.

THEOREM 19 ([21],[30]). Let X be an abelian group and $q: X \to \mathbb{R}$ a quadratic form on X. If $q(x) \geq 0$ $(x \in X)$, then

1. For any system of elements $x_1,\dots,x_k \in X$ the matrix

(4) $$G(x_1,\dots,x_k) = \begin{bmatrix} (x_1|x_1) & (x_1|x_2)\dots & (x_1|x_k) \\ (x_2|x_1) & (x_2|x_2)\dots & (x_2|x_k) \\ \vdots & & \\ (x_k|x_1) & (x_k|x_2)\dots & (x_k|x_k) \end{bmatrix}$$

is positive semidefinite;

2. The mapping $x \to |x| = (q(x))^{1/2}$ possesses the following properties

(5) $$|(x|y)| \leq |x|\ |y|\ ,\quad |x+y| \leq |x|+|y| \qquad (x,y \in X).$$

3. The set

$$X_0 = \{\ x \in X: q(x) = 0\ \}$$

is a subgroup of X, and the form

$$\hat{q}: X/_{X_0} \to \mathbb{R}$$

defined by

(6) $$\hat{q}(x+X_0) = q(x) \qquad (x \in X)$$

is quadratic on $X/_{X_0}$, $\hat{q}(x+X_0) \geq 0$ for any $x \in X$, and

$$\hat{q}(x+X_0) = 0 \Rightarrow x \in X_0.$$

Proof. Since $q(px) = p^2q(x)$ holds for any $x \in X$ and any integer p, by adding equalities:

$$\frac{1}{2}(q(px+y)+q(px-y)) = q(px)+q(y) = p^2q(x)+q(y)$$

$$\frac{1}{2}(q(px+y)-q(px-y)) = 2(px|y) = 2p(x|y)$$

we find

(7) $$q(px+y) = p^2q(x)+2p(x|y) + q(y)\ .$$

If $x_1,\dots,x_k$ are elements of X and $p_1,\dots,p_k$ integers, then by setting $p=p_1$, $x=x_1$ and $y=p_2x_2+\dots+p_kx_k$ in (7) we get

$$q(\sum_{i=1}^{k} p_i x_i) = p_1^2 q(x_1)+2\sum_{i=1}^{k} p_1 p_i (x_1|x_i) + q(\sum_{i=2}^{k} p_i x_i) \Longrightarrow$$

(8) $$q(\sum_{i=1}^{k} p_i x_i) = \sum_{i,j=1}^{k} p_i p_j (x_i|x_j) \geq 0.$$

Now, suppose that $r_1,\dots,r_k$ are rational numbers. Writting $r_i=p_i/p$ with integers p_i and a natural number p, we have

(9) $$\sum_{i,j=1}^{k} r_i r_j (x_i|x_j) = \frac{1}{p^2} \sum_{i,j=1}^{k} p_i p_j (x_i|x_j) \geq 0$$

and this implies

$$\sum_{i,j=1}^{k} t_i t_j (x_i|x_j) \geq 0 \qquad (t_i,t_j \in \mathbb{R}) .$$

Thus the matrix (4) is positive semidefinite.

Since

$$\det G(x,y) = q(x)q(y)-(x|y)(y|x) \geq 0 \Longrightarrow$$

$$(x|y)^2 \leq (x|x)(y|y) \text{ , i.e. } |(x|y)| \leq |x|\ |y|.$$

Now,

$$|x+y|^2 = q(x+y) = q(x)+q(y)+2(x|y) \leq |x|^2+|y|^2 + 2|x|\ |y| \Longrightarrow$$

$$|x+y| \leq |x|+|y|.$$

In order to prove the third part of Theorem 19 we note that $x_0 \in X_0$ and $|(x_0|y)| \leq |x_0|\ |y|$ imply $(x_0|y)=0$ for any $y \in X$. Hence

$$q(x_0+y) = q(x_0-y) \qquad (x_0 \in X_0,\ y \in X) .$$

Furthermore, $x_0,y_0 \in X_0$ and $q(x) \geq 0$ for any $x \in X$ imply

$$q(x_0+y_0)+q(x_0-y_0) = 2q(x_0)+2q(y_0) \Longrightarrow$$

$$q(x_0+y_0) = q(x_0-y_0) = 0 .$$

Thus $x_0,y_0 \in X_0 \Longrightarrow x_0+y_0 \in X_0,\ x_0-y_0 \in X_0$, i.e. X_0 is a subgroup of X. If $x,x' \in X$ are such that $x_0 = x'-x \in X_0$, then

$$q(x') = q(x_0+x) = q(x_0) + q(x)+2(x_0|x) = q(x)$$

implies that by (6) a function $\hat{q}$ is well defined. It is obvious that $\hat{q}$ is possitive semidefinite and that $\hat{q}(x+X_0)=0$ implies $x \in X_0$. □

COROLLARY 10. Let X be a vector space over a field Φ and $q\colon X \to \mathbb{R}$ a quadratic form. If

(i) $q(x) \geq 0 \quad (x \in X)$,

(ii) $\lim_{t\to 0} q(tx) = 0 \quad (x \in X,\ t \in \mathbb{R})$ and

(iii) $q(ix) = q(x) \quad (x \in X)$ in the case $\Phi = C$,

then the function (3) is bilinear if $\Phi = \mathbb{R}$ and the function

(10) $$(x|y) = \frac{1}{4}(q(x+y)-q(x-y)) + \frac{i}{4}(q(x+iy)-q(x-iy))$$

is sesquilinear if $\Phi = C$. In any case

$$q(x) = (x|x) \quad (x \in X).$$

If in addition $q(x) = 0$ implies $x=0$, then $(x,y) \to (x|y)$ is a scalar product in X.

Proof. From (5), $|(tx|y)| \leq |tx|\cdot|y| = \sqrt{q(tx)}\cdot|y|$ and (ii) we find $\lim_{t\to 0} (tx|y) = 0$. Thus $(tx|y) = t(x|y)$.□

Corollary 10 is the P.Jordan and J.von Neumann characterization of an inner product space quoted in Remark 2 of section 2.

COROLLARY 11. Let X be a vector space over a field Φ and $q: X \to \mathbb{R}$ a quadratic form on X. If

(i) $q(x) \geq 0 \quad (x \in X)$ and

(ii) there is an algebraic basic set $(e_i, i \in I)$

of X such that $q(te_i) = t^2 q(e_i)$ $(i \in I)$, then the function (3) is bilinear in the case $\Phi = \mathbb{R}$ and the function (10) is sesquilinear in the case $\Phi = C$.

Proof. For $x = \Sigma\, \xi_i e_i \in X$, $y \in X$ and $t \in \mathbb{R}$ (3) and (5) imply

$$|(tx|y)| \leq |t\ \Sigma\xi_i e_i|\ |y| \leq \Sigma\ |t\xi_i e_i|\ |y| = |t|\Sigma|\xi_i e_i|\ |y|.$$

Hence $\lim_{t\to 0} (tx|y) = 0$. Since an additive function $t \to (tx|y)$ is continuous in $t=0$ we find $(tx|y) = t(x|y)$.□

THEOREM 20 ([21]). Let X be an abelian group and Y an inner product space. If $f: X \to Y$ is such a function that

(11) $$|f(x+y)| = |f(x)+f(y)| \quad (x,y \in X) ,$$

then f is an additive function.

Proof. Set

(12) $$q(x) = |f(x)|^2 \quad (x \in X) .$$

Then

(13) $$q(x+y) = q(x)+q(y) + (f(x)|f(y)) + (f(y)|f(x)) .$$

From (11) for $x=y=0$ we get $f(0) = 0$, for $x=-y$ we find $f(-x) = -f(x)$. This and (13) imply that q is a quadratic form. But then the function $q(x+y)-q(x-y)$ is biadditive, i.e. the function

$$(f(x)|f(y)) + (f(y)|f(x))$$

is biadditive. Thus

(14) $$(f(x+x')-f(x')-f(x)|f(y)) + (f(y)|f(x+x')-f(x)-f(x')) = 0 .$$

If in (14) we take $y = x+x'$, $y=-x'$, $y=-x$ and sum up we get

$$2(f(x+x')-f(x')-f(x)|f(x+x')-f(x')-f(x)) = 0$$

from which the additivity of f follows. □

THEOREM 21 ([21]). Let X be an abelian group and $q: X \to \mathbb{R}$ a quadratic form. Then an additive form $g: X \to \mathbb{R}$ such that

(15) $$q(x) = (g(x))^2 \quad (x \in X)$$

exists if and only if q satisfies the following subsidiary condition:

(16) $$(q(x+y)-q(x-y))^2 = 16q(x)q(y) \quad (x,y \in X).$$

Proof. Suppose that $q(x) \geq 0$ $(x \in X)$ and that it satisfies (16), i.e.

(17) $$(x|y)^2 = q(x)q(y) .$$

If $q=0$ we can take $g=0$. If $q\neq 0$, then $q(y) > 0$ for at least one $y \in X$. Now, we define

$$g(x) = \frac{1}{\sqrt{q(y)}}(x|y) = \frac{(x|y)}{|y|} .$$

Then g is additive and (17) implies $q(x) = (g(x))^2$. □

THEOREM 22 ([22]). Suppose that $q: \mathbb{R} \to \mathbb{R}$ is a function such that

(18) $$(q(x+y)-q(x-y))^2 = 16q(x)q(y) \quad (x,y \in \mathbb{R}).$$

Then

(19) $$q(rx) = r^2q(x) \quad (x \in R : r \in Q) .$$

If q is continuous, then $q(x) = x^2q(1)$ $(x \in \mathbb{R})$.

Remark. The relation (18) can be written in the form

$$\det\begin{pmatrix} 4q(x) & S(x,y) \\ S(y,x) & 4q(y) \end{pmatrix} = 0, \quad S(x,y)=q(x+y)-q(x-y)$$

which can be generalized to a functional equation with the corresponding determinant of order $n \geq 2$:

$$(20) \qquad \det\begin{pmatrix} 4q(x_1) & S(x_1,x_2) & \dots & S(x_1,x_n) \\ S(x_2,x_1) & 4q(x_2) & \dots & S(x_2,x_n) \\ \vdots & & & \\ S(x_n,x_1) & S(x_n,x_2) & \dots & 4q(x_n) \end{pmatrix} = 0 \quad (x_1,\dots,x_n \in X)$$

where $q\colon \mathbb{R} \to \mathbb{R}$ is an unknown function and

$$S(x,y) = q(x+y) - q(x-y) .$$

From (20) one can deduce : $q(0) = 0$, $q(-x) = q(x)$. If q is differentiable function then $q(x) = x^2 q(1)$ $(x \in \mathbb{R})$.

THEOREM 23. ([22]). Let X be an abelian group and $q\colon X \to \mathbb{R}$ a given function. Then, the function q is of the form

$$q(x) = \sum_{i,j=2}^{n} b_{ij} f_i(x) f_j(x) \qquad (x \in X)$$

with additive functions $f_i\colon X \to \mathbb{R}$ $(i=2,\dots,n)$ and real numbers $b_{ij}=b_{ji}$ $(i,j=2,\dots,n)$ if and only if q is quadratic and it satisfies the condition (20).

This theorem answers the question raised by Prof. A.Ostrowski in Oberwolfach "Tagung über Funktionalgleichungen, 1965".By the use of a Hamel base (H_α) of $\mathbb{R}$ as a vector space over Q, for a quadratic form q and $x = \Sigma r_\alpha H_\alpha$ we have

$$q(x) = \Sigma b_{\alpha\beta} r_\alpha r_\beta \ , \ b_{\alpha\beta} = S(H_\alpha, H_\beta) .$$

By $x \to r_\alpha = f_\alpha(x)$ additive functions are defined so that [2] :

$$q(x) = \Sigma\, b_{\alpha\beta} f_\alpha(x) f_\beta(x) \qquad (x \in \mathbb{R})$$

In Theorem 23 only a finite number of additive functions appears.

10. QUADRATIC FORMS ON GROUPS

DEFINITION. A function $q\colon G \to A$ from a group G into an abelian group A is quadratic if

$$(1) \qquad q(xy) + q(xy^{-1}) = 2q(x)+2q(y) \qquad (x,y \in G).$$

We assume that $a \in A$, $2a = 0$ implies $a=0$.

The main question which is raised now is whether the function

$$(2) \qquad S(x,y) = q(xy)-q(xy^{-1}) \qquad (x,y \in G)$$

is an additive biomorphism, i.e.

$$S(xy,z) = S(x,z) + S(y,z)$$

$$S(z,xy) = S(z,x) + S(z,y)$$

for all $a,y,z \in G$.

THEOREM 24 (S.Kurepa [24], 1971). If $q: G \to A$ satisfies (1), then the restriction of the function (2) to any subgroup G_2 of G is an additive bimorphism provided that G_2 is generated by two elements.

THEOREM 25 ([24]). If in Theorem 24 , $A = \mathbb{R}$ is the additive group of real numbers and $q(x) \geq 0$ for any $x \in G$, then the function S defined by (2) is an additive bimorphism of a group G into the additive group of real numbers.

In the case of Theorem 25 one can extend results of Theorem 19 to this situation. For proofs and details see [24].

REFERENCES:

[1] J.Aczél, Lectures on functional equations and their applications, Academic Press, New York, London, 1966.

[2] J.Aczél, The general solution of two functional equations by reducing to functions additive in two variables and with the aid of Hamel basis, Glasnik mat.fiz. i astr. 20 (1965), 65-72.

[3] J.Aczél and J.Dhombres, Functional equations containing several variables, Addison-Wesley, Reading, Mass., 1985.

[4] J.A.Baker, On quadratic functionals continuous along rays, Glasnik mat. 3 (23) (1968), 215-229.

[5] J.A.Baker and K.R.Davidson, Cosine, exponential and quadratic functions, Glasnik mat. 16 (36)(1981), 269-274.

[6] A.Clarkson, The von Neumann-Jordan constants for the Lebesgue spaces, Ann. of Math. 38 (1937), 114-115.

[7] M.Fréchet, Sur la definition axiomatique d'une classe d'espaces vectoriels distanciés applicables vectoriellement sur l'espace de Hilbert, Annals of Math. 36 (1935), 705-718.

[8] A.M.Gleason, The definition of a quadratic form, Amer.Math. Monthly 73 (1966), 1049-1056.

[9] A.Grząślewicz, On the solution of the system of functional equations related to quadratic functional, Glasnik mat. 14 (1979), 77-82.

[10] P.Jordan and J.von Neumann, On inner products in linear, metric space, Annals of Math. 36 (1935), 719-723.

[11] W.B.Jurkat, On Cauchy's functional equation, Proc. Amer.Math. Soc. 16 (1965), 683-686.

[12] P.L.Kannappan and S.Kurepa, Some relations between additive functions I, Aequationes Math. 4 (1970), 163-175.

[13] P.L.Kannappan and S.Kurepa, Some relations between additive functions II, Aequationes Math. 6 (1971), 46-58.

[14] P.L.Kannappan and C.T.Ng, On a generalized fundamental equation of information, Can. J.Math. 35 (1983), 863-872.

[15] M.Kuczma, An Introduction to the Theory of Functional Equations and Inequalities, Panstwowe Wydawnictwo Naukowe, Warszawa-Kraków-Katowice, 1985.

[16] S.Kurepa, Convex functions, Glasnik mat.fiz. astr. 11 (1956), 89-94.

[17] S.Kurepa, On the quadratic functional, Publ.Inst.Math.Acad. Serbe.Sci. Beograd, 13 (1959), 57-72.

[18] S.Kurepa, Functional equations for invariants of a matrix, Glasnik mat.fiz.astr. 14 (1959), 97-113.

[19] S.Kurepa, A cosine functional equation in Hilbert space, Can. J.Math. 12 (1960), 45-50.

[20] S.Kurepa, The Cauchy functional equation and scalar product in vector spaces, Glasnik mat.fiz.astr. 19 (1964), 23-36.

[21] S.Kurepa, Quadratic and sesquilinear functionals, Glasnik mat. fiz.astr. 20 (1965), 79-92.

[22] S.Kurepa, On a nonlinear functional equation, Glasnik mat.fiz. astr. 20 (1965), 243-249.

[23] S.Kurepa, Quadratic functionals conditioned on an algebraic basic set, Glasnik mat. 6 (26)(1971), 265-275.

[24] S.Kurepa, On Bimorphisms and Quadratic Forms on Groups, Aequationes Math. 9 (1973), 30-45.

[25] S.Kurepa, Semigroups and cosine functions, Lecture Notes in Math. 948 (Functional Analysis,Proceedings, Dubrovnik(1981), p.p. 47-72, Springer-Verlag, Berlin, Heidelberg, New York, 1982.

[26] A.Nishiyama and S.Horinouchi, On a System of Functional Equations, Aequationes Math. 1 (1968), 1-5.

[27] C.T.Ng, The equation F(x)+M(x)G(1/x)=0 and homogeneous biadditive forms, J.Linear Algebra (in press).

[28] L.Paganoni and S.Paganoni Marzegalli, Cauchy's functional equation on semigroups, Fund.Math. 110 (1980), 63-74.

[29] J.Rätz, On the homogeneity of additive mappings, Aequationes Math. (1976), 67-71.

[30] H.Rubin and M.H.Stone, Postulates for generalizations of Hilbert space, Proc.Amer.Math.Soc. 4 (1953), 611-616.

[31] P.Vrbová, Quadratic functionals and bilinear forms, Časopis pro pestovani matematiky 98 (1973), 159-161, Praha.

[32] J.Vukman, A result concerning additive functions in hermitian Banach *-algebras and an application, Proc.Amer.Math.Soc. 91 (1984), 367-372.

[33] J.Vukman, Some results concerning the Cauchy functional equation in certain Banach algebras, Bull.Austral.Math.Soc. 31 (1985), 137-144.

[34] O.Zariski and P.Samuel, Commutative algebra, Van Nostrand Comp. Inc. New York, 1958. pp. 120-131.

THE GENERAL MARGINAL PROBLEM

J. Hoffmann-Jørgensen

Contents

1. Introduction

Let $(T,\mathcal{B})$ be a measurable space, i.e. a set T with a σ-algebra $\mathcal{B}$ of subsets of T, and suppose that q_γ is a measurable map from $(T,\mathcal{B})$ into a probability space $(T_\gamma,\mathcal{B}_\gamma,P_\gamma)$ for all γ in a certain set Γ. Then the general marginal problem is the problem of finding necessary and/or sufficient conditions for the existence of a probability measure P on $(T,\mathcal{B})$, such that the law of q_γ under P equals P_γ for all $\gamma \in \Gamma$, i.e. such that $P_\gamma = q_\gamma P \;\forall \gamma$, where $q_\gamma P$ is the image measure of P under q_γ.

This problem has many aspects, and special cases has been studied by many authors in many different contexts. The classical paper of V. Strassen [22] treats the case where $\Gamma = \{1,2\}$ is a two point-set, and later this case has been taken up by many others, see [3] where further references can be found. Projective limits of probability measures is another special case which has drawn the attention of many authors, see e.g. [1],[13],[14],[15], [19] and [24]. Another special case is simultaneous extensions of probability measures, in this case $T_\gamma = T$ and q_γ = the identity map for all γ. Another special case is the problem of a.s.-realization of weak convergency, in this case we have a topological space S and a net $\{P_\lambda \mid \lambda \in \Lambda\}$ of Baire probability measures on S such that $\{P_\lambda\}$ converges weakly to a Baire probability P_0, and the problem is to find a probability measure P on

$$T = \{(s_0,(s_\lambda)) \in S \times S^\Lambda \mid s_0 = \lim_\lambda s_\lambda\}$$

such that the marginals of P equal P_λ for $\lambda \in \{0\} \cup \Lambda$.

To solve the general marginal problem, we first solve it for finitely additive probabilities (the socalled probability contents, see section 4), where the solution follows easily from Hahn-Banach theorem. Section 2 and 4 contain a detailed study of probability contents per se without reference to the general marginal problem. In section 3 we study smoothness of functionals defined on function spaces, and this study is carried on in section 5 for a certain class of functionals, the socalled outer probability contents. The smoothness results of section 3 and 5, are then used in section 6 to obtain σ-additivity of the finitely additive solution the general marginal problem, and in section 7 we specialize to the case where T is a product space and the $\{q_\gamma\}$ are the projections. Finally, in the appendices I list a sequence of wellknown results from topology and measure theory needed in the previous sections.

The subject of this exposition was the theme of a series of lectures given at the Postgraduate School in Functional Analysis, Dubrovnik, November 1985, and I would like to thank the organizers D. Butković, H. Kraljević and S. Kurepa for giving me the opportunity to lecture on this subject for a lively and engaged audience, and for the opportunity to collect the material in this exposition.

2. Function spaces. We shall in this section study the topology and the σ-algebra generated by a set of $\overline{\mathbb{R}}$-valued functions on a set T.

Let T be a set and let F be a paving on T, then a map $f:T \to \overline{\mathbb{R}}$ is called an upper (resp. a lower) F-function. if $\{f \geq a\} \in F$ (resp. if $\{f \leq a\} \in F$) for all $a \in \mathbb{R}$. And we put

$$U(T,F) = \{f \in \overline{\mathbb{R}}^T \mid f \text{ is an upper } F\text{-function}\}$$
$$L(T,F) = \{f \in \overline{\mathbb{R}}^T \mid f \text{ is a lower } F\text{-function}\}$$

Clearly we have

$$(2.1) \qquad L(T,F) = -U(T,F)$$

and if T is a topological space and $F = F(T)$, then

$$(2.2) \qquad U(T,F) = \mathrm{Usc}(T), \quad L(T,F) = \mathrm{Lsc}(T)$$

Let us recall the definition of the Souslin operation. As usual we let $\mathbb{N} = \{1,2,\dots\}$ denote the set of positive integers and we put

$$\mathbb{N}^{(\mathbb{N})} = \bigcup_{k=1}^{\infty} \mathbb{N}^k$$

If $j \in \mathbb{N}$ and $\sigma \in \mathbb{N}^{\mathbb{N}}$ or $\sigma \in \mathbb{N}^k$ for some $k \geq j$, then we put

$$\sigma|j = (\sigma(1),\dots,\sigma(j)) \in \mathbb{N}^j$$

Let T be a set and let F be a subset of $\overline{\mathbb{R}}^T$, then an F-Souslin scheme is a map $\phi:\mathbb{N}^{(\mathbb{N})} \to F$. If Φ is an F-Souslin scheme, then we put

$$S(\Phi) = \sup_{\sigma \in \mathbb{N}^{\mathbb{N}}} \{ \inf_{j \in \mathbb{N}} \Phi(\sigma|n)\} \in \overline{\mathbb{R}}^T$$

And we define

$$S(F) = \{S(\Phi) \mid \Phi \text{ is an F-Souslin scheme}\}$$

$$\mathcal{S}(F) = \{B \subseteq T \mid 1_B \in S(F)\} = 2^T \cap S(F)$$

cf. [20; p. 10 and p. 201].

Let T be a set and let F be a subset of $\overline{\mathbb{R}}^T$, then we define the <u>weight</u> of F by

$$\text{weight}(F) = \min\{\text{card } H \mid H \subseteq F,\ \tau(H) = \tau(F)\}$$

And if T is a topological space we define the <u>weight</u> of T (see [5; p. 27]) by

$$\text{weight}(T) = \min\{\text{card}(\mathcal{G}) \mid \mathcal{G} \text{ is an open base for } T\}$$

Note that if $H \subseteq F \subseteq \mathbb{R}^T$, then we have

(2.3) $\quad \tau(H) = \tau(F) \iff F \subseteq C(T, \tau(H))$

Let T be a set, and F a subset of $\mathbb{R}^T$, then we say that F is a <u>cone</u>, <u>linear space</u>, <u>algebra</u> or <u>convex</u> if F is so with respect to the usual addition, scalar multiplication and multiplication of real valued functions. And we say that the pair (T,A) is <u>an algebraic function space</u>, if T is set and A is an algebra, such that $1_T \in A \subseteq B(T)$.

<u>Proposition 2.1.</u> <u>Let</u> F <u>be a</u> $(\cup f)$<u>-stable paving on</u> T, <u>and let</u> D <u>be a</u> $(\vee f)$<u>-stable subset of</u> $\overline{\mathbb{R}}^Q$ <u>satisfying</u>

(2.1.1) $\quad F$ <u>is</u> $(\cap\xi)$<u>-stable</u>

(2.1.2) $\quad \forall \mathcal{G} \subseteq \mathcal{G}(D)\ \exists\, \mathcal{G}_o \subseteq \mathcal{G}:\ \text{card}(\mathcal{G}_o) \leq \xi$ <u>and</u> $\cup\mathcal{G}_o = \cup\mathcal{G}$

<u>where</u> ξ <u>is an infinite cardinal, and</u> D <u>has its product topology.</u> <u>If</u> $\varphi: D \to \overline{\mathbb{R}}$ <u>is increasing and upper semicontinuous, and if</u> $f = (f_q)_{q \in Q}$

<u>is a map from</u> T <u>into</u> D, <u>such that</u> $f_q \in U(T,F)$ <u>for all</u> $q \in Q$, <u>then we have that</u> $\varphi \circ f \in U(T,F)$. <u>I.e.</u> $U(T,F)$ <u>is</u> φ<u>-stable.</u>

<u>Remarks</u> (1): Of course we have a similar stability result for lower F-functions, in that case D should be $(\wedge f)$-stable and φ should be increasing and lower semicontinuous

(2): If weight(D) $\leq \xi$ then clearly (2.1.2) hold. Note that weight(D) $\leq \aleph_0 \vee$ card(Q) for all $D \subseteq \overline{\mathbb{R}}^Q$. A topological space D satisfying (2.1.2) is usually called <u>hereditarily</u> ξ-<u>Lindelöf</u>.

<u>Proof</u>. Since $\overline{\mathbb{R}}$ is homeomorphic and order isomorphic to [0,1], it follows easily that it is no loss of generality to assume that $D \subseteq \mathbb{R}^Q$ and that φ maps D into $\mathbb{R}$.

If σ is a finite subset of Q, and $\omega' \in \mathbb{R}^\sigma$ and $a \in \mathbb{R}$ we put

$$D(\omega',\sigma) = \{\omega \in D \mid \omega(q) < \omega'(q) \ \forall q \in \sigma\}$$

$$\Delta(a) = \left\{ (\omega',\sigma) \,\middle|\, \begin{array}{l} \sigma \text{ is a finite subset of } Q \\ \omega' \in \mathbb{R}^\sigma \text{ and } \varphi(\omega) < a \ \forall \omega \in D(\omega',\sigma) \end{array} \right\}$$

Then I claim, that we have

$$\text{(i)} \qquad \{\omega \in D \mid \varphi(\omega) \geq a\} = D \setminus \bigcup_{(\omega',\sigma)\in\Delta(a)} D(\omega',\sigma)$$

So let ω_0 belong to the right hand side of (i), and let $b > \varphi(\omega_0)$. Then by upper semicontiuity of φ there exist a finite subset σ of Q, and $\varepsilon > 0$ such that $\varphi(\omega) < b$ for all $\omega \in G \cap D$ where

$$G = \{\omega \in \mathbb{R}^Q \,|\, |\omega(q) - \omega_0(q)| < 2\varepsilon \ \ \forall q \in \sigma\}$$

Now let $\omega'(q) = \omega_0(q) + \varepsilon$ for $q \in \sigma$. Then $\omega' \in \mathbb{R}^\sigma$, and if $\omega \in D(\omega',\sigma)$, then $\omega_1 = \omega \vee \omega_0 \in D$ by $(\vee f)$-stability of D, and

$$\omega_o(q) \le \omega_1(q) \le \omega_o(q) + \varepsilon \qquad \forall q \in \sigma$$

Hence $\omega_1 \in G \cap D$, and so $\varphi(\omega) \le \varphi(\omega_1) < b$. Thus $(\omega',\sigma) \in \Delta(b)$ and $\omega_o \in D(\omega',\sigma)$. By assumption we know that ω_o belongs to the right hand side of (i), and so $(\omega',\sigma) \in \Delta(b) \setminus \Delta(a)$. Hence $b > a$ for all $b > \varphi(\omega_o)$, thus $\varphi(\omega_o) \ge a$ and ω_o belongs to the left hand side of (i).

Now suppose that $\omega_o \in D$ and $\varphi(\omega_o) \ge a$. Let $(\omega',\sigma) \in \Delta(a)$, then $\varphi(\omega) < a \le \varphi(\omega_o)$ for all $\omega \in D(\omega',\sigma)$, and so $\omega_o \notin D(\omega',\sigma)$. Hence the converse inclusion holds, and (i) is proved.

Now note that $D(\omega',\sigma)$ is open in D, and so by (2.1.2) and (i), there exist $\Gamma(a) \subseteq \Delta(a)$ such that

(ii) $\quad \mathrm{card}(\Gamma(a)) \le \xi$

(iii) $\quad \{\omega \in D \mid \varphi(\omega) \ge a\} = D \setminus \bigcup_{(\omega',\sigma)\in\Gamma(a)} D(\omega',\sigma)$

And if $g = \varphi \circ f$, then by (iii) we have

(iv) $\quad \{g \ge a\} = \bigcap_{(\omega',\sigma)\in\Gamma(a)} f^{-1}(D \setminus D(\omega',\sigma))$

(v) $\quad f^{-1}(D \setminus D(\omega',\sigma)) = \bigcup_{q\in\sigma} \{f_q \ge \omega'(q)\}$

Hence g is an upper F-function by $(\cup f,\cap\xi)$-stability of F, since $f_q \in U(T,F)$ for all $q \in Q$.

□

<u>Proposition 2.2</u>. <u>Let</u> $F \subseteq \mathbb{R}^T$ <u>be a convex cone containing</u> $\pm 1_T$, <u>and let</u> $\mathcal{F} \subseteq 2^T \cap F$, <u>then we have</u>

(2.2.1) $\quad \forall f \in B(T) \cap U(T,\mathcal{F})\ \exists g_n, h_n \in F\colon g_n \uparrow\uparrow f,\ h_n \downarrow\downarrow f$

(2.2.2) $\quad \forall f \in B^*(T) \cap U(T,\mathcal{F})\ \exists h_n \in F\colon h_n \downarrow f$

(2.2.3) $\quad \forall f \in B_*(T) \cap U(T,\mathcal{F})\ \exists g_n \in F\colon g_n \uparrow f$

In particular we have

(2.2.4) $U(T,F) \cap B(T) \subseteq \bar{F}$,

(2.2.5) $U(T,F) \cap B^*(T) \subseteq F_\delta$, $U(T,F) \cap B_*(T) \subseteq F_\sigma$

(2.2.6) $U(T,F) \subseteq F_{\delta\sigma} \cap F_{\sigma\delta}$ if $\emptyset \in F$ and $T \in F$

Proof. (2.2.3): Let $f \in U(T,F) \cap B_*(T)$ and choose $a \in \mathbb{R}$ so that $f \geq a$. Then we put

$$F_{jn} = \{f \geq j2^{-n} - a\} = \{2^n(f+a) \geq j\}$$

$$g_n = -a + \sum_{j=1}^{n2^n} 2^{-n} 1_{F_{jn}}$$

Since $F_{jn} \in F \subseteq F$, we have that $g_n \in F$, and it is easily checked that

$$g_n = -a + 2^{-n}\text{int}[2^n(n \wedge (f+a))]$$

From which it follows that $g_1 \leq g_2 \leq \dots$ and

$$f \wedge (n-a) - 2^{-n} \leq g_n \leq f \wedge (n-a)$$

Hence $g_n \uparrow f$ and if f is bounded then $\|g_n - f\| \leq 2^{-n}$ for $n \geq a + \|f\|_T$. This proves (2.2.3) and the first part of (2.2.1).

(2.2.2): If $f \in B^*(T) \cap U(T,F)$ we choose $b \in \mathbb{R}$ so that $f \leq b$, and we put

$$H_{jn} = \{f \geq b - j2^{-n}\} = \{j \geq 2^n(b-f)\}$$

$$h_n = b - n + 2^{-n} + 2^{-n} \sum_{j=1}^{n2^n - 1} 1_{H_{jn}}$$

Then exactly as above one shows that $h_n \downarrow f$ and $h_n \downarrow\downarrow f$ if f is bounded. And since $H_{jn} \in F \subseteq F$, we see that $h_n \in F$ and so the proposition is proved. □

Proposition 2.3. Let F be a $(\downarrow c)$-stable subset of $\overline{\mathbb{R}}^T$ satisfying

(2.3.1) $f+a,\ f\wedge a$ and $f\vee a$ belongs to $F \quad \forall f\in F \ \ \forall a\in\mathbb{R}$

(2.3.2) $nf\in F \quad \forall f\in F \quad \forall n\in\mathbb{N}$

Then we have that $F\subseteq U(T,\mathcal{F})$, where $\mathcal{F}=F\cap 2^T$.

Proof. Let $f\in F$ and $a\in\mathbb{R}$, then

$$f_n=(nf-na+1)^+\wedge 1\in F$$

by (2.3.1) and (2.3.2), and $f_n\downarrow 1_{\{f\geq a\}}$. Hence $\{f\geq a\}\in\mathcal{F}$ by $(\downarrow c)$-stability of F. Thus f is an upper $\mathcal{F}$-function for all $f\in F$.

□

Proposition 2.4. Let F and G be subsets of $\overline{\mathbb{R}}^T$, then we have

(2.4.1) $F\subseteq S(F)=S(S(F))$

(2.4.2) $S(F)\cap G=S(S(F)\cap G)\cap G$

(2.4.3) $S(F)$ is $(\wedge c,\vee c)$-stable

And if $n\in\mathbb{N}$ and $\varphi:\overline{\mathbb{R}}^n\times T\to\overline{\mathbb{R}}$ is a map satisfying

(2.4.4) $\varphi(\cdot,t)$ is increasing and continuous $\forall t\in T$

(2.4.5) $\varphi(f_1(\cdot),\dots,f_n(\cdot),\cdot)\in S(F) \quad \forall f_1,\dots,f_n\in F$

Then we have

(2.4.6) $\varphi(f_1(\cdot),\dots,f_n(\cdot),\cdot)\in S(F) \quad \forall f_1,\dots,f_n\in S(F)$

<u>Proof</u> (2.4.1): Let $f \in F$, then $\Phi(\alpha) \equiv f$ for $\alpha \in \mathbb{N}^{(\mathbb{N})}$ is an F-Souslin scheme with $S(\Phi) = f$. Thus we have

$$F \subseteq S(F) \subseteq S(S(F))$$

And the inclusion: $S(S(F)) \subseteq S(F)$, follows in exactly the same way as the corresponding inclusion for sets, see [20; p. 12-16].

(2.4.2): Let $H = S(F) \cap G$, then by (2.4.1) we have

$$H \subseteq G \cap S(H) \subseteq G \cap S(S(F)) = H$$

and so (2.4.2) holds.

(2.4.3) Let $f_n \in S(F)$, and put

$$\phi(\alpha) = f_{\alpha(1)} \qquad \Psi(\alpha) = f_{\alpha(k)} \qquad \forall \alpha \in \mathbb{N}^k$$

Then we have

$$S(\phi) = \sup_n f_n , \qquad S(\Psi) = \inf_n f_n$$

and so $S(F)$ is $(\wedge c, \vee c)$-stable by (2.4.1).

(2.4.6): The proof goes by induction in n. Suppose that $n = 1$, and let Φ be an F-Souslin scheme, then

$$\Psi(\alpha) = \varphi(\Phi(\alpha), \cdot) \qquad \text{for} \quad \alpha \in \mathbb{N}^{(\mathbb{N})}$$

is an $S(F)$-Souslin scheme by (2.4.5), and if $f = S(\Phi)$, then by (2.4.4) we have that

$$\varphi(f(\cdot), \cdot) = S(\Psi)$$

Thus (2.4.6) holds by (2.4.1) if $n = 1$.

Now suppose that (2.4.6) holds for some $n \geq 1$, and suppose that $\varphi: \overline{\mathbb{R}}^{n+1} \times T \to \overline{\mathbb{R}}$ satisfies (2.4.4) and (2.4.5). Let $f \in F$ be given and put

$$\varphi_o(x_1,\ldots,x_n,t) = \varphi(x_1,\ldots,x_n,f(t),t) \quad \forall (x_1\ldots x_n,t) \in \overline{\mathbb{R}}^n \times T$$

Then φ_0 satisfies (2.4.4+5), and thus also (2.4.6) by induction hypothesis. Let $f_1,\ldots,f_n \in S(F)$ be given and put

$$\psi(x,t) = \varphi(f_1(t),\ldots,f_n(t),x,t) \quad \forall (x,t) \in \overline{\mathbb{R}} \times T$$

Then ψ satisfies (2.4.4) and (2.4.5) with $n = 1$. And we have just shown that (2.4.6) holds for $n = 1$, we see that (2.4.6) holds for $n + 1$. Thus the induction step is completed and (2.4.6) is proved.

□

Theorem 2.5. *Let* $F \subseteq \mathbb{R}^T$ *be a convex cone containing* $\pm 1_T$, *and put*

$$\mathcal{B} = \{B \subseteq T \mid B \in \mathcal{S}(F),\ T \setminus B \in \mathcal{S}(F)\}$$

If $f \in \overline{\mathbb{R}}^T$ *then we have*

(2.5.1) $S(F) = U(T, \mathcal{S}(F))$

(2.5.2) $\mathcal{B}$ *is a* σ*-algebra*

(2.5.3) f *is* $\mathcal{B}$*-measurable* $\Longleftrightarrow f \in S(F)$ *and* $-f \in S(F)$

(2.5.4) $\sigma(F) \subseteq \mathcal{B}$ *if* $(-F) \subseteq S(F)$

Moreover if $\mathcal{F}$ *is a paving on* T, *so that* $\mathcal{F} \subseteq S(F)$ *and* $F \subseteq U(T,\mathcal{F})$, *then we have*

(2.5.5) $\mathcal{S}(F) = \mathcal{S}(\mathcal{F})$

Proof (2.5.1): Let $H = S(F) \cap \mathbb{R}^T$, then by applying (2.4.6) to $\varphi(x,t) = x + f(t)$ and $\varphi(x) = ax$ it follows that H is a convex cone containing F, and thus $\pm 1_T \in H$. Moreover $\mathcal{S}(F) = H \cap 2^T$, so by

Proposition 2.2 we have that

$$U(T,\mathcal{S}(F)) \subseteq H_{\delta\sigma} \subseteq S(F)$$

since $S(F)$ is $(\wedge c,\vee c)$-stable by Propositon 2.4. The converse inclusion follows easily from Propositions 2.3 and 2.4

(2.5.2): Trivial consequence of (2.4.3).

(2.5.3): If f is $\mathcal{B}$-measurable, then f and $(-f)$ are upper $\mathcal{S}(F)$-functions by definition of $\mathcal{B}$ and so f and $(-f)$ belongs to $S(F)$ by (2.5.1). Conversely if f and $(-f)$ are upper $\mathcal{S}(F)$-functions then $\{f \geq a\} \in \mathcal{S}(F)$ and

$$\{f < a\} = \bigcup_{n=1}^{\infty} \{-f \geq 2^{-n}-a\} \in \mathcal{S}(F)$$

since $\mathcal{S}(F)$ is $(\cup c)$-stable. Thus f is $\mathcal{B}$-measurable.

(2.5.4) follows from (2.5.3).

(2.5.5): Since $F \subseteq \mathcal{S}(F)$ we have by (2.4.2) that $\mathcal{S}(F) \subseteq \mathcal{S}(\mathcal{S}(F)) = \mathcal{S}(F)$. If $B \in \mathcal{S}(F)$, then there exist an F-Souslin scheme Φ, such that $1_B = S(\Phi)$. Now put

$$Q(\alpha) = \{\Phi(\alpha) \geq \tfrac{1}{2}\} \qquad \forall \alpha \in \mathbb{N}^{(\mathbb{N})}$$

Then $\{Q(\alpha) \mid \alpha \in \mathbb{N}^{(\mathbb{N})}\}$ is an $\mathcal{F}$-Souslin scheme since $\Phi(\alpha) \in F \subseteq U(T,\mathcal{F})$. And the reader easily verifies that $S(Q) = B$. Hence $B \in \mathcal{S}(\mathcal{F})$ and so (2.5.5) follows.

Proposition 2.6. Let T be a topological space and let F be a subset of $\overline{\mathbb{R}}^T$ inducing the topology on T. Then we have

(2.6.1) $\aleph_0 \vee \text{weight}(F) = \aleph_0 \vee \text{weight}(T)$

(2.6.2) T is separable and pseudo-metrizable, if and only if $\text{weight}(F) \leq \aleph_0$.

<u>Proof</u>. Let $H \subseteq F$ so that $\mathrm{card}(H) = \mathrm{weight}(F)$ and $\tau(H) = \tau(F)$. Let $\mathcal{G}$ be a countable open base for the topology on $\overline{\mathbb{R}}$. Then

$$\mathcal{G}_o = \{\bigcap_{j=1}^{n} h_j^{-1}(G_j) \mid n \in \mathbb{N}, G_1 \ldots G_n \in \mathcal{G}, h_1, \ldots, h_n \in H\}$$

is a base for the topology on T, and so

$$\begin{aligned} \mathrm{weight}(T) &\leq \aleph_o \vee \mathrm{card}(\mathcal{G}_o) \leq \aleph_o \vee \mathrm{card}\ H \\ &= \aleph_o \vee \mathrm{weight}(F) \end{aligned}$$

And by [5; Theorem 1.1.15] there exist a base $\mathcal{U}$ for the topology on T, such that $\mathcal{U} \subseteq \mathcal{G}_o$ and $\mathrm{card}(\mathcal{U}) = \mathrm{weight}(T)$. If $U \in \mathcal{U}$, then there exist $n(U) \in \mathbb{N}$ and $G_{jU} \in \mathcal{G}$ and $h_{jU} \in H$ for $1 \leq j \leq n(U)$ so that

$$U = \bigcap_{j=1}^{n(U)} h_{jU}^{-1}(G_{jU})$$

Now let $H_o = \{h_{jU} \mid U \in \mathcal{U}, 1 \leq j \leq n(U)\}$, then $H_o \subseteq H$ and $\tau(H_o) = \tau(F)$. Hence we have

$$\begin{aligned} \mathrm{weight}(F) &\leq \mathrm{card}\ H_o \leq \aleph_o \vee \mathrm{card}\ \mathcal{U} \\ &= \aleph_o \vee \mathrm{weight}(T) \end{aligned}$$

Thus (2.6.1) is proved.

(2.6.2) follows from (2.6.1) and [5; Theorem 4.2.9].

□

<u>Proposition 2.7</u>. <u>Let</u> (T,A) <u>be an algebraic function space, let</u> Q <u>be a set, and let</u> D <u>be a closed subset of</u> $\mathbb{R}^Q$. <u>Then</u> $\bar{A}$ <u>is</u> φ-<u>stable for any continuous map</u> $\varphi: D \to \mathbb{R}$.

<u>Proof</u>. Let $K = \{\omega \in D \mid |\omega(q)| \leq \|f_q\| \ \forall q \in Q\}$, then K is compact, and the set

$$\{\varphi \in C(K) \mid \varphi \circ f \in \overline{A}\} = C$$

is a $\|\cdot\|_K$-closed algebra, containing 1 and separating points in K. Thus $C = C(K)$ by Stone-Weierstrass theorem [5; Theorem 3.2.21]

□

Proposition 2.8. Let T be a set, let $F \subseteq \overline{\mathbb{R}}^T$, and let $\{f_\gamma \mid \gamma \in \Gamma\}$ be a decreasing net in $\overline{\mathbb{R}}^T$, so that $f_\gamma \downarrow \varphi$.

(1): If $f_\gamma + a1_T \in \overline{F}$ for all $\gamma \in \Gamma$ and all $a \in \mathbb{R}_+$ then there exist a decreasing net $\{g_\lambda \mid \lambda \in \Lambda\}$ in F and an increasing cofinal map $\Lambda \to \Gamma$ satisfying

(2.8.1) $g_\lambda \downarrow \varphi$ and $g_\lambda \geq f_{\sigma(\lambda)}$ $\forall \lambda \in \Lambda$

(2.8.2) $\|g_\lambda - f_{\sigma(\lambda)}\| \to 0$ and $\text{card}(\Lambda) \leq \aleph_o \vee \text{card}(\Gamma)$

And if $\Gamma = \mathbb{N}$ we may take $\Lambda = \mathbb{N}$ and $\sigma(n) = n$, $\forall n \in \mathbb{N}$.

(2): Suppose that $f + a1_T$ and $f \wedge g$ belongs to $\overline{F}$ for all $f, g \in F$ and all $a \in \mathbb{R}_+$. If $f_\gamma \in \overline{F}_\xi$ for all $\gamma \in \Gamma$, where ξ is an infinite cardinal, then there exist a decreasing net $\{g_\lambda \mid \lambda \in \Lambda\}$ in F satisfying (2.8.1), such that $\text{card}(\Lambda) \leq \xi \vee \text{card}(\Gamma)$. Moreover if Γ and ξ are countable, and if Γ has no maximal element, then we may take $\Lambda = \Gamma$ and $\sigma(\gamma) = \gamma$ for all $\gamma \in \Gamma$.

(3): If F is $(\wedge f)$-stable and $f_\gamma \in F_\xi$ for all $\gamma \in \Gamma$, then there exist a decreasing net $\{g_\lambda \mid \lambda \in \Lambda\}$ in F, and an increasing cofinal map $\sigma: \Lambda \to \Gamma$ satisfying (2.8.1), such that $\text{card}(\Lambda) \leq \xi \vee \text{card}(\Gamma)$. Moreover if Γ and ξ are countable, and Γ has no maximal element, then we may take $\Lambda = \Gamma$ and $\sigma(\gamma) = \gamma$ for all $\gamma \in \Gamma$.

Remarks (1): Of course we have an analogue result for increasing nets

(2): If (T,A) is an algebraic function space, then the hypothesis of (1) holds, if $F = A$ and $f_\gamma \in \overline{A}$ $\forall \gamma \in \Gamma$, and the hypothesis of (2) holds, if $F = A$.

<u>Proof</u> (1): Let Λ be the set of all finite non-empty subsets of Γ ordered by inclusion. Then Λ is a directed set. Now put

$$\Lambda_n = \{\lambda \in \Lambda \mid \text{card } \lambda \leq n\} \qquad \forall n = 1,2,\ldots$$

Then $\Lambda_n \uparrow \Lambda$. If $\lambda \in \Lambda_1$ then $\lambda = \{\gamma\}$ for some $\gamma \in \Gamma$ and we put $\sigma(\lambda) = \gamma$, now suppose that $\sigma : \Lambda_n \to \Gamma$ has been defined, such that σ is increasing. If $\lambda \in \Lambda_{n+1} \setminus \Lambda_n$, then there is atmost 2^{n+1} sets $\mu \in \Lambda_n$, so that $\mu \subseteq \lambda$. And since Γ is upwards directed we may choose $\sigma(\lambda) \in \Gamma$, such that $\sigma(\lambda) \geq \sigma(\mu)$ if $\mu \subseteq \lambda$, $\mu \neq \lambda$. Thus by induction in n, there exist an increasing map σ from Λ into Γ, such that

(i) $\qquad \sigma(\lambda) = \gamma \quad \text{if} \quad \lambda = \{\gamma\}$

Since $\sigma(\lambda) \geq \gamma$ for all $\lambda \geq \{\gamma\}$, we see that σ is an increasing cofinal map from Λ into Γ.

By assumption there exist $g_\lambda \in F$, such that

(ii) $\qquad \|g_\lambda - (f_{\sigma(\lambda)} + 3\cdot 2^{-\text{card }\lambda})\| \leq 2^{-\text{card }\lambda} \qquad \forall \lambda \in \Lambda$

Let $\lambda, \mu \in \Lambda$ so that $\mu \subseteq \lambda$ and $\mu \neq \lambda$. Let $n = \text{card}(\lambda)$ and $k = \text{card}(\mu)$. Then

$$g_\lambda \leq f_{\sigma(\lambda)} + 4\cdot 2^{-n} \leq f_{\sigma(\mu)} + 3\cdot 2^{-k} - 2^k \leq g_\mu$$

since $f_{\sigma(\lambda)} \leq f_{\sigma(\mu)}$ and $k+1 \leq n$. Thus $\{g_\lambda \mid \lambda \in \Lambda\}$ is a decreasing net satisfying (2.8.1), since σ is cofinal and so $f_{\sigma(\lambda)} \downarrow \varphi$. Moreover

$$\text{card}(\Lambda) \leq \aleph_0 \vee \text{card}(\Gamma)$$
$$\|g_\lambda - f_{\sigma(\lambda)}\| \leq 4\cdot 2^{-\text{card }\lambda}$$

and so (2.8.2) holds.

Now suppose that $\Gamma = \mathbb{N}$, then we put $\Lambda = \mathbb{N}$ and choose $g_n \in F$, so that

$$\|g_n - (f_n + 3\cdot 2^{-n})\| \leq 2^{-n}$$

Then as above one finds that $\{g_n\}$ is a decreasing sequence in F satisfying (2.8.1) and (2.8.2) with $\sigma(n) = n$ for all $n \in \mathbb{N}$.

(2): From the assumptions it follows easily that $\overline{F}$ is (vf)-stable and $f + a1_T \in \overline{F}$ for all $f \in \overline{F}$ and all $a \in \mathbb{R}_+$. Since $f_\gamma \in \overline{F}_\xi$ there exist functions $f_{\alpha\gamma} \in \overline{F}$ for $\alpha \in \xi$ and $\gamma \in \Gamma$ so that

$$f_\gamma = \inf_{\alpha\in\xi} f_{\alpha\gamma} \qquad \forall\, \gamma \in \Gamma$$

Now let Λ be the set of all non-empty finite subsets of $\Gamma \times \xi$ ordered by inclusion, and put

$$f_\lambda = \min_{(\alpha,\gamma)\in\lambda} f_{\alpha\gamma}$$

Then $\{f_\lambda \mid \lambda \in \Lambda\}$ is a decreasing net in $\overline{F}$, so that $f_\lambda \downarrow \varphi$, and $f_\lambda + a1_T \in \overline{F}$ for all $\lambda \in \Lambda$ and all $a \in \mathbb{R}_+$. And above there exist an increasing cofinal map σ from Λ into Γ, so that

(iii) $\qquad \sigma(\lambda) = \gamma \qquad \text{if} \qquad \gamma = \{(\alpha,\gamma)\}$

Then there exist a net $\{g_\lambda \mid \lambda \in \Lambda\}$ in F so that

(iv) $\qquad \|g_\lambda - (f_\lambda + 3\cdot 2^{-\operatorname{card}\lambda})\| \leq 2^{-\operatorname{card}\lambda} \qquad \forall\, \lambda \in \Lambda$

And so as above we have that $\{g_\lambda\}$ is a decreasing net in F so that $g_\lambda \downarrow \varphi$ and $g_\lambda \geq f_\lambda$ for all $\lambda \in \Lambda$. Now let $\lambda \in \Lambda$, and let μ be the projection of λ onto Γ, then by (iii) and monotonicity of σ we have that $\sigma(\lambda) \geq \gamma$ for all $\gamma \in \mu$, and so we have

$$g_\lambda \geq f_\lambda \geq \min_{\gamma\in\mu} \inf_{\alpha\in\xi} f_{\alpha\gamma} \geq \min_{\gamma\in\mu} f_\gamma \geq f_{\sigma(\lambda)}$$

Thus (2.8.1) holds, and clearly $\operatorname{card} \Lambda \leq \xi \vee \operatorname{card} \Gamma$.

Now suppose that Γ and ξ are atmost countable. Then Λ is countable and so there exist an increasing cofinal map θ from $\mathbb{N}$ into Λ. And we put

$$\eta(\gamma) = \sup\{k \in \mathbb{N} \mid \gamma \geq \sigma(\theta(k))\}$$

with the convention that $\sup \emptyset = 1$. Since $\sigma \circ \theta$ is increasing and cofinal from $\mathbb{N}$ into Γ, and Γ has no maximal element, we see that $\eta(\gamma) \in \mathbb{N}$ for all $\gamma \in \Gamma$. And clearly η is increasing and cofinal from Γ into $\mathbb{N}$, since $\eta(\gamma) \geq k$ for all $\gamma \geq \sigma(\theta(k))$.

Now put $\tau = \theta \circ \eta$ and $\hat{g}_\gamma = g_{\tau(\gamma)}$. Then τ is an increasing cofinal map from Γ into Λ, and so $\{\hat{g}_\gamma \mid \gamma \in \Gamma\}$ is a decreasing net in F, such that $\hat{g}_\gamma \downarrow \varphi$. Moreover since $\sigma(\tau(\gamma)) \leq \gamma$ by definition of η we have by (2.8.1)

$$\hat{g}_\gamma = g_{\tau(\gamma)} \geq f_{\sigma(\tau(\gamma))} \geq f_\gamma$$

Thus $\{\hat{g}_\gamma\}$ is a decreasing net in F satisfying (2.8.1) with $\Lambda = \Gamma$ and $\sigma(\gamma) = \gamma$ for all $\gamma \in \Gamma$.

(3) In this case we proceed as in (2) and we put $g_\lambda = f_\lambda$ for $\lambda \in \Lambda$, then it is easily checked that $\{g_\lambda \mid \lambda \in \Lambda\}$ is a decreasing net in F satisfying (2.8.1), and the last part of (3) follows as above.

□

<u>Proposition 2.9</u>. <u>Let</u> (T,A) <u>be an algebraic function space, and let</u> ξ <u>be an infinite cardinal. Then we have</u>

(2.9.1) $$A_\xi = \overline{A}_\xi = B^*(T) \cap U(T, F_\xi(A))$$
$$= B^*(T) \cap \bigcup\{\operatorname{Usc}(T,\tau(Q)) \mid Q \subseteq A,\ \operatorname{card}(Q) \leq \xi\}$$
$$= B^*(T) \cap \bigcup\{\operatorname{Usc}(T,\tau(Q)) \mid Q \subseteq A,\ \operatorname{weight}(Q) \leq \xi\}$$

(2.9.2) $\mathcal{F}_\xi(A) = 2^T \cap A_\xi =$ <u>the</u> $(\cap\xi)$<u>-closure of</u> $\mathcal{F}_o(A)$

$= \cup\{\mathcal{F}(Q) \mid Q \subseteq A \quad \text{weight}(Q) \subseteq A\}$

$$= \left\{ \bigcap_{\gamma\in\Gamma} \{\varphi_\gamma \geq 0\} \;\middle|\; \begin{array}{l} \text{card}(\Gamma) \leq \xi \ \ \underline{\text{and}} \ \ \{\varphi_\gamma\} \ \underline{\text{is}} \\ \underline{\text{a decreasing net in}} \ \ A \end{array} \right\}$$

(2.9.3) $\sigma(A_\xi) = \sigma(A^\xi) = \sigma(\mathcal{F}_\xi(A)) = \sigma(\mathcal{G}_\xi(A))$

<u>Moreover</u> $\bar{A}$ <u>equals the set of all</u> $f \in \mathbb{R}^T$, <u>which are uniformly continuous with respect to the uniformity induced by the family of pseudo-metrics</u>: $\{d_\varphi \mid \varphi \in A\}$, <u>where</u>

$$d_\varphi(u,v) = |\varphi(u)-\varphi(v)| \qquad \forall\, u,v \in T \quad \forall\, \varphi \in A$$

<u>In particular if</u> $(T,\tau(A))$ <u>is hereditarily ξ-Lindelöf, (e.g. if</u> $\text{weight}(A) \leq \xi$), <u>then</u>

(2.9.4) $\mathcal{F}(A) = \mathcal{F}_\xi(A)$, $\quad A_\xi = B^*(T) \cap \text{Usc}(T,\tau(A))$

(2.9.5) $\mathcal{G}(A) = \mathcal{G}_\xi(A)$, $\quad A^\xi = B_*(T) \cap \text{Lsc}(T,\tau(A))$

(2.9.6) $\mathcal{K}(A) = \mathcal{K}_\xi(A)$, $\quad \mathcal{B}(A) = \sigma(A_\xi) = \sigma(A^\xi)$

<u>Remark</u>. Of course we have complete analogues of (2.9.1) and (2.9.2) for A^ξ and $\mathcal{G}_\xi(A)$.

<u>Proof</u> (2.9.1): By Proposition 2.8 we have that $\bar{A} \subseteq A_\xi$ and so $A_\xi = \bar{A}_\xi$. Since $\mathcal{F}_\xi(A)$ is $(\cap\xi, \cup f)$-stable and $A \subseteq U(T,\mathcal{F}_\xi(A))$ we have by Proposition 2.1 that

$$A_\xi \subseteq B^*(T) \cap U(T,\mathcal{F}_\xi(A))$$

And if U_o and U_1 denote the two last sets in (2.9.1) then clearly we have

(i) $\quad A_\xi \subseteq B^*(T) \cap U(T,\mathcal{F}_\xi(A)) \subseteq U_o \subseteq U_1$

Now let $f \in U_1$, then there exist $Q \subseteq A$ so that weight$(Q) \leq \xi$ and f is upper $\tau(Q)$-continuous. Let B be the algebra generated by $Q \cup \{1_T\}$, and put

$$\Phi = \{\varphi \in B \mid \varphi \geq f\}, \qquad f_o = \inf_{\varphi \in \Phi} \varphi$$

If $t_o \in T$ and $a \in \mathbb{R}$ so that $f(t_o) < a$. Then there exist $\varphi_1, \ldots, \varphi_n \in Q$ and $\delta > 0$, such that

$$\text{(ii)} \qquad f(t) < a \quad \text{if} \quad \sum_{j=1}^{n} |\varphi_j(t) - \varphi_j(t_o)|^2 < \delta$$

Now choose $b \in \mathbb{R}$, so that $b > a$ and $f \leq b$ (recall that $f \in B^*(T)$), and put

$$\varphi(t) = a + (b-a)\delta^{-1} \sum_{j=1}^{n} (\varphi_j(t) - \varphi_j(t_o))^2$$

Then $\varphi \in B$ and $\varphi(t_o) = a$. Moreover since $f \leq b$ it follows easily from (ii) that $\varphi \geq f$. Hence $\varphi \in \Phi$ and so $f_o(t_o) \leq a$. Thus $f_o \leq f$ and since the converse inequality is evident, we have that $f_o = f$. Moreover since $B \subseteq C(T, \tau(Q))$, and $(T, \tau(Q))$ is heriditarily ξ-Lindelöf there exist $\Psi \subseteq \Phi$ so that card$(\Psi) \leq \xi$, and

$$f = f_o = \inf_{\psi \in \Psi} \psi$$

Thus $f \in A_\xi$ since $\Psi \subseteq B \subseteq A$. Hence $U_1 \subseteq A_\xi$ and so (2.9.1) follows from (i).

(2.9.2): By (2.9.1) we have that $A_\xi \cap 2^T = F_\xi(A)$. If weight$(Q) \leq \xi$, then there exist $Q_o \subseteq Q$ so that card$(Q_o) \leq \xi$ and $F(Q) = F(Q_o)$, hence

$$F_\xi(A) \supseteq \cup \{F(Q) \mid Q \subseteq A, \text{ weight}(Q) \leq \xi\}$$

and the converse inclusion is evident.

Let F_o be the $(\cap\xi)$-closure of $F_o(A)$, and let F_1 denote the last set in (2.9.2). Since $F_\xi(A)$ is $(\cap\xi)$-stable and $F_\xi(A) \supseteq F_o(A)$ we have

(iii) $\quad F_1 \subseteq F_o \subseteq F_\xi(A)$

And if $F \in F_\xi(A)$, then by (2.9.1) and Proposition 2.8 there exist a decreasing net $\{\varphi_\gamma \mid \gamma \in \Gamma\}$ in A, such that $\mathrm{card}(\Gamma) \leq \xi$ and $\varphi_\gamma \downarrow 1_F$, hence

$$F = \bigcap_\gamma \{\varphi_\gamma \geq 1\} \in F_1$$

and so (2.9.2) follows from (iii).

(2.9.3) follows easily from (2.9.1) and (2.9.2).

Let $f \in \mathbb{R}^T$ be uniformly continuous with respect to the uniformity generated by $\{d_\varphi \mid \varphi \in A\}$ and let $\varepsilon > 0$. Then there exist $\varphi_1, \ldots, \varphi_n \in A$ and $\delta > 0$ such that

(iv) $\quad |f(u) - f(v)| \leq \varepsilon \quad$ if $\quad \sum_1^n |\varphi_j(u) - \varphi_j(v)| \leq \delta$

Then f is bounded since $\varphi_1, \ldots, \varphi_n$ are so, and if

$$F(x) = \inf_{t \in T}\{f(t) + a \sum_{j=1}^n |\varphi_j(t) - x_j|\} \qquad \forall x = (x_1, \ldots, x_n) \in \mathbb{R}^n$$

where $a = 2\delta^{-1}\|f\|$, then F maps $\mathbb{R}^n$ into $\mathbb{R}$ and

$$F(x) \leq f(t) + a \sum_1^n |\varphi_j(t) - y_j| + a \sum_1^n |x_j - y_j|$$

for all $x, y \in \mathbb{R}^n$ and all $t \in T$. So taking infimum over $t \in T$, we find

$$|F(x) - F(y)| \leq a \sum_{j=1}^n |x_j - y_j| \qquad \forall x, y \in \mathbb{R}^n$$

Hence F is continuous from $\mathbb{R}^n$ into $\mathbb{R}$, and so by Proposition 2.7, we have that $\varphi = F(\varphi_1, \ldots, \varphi_n)$ belongs to $\bar{A}$. Putting $t = u$ in the infimum defining $\varphi(u)$ we see that $\varphi \leq f$. Moreover by (iv) we have

$$f(t) + a \sum_{j=1}^{n} |\varphi_j(t) - \varphi_j(u)| \geq (f(u)-\varepsilon) \wedge (f(t) + 2\|f\|)$$

$$\geq f(u) - \varepsilon$$

since $f(t) + 2\|f\| \geq f(u)$. Hence $f - \varepsilon \leq \varphi \leq f$, and so $f \in \bar{A}$.

Conversely if $f \in \bar{A}$, then evidently f is uniformly continuous with respect to the uniformity induced by $\{d_\varphi | \varphi \in A\}$, and so the proposition is proved.

□

<u>Proposition 2.10</u>. <u>Let</u> (T,A) <u>be an algebraic function space, let</u> ξ <u>be an infinite cardinal, and let</u> K <u>be a subset of</u> T. <u>Then the following five statements are equivalent</u>

(2.10.1) $K \in \mathcal{K}_\xi(A)$

(2.10.2) $p_Q(K) \in \mathcal{K}(\mathbb{R}^Q)$ $\forall Q \subseteq A$ <u>with</u> $\mathrm{weight}(Q) \leq \xi$

(2.10.3) $p_Q(K) \in \mathcal{K}(\mathbb{R}^Q)$ $\forall Q \subseteq A$ <u>with</u> $\mathrm{card}(Q) \leq \xi$

(2.10.4) $\{K \cap F | F \in \mathcal{F}_\xi(A)\}$ <u>is</u> ξ<u>-compact</u>

(2.10.5) $\{K \cap \{\varphi \geq 0\} | \varphi \in A\}$ <u>is mono-</u>ξ<u>-compact</u>

<u>where</u> $p_Q(t) = (q(t))_{q \in Q}: T \to \mathbb{R}^Q$ <u>for</u> $Q \subseteq \mathbb{R}^T$.

<u>Moreover if</u> $K \in \mathcal{K}_\xi(A)$, <u>then</u> $K \in \mathcal{F}_\xi(A)$, <u>if and only if</u> K <u>is</u> Q<u>-saturated for some</u> $Q \subseteq A$ <u>with</u> $\mathrm{card}(Q) \leq \xi$

<u>Proof</u> (2.10.1) ⇒ (2.10.2): Let $Q \subseteq A$ with $\mathrm{weight}(Q) \leq \xi$. Then there exist $R \subseteq Q$ with $\mathrm{card}(R) \leq \xi$ and $\tau(R) = \tau(Q)$. Then q is $\tau(R)$-continuous for all $q \in Q$, and so p_Q is $\tau(R)$-continuous from T into $\mathbb{R}^Q$, and K is $\tau(R)$-compact by assumption. Hence (2.10.2) follows

(2.10.2) ⇒ (2.10.3): Evident!

(2.10.3) ⇒ (2.10.4): Let $\{F_i \mid i \in I\} \subseteq F_\xi(A)$ so that $\cap(F_i \cap K) = \emptyset$, and $\operatorname{card} I \leq \xi$. Then there exist a set $Q \subseteq A$ so that $\operatorname{card}(Q) \leq \xi$ and $F_i \in F(Q)$ for all $i \in I$. By (F.6) there exist $E_i \in F(\mathbb{R}^Q)$ so that $F_i = p_Q^{-1}(E_i)$ for all $i \in I$. But then the reader easily verifies that we have

$$\text{(i)} \qquad p_Q(K) \cap \bigcap_{j \in J} E_j = p_Q(K \cap \bigcap_{j \in J} F_j) \qquad \forall J \subseteq I$$

Putting $J = I$ in (i) we see that $p_Q(K) \cap \cap E_i = \emptyset$. So by compactness of $p_Q(K)$ and closeness of E_i there exist a finite set $J \subseteq I$, such that

$$p_Q(K \cap \bigcap_{j \in J} F_j) = p_Q(K) \cap \bigcap_{j \in J} E_j = \emptyset$$

But then $K \cap \cap_{j \in J} F_j = \emptyset$, and so (2.10.4) holds.

(2.10.4) ⇒ (2.10.5): Evident!

(2.10.5) ⇒ (2.10.6): Let $Q \subseteq A$ so that $\operatorname{card}(Q) \leq \xi$ and let $\{F_i \mid i \in I\}$ be a family in $F(Q)$, so that $F \cap K = \emptyset$ where $F = \cap F_i$. Since $(T, \tau(Q))$ is hereditarily ξ-Lindelöf there exist $J \subseteq I$ so that $\operatorname{card}(J) \leq \xi$ and

$$F = \bigcap_{j \in J} F_j$$

Now let Γ be the set of all nonempty finite subsets of Γ,

$$F_\gamma = \bigcap_{j \in \gamma} F_j \qquad \forall \gamma \in \Gamma$$

Then $F_\gamma \in F(Q) \subseteq F_\xi(A)$ and $F_\gamma \downarrow F$. Since $\operatorname{card}(\Gamma) \leq \xi$, it follows from Propositions 2.8 and 2.9, that there exist a net $\{\varphi_\lambda \mid \lambda \in \Lambda\}$ and an increasing cofinal map $\sigma : \Lambda \to \Gamma$ such that

(ii) $\varphi_\lambda \downarrow 1_F$ and $\varphi_\lambda \geq 1_{F_{\sigma(\lambda)}}$ $\forall \lambda \in \Lambda$

(iii) $\text{card}(\Lambda) \leq \xi$

Then $K \cap \{\varphi_\lambda \geq 1\} \downarrow \emptyset$, and so by assumption there exist $\lambda \in \Lambda$ so that $K \cap \{\varphi_\lambda \geq 1\} = \emptyset$. Moreover if $\gamma = \sigma(\lambda)$ then $F_\gamma \subseteq \{\varphi_\lambda \geq 1\}$ by (ii) and so we have

$$K \cap F_\gamma = K \cap \bigcap_{j \in \gamma} F_j = \emptyset$$

and since $\gamma \subseteq I$ is finite we see that $K \in K(Q)$, and so (2.10.1) holds.

Now suppose that $K \in K_\xi(A)$ and K is Q-saturated for some $Q \subseteq A$ with $\text{card}(Q) \leq \xi$. Then by (C.3) we have that $K = p_Q^{-1}(C)$, where $C = p_Q(K)$. By (2.10.3) we know that C is compact and thus closed in $\mathbb{R}^Q$, since $\mathbb{R}^Q$ is Hausdorff. Thus $K \in F(Q) \subseteq F_\xi(A)$. Conversely every set in $F_\xi(A)$ is evidently Q-saturated for some $Q \subseteq A$ with $\text{card}\, Q \leq \xi$.

□

<u>Proposition 2.11</u>. <u>Let</u> ξ <u>be an infinite cardinal number, let</u> F <u>be a</u> $(\cup f, \cap \xi)$<u>-stable paving, and let</u> φ <u>be a map from</u> $\mathbb{R}_+^Q \times T$ <u>into</u> $\overline{\mathbb{R}}$, <u>where</u> $\text{card}(Q) \leq \xi$, <u>such that</u>

(2.11.1) $\varphi(\cdot, t)$ <u>is increasing and upper semicontinuous</u> $\forall t \in T$

(2.11.2) $\varphi(\omega, \cdot) \in U(T, F)$ $\forall \omega \in \mathbb{R}_+^Q$

<u>Now let</u> $\hat{f} = (f_q)$ <u>be a map from</u> T <u>into</u> $\mathbb{R}^Q$, <u>such that</u> $f_q \in U(T,F) \cap B^+(T)$ <u>for all</u> $q \in Q$. <u>Then the map</u>

$$f(t) = \varphi(\hat{f}(t), t) \quad \text{for} \quad t \in T$$

<u>belongs to</u> $U(T, F)$.

<u>Proof</u>. If $Q = \{\emptyset\}$, then the proposition is evident. Now suppose that $Q = \{1\}$. Then $f_1 \in U(T,F) \cap B^+(T)$ and so $\emptyset$, $T \in F$ and $0 \le f_1 < a$ for some $a \in \mathbb{R}_+$. Let $F_{jn} = \{f_1 \ge j\, n^{-1} a\}$ for $j = 0,1,\dots,n$, and put

$$h_n = n^{-1} a \sum_{j=0}^{n} 1_{F_{jn}}$$

Then $f_1 \le h_n \le n^{-1} a + f_1$ and $T = F_{on} \supseteq \dots \supseteq F_{nn} = \emptyset$. Put

$$g_n(t) = \varphi(h_n(t),t)$$

Then we have

$$\text{(i)} \qquad \{g_n \ge c\} = \bigcup_{j=0}^{n} \{t \in T \mid \varphi(n^{-1}(j+1)a,t) \ge c\} \cap F_{jn}$$

To see this let $g_n(t) \ge c$. Then $t \in F_{jn} \setminus F_{j+1n}$ for some $0 \le j \le n-1$, and so $h_n(t) = n^{-1}(j+1)a$ and

$$g_n(t) = \varphi(n^{-1}(j+1)a,t) \ge c$$

Conversely if t belongs to the right hand side of (i) then $t \in F_{jn}$ and $\varphi(n^{-1}(j+1)a,t) \ge c$ for some $0 \le j \le n$. But then $h_n(t) \ge n^{-1}(j+1)a$ and since $\varphi(\cdot,t)$ is increasing we have

$$g_n(t) \ge \varphi(n^{-1}(j+1)a,t) \ge c$$

Thus (i) holds.

Now by (i), (2.11.2) and $(\cup f,\cap f)$-stability of F we have that $g_n \in U(T,F)$. Since $h_n \ge f_1$ and $h_n \to f_1$ we have by (2.11.1)

$$\varphi(f_1(t),t) \le \inf_n \varphi(h_n(t),t) \le \limsup_{n\to\infty} \varphi(h_n(t),t)$$
$$\le \varphi(f_1(t),t)$$

I.e. $f = \inf_n g_n$ and so $f \in U(T,F)$ by Proposition 2.1. (Recall that ξ is infinite and so F is $(\cup f,\cap c)$-stable).

Now let $n \in \mathbb{N}$ and suppose that the proposition holds whenever $\mathrm{card}(Q) < n$. Let $Q = \{1,\dots,n\}$ and put

$$\hat{\varphi}(x_1,\dots,x_{n-1},t) = \varphi(x_1,\dots,x_{n-1},f_n(t),t)$$

where $f_1,\dots f_n \in U(T,F) \cap B^+(T)$. By the argument above we have that $\hat{\varphi}$ satisfies (2.11.1) and (2.11.2), and so by induction hypothesis we have that

$$\varphi(f_1(t),\dots,f_n(t),t) \in U(T,F)$$

Thus by induction we see that the proposition holds whenever Q is finite.

Now let Q be arbitrary, and let $a_q = \|f_q\|$. If π is finite subset of Q, we put

$$(\sigma_\pi \omega)(q) = \begin{cases} a_q \wedge \omega(q) & \text{if } q \in \pi, \omega \in \mathbb{R}_+^\pi \\ a_q & \text{if } q \in Q \setminus \pi, \omega \in \mathbb{R}_+^\pi \end{cases}$$

$$\varphi_\pi(\omega,t) = \varphi(\sigma_\pi\omega,t) \qquad \forall\, \omega \in \mathbb{R}_+^\pi \quad \forall\, t \in T$$

$$f^\pi = (f_q)_{q\in\pi}, \qquad f_\pi = \varphi_\pi(f^\pi(t),t)$$

Then by the finite case we have that $f_\pi \in U(T,F)$ for all finite sets. And since $\sigma_\pi f^\pi \downarrow \hat{f}$ we have

$$f(t) = \varphi(\hat{f}(t),t) \leq \inf_\pi \varphi(\sigma_\pi f^\pi(t),t)$$

$$= \inf_\pi f_\pi(t) \leq \lim_\pi \sup \varphi(\sigma_\pi f^\pi(t),t)$$

$$\leq \varphi(\hat{f}(t),t) = f(t)$$

and so $f = \inf_\pi f_\pi$, and the cardinality of the set of all finite subsets of Q is $\leq \xi$, so it follows from Proposition 2.1 that $f \in U(T,F)$, since F is $(\cup f, \cap\xi)$-stable. □

<u>Proposition 2.12</u>. <u>Let</u> (T,A) <u>be an algebraic function space</u>, <u>and let</u> φ <u>be a map from</u> $D \times T$ <u>into</u> $\mathbb{R}$, <u>where</u> D <u>is a closed subset of</u> $\mathbb{R}^Q$, <u>such that</u>

(2.12.1) $\{\varphi(\cdot,t) \mid t \in T\}$ <u>is equicontinuous on</u> K <u>for all compact sets</u> $K \subseteq D$

(2.12.) $\varphi(\omega,\cdot) \in \bar{A} \quad \forall \omega \in D$

<u>If</u> $\hat{f} = (f_q)$ <u>is a map from</u> T <u>into</u> D, <u>such that</u> $f_q \in \bar{A}$, <u>then</u> $f(t) = \varphi(\hat{f}(t),t)$ <u>belongs to</u> $\bar{A}$.

<u>Proof</u>. Let $K = \{\omega \in D \mid |\omega(q)| \leq \|f_q\| \ \forall q \in Q\}$, then K is a compact subset of D. Now let $\varepsilon > 0$ and $\omega_0 \in K$ be given, then by (2.12.1) there exist a neighborhood $U(\omega_0)$ of ω_0 in $\mathbb{R}^Q$, such that

(i) $|\varphi(\omega,t) - \varphi(\omega_0,t)| < \varepsilon/4 \quad \forall \omega \in K \cap U(\omega_0) \quad \forall t \in T$

By definition of the product topology, there exist a finite set $\pi(\omega_0) \subseteq Q$ and $\delta(\omega_0) > 0$, so that $V(\omega_0) \subseteq U(\omega_0)$ where

$$V(\omega_0) = \{\omega \mid |\omega(q) - \omega_0(q)| < 2\delta(\omega_0) \quad \forall q \in \pi(\omega_0)\}$$

Now let

$$W(\omega_0) = \{\omega \mid |\omega(q) - \omega_0(q)| < \delta(\omega_0) \quad \forall q \in \pi(\omega_0)\}$$

Then by compactness of K there exist $\omega_1,\ldots,\omega_N \in K$, so that

(ii) $K \subseteq \bigcup_{j=1}^{N} W(\omega_j)$

Now put $\delta = (\varepsilon/2) \wedge \min \delta(\omega_j)$ and $\pi = \cup \pi(\omega_j)$. If $u,v \in T$ we define

$$\rho(u,v) = \sum_{j=1}^{N} |\varphi(\omega_j,u) - \varphi(\omega_j,v)| + \sum_{q\in\pi} |f_q(u)-f_q(v)|$$

then ρ is a pseudo-metric on T. Suppose that $\rho(u,v) < \delta$, and choose $j \in \{1,\ldots,N\}$ so that $\hat{f}(u) \in W(\omega_j)$ (note that $\hat{f}(u) \in K$). Then we have

$$|f(u)-f(v)| = |\varphi(\hat{f}(u),u)-\varphi(\hat{f}(v),v))$$

$$\leq |\varphi(\hat{f}(u),u)-\varphi(\omega_j,u)| + \rho(u,v) + |\varphi(\omega_j,v)-\varphi(\hat{f}(v),v)|$$

Then $\hat{f}(u) \in W(\omega_j) \subseteq U(\omega_j)$ and since $\delta \leq \delta(\omega_j)$ we have

$$|\omega_j(q) - f_q(v)| \leq |\omega_j(q) - f_q(u)| + |f_q(u) - f_q(v)|$$

$$< \delta(\omega_j) + \rho(u,v) \leq 2\delta(\omega_j)$$

for all $q \in \pi(\omega_j)$. Hence $\hat{f}(v) \in V(\omega_j) \subseteq U(\omega_j)$, and so by (i) we have

$$|f(u) - f(v)| \leq \varepsilon \quad \text{if} \quad \rho(u,v) < \delta$$

since $\delta < \varepsilon/2$. Since $f_q \in \bar{A}$ for all $q \in \pi$, and since $\varphi(\omega_j,\cdot) \in \bar{A}$ by (2.12.2) for all $j = 1,\ldots,N$, we see that f is uniformly continuous with respect to the uniformity defined in Proposition 2.9, and so $f \in \bar{A}$ by Proposition 2.9. □

3. Positive functionals.

We shall in this section study <u>functionals</u> on T, i.e. functions μ from a subset, denoted $D(\mu)$, of $\bar{\mathbb{R}}^T$ into $\bar{\mathbb{R}}$. We shall restrict ourselves to <u>increasing functionals</u>, i.e. functionals μ, such that $\mu(f) \leq \mu(g)$, whenever $f,g \in D(\mu)$ and $f \leq g$, or even <u>positive functionals</u>, i.e. increasing functionals μ, such that $0 \in D(\mu)$ and $\mu(0) = 0$. A functional defined on all of $\bar{\mathbb{R}}^T$ is said to be <u>every where defined</u>.

Now let μ be an incrasing functional on T, and let F be a subset of $D(\mu)$, then we define <u>the upper</u> and <u>lower functionals</u> μ^F and μ_F by

$$\mu_F(h) = \sup\{\mu(f) \mid f \in F,\ f \leq h\} \qquad \forall h \in \bar{\mathbb{R}}^T$$

$$\mu^F(h) = \inf\{\mu(f) \mid f \in F,\ f \geq h\} \qquad \forall h \in \bar{\mathbb{R}}^T .$$

If $F = D(\mu)$ we write μ^* and μ_* in place of μ^F and μ_F. Clearly μ^F and μ_F are increasing everywhere defined functional satisfying

$$\text{(3.1)} \qquad \mu_H \leq \mu_F \leq \mu_* \leq \mu^* \leq \mu^F \leq \mu^H \qquad \forall H \subseteq F \subseteq D(\mu)$$

$$\text{(3.2)} \qquad \mu^F = \mu_F = \mu \quad \text{on} \quad F$$

And we define <u>the dual functional</u> μ^0 by

$$D(\mu^0) = -D(\mu), \quad \mu^0(f) = -\mu(-f) \quad \forall f \in D(\mu^0).$$

Then μ^0 is an increasing functional on T, and we have the following duality relations:

$$\text{(3.3)} \qquad \mu = \mu^{00}, \quad (\mu^F)^0 = (\mu^0)_{-F}, \quad (\mu_F)^0 = (\mu^0)^{-F}$$

for all $F \subseteq D(\mu)$.

Now let μ be an increasing functional on T. If F and H are subsets of $D(\mu)$, $\mathcal{K}$ is a paving on T and $\xi \geq 1$ is a cardinal number, then we say that μ is

ξ-<u>subadditive</u> resp. ξ-<u>superadditive</u> on F, if we have $\mu_F(\Sigma^* f_q) \leq \Sigma^* \mu(f_q)$ resp. $\mu^F(\Sigma_* f_q) \geq \Sigma_* \mu(f_q)$, whenever $\{f_q \mid q \in Q\} \subseteq F$, and $\text{card}(Q) \leq \xi$

<u>submodular</u> resp. <u>supermodular</u> on F, if we have $\mu_F(f \wedge g) \dotplus \mu_F(f \vee g) \leq \mu(f) \dot{+} \mu(g)$ resp. $\mu(f) \dotplus \mu(g) \leq \mu^F(f \wedge g) \dot{+} \mu^F(f \vee g)$, whenever $f, g \in F$

ξ-<u>subsmooth</u> resp. ξ-<u>supersmooth at</u> H <u>along</u> F, if we have that $\mu(h) = \lim \mu(f_\gamma)$, whenever $h \in H$ and $\{f_\gamma \mid \gamma \in \Gamma\}$ is a increasing resp. decreasing net in F such that $h = \lim f_\gamma$ and $\text{card}(\Gamma) \leq \xi$.

K-<u>tight along</u> F, if $\inf_{\delta>0} \inf_{K \in K} \mu_F(1_{T \setminus K} + \delta 1_T) \leq 0$.

If μ is ξ-subadditive and ξ-superadditive on F we say that μ is ξ-<u>additive</u> on F, and we define <u>modularity</u> and ξ-<u>smoothness</u> similarly. We say that μ is <u>additive</u> resp. <u>finitely additive</u> on F, if μ is 2-additive resp. n-additive $\forall n \in \mathbb{N}$ on F, and we define <u>sub/superadditivity</u> and <u>finite sub/superadditivity</u> similarly. We shall use the symbol σ for the countable cardinal $\aleph_0$, e.g. σ-<u>additive</u> etc., and we shall use the symbol τ to express that the property holds for all cardinals, e.g. τ-<u>additive</u> etc.

Note that μ has one of the subproperties, if and only if the dual functional μ^0 has the corresponding superproperty on the reflected set. This observation allows a dualization of most of the results in this section.

Let μ be an increasing every where defined functional on T, and let $F \subseteq \bar{\mathbb{R}}^T$, then we put

$$R^*(\mu,F) = \{h \in \bar{\mathbb{R}}^T \mid \inf\{\mu(f-h) \mid f \in F,\ f \geq h\} \leq 0\}$$

$$R_*(\mu,F) = \{h \in \bar{\mathbb{R}}^T \mid \inf\{\mu(h-f) \mid f \in F,\ f \leq h\} \leq 0\}.$$

Hence if $F - h = \{f-h \mid f \in F\}$, $h - F = \{h-f \mid f \in F\}$, then

(3.4) $h \in R^*(\mu,F) \iff \mu^{F-h}(0) \leq 0$

(3.5) $h \in R_*(\mu,F) \iff \mu^{h-F}(0) \leq 0$

(3.6) $R_*(\mu,F) = -R^*(\mu,-F)$

(3.7) $F \subseteq R_*(\mu,F) \cap R^*(\mu,F)$ if $\mu(0) \leq 0$.

Moreover if μ is increasing and subadditive on all of $\bar{\mathbb{R}}^T$; then we have

(3.8) $\mu^0(f \dot{+} g) \leq \mu^0(f) \dot{+} \mu(g), \quad \mu^0(f) \underset{\bullet}{+} \mu(g) \leq \mu(f \underset{\bullet}{+} g)$

(3.9) $\mu^0 \leq \mu$ if $\mu(0) > -\infty$

(3.10) $R^*(\mu,F) \subseteq R^*(\mu,G)$ if $F \subseteq R^*(\mu,G)$

(3.11) $R_*(\mu,F) \subseteq R_*(\mu,G)$ if $F \subseteq R_*(\mu,G)$

(3.12) $\mu = \mu^F$ on $R^*(\mu,F)$ and $\mu = \mu_F$ on $R_*(\mu,F)$

(3.13) $h \in R^*(\mu^0,F)$ if $\mu(h) = \mu^F(h) \neq \pm\infty$

(3.14) $h \in R_*(\mu^0,F)$ if $\mu(h) = \mu_F(h) \neq \pm\infty$.

For the rest of this section we let T denote a given set and μ, ν, μ_q, ν_q will denote everywhere defined increasing functionals on T, unless otherwise stated.

<u>Lemma 3.1</u>. <u>Let</u> Q <u>be a set, and let</u> $\psi: \bar{\mathbb{R}}^Q \times T \to \bar{\mathbb{R}}$ <u>be a function, and let</u> $\alpha_q \in \bar{\mathbb{R}}_+^T$ $\forall q \in Q$. <u>Let</u> G <u>and</u> G_q <u>for</u> $q \in Q$ <u>be subsets</u> of $\bar{\mathbb{R}}^T$, <u>such that</u>

(3.1.1) $\psi(\hat{g}(t),t) \in G$ <u>if</u> $\hat{g} = (g_q): T \to \bar{\mathbb{R}}^Q$ <u>and</u> $g_q \in G_q$ $\forall q$

(3.1.2) $0 \leq \psi(\omega',t) - \psi(\omega'',t) \leq \sum_{q \in Q} \alpha_q(t)(\omega'_q - \omega''_q)$

<u>whenever</u> $t \in T$ <u>and</u> $\omega' \geq \omega''$. <u>Now let</u> $\hat{h} = (h_q)$ <u>be a map from</u> T <u>into</u> $\bar{\mathbb{R}}^Q$, <u>and suppose that there exist</u> $a \in \bar{\mathbb{R}}$ <u>and a countable set</u>

$N \subseteq Q$, <u>such that</u>

(3.1.3) $$h_q \in R^*(\nu_q, G_q) \quad \forall q \in N$$

(3.1.4) $$\forall q \in Q \setminus N \ \exists \ g_q \in G_q : g_q \geq h_q \ \underline{\text{and}} \ \nu_q(g_q - h_q) \leq 0$$

(3.1.5) $$\nu(h \dot{+} \sum_{q \in Q}^{*} \alpha_q f_q) \leq a \dot{+} \sum_{q \in Q}^{*} \nu_q(f_q) \quad \forall \{f_q\} \subseteq \bar{\mathbb{R}}_+^T$$

<u>where</u> $h(t) = \psi(\hat{h}(t), t)$. <u>Then we have, that</u> $\nu^G(h) \leq a$.

<u>Remark</u>. Of course (3.1.4) implies that $h_q \in R^*(\nu_q, G_q)$ for $q \in Q \setminus N$. Conversely if $h_q \in R^*(\nu_q, G_q)$ and G_q is $(\wedge c)$-stable for all $q \in Q \setminus N$, then (3.1.4) holds.

<u>Proof</u>. Let $\varepsilon > 0$ be given, then there exist $\varepsilon_q \geq 0$ for $q \in Q$ such that

(i) $$\sum_{q \in Q} \varepsilon_q < \varepsilon, \quad \varepsilon_q > 0 \ \forall q \in N, \quad \varepsilon_q = 0 \quad \forall q \in Q \setminus N.$$

And by (3.1.3) and (3.1.4) there exist $g_q \in G_q$ for all $q \in Q$ such that

(ii) $$g_q \geq h_q \quad \text{and} \quad \nu_q(g_q - h_q) \leq \varepsilon_q \quad \forall q \in Q.$$

Let $\hat{g} = (g_q)$ and $g(t) = \psi(\hat{g}(t), t)$, then $g \in G$ and $g \geq h$ by (3.1.1) and (3.1.2), and by (3.1.2) and (A.2) we have

$$g \leq h \dot{+} (g-h) \leq h \dot{+} \sum_{q \in Q} \alpha_q (g_q - h_q) .$$

So by (3.1.5) and (i) we have

$$\nu^G(h) \leq \nu(g) \leq a \dot{+} \sum_q^{*} \nu_q(g_q - h_q) \leq a + \varepsilon$$

and so the lemma is proved. □

<u>Corollary 3.2</u>. <u>Let</u> $\varphi: \bar{\mathbb{R}}^Q \to \mathbb{R}$ <u>be map, let</u> $\alpha_q \in \bar{\mathbb{R}}_+$ $\forall q \in Q$ <u>and let</u> F <u>be a subset of</u> $\bar{\mathbb{R}}^T$ <u>so that</u>

(3.2.1) $$0 \le \varphi(\omega') - \varphi(\omega'') \le \sum_{q\in Q} \alpha_q(\omega'_q - \omega''_q) \quad \text{if} \quad \omega' \ge \omega''$$

(3.2.2) μ is ξ-subadditive on F, where $\xi = \mathrm{card}(Q)$.

Now let U and V be φ-stable subsets of $\bar{\mathbb{R}}^T$, such that

(3.2.3) $$v - u \in F \quad \forall u \in U \quad \forall v \in V \quad \text{so that} \quad u \le v$$

Then we have that $U \cap R^*(\mu,V)$ (resp. $V \cap R_*(\mu,U)$) is φ-stable in either of the following two cases

Case 1°: Q is countable and $\forall \varepsilon > 0 \quad \forall q \in Q \ \exists \ \delta > 0$, such that $\mu^F(\alpha_q f) < \varepsilon$ if $f \in F$ and $\mu(f) < \delta$.

Case 2°: V is $(\wedge c)$-stable (resp. U is $(\vee c)$-stable) and $\mu^F(\alpha_q f) \le 0$ whenever $q \in Q$, $f \in F$ and $\mu(f) \le 0$.

Remark. Each of the following four maps satisfies condition (3.2.1)

$$\varphi(\omega) = \sup_{q\in Q} \alpha_q \omega_q, \qquad \varphi(\omega) = \inf_{q\in Q} \alpha_q \omega_q$$

$$\varphi(\omega) = \sum_{q\in Q}^{*} \alpha_q \omega_q, \qquad \varphi(\omega) = \sum_{*\, q\in Q} \alpha_q \omega_q$$

Proof. Let $h_q \in U \cap R^*(\mu,V)$ for all $q \in Q$, and put $\hat{h} = (h_q)$ and $w = \varphi \circ \hat{h}$. Then $w \in U$ by φ-stability of U, so by (3.4) it suffices to show that $\mu^G(0) \le 0$, where $G = \{v-w \mid v \in V, \ v \ge w\}$. To do this we shall apply Lemma 3.1 with

$$\psi(\omega,t) = \varphi(\omega) - w(t), \quad \alpha_q(t) = \alpha_q, \ h_q = h_q, \ a = 0$$

$$G = G, \quad G_q = \{v \in V \mid v \ge h_q\} \quad \forall q \in Q$$

$$\nu = \mu_F \quad \text{and} \quad \nu_q(f) = \mu^F(\alpha_q f) \quad \forall f \in \bar{\mathbb{R}}^T \quad \forall q \in Q$$

Then by (A.2) and (A.3) it follows easily that (3.1.1), (3.1.2) and (3.1.5) holds, since $\psi(\hat{h}(t),t) \equiv 0$. In case 1° we put $N = Q$, then

(3.1.4) holds trivially, and (3.1.3) follows easily from (3.2.3) and the assumption of case 1°. In case 2° we put $N = \emptyset$, then (3.1.3) holds trivially and (3.1.4) follows easily from (3.2.3) and the assumption of case 2° (see also the remark to Lemma 3.1).

Hence by Lemma 3.1 we have that $\nu^G(0) \leq 0$, and so $h \in R^*(\nu,V) \cap U$, however since $\mu = \mu_F$ on F it follows from (3.2.3) that $R^*(\nu,V) \cap U = R^*(\mu,V) \cap U$ and so $R^*(\mu,V) \cap U$ is φ-stable.

Applying this to $\varphi^o(\omega) = -\varphi(-\omega)$, $U^o = -V$ and $V^o = -U$, it follows from (3.6) that $V \cap R_*(\mu,U)$ is φ-stable in both cases. □

Corollary 3.3. *Let μ be an everywhere defined positive functional on T, and let $F \subseteq \bar{\mathbb{R}}^T$, and put $\mathcal{F} = 2^T \cap R^*(\mu,F)$. Then we have*

(1): *If F is $(\wedge c)$-stable, and if $\mu(\infty f) = 0$ whenever $f \in \bar{\mathbb{R}}_+^T$ and $\mu(f) = 0$, then $R^*(\mu,F)$ is φ-stable for every increasing map $\varphi\colon \bar{\mathbb{R}} \to \bar{\mathbb{R}}$, such that F is φ-stable.*

(2): *If $F \subseteq \mathbb{R}^T$ is a convex cone containing $\pm 1_T$, and if μ is subadditive on $\mathbb{R}_+^T$, then we have*

(3.3.1) $\qquad U(T,\mathcal{F}) \cap B(T) \subseteq \overline{R^*(\mu,F)}$

(3.3.2) $\qquad U(T,\mathcal{F}) \cap B^*(T) \subseteq R^*(\mu,F)_\delta$

(3.3.3) $\qquad U(T,\mathcal{F}) \cap B_*(T) \subseteq R^*(\mu,F)_\sigma$

(3.3.4) $\qquad R^*(\mu,F)$ *is $\|\cdot\|$-closed if* $\lim_{\varepsilon \downarrow 0} \mu(\varepsilon 1_T) = 0$

(3): *If μ is ξ-subadditive on $\bar{\mathbb{R}}_+^T$ for some cardinal number ξ, then we have*

(3.3.5) $\qquad R^*(\mu,F_\xi)$ *is $(\wedge\xi)$-stable,* $R_*(\mu,F^\xi)$ *is $(\vee\xi)$-stable*

(3.3.6) $\qquad$ *If $\xi \leq \aleph_0$, then $R^*(\mu,F^\xi)$ is $(\vee\xi)$-stable and $R_*(\mu,F_\xi)$ is $(\wedge\xi)$-stable.*

Proof. (1): Follows from Corollary 3.2 with $Q = \{1\}$, $\alpha_1 = \infty$, $F = \bar{\mathbb{R}}^T$, $U = \bar{\mathbb{R}}^T$ and $V = F$.

(2): By Corollary 3.2 if follows that $\mathbb{R}^T \cap R^*(\mu,F)$ is a convex cone, and by (3.7) it contains $\pm 1_T$. But then (3.3.1)-(3.3.3) follows from Proposition 2.2. Now let h belong to the $\|\cdot\|$-closure of $R^*(\mu,F)$, then there exists $g \in R^*(\mu,F)$ and $f \in F$ such that

(i) $$\|h-g\| \leq \varepsilon, \quad f \geq g \quad \text{and} \quad \mu(f-g) < \varepsilon.$$

Let $f_0 = f + \varepsilon 1_T$, then $f_0 \in F$ and by (A.2) we have

$$0 \leq f_0 - h \leq \varepsilon 1_T \dot{+} (f-g) \dot{+} (g-h) \leq 2\varepsilon 1_T + (f-h)$$

so by subadditivity of μ we have

$$\mu(f_0-h) < \mu(2\varepsilon 1_T) \dot{+} \mu(f-h) < \mu(2\varepsilon 1_T) + \varepsilon.$$

Hence letting $\varepsilon \to 0$ we see that $h \in R^*(\mu,F)$.

(3): Follows easily from Corollary 3.2. □

Proposition 3.4. *Let* F *and* G *be subsets of* $\bar{\mathbb{R}}^T$, *and let* $E = \{f \in \bar{\mathbb{R}}^T \mid \mu(f) \leq \nu(f)\}$. *Then we have*

(3.4.1) $\mu_E \leq \nu_E \leq \nu$ *and* $\mu \leq \mu^E \leq \nu^E$

(3.4.2) $R^*(\nu^o,E) \subseteq E$ *if* ν *is superadditive on* $\bar{\mathbb{R}}^T$

(3.4.3) $R_*(\mu,E) \subseteq E$ *if* μ *is subadditive on* $\bar{\mathbb{R}}^T$

(3.4.4) $G \cap F^\xi \subseteq E$ *if* $F \subseteq E$ *is* ($\vee$f)-*stable and* μ *is* ξ-*subsmooth at* G *along* F

(3.4.5) $G \cap F_\xi \subseteq E$ *if* $F \subseteq E$ *is* ($\wedge$f)-*stable and* ν *is* ξ-*supersmooth at* G *along* F

If μ *is subadditive on* G *and* ν *is superadditive on* F,

<u>then</u> $h \in E$ <u>whenever</u> $h \in \bar{\mathbb{R}}^T$ <u>is a function, such that for all</u> $\varepsilon > 0$ <u>there exist</u> $g_0, g_1 \in G$ <u>and</u> $f_0, f_1 \in F$ <u>satisfying</u>

(3.4.6) $$f_0 \dot{+} f_1 \leq h \leq g_0 \dot{+} g_1, \quad f_0 \dot{+} f_1 \in F \quad \text{and} \quad g_0 \dot{+} g_1 \in G$$

(3.4.7) $$\text{either } \mu(g_j) < \infty \text{ for } j = 0,1 \text{ or } \nu(f_j) > -\infty \text{ for } j = 0,1$$

(3.4.8) $$\mu(g_j) \leq \varepsilon + \nu(f_j) \quad \text{for } j = 0,1.$$

<u>Now suppose that</u> $F \subseteq E$ <u>is</u> $(\vee f, \wedge f)$<u>-stable, and that</u> μ <u>is</u> σ<u>-subsmooth at</u> G, <u>along</u> G <u>and that</u> ν <u>is</u> σ<u>-supersmooth at</u> $G \cap F_\delta$ <u>along</u> F. <u>If</u> G <u>satisfies</u>

(3.4.9) $$\forall g \in G \; \exists \varphi \in F_\delta : \varphi \leq g \quad \text{and} \quad G \supseteq \{f \in S(F) \mid \varphi \leq f \leq g\}.$$

<u>Then, if</u> $H = G \cap F_\delta$, <u>we have</u>

(3.4.10) $$\mu(g) \leq \nu_H(g) \leq \nu(g) \quad \forall g \in G \cap S(F)$$

(3.4.11) $$G \cap S(F) \subseteq E.$$

<u>Proof</u> (3.4.1): If $f \in E$ and $f \leq h$, then we have that $\mu(f) \leq \nu(f) \leq \nu_E(h)$ and so $\mu_E \leq \nu_E \leq \nu$. Similar we have that $\mu \leq \mu^E \leq \nu^E$.

(3.4.2): Let $h \in R^*(\nu^o, E)$, then there exist $f \in E$ so that $f \geq h$ and $\nu^o(f-h) < \varepsilon$. And by (3.8) we have

$$\mu(h) \leq \mu(f) \leq \nu(f) \leq \nu(h) \dot{+} \nu^o(f-h) \leq \nu(h) + \varepsilon$$

since $f \leq h \dot{+} (f-h)$ by (A.2). Thus $h \in E$

(3.4.3) is proved similarly.

(3.4.4). Let $g \in G \cap F^\xi$, then there exist an increasing net $\{f_\gamma \mid \gamma \in \Gamma\}$ in F, so that $f_\gamma \uparrow g$ and $\text{card}(\Gamma) \leq \xi$. Then we have

$$\mu(g) = \lim_\gamma \mu(f_\gamma) \leq \lim_\gamma \nu(f_\gamma) \leq \nu(g)$$

and so $g \in E$.

(3.4.5) is proved similarly.

Now suppose that $h \in \bar{\mathbb{R}}^T$ satisfies (3.4.6) and (3.4.7) and that μ is subadditive on G and ν is superadditive on F. If $\mu(g_j) < \infty$ for $j = 0,1$, then we have

$$\mu(h) \leq \mu(g_0 \dot{+} g_1) \leq \mu(g_0) \dot{+} \mu(g_1) = \mu(g_0) \underset{\bullet}{+} \mu(g_1)$$

$$\leq 2\varepsilon + \nu(f_0) \underset{\bullet}{+} \nu(f_1) \leq 2\varepsilon + \nu(f_0 \underset{\bullet}{+} f_1) \leq 2\varepsilon + \nu(h)$$

by (3.4.6) and (3.4.8).

Similar if $\nu(f_j) > -\infty$ for $j = 0,1$, then we have

$$\mu(h) \leq \mu(g_0 \dot{+} g_1) \leq \mu(g_0) \dot{+} \mu(g_1) \leq 2\varepsilon + \nu(f_0) \dot{+} \nu(f_1)$$

$$= 2\varepsilon + \nu(f_0) \underset{\bullet}{+} \nu(f_1) \leq 2\varepsilon + \nu(f_0 \underset{\bullet}{+} f_1) \leq 2\varepsilon + \nu(h).$$

Letting $\varepsilon \to 0$ we see that $h \in E$ in both cases.

(3.4.10): Let $g \in G \cap S(F)$, and choose $\varphi \in F_\delta$ according to (3.4.9). Let Φ be an F-Souslin scheme, such that $g = S(\Phi)$, since F is $(\wedge f)$-stable we may assume that Φ is decreasing (i.e. $\phi(\sigma|n+1) \leq \phi(\sigma|n)$ for all $\sigma \in \mathbb{N}^{\mathbb{N}}$ and all $n \geq 1$). Moreover there exist $\varphi_k \in F$ such that $\varphi_k \downarrow \varphi$. If $\alpha \in \mathbb{N}^k$ for some $k \geq 1$ we put

$$N(\alpha) = \{\beta \in \mathbb{N}^k \mid \beta(j) \leq \alpha(j) \quad \forall 1 \leq j \leq k\}$$

$$\hat{\Phi}(\alpha) = \varphi_k \vee \max_{\beta \in N(\alpha)} \Phi(\beta)$$

$$\Phi^*(\alpha) = \sup_{\sigma \in \mathbb{N}^{\mathbb{N}}} \inf_{n \geq 1} \hat{\Phi}(\alpha, \sigma|n)$$

where $(\alpha,\beta) = (\alpha_1,\dots,\alpha_k, \beta_1 \dots \beta_n)$ if $\alpha \in \mathbb{N}^k$ and $\beta \in \mathbb{N}^n$. If $\alpha \in \mathbb{N}^{(\mathbb{N})}$ and $\sigma \in \mathbb{N}^{\mathbb{N}}$ then clearly we have

(i) $\hat{\Phi}(\alpha) \in F, \quad \Phi^*(\alpha) \in S(F)$

(ii) $\hat{\Phi}(\sigma|n+1) \leq \hat{\Phi}(\sigma|n) \qquad \forall n \geq 1$

(iii) $\varphi \leq \Phi^*(\alpha) \leq \hat{\Phi}(\alpha)$ and $\Phi(\alpha) \leq \hat{\Phi}(\alpha)$.

And we shall show that

(iv) $g = S(\hat{\Phi})$.

So let $h = S(\hat{\Phi})$, let $t \in T$ and let $a < h(t)$. Then there exist $\sigma \in \mathbb{N}^{\mathbb{N}}$ so that $a < \hat{\Phi}(\sigma|n)(t)$ for all $n \geq 1$. Hence if

$$S_0 = \bigcup_{k=1}^{\infty} \{\alpha \in N(\sigma|k) \mid \Phi(\alpha)(t) \vee \varphi_k(t) > a\}$$

then $S_0 \cap N(\sigma|k) \neq \emptyset$ for all $k \geq 1$. Thus S_0 is infinite and since $\alpha(1) \in N(\sigma(1))$ for all $\alpha \in S_0$, and $N(\sigma(1))$ is finite, there exist $\tau(1) \in N(\sigma(1))$ so that

$$S_1 = \bigcup_{k=2}^{\infty} \{\alpha \in N(\sigma|k) \mid \alpha(1) = \tau(1) \text{ and } \alpha \in S_0\}$$

is infinite, and so we may proceed as above. Continuing this procedure we obtain $\tau \in \mathbb{N}^{\mathbb{N}}$, such that $\tau(j) \leq \sigma(j)$ for all $j \geq 1$ and such that

$$S_n = \bigcup_{k=n+1}^{\infty} \{\alpha \in N(\sigma|k) \mid \alpha|n = \tau|n \text{ and } \alpha \in S_{n-1}\}$$

is infinite for all $n \geq 1$. Now let $n \in \mathbb{N}$ be given, then there exist an $\alpha \in S_n$, and so $\alpha \in \mathbb{N}^k$ for some $k \geq n+1$. Since $\alpha|n = \tau|n$ $\alpha \in S_{n-1} \subseteq S_0$ we have

$$\Phi(\tau|n)(t) \vee \varphi_n(t) \geq \Phi(\alpha)(t) \vee \varphi_k(t) > a.$$

Thus we have

$$a \leq \varphi(t) \vee \inf_{n \geq 1} \Phi(\tau|n)(t) \leq g(t)$$

and so $g \geq h$. But the converse inequality follows from (iii) and so (i) is proved.

From (iv) and the definition of Φ^* we have

(v) $\Phi^*(n) \uparrow g$ as $n \uparrow \infty$ in $\mathbb{N}$

(vi) $\Phi^*(\alpha,n) \uparrow \Phi^*(\alpha)$ as $n \uparrow \infty$ in $\mathbb{N}$, $\forall \alpha \in \mathbb{N}^{(\mathbb{N})}$

Now let $a < \mu(g)$. By (i), (iii), (iv) and (3.4.9) we have that $\Phi^*(\alpha) \in G$, so by (v) and σ-subsmoothness of μ there exist $\sigma(1) \in \mathbb{N}$ so that $\mu(\Phi^*(\sigma(1))) > a$. And by succesive use of (vi) and σ-subsmoothness of μ, there exist a $\sigma \in \mathbb{N}^{\mathbb{N}}$ so that

(vii) $\mu(\Phi^*(\sigma|n)) > a \quad \forall n \geq 1.$

Now let $f_n = \hat{\Phi}(\sigma|n)$ and $f = \inf_n f_n$. Then $f_n \in F$ by (i) and $f_n \downarrow f$ by (ii). Hence $f \in F_\delta \subseteq S(F)$ and since $\varphi \leq f \leq g$ by (iv), we have that $f \in G \cap F_\delta$ by (3.4.9). Thus we find by (vii) and σ-supersmoothness of ν:

$$\nu(g) \geq \nu_H(g) \geq \nu(f) = \lim_{n\to\infty} \nu(f_n)$$

$$\geq \lim_{n\to\infty} \mu(f_n) \geq \lim_{n\to\infty} \mu(\Phi^*(\sigma|n)) \geq a$$

since $f_n \in F \subseteq E$ and $f_n \geq \Phi^*(\sigma|n)$ by (iii). Letting $a \uparrow \mu(g)$ we see that (3.4.10) holds.

(3.4.11) is a trivial consequence of (3.4.10). □

<u>Proposition 3.5</u>. <u>Let</u> ξ <u>be an infinite cardinal number and suppose that</u> μ <u>is</u> ξ-<u>supersmooth at</u> H <u>along</u> F. <u>Then we have</u>

(1): <u>If</u> $f + a1_T \in \bar{F}$ <u>for all</u> $f \in F$ <u>and all</u> $a \in \mathbb{R}_+$, <u>then</u> μ^F <u>and</u> μ <u>are</u> ξ-<u>supersmooth at</u> H <u>along</u> $\bar{F}$.

(2): <u>If</u> $f + a1_T$ <u>and</u> $f \wedge g$ <u>belongs to</u> $\bar{F}$ <u>for all</u> $f,g \in F$ <u>and all</u> $a \in \mathbb{R}_+$, <u>then</u> μ^F, μ <u>and</u> $\mu_{\bar{F}_\xi}$ <u>are</u> ξ-<u>supersmooth at</u> H <u>along</u> $\bar{F}_\xi$.

(3): <u>If</u> F <u>is</u> $(\wedge f)$-<u>stable, then</u> μ^F, μ <u>and</u> μ_{F_ξ} <u>are</u> ξ-<u>supersmooth of</u> H <u>along</u> F_ξ.

(4) <u>If</u> $G = \{g \in \bar{\mathbb{R}}^T \mid g \vee h \in F \cap H \quad \forall h \in H\}$ <u>then</u> μ^H <u>is</u> ξ-<u>super</u>-

<u>smooth at</u> $\bar{\mathbb{R}}^T$ <u>along</u> G.

<u>Proof</u> (1)-(3): Let $E = \bar{F}, \bar{F}_\xi$ or F_ξ according to case (1), (2) or (3), and let $\{g_\gamma \mid \gamma \in \Gamma\}$ be a decreasing net in E, so that $g_\gamma \downarrow h$, where $h \in H$ and $\text{card}(\Gamma) \leq \xi$. Then by Proposition 2.8 there exist a net $\{f_\lambda \mid \lambda \in \Lambda\} \subseteq F$, and increasing cofinal map $\sigma: \Lambda \to \Gamma$ so that

(i) $$f_\lambda \downarrow h, \quad f_\lambda \geq g_{\sigma(\lambda)} \quad \forall \lambda \in \Lambda, \quad \text{card}(\Lambda) \leq \xi.$$

And so we have

$$\begin{aligned}\mu(h) &\leq \lim_\gamma \mu_E(g_\gamma) \leq \lim_\gamma \mu(g_\gamma) \leq \lim_\gamma \mu^F(g_\gamma) \\ &= \lim_\lambda \mu^F(g_{\sigma(\lambda)}) \leq \lim_\lambda \mu(f_\lambda) = \mu(h) \\ &\leq \mu^F(h) \leq \lim_\gamma \mu^F(g_\gamma).\end{aligned}$$

Thus μ^F and μ are ξ-supersmooth at H along E, and in case (2) and (3) we have that $h \in E$ and so in case (2) and (3) we have that μ_E is ξ-supersmooth at H along E.

(4): Let $\{g_\gamma \mid \gamma \in \Gamma\}$ be a decreasing net in G, so that $\text{card}(\Gamma) \leq \xi$ and $g_\gamma \downarrow g$. Let $a > \mu^H(g)$ then there exist $h \in H$ such that $\mu(h) < a$, and $h \geq g$. Hence we have that $f_\gamma = g_\gamma \vee h \in F$, and $f_\gamma \downarrow g \vee h = h$. Thus we have

$$\mu^H(g) \leq \lim_\gamma \mu^H(g_\gamma) \leq \lim_\gamma \mu(f_\gamma) = \mu(h) \leq a$$

since $g_\gamma \leq f_\gamma$ and $f_\gamma \in H$. Hence letting $a \downarrow \mu^H(g)$ we see that μ^H is ξ-supersmooth of $\bar{\mathbb{R}}^T$ along G. □

<u>Proposition 3.6</u>. <u>Let</u> U, V, F <u>and</u> H <u>be subsets of</u> $\bar{\mathbb{R}}^T$, <u>let</u> $G \subseteq]-\infty,\infty]^T$, <u>and let</u> ξ <u>be an infinite cardinal, so that</u>

(3.6.1) ν <u>is</u> ξ-<u>supersmooth at</u> 0 <u>along</u> F <u>and</u> $\nu(0) \leq 0$

(3.6.2) $\mu(u) \leq \lambda(v) \dot{+} \nu((u-v)^+) \quad \forall u \in U \quad \forall v \in V$

(3.6.3) $(h-g)^+ \in F \quad \forall h \in H \quad \forall g \in G.$

Let $\psi \in \bar{\mathbb{R}}^T$ and let $\{\varphi_\gamma \mid \gamma \in \Gamma\}$ be a decreasing net in $\bar{\mathbb{R}}^T$, such that $\text{card}(\Gamma) \leq \xi$ and $\varphi_\gamma \downarrow \varphi$. Then we have

(3.6.4) $$\lim_\gamma \mu_U(\varphi_\gamma - \psi) \leq \inf\{\lambda(v) \mid v \in V \quad \exists g \in G: g \geq \varphi,\ v \geq g-\psi\}$$

in either of the following two cases

Case 1°. $\varphi_\gamma \in H \quad \forall \gamma \in \Gamma$

Case 2°. ν is subadditive on $\bar{\mathbb{R}}^T_+$, F is $(\wedge f)$-stable and for all $\varepsilon > 0$ there exist $h_\gamma \in H_\xi$ such that $h_\gamma \leq \varphi_\gamma$ and $\sum_{\gamma \in \Gamma} \nu(\varphi_\gamma - h_\gamma) < \varepsilon$.

In particular, if $L = \{\varphi \in \bar{\mathbb{R}}^T \mid \mu(\varphi) \geq \lambda^{V\cap G}(\varphi)\}$ then, we have that μ is ξ-supersmooth at L along $U \cap H$.

Proof. Let $v \in V$ and $g \in \bar{\mathbb{R}}^T_+$, and let $u \in U$ such that $u \leq v \dot{+} g$, then $g \geq (u \dot{-} v)^+ = (u-v)^+$ by (A.2) and so

$$\mu(u) \leq \lambda(v) \dot{+} \nu((u-v)^+) \leq \lambda(v) \dot{+} \nu(g)$$

by (3.6.2). Hence we have

(i) $$\mu_U(v \dot{+} g) < \lambda(v) \dot{+} \nu(g) \quad \forall v \in V \quad \forall g \in \bar{\mathbb{R}}^T_+$$

Now let $a > b$ where b denotes the right hand side of (3.6.4). Then there exist $v \in V$ and $g \in G$ so that

(ii) $$g \leq \varphi, \quad v \geq g - \psi \quad \text{and} \quad \lambda(v) < a$$

Case 1°: In this case we put $f_\gamma = (\varphi_\gamma - g)^+$. Then $f_\gamma \in F$ by (3.6.3), and since $g > -\infty$ and $g \geq \varphi$ we have that $f_\gamma \downarrow 0$. Hence $\lim \nu(f_\gamma) \leq 0$ by (3.6.1) and by (i) we have

$$\mu_U(\varphi_\gamma - \psi) \le \mu_U(v \dot{+} f_\gamma) \le \lambda(v) \dot{+} \nu(f_\gamma) \le a \dot{+} \nu(f_\gamma)$$

since $\varphi_\gamma - \psi \le v \dot{+} f_\gamma$ by (ii) and (A.2). Hence we see that

$$\lim_\gamma \mu_U(\varphi_\gamma - \psi) \le a$$

and so (3.6.4) follows.

Case 2°. Let $h_\gamma \in H_\xi$ be chosen according to the assumption of case 2°, where $\varepsilon > 0$ is any given number. Let Λ be the set of all non-empty finite subsets of Γ, and let σ be an increasing cofinal map from Λ into Γ such that (see the proof of Proposition 2.8):

(iii) $\sigma(\lambda) = \gamma$ if $\lambda = \{\gamma\}$ for some $\gamma \in \Gamma$.

Now put

$$h_\lambda = \min_{\gamma \in \lambda} h_\gamma, \quad f_\lambda = (h_\lambda - g)^+ \quad \forall \lambda \in \Lambda$$

Then $h_\lambda \in H_\xi$ and $h_\lambda \downarrow h$ for some h with $h \le \varphi$. As above we have that $f_\lambda \downarrow 0$, and by (3.6.3) we have that $f_\lambda \in F_\xi$ for all $\lambda \in \Lambda$. Hence by Proposition 3.5.(3) and (3.6.1) we have

(iv) $$\lim_\lambda \nu(f_\lambda) \le 0 .$$

Moreover by (A.2) and (ii) we have

$$\varphi_{\sigma(\lambda)} - \psi \le (\varphi_{\sigma(\lambda)} - h_\lambda) \dot{+} (h_\lambda - g) \dot{+} (g - \psi)$$
$$\le v \dot{+} f_\lambda \dot{+} \sum_{\gamma \in \lambda} (\varphi_\gamma - h_\gamma)$$

since $\varphi_\gamma - h_\gamma \ge (\varphi_{\sigma(\lambda)} - h_\gamma)^+$ for $\gamma \in \lambda$ by (iii) and monotonicity of σ. Thus by (i) and subadditivity of ν we find that

$$\mu_U(\varphi_{\sigma(\lambda)} - \psi) \le \lambda(v) \dot{+} \nu(f_\lambda) \dot{+} \sum_{\gamma \in \lambda} \nu(\varphi_\gamma - h_\gamma)$$
$$\le a \dot{+} \varepsilon \dot{+} \nu(f_\lambda)$$

and so (3.6.4) follows from (iv) since σ is cofinal and $\varepsilon > 0$ and $a > b$ are arbitrary. □

<u>Proposition 3.7</u>. <u>Let</u> $F \subseteq \bar{\mathbb{R}}^T$ <u>and let</u> $\xi \geq 1$ <u>be a cardinal number. Then we have</u>

(1): <u>If</u> μ <u>is supermodular on</u> F, <u>then we have</u>

$$\mu_F(f) \dot{+} \mu_F(g) \leq \mu^F(f \wedge g) \dot{+} \mu^F(f \vee g) \quad \forall f,g \in \bar{\mathbb{R}}^T \tag{3.7.1}$$

<u>And if</u> μ <u>is supermodular on</u> F, <u>and</u> F <u>is</u> $(\vee f, \wedge f)$<u>-stable then</u> μ_F <u>is supermodular on</u> $\bar{\mathbb{R}}^T$.

(2): <u>If</u> μ <u>is</u> ξ<u>-superadditive on</u> F, <u>then we have</u>

$$\sum_{q \in Q}{}^{*} \mu_F(f_q) \leq \mu^F\Big(\sum_{q \in Q}{}^{*} f_q\Big) \quad \forall \{f_q\} \subseteq \bar{\mathbb{R}}^T \tag{3.7.2}$$

<u>provided that</u> $\text{card}(Q) \leq \xi$. <u>And if</u> μ <u>is</u> ξ<u>-superadditive on</u> F <u>and</u> F <u>is</u> $(\Sigma_*\xi)$<u>-stable, then</u> μ_F <u>is</u> ξ<u>-superadditive on</u> $\bar{\mathbb{R}}^T$.

(3): <u>If</u> μ_F <u>is supermodular on</u> $\bar{\mathbb{R}}^T$ <u>and</u> μ_F <u>is</u> σ<u>-supersmooth at</u> F_δ <u>along</u> F, <u>then</u> μ_F <u>is</u> σ<u>-supersmooth at</u> $\bar{\mathbb{R}}^T$ <u>along</u> $H = \{h \in \bar{\mathbb{R}}^T \mid \mu_F(h) < \infty\}$.

<u>Proof</u>. (1): Note that (3.7.1) follows directly from the definition of supermodularity and (A.13). Moreover if μ is supermodular and F is $(\wedge f, \vee f)$-stable then

$$\mu(u) \dot{+} \mu(v) \leq \mu(u \wedge v) \dot{+} \mu(u \vee v)$$

$$\leq \mu_F(f \wedge g) \dot{+} \mu_F(f \vee g)$$

whenever $u, v \in F$, $f,g \in \bar{\mathbb{R}}^T$ and $u \leq f$, $v \leq g$. Hence we see that μ_F is supermodular on $\bar{\mathbb{R}}^T$ by (A.13).

(2): Follows similarly.

(3): Let $h_n \in H$ so that $h_n \downarrow h$. If $\mu_F(h_n) = -\infty$ for some $n \in \mathbb{N}$, then clearly we have

$$\mu_F(h) = -\infty = \lim_{n\to\infty} \mu_F(h_n).$$

Thus it is no loss of generality to assume that $\mu_F(h_n) \in \mathbb{R}$ for all $n \in \mathbb{N}$.

Let $\varepsilon > 0$ be given, since $\mu_F(h_1) \in \mathbb{R}$, there exist $f_1 \in F$, such that

$$\text{(i)} \qquad f_1 \leq h_1 \quad \text{and} \quad \mu(f_1) \leq \mu_F(h_1) \leq \tfrac{1}{2}\varepsilon + \mu(f_1).$$

Then $\mu(f_1) \in \mathbb{R}$, and since μ_F is supermodular on $\bar{\mathbb{R}}^T$ we have

$$\begin{aligned} -\infty < \mu_F(h_2) + \mu(f_1) &\leq \mu_F(h_2 \wedge f_1) \dot{+} \mu_F(h_2 \vee f_1) \\ &\leq \mu_F(h_2 \wedge f_1) \dot{+} \mu_F(h_1) \\ &\leq \mu_F(h_2) \dot{+} \mu_F(h_1) < \infty \end{aligned}$$

since $h_2 \wedge f_1 \leq h_2$ and $h_2 \vee f_1 \leq h_1$. Hence $\mu_F(h_2 \vee f_1)$ and $\mu_F(h_2 \wedge f_1)$ are finite. And so there exist $f_2 \in F$, such that

$$f_2 \leq f_1 \wedge h_2 \quad \text{and} \quad \mu(f_2) \leq \mu_F(f_1 \wedge h_2) \leq h(f_2) + \varepsilon/4.$$

Continuing like this we may construct $f_n \in F$, such that

$$\text{(ii)} \qquad \mu(f_n),\ \mu_F(f_n \wedge h_{n+1}) \quad \text{and} \quad \mu_F(f_n \vee h_{n+1}) \quad \text{are finite} \quad \forall n \geq 1$$

$$\text{(iii)} \qquad f_{n+1} \leq f_n \wedge h_{n+1} \qquad \text{and} \quad \mu_F(f_n \wedge h_{n+1}) \leq 2^{-n-1}\varepsilon + \mu(f_{n+1}).$$

By supermodularity of μ_F we have

$$\begin{aligned} \mu_F(h_{n+1}) + \mu(f_n) &\leq \mu_F(f_n \wedge h_{n+1}) + \mu_F(f_n \vee h_{n+1}) \\ &\leq 2^{-n-1}\varepsilon + \mu(f_{n+1}) + \mu_F(h_n) \end{aligned}$$

by (iii) since $f_n \vee h_{n+1} \leq h_n$. Hence by (i), (ii) and a simple induction in n, we find

$$\mu_F(h_n) \leq \mu(f_n) + \sum_{j=1}^{n} 2^{-j}\varepsilon \leq \mu(f_n) + \varepsilon \qquad \forall n \geq 1.$$

Since μ_F is σ-supersmooth at F_δ along F, and since $f_n \downarrow f$ for some $f \in F_\delta$ with $f \leq h$ we have

$$\mu_F(h) \leq \lim_{n\to\infty} \mu_F(h_n) \leq \lim_{n\to\infty} \mu(f_n) + \varepsilon$$
$$= \mu_F(f) + \varepsilon \leq \mu_F(h) + \varepsilon$$

and so (3) is proved. □

<u>Proposition 3.8</u>. <u>Let</u> L <u>be a linear space with a locally convex topology</u> η, <u>let</u> K <u>be a subset of</u> L <u>and let</u> $\lambda: K \to \bar{\mathbb{R}}$ <u>be a map satisfying</u>

(3.8.1) K <u>is convex and</u> (K,η) <u>is pseudo-metrizable</u>

(3.8.2) K <u>is countably compact in the weak topology on</u> L

(3.8.3) $\{u \in K \mid \lambda(u) \leq a\}$ <u>is convex</u> $\forall a \in \mathbb{R}$.

<u>Let</u> $F \subseteq \bar{\mathbb{R}}^T$, <u>let</u> θ <u>be a map</u>: $K \to \bar{\mathbb{R}}^T$, <u>and let</u> μ <u>be an increasing map</u>: $F \to \bar{\mathbb{R}}$ <u>satisfying</u>

(3.8.4) $\{u \in K \mid \theta(u,t) \geq a\}$ <u>is convex</u> $\forall a \in \mathbb{R}$ $\forall t \in T$

(3.8.5) $\mu(f) = \inf\{\lambda(u) \mid u \in K, \ \theta u \geq f\}$ $\forall f \in F$

(3.8.6) $\mu^F(\liminf_{n\to\infty} \theta u_n) \leq \sup_{n\in\mathbb{N}} \lambda(u_n)$ <u>if</u> $u_n \in K$ $\forall n \geq 1$ <u>and</u> $u_n \to u$ <u>in</u> (L,η) <u>for some</u> $u \in K$

<u>where</u> $\theta(u,t) = (\theta u)(t)$ <u>for all</u> $u \in K$ <u>and</u> $t \in T$. <u>Then</u> μ^F <u>is</u> σ-<u>subsmooth at</u> $\bar{\mathbb{R}}^T$ <u>along</u> $\bar{\mathbb{R}}^T$.

<u>Remarks</u>. A map λ satisfying (3.8.3) is usually called <u>quasiconvex</u>, and <u>quasiconcave</u> functions are defined similarly. Thus (3.8.4) just states that $\theta(\cdot,t)$ is quasiconcave for all $t \in T$. Note that a convex (resp. concave) function is quasiconvex (resp. quasiconcave).

Also note that, if $\mathcal{C}$ is the set of all convex subsets of K, then $\lambda: K \to \bar{\mathbb{R}}$ is quasiconvex (resp. quasiconcave) if and only if λ is a lower (resp. upper) $\mathcal{C}$-function.

Proof. Let $f_n \in \bar{\mathbb{R}}^T$ for $n \geq 1$ so that $f_n \uparrow f$, and let $a \in \mathbb{R}$ so that $a > \lim \mu^F(f_n)$. Since $\mu^F(f_n) < a$ we have by (3.8.5) that there exist $u_n \in K$ so that

(i) $$\theta u_n \geq f_n \quad \text{and} \quad \lambda(u_n) \leq a \quad \forall n \geq 1.$$

Let K_n be the convex hull of $\{u_j \mid j \geq n\}$ and let ρ be a pseudometric for (K,η). Then $K_n \subseteq K$, and since (L,η) is locally convex we have the weak closure of K_n equals the η-closure of K_n. By (3.8.2) there exist a weak limit point $u \in K$ of $\{u_n\}$ and so u belongs to the closure of K_n in (K,η). Hence there exist $v_n \in K$ satisfying

(ii) $$v_n \in K_n \quad \forall n \geq 1$$

(iii) $$\rho(u,v_n) \leq 2^{-n} \ \forall n \geq 1.$$

Since $\theta(u_j,t) \geq f_j(t) \geq f_n(t)$ for all $t \in T$ and all $j \geq n$ we have by (ii) and (3.8.4) that $\theta v_n \geq f_n$ for all $n \geq 1$. Hence

$$f = \lim_{n\to\infty} f_n \leq \liminf_{n\to\infty} \theta v_n$$

And by (iii) and (3.8.6) we have

$$\mu^F(f) \leq \mu^F(\liminf_{n\to\infty} \theta v_n) \leq \sup_{n\geq 1} \lambda(v_n)$$

Since $\lambda(u_j) \leq a$ for all $j \geq 1$, then by (i) and (3.8.3) we have that $\lambda(v_n) \leq a$. Thus we conclude that $\mu^F(f) \leq a$ for all $a > \lim \mu^F(f_n)$ and so

$$\mu^F(f) \leq \lim \mu^F(f_n) \leq \mu^F(f).$$

Hence the proposition is proved. □

Corollary 3.9. Let $(S_j, \mathcal{S}_j, \mu_j)$ be finite positive measure spaces for $j = 1,\dots,k$, and let B_j be a convex set of non-negative μ_j-measurable functions, satisfying

(3.9.1) $\quad \varphi \wedge a \in B_j \quad \forall \varphi \in B_j$

(3.9.2) $\quad \forall \varphi \in \hat{B}_j(a)\ \exists \psi \in B_j\colon \psi \geq \varphi \wedge a$ and $\psi = \varphi$ μ-a.e.

for all $1 \leq j \leq k$ and all $a \in \mathbb{R}_+$, where $\hat{B}_j(a)$ is the closure of $B_j(a) = \{\varphi \in B_j \mid \varphi \leq a\}$ in $L^1(\mu_j)$.

Now let T be a set and let $\pi_j\colon T \to S_j$ be maps for $j = 1,\dots,k$ and put

$$\mu(f) = \inf\left\{ \sum_{j=1}^{k} \int \varphi_j \, d\mu \;\middle|\; \begin{array}{l} \varphi_j \in B_j \quad \text{for} \quad 1 \leq j \leq k \\ \text{and} \quad f \leq \sum_j \varphi_j \circ \pi_j \end{array} \right\}$$

for $f \in \overline{\mathbb{R}}^T$. Then μ is σ-subsmooth at $B^*(T)$ along $\overline{\mathbb{R}}^T$.

Proof. So let $f_n \in \overline{\mathbb{R}}^T$ for $n \geq 1$, so that $f_n \uparrow f$ where $f \in B^*(T)$. Then there exist $a \in \mathbb{R}_+$, such that $f(t) \leq a$ for all $t \in T$. We shall apply Proposition 3.8 with

$$L = L^1(\mu_1) \times \dots \times L^1(\mu_k)$$

$$K = B_1(a) \times \dots \times B_k(a)$$

$$F = \{h \in \overline{\mathbb{R}}^T \mid h(t) \leq a \ \forall t \in T\}$$

$$\eta = \text{the product of the } \|\cdot\|_1\text{-topogies on } L^1(\mu_j)$$

$$\theta u = \sum_j u_j \quad \text{and} \quad \lambda(u) = \sum_j \int u_j \, d\mu_j$$

for $u = (u_1,\dots,u_k) \in K$. Then the weak topology on L is the product of the weak topologies on $L^1(\mu_j)$. By [4; Theorem IV.8.9] we have that $\hat{B}_j(a)$ is weakly sequentially compact, so by (3.9.1) and (3.9.2) we have that $B_j(a)$ is weakly countably compact. Thus (3.8.1)-(3.8.3) holds. Moreover (3.8.4) is evident and (3.8.5) follows from (3.9.1) and the definition of μ by noting that if

$$h \leq a \quad \text{and} \quad h \leq \sum_j \varphi_j \circ \pi_j$$

where $\varphi_j \geq 0$, then $h \leq \Sigma\psi_j \circ \pi_j$ where $\psi_j = \varphi_j \wedge a$.

Now suppose that $u_n = (u_{1n},\dots,u_{kn}) \in K$ converge to $v = (v_1,\dots,v_k) \in K$ in η. Then there exist integers $\sigma(1) < \sigma(2) < \dots$ such that $u_{j\sigma(n)} \to v_j$ μ_j - a.e. Now put

$$h(t) = \liminf_{n\to\infty} \sum_{j=1}^{k} u_{jn}(\pi_j(t)) \qquad \forall t \in T$$

$$\varphi_j(s_j) = \limsup_{n\to\infty} u_{j\sigma(n)}(s_j) \quad \forall s_j \in S_j \quad \forall 1 \leq j \leq k$$

Then $\varphi_j = v_j$ μ_j - a.e. and $\varphi_j \leq a$. So by (3.9.1) and (3.9.2) there exist $\psi_j \in B_j(a)$ so that $\psi_j \geq \varphi_j$ and $\psi_j = v_j$ μ_j - a.e., and then we have

$$h \leq \limsup \sum_{j=1}^{k} u_{j\sigma(n)} \circ \pi_j \leq \sum_{j=1}^{k} \varphi_j \circ \pi_j \leq \sum_{j=1}^{k} \psi_j \circ \pi_j$$

And so

$$\mu(h) \leq \sum_{j=1}^{k} \int \psi_j \, d\mu_j = \lim_{n\to\infty} \sum_{j=1}^{k} \int u_{jn} \, d\mu_j$$

$$\leq \sup_n \lambda(u_n)$$

since $u_{jn} \to \psi_j$ in $L^1(\mu_j)$ for all $1 \leq j \leq k$. Thus (3.8.6) holds and so the corollary follows from Proposition 3.8. □

<u>Proposition 3.10</u>. <u>Let</u> $F \subseteq \overline{\mathbb{R}}^T$ <u>and suppose that</u> μ <u>is superadditive on</u> F. <u>Let</u> $\varphi \in \overline{\mathbb{R}}^T$, <u>such that for some</u> $f \in F$ <u>we have</u>

(3.10.1) $f \geq \varphi$ and $\mu_F(f \dot{-} \varphi) \neq \pm\infty$

(3.10.2) $\mu(f) \leq \nu(\varphi) \dotplus \mu_F(f \dot{-} \varphi)$

(3.10.3) $\forall f_1 \in F$ <u>with</u> $\varphi \dotplus f_1 \leq f$ $\exists f_0 \in F$ such that $f_0 \geq \varphi$ <u>and</u> $f_0 \dotplus f_1 \leq f$.

<u>Then we have that</u> $\mu^F(\varphi) \le \nu(\varphi)$.

<u>Remark</u>. Note that if $\varphi = 1_C$ and $F = \{1_E \mid E \in \mathcal{F}\}$ for some paving $\mathcal{F}$, then (3.10.3) is equivalent to

(3.10.4) $\quad \forall E_1 \in \mathcal{F}$ with $E_1 \subseteq E \setminus C \quad \exists \quad E_0 \in \mathcal{F}$ such that $C \subseteq E_0 \subseteq E \setminus E_1$

where $f = 1_E$.

<u>Proof</u>. By (3.10.1) there exist $f_1 \in F$, such that $f_1 \le f \dot{-} \varphi$ and $\mu_F(f \dot{-} \varphi) \le \varepsilon + \mu(f_1)$, where $\varepsilon > 0$ is a given positive number. Then $\varphi \underset{\bullet}{+} f_1 \le f$ by (A.2), and so by (3.10.3) there exist $f_0 \in F$, such that $f_0 \ge \varphi$ and $f_0 \underset{\bullet}{+} f_1 \le f$. Then by (3.10.2) and superadditivity of μ we have

$$\begin{aligned} \mu(f_0) \underset{\bullet}{+} \mu(f_1) &\le \mu^F(f_0 \underset{\bullet}{+} f_1) \le \mu(f) \\ &\le \nu(\varphi) \overset{\bullet}{+} \mu_F(f \dot{-} \varphi) \\ &\le \nu(\varphi) \overset{\bullet}{+} \varepsilon \overset{\bullet}{+} \mu(f_1) \end{aligned}$$

Moreover by (3.10.1) we have

$$-\infty < \mu_F(f \dot{-} \varphi) \le \varepsilon + \mu(f_1) \le \varepsilon + \mu_F(f \dot{-} \varphi) < \infty$$

and so $\mu(f_1) \in \mathbb{R}$. Hence we have

$$\mu^F(\varphi) \le \mu(f_0) \le \nu(\varphi) + \varepsilon$$

and since $\varepsilon > 0$ is arbitrary the proposition is proved. □

Our next theorem is fundamental for the rest of this exposition. It is a simultaneous extension of the Riesz-Daniell representation theorem for positive linear and the Caracthéodory extension theorem for set functions.

Theorem 3.11. Let $V \subseteq \bar{\mathbb{R}}^T$, and let $\lambda: V \to \bar{\mathbb{R}}$ be a positive functional. Let U be $(\wedge f, \vee f)$-stable subset of $\bar{\mathbb{R}}_+^T$ containing 0, and let F be a $(\wedge f)$-stable subset of $\bar{\mathbb{R}}_+^T$ satisfying

(3.11.1) $\lambda^*(u) \leq \lambda(v) + \lambda^*((u-v)^+) \quad \forall u \in U \quad \forall v \in V$

(3.11.2) $(u-v)^+ \in U_\xi \quad \forall u \in U \quad \forall v \in V$

(3.11.3) λ^* is supermodular on U and λ^* is ξ-supersmooth at 0 along U

(3.11.4) $\lambda^*(f) = \sup\{\lambda^*(u) \mid u \in U_\xi,\ u \leq f\} < \infty \quad \forall f \in F$

(3.11.5) $\lambda^*(u) \leq \sup\{\lambda^*(f) \mid f \in F,\ f \leq 1_{\{u>0\}}\} \quad \forall u \in U_\xi \cap [0,1]^T$

where ξ is an infinite cardinal number. Let C be the paving of all sets of the form $\{f \geq a\}$ where $f \in F$ and $0 < a < \infty$, and suppose that $C \subseteq F$. Now put

$$\hat{C} = \{\hat{C} \subseteq T \mid \hat{C} \cap C \in U_\xi \quad \forall C \in C\}$$

$$R = \{R \subseteq T \mid \lambda^*(R) = \sup\{\lambda^*(u) \mid u \in U_\xi,\ u \leq 1_R\}\}$$

Then there exist a σ-algebra B on T and a positive measure μ on (T,B) satisfying

(3.11.6) $C \subseteq B$ and $\mu(C) = \lambda^*(C) < \infty \quad \forall C \in C$

(3.11.7) $B \in B \iff B \cap C \in B \quad \forall C \in C$

(3.11.8) If $B \cap C \in R \quad \forall C \in C$, then $B \in B$

(3.11.9) $\mu_*(A) = \sup\{\lambda^*(u) \mid u \in U_\xi,\ u \leq 1_A\} \quad \forall A \subseteq T$

(3.11.10) $\mu_*(A) = \sup\{\lambda^*(C) \mid C \in C,\ C \subseteq A\} \quad \forall A \subseteq T$

(3.11.11) $\hat{C}$ is $(\cap\xi, \cup f)$-stable, and $\hat{C} \subseteq B$

(3.11.12) μ is ξ-supersmooth at B along $\{\hat{C} \in \hat{C} \mid \mu(\hat{C}) < \infty\}$

If $\lambda^*(C) \geq n\,\lambda^*(n^{-1}1_C)$ *for all* $C \in \mathcal{C}$ *and all* $n \in \mathbb{N}$, *then we have*

$$(3.11.13) \qquad \lambda^*(f) \leq \int_* g\,d\mu \quad \forall g \in \bar{\mathbb{R}}_+^T \quad \forall f \in F \cap \mathbb{R}^T \quad \text{with} \quad f \leq g$$

If U *is a convex cone in* $\mathbb{R}_+^T$, *if* λ^* *is superadditive on* U *and if* λ^* *is positively homogenuous on* U, (i.e. $\lambda^*(au) = a\,\lambda^*(u)$ $\forall a \in \mathbb{R}_+$ $\forall u \in U$), *then we have*

$$(3.11.14) \qquad \int_* g\,d\mu \leq \sup\{\lambda^*(u) \mid u \in U_\xi,\ u \leq g\} \quad \forall g \in \bar{\mathbb{R}}_+^T$$

Proof. By Proposition 3.5 and (3.11.3) we have that λ^* is ξ-supersmooth at 0 along U_ξ. Putting $\mu = \nu = \lambda = \lambda^*$, $U = U$, $V = V$, $F = U_\xi$, $H = U$, $G = V$ and $\psi = 0$, in Proposition 3.6 we see that λ^* is ξ-supersmooth at $\bar{\mathbb{R}}^T$ along U. Thus by Proposition 3.5 we have

(i) λ^* is ξ-supersmooth at $\bar{\mathbb{R}}^T$ along U_ξ.

So by supermodularity of λ^* on U, and $(\wedge f, \vee f)$-stability of U we find

(ii) U_ξ is $(\wedge\xi, \vee f)$-stable, and λ^* is supermodular on U_ξ.

Now let us define

$$\nu(h) = \sup\{\lambda^*(u) \mid u \in U_\xi,\ u \leq h\} \quad \forall h \in \bar{\mathbb{R}}^T .$$

Then by (ii) and Proposition 3.7 we have

(iii) ν is supermodular on $\bar{\mathbb{R}}^T$ and $\nu \leq \lambda^*$

(iv) $\nu(u) = \lambda^*(u) \quad \forall u \in U_\xi$.

Next we show

(v) $\nu(f+g) \leq \lambda^*(f) + \nu(g) \quad \forall f, g \in \bar{\mathbb{R}}_+^T$

So let $f, g \in \bar{\mathbb{R}}_+^T$ be given an choose $u \in U_\xi$ and $v \in V$, so that $u \leq f+g$ and $g \leq v$. Let $\{u_\gamma \mid \gamma \in \Gamma\}$ be a decreasing in U so that $u_\gamma \downarrow u$ and $\mathrm{card}(\Gamma) \leq \xi$. Since $v \geq 0$ we have that $(u_\gamma - v)^+ \downarrow (u-v)^+$, and so by (i), (3.11.1) and (3.11.2) we have

$$\lambda^*(u) = \lim \lambda^*(u_\gamma) \leq \lim\{\lambda(v) + \lambda^*(u_\gamma - v)^+)\}$$

$$= \lambda(v) + \lambda^*((u-v)^+)$$

$$\leq \lambda(v) + \nu(f)$$

since $(u-v)^+ \in U_\xi$ and (see (A.2))

$$(u-v)^+ = (u \dot{-} v)^+ \leq (f \dot{+} g) \dot{-} g \leq f.$$

But then (v) follows by taking infinium over v and supremum over u.

Since ν is supermodular on U_ξ and σ-supersmooth at U_ξ along U_ξ by (i)-(iv), we have by Proposition 3.7 that

(vi) ν is σ-supersmooth at $\bar{\mathbb{R}}^T$ along H

where $H = \{h \in \bar{\mathbb{R}}_+^T \mid \nu(h) < \infty\}$.

Now we are able to define the measure μ satisfying the hypothesis of the theorem. Let

$$\mathcal{B} = \{B \subseteq T \mid \nu(A) = \nu(A \cap B) + \nu(A \setminus B) \quad \forall A \subseteq T\}$$

$$\mu(B) = \nu(B) \qquad \forall B \in \mathcal{B}.$$

I.e. $\mathcal{B}$ is the set of all Caracthéodory ν-measurable sets, and so by [4; Lemma III.5.2] we have

(vii) $\mathcal{B}$ is an algebra and μ is finitely additive on $\mathcal{B}$

(Note that $\mu(\phi) = \nu(0) = \lambda^*(0) = 0$ since $0 \in U$). Next we show

(viii) $\nu(A) = \sup\{\nu(C) \mid C \in \mathcal{C},\ C \subseteq A\} \quad \forall A \subseteq T$

So let $A \subseteq T$ and let $a < \nu(A)$, then there exist $u \in U_\xi$ so that $u \leq 1_A$ and $\lambda^*(u) > a$. Let $D = \{u > 0\}$ then by (3.11.5) there exist $f \in F$ so that $0 \leq f \leq 1_D$ and $\lambda^*(f) > a$. Now let $C_n = \{f \geq 1/n\}$, and put

$$f_n(t) = \begin{cases} f(t) & \text{if} \quad t \notin C_n \\ 0 & \text{if} \quad t \in C_n \end{cases}$$

Then $f \leq f_n + 1_{C_n}$, and since $\lambda^*(f) = \nu(f)$ by (3.11.4) it follows from (v) that

$$a < \lambda^*(f) = \nu(f) \leq \nu(f_n) + \lambda^*(C_n).$$

Now $f_n \downarrow 0$ and $\nu(f_n) \leq \nu(f) = \lambda^*(f) < \infty$, so by (vi) we have that $\nu(f_n) \to 0$, and since $C_n \in \mathcal{C} \subseteq F$, and thus $\lambda^*(C_n) = \nu(C_n)$ by (3.11.4), we have

$$\lambda^*(C_n) = \nu(C_n) > a$$

for n sufficiently large. Thus (viii) follows.

By (iii) we have, that ν is superadditive on 2^T, and so by (viii) we have

(ix) $$B \in \mathcal{B} \iff \nu(C) \leq \nu(C \cap B) + \nu(C \setminus B) \quad \forall C \in \mathcal{C}.$$

Next we show

(x) $$\lambda^*(C) = \nu(C) < \infty \quad \forall C \in \mathcal{C}.$$

So let $C \in \mathcal{C}$, then $C = \{f \geq a\}$ for some $f \in F$ and some $0 < a < \infty$. Then $C \in F$ and so $\lambda^*(C) = \nu(C)$ by (3.11.4), and $1_C \leq nf$, where $n \in \mathbb{N}$ is chosen so that $n \geq a^{-1}$. Hence by (iv) and (v) we have

$$\lambda^*(C) = \nu(C) \leq \nu(nf) \leq (n-1)\lambda^*(f) + \nu(f) \leq n\lambda^*(f) < \infty$$

since $\lambda^*(f) < \infty$ by (3.11.4).

Next we show

(xi) $\mathcal{B}$ is a σ-algebra and μ is a measure on $(T,\mathcal{B})$.

So let $B_n \in \mathcal{B}$ so that $B_n \uparrow B$, and let $C \in \mathcal{C}$ be given. Then $C \setminus B_n \downarrow C \setminus B$ and $\nu(C \setminus B_n) \leq \nu(C) < \infty$. Hence by (vi) we have

$$\begin{aligned} \nu(C) &= \lim_{n\to\infty} [\nu(C \cap B_n) + \nu(C \setminus B_n)] \\ &\leq \lim_{n\to\infty} \nu(C \cap B_n) + \nu(C \setminus B) \\ &\leq \nu(C \cap B) + \nu(C \setminus B) \end{aligned}$$

and so $B \in \mathcal{B}$ by (ix). Moreover since the inequalities above holds for all $C \in \mathcal{C}$ find by (viii) that

$$\nu(B) \leq \lim \nu(B_n) \leq \nu(B).$$

Thus (xi) follows from (vii).

We have now shown that μ is a measure on $(T,\mathcal{B})$ so let us now show that μ satisfies (3.11.6)-(3.11.14).

(3.11.6): Let $C_0, C_1 \in \mathcal{C}$, then $C_0 \cap C_1 \in F$ by $(\wedge f)$-stability of F, since $\mathcal{C} \subseteq F$. Hence by (3.11.4) and (v) we have

$$\nu(C_0) \leq \nu(C_0 \setminus C_1) + \lambda^*(C_0 \cap C_1) = \nu(C_0 \setminus C_1) + \nu(C_0 \cap C_1)$$

and so $C_1 \in \mathcal{B}$ by (ix). Hence $\mathcal{C} \subseteq \mathcal{B}$, and the second part of (3.11.6) follows from (x).

(3.11.7). Since $\mathcal{C} \subseteq \mathcal{B}$ we have that $B \cap C \in \mathcal{B}$ for all $B \in \mathcal{B}$ and all $C \in \mathcal{C}$. Now suppose that $B \cap C \in \mathcal{B}$ for all $C \in \mathcal{C}$. Then

$$\nu(C) = \nu(C \cap (B \cap C)) + \nu(C \setminus (B \cap C)) = \nu(C \cap B) + \nu(C \setminus B)$$

for all $C \in \mathcal{C}$, so $B \in \mathcal{B}$ by (ix)

(3.11.8): Suppose that $B \cap C \in R$ for all $C \in \mathcal{C}$, then $\lambda^*(B \cap C) = \nu(B \cap C)$, and so by (v) we have

$$\nu(C) \leq \lambda^*(C \cap B) + \nu(C \setminus B) = \nu(C \cap B) + \nu(C \setminus B)$$

for all $C \in \mathcal{C}$, thus $B \in \mathcal{B}$ by (ix).

(3.11.9+10): Since $\mathcal{C} \subseteq \mathcal{B}$ then by (viii) we have that $\nu = \mu_*$ and (3.11.9) and (3.11.10) follows from the definition of ν and (viii) and (x).

(3.11.11): Since U_ξ is $(\wedge\xi, \vee f)$-stable it follows easily that $\hat{\mathcal{C}}$ is $(\cap\xi, \cup f)$-stable, and by (3.11.8) we have that $\hat{\mathcal{C}} \subseteq \mathcal{B}$.

(3.11.12): Let $\{\hat{C}_\gamma \mid \gamma \in \Gamma\}$ be a decreasing net in $\hat{\mathcal{C}}$, such that $\text{card}(\Gamma) \leq \xi$, $\mu(\hat{C}_\gamma) < \infty$ and $\hat{C}_\gamma \downarrow \hat{C}$. Now fix a $\beta \in \Gamma$ and let $\varepsilon > 0$ be given. Since $\mu(\hat{C}_\beta) < \infty$, then by (3.11.10) there exist $C \in \mathcal{C}$, such that $C \subseteq \hat{C}_\beta$ and $\mu(\hat{C}_\beta \setminus C) < \varepsilon$. Then we have

$$\mu(\hat{C}_\gamma) \leq \mu(C \cap \hat{C}_\gamma) + \mu(\hat{C}_\beta \setminus C) \leq \mu(C \cap \hat{C}_\gamma) + \varepsilon$$

for all $\gamma \geq \beta$. And since $\hat{C}_\gamma \cap C \in U_\xi$ and $\hat{C}_\gamma \cap C \downarrow \hat{C} \cap C$ it follows from (i) and (iv) that

$$\lim_\gamma \mu(\hat{C}_\gamma) \leq \mu(C \cap \hat{C}) + \varepsilon \leq \mu(\hat{C}) + \varepsilon$$

and so (3.11.12) follows.

(3.11.13): Let $f \in F \cap \mathbb{R}_+^T$ and $g \in \overline{\mathbb{R}}_+^T$ so that $f \leq g$. If $\int_* g \, d\mu = \infty$ there is nothing to prove, so suppose that $\int_* g \, d\mu < \infty$, and put

$$C_{jn} = \{f \geq j/n\}, \quad C_n = \{f \geq n\}, \quad D_n = \{f < 1/n\}$$

$$f_n = \sum_{j=1}^{n^2} n^{-1} 1_{C_{jn}}, \quad f_n' = n^{-1} 1_{C_{1n}}$$

$$f_n'' = f(1_{C_n} + 1_{D_n}) .$$

Then $f_n \leq f \leq f_n + f_n' + f_n''$, and by (v), (3.11.4) and (3.11.6) we have

$$\begin{aligned}\lambda^*(f) &= \nu(f) \le \nu(f_n + f_n' + f_n'') \\ &\le \nu(f_n'') + n^{-1}\lambda^*(C_{1n}) + \sum_{j=1}^{n^2} n^{-1}\lambda^*(C_{jn}) \\ &= \nu(f_n'') + n^{-1}\mu(f \ge 1/n) + \sum_{j=1}^{n^2} n^{-1}\mu(C_{jn}) \\ &= \nu(f_n'') + n^{-1}\mu(f \ge 1/n) + \int f_n\, d\mu \\ &\le \nu(f_n'') + n^{-1}\mu(f \ge 1/n) + \int_* g\, d\mu\end{aligned}$$

since $f \in F$, $C_{jn} \in C$, $f_n \le f \le g$ and $\lambda^*(n^{-1}1_C) \le n^{-1}\mu(C)$ for all $C \in C$ and $n \in \mathbb{N}$. Now observe that

(xii) $$\lim_{\varepsilon\to 0} \varepsilon\,\mu(|h| \ge \varepsilon) = 0 \quad \forall h \in L^1(\mu)$$

since

$$\begin{aligned}\varepsilon\,\mu(|h| \ge \varepsilon) &= \varepsilon\,\mu(|h|) \ge \sqrt{\varepsilon}) + \varepsilon\,\mu(\varepsilon \le |h| < \sqrt{\varepsilon}) \\ &\le \sqrt{\varepsilon}\int |h|d\mu + \int_{\{|h| \le \sqrt{\varepsilon}\}} |h|d\mu\end{aligned}$$

for all $0 < \varepsilon < 1$.

Since $\int_* g\, d\mu < \infty$ we have that $f \in L^1(\mu)$, so by (xii) we have that

$$\lim_{n\to\infty} n^{-1}\mu(f \ge 1/n) = 0$$

And since $f < \infty$ we have that $f_n'' \downarrow 0$, and $\nu(f_n'') \le \nu(f) < \infty$, so by (vi) we find

$$\lim_{n\to\infty} \nu(f_n'') = 0 .$$

Thus (3.11.13) follows.

(3.11.14): Let $a < \int_* g\, d\mu$, then there exist a function $h \in B^+(T) \cap L^1(\mu)$ so that $h \le g$ and $\int h\, d\mu > a$. Since U is a convex cone in $\mathbb{R}_+^T$, then evidently U_ξ is so, and by (i) it follows that λ^* is superadditive on U_ξ and positively homogenuous on U_ξ.

Now let $b \in \mathbb{N}$ be chosen so that $h \le b$, and put

$$C_{jn} = \{h \geq bj/n\}, \quad D_n = \{h < b/n\}$$

$$h_n = \sum_{j=1}^{n} bn^{-1} 1_{C_{jn}}, \quad h'_n = h\, 1_{D_n}, \quad h''_n = bn^{-1} 1_{C_{1n}}$$

Then $h_n \leq h \leq h_n + h'_n + h''_n$ and so

$$\int h\, d\mu \leq bn^{-1} \sum_{j=1}^{n} \mu(C_{jn}) + \int h'_n\, d\mu + \int h''_n\, d\mu.$$

By (3.11.9) there exist $u_{jn} \in U$, so that

$$u_{jn} \leq 1_{C_{jn}} \quad \text{and} \quad \mu(C_{jn}) \leq n^{-1} + \lambda^*(u_{jn}).$$

Since λ^* is superadditive and positively homogenuous on U_ξ we find

$$\int h\, d\mu \leq bn^{-1} \sum_{j=1}^{n} \lambda^*(u_{jn}) + bn^{-1} + \int h'_n\, d\mu + \int h''_n\, d\mu$$

$$\leq \overset{*}{\lambda}(u_n) + bn^{-1} + \int h'_n\, d\mu + \int h''_n\, d\mu$$

where

$$u_n = \sum_{j=1}^{n} bn^{-1} u_{jn}$$

Now $h'_n \downarrow 0$ and so $\int h'_n\, d\mu \to 0$, and by (xii) we have that $\int h''_n\, d\mu \to 0$. Thus we conclude that

$$a < \int h\, d\mu \leq \liminf \lambda^*(u_n)$$

and since $u_n \in U_\xi$ and $u_n \leq h_n \leq h$, we see that (3.11.14) holds and the theorem is proved. □

<u>Proposition 3.12</u>. <u>Let</u> $\mathcal{K}$ <u>be a paving on</u> T <u>and let</u> F <u>and</u> G <u>be subsets of</u> $\bar{\mathbb{R}}^T$. <u>Now put</u>

$$F^* = \{h \in \bar{\mathbb{R}}^T \mid \sup\{\mu(f) \mid f \in F,\ f \leq h + \varepsilon\} \geq \nu(h) \quad \forall \varepsilon > 0\}$$

$$F_* = \{h \in \bar{\mathbb{R}}^T \mid \sup\{\mu(f) \mid f \in F,\ \exists \varepsilon > 0:\ f + \varepsilon \leq h\} \geq \nu(h)\}$$

<u>Then we have</u>

(3.12.1) <u>If</u> μ <u>is</u> $\mathcal{K}$-<u>tight along</u> F, <u>then</u> ν <u>is</u> $\mathcal{K}$-<u>tight along</u> F

Now suppose that μ satisfies

(3.12.2) $$\inf_{K \in K} \mu_F(1_{T\setminus K}) \leq 0$$

Then ν is K-tight along F_*. Moreover if μ satisfies (3.12.2) and the following two conditions:

(3.12.3) $$\nu(g) \leq \nu(g1_K) \dot{+} \mu_F(g1_{T\setminus K}) \quad \forall g \in G \quad \forall K \in K$$

(3.12.4) $$\mu_F(a1_{T\setminus K}) \leq b\, \mu_F(1_{T\setminus K}) \quad \forall K \in K$$

where $a, b \in \mathbb{R}_+$ are given numbers, then we have

(3.12.5) $$\nu_G(\varphi) = \sup\{\nu(g1_K) \mid g \in G,\ K \in K,\ g \leq \varphi\}$$

for all $\varphi \in \bar{\mathbb{R}}^T$ with $0 \leq \varphi \leq a$.

Proof (3.12.1): Let $\varepsilon > 0$ be given and choose $\delta > 0$ and $K \in K$ such that $\mu_F(1_{T\setminus K} + 2\delta 1_T) < \varepsilon$. Now let $h \in F^*$ such that $h \leq 1_{T\setminus K} + \delta$ and let $a < \nu(h)$. Then by definition of F^* there exist $f \in F$, such that $f \leq h + \delta$ and $\mu(f) > a$. But then

$$f \leq h + \delta \leq 1_{T\setminus K} + 2\delta$$

and so $a < \mu(f) < \varepsilon$, for all $a < \nu(h)$. Thus $\nu(h) \leq \varepsilon$ and so we see that ν is K-tight along F^*.

Now suppose that (3.12.2) holds. Then exactly as above one show that ν is K-tight along F_*.

Now suppose that (3.12.2)-(3.12.4) holds, and let φ be a function such that $0 \leq \varphi \leq a$. Now let $c < \nu_G(\varphi)$ then there exist $\varepsilon > 0$ and $g \in G$ so that $c + \varepsilon \leq \nu(g)$ and $g \leq \varphi$. And by (3.12.2) there exist $K \in K$ such that $\mu_F(1_{T\setminus K}) < \varepsilon/b$. Then by (3.12.3) and (3.12.4) we have

$$c + \varepsilon \leq \nu(g) \leq \nu(g1_K) \dot{+} \mu_F(g1_{T\setminus K})$$

$$\leq \nu(g1_K) \dot{+} \mu_F(a1_{T\setminus K})$$

$$\leq \nu(g1_K) \dotplus b\ \mu_F(1_{T\setminus K})$$

$$\leq \nu(g1_K) + \varepsilon$$

since $g \leq \varphi \leq a$. Thus we see that we have $\leq$ in (3.13.5). Now if $g \leq \varphi$ then $g \cdot 1_K \leq \varphi$ since $\varphi \geq 0$ and so the converse inquality holds. □

Proposition 3.13. *Let K be a paving on T and $F \subseteq \bar{\mathbb{R}}^T$, such that μ is K-tight along F, and let $a, b \in \mathbb{R}_+$ so that*

(3.13.1) $$\mu_F(a1_{T\setminus K} + a\delta) \leq b\ \mu_F(1_{T\setminus K} + \delta) \quad \forall K \in K \quad \forall \delta \in \mathbb{R}_+$$

Let ξ be an infinite cardinal such that the paving:

$$L(K,\varepsilon) = \{K \cap \{f \geq \varepsilon\} \mid f \in F\}$$

is mono-ξ-compact for all $K \in K$ and all $\varepsilon > 0$. If $\mu(0) = 0$ then μ is ξ-supersmooth at 0 along $F_a = \{f \in F \mid f \leq a\}$.

Proof. Let $\{f_\gamma \mid \gamma \in \Gamma\}$ be a decreasing net in F_a, such that $f_\gamma \downarrow 0$ and $\mathrm{card}(\Gamma) \leq \xi$. Let $\varepsilon > 0$ be given and choose $K \in K$ and $\delta > 0$, so that $\mu_F(1_{T\setminus K} + \delta) < \varepsilon/b$. If $a = 0$ then $f_\gamma \equiv 0$ and there is nothing to prove. So suppose that $a > 0$, then $L_\gamma = K \cap \{f_\gamma \geq a\delta\} \in L(K,a\delta)$ and $L_\gamma \downarrow \emptyset$. Hence by mono-$\xi$-compactness of $L(K,a\delta)$ there exist $\beta \in \Gamma$, such that $L_\beta = \emptyset$. Hence we have

$$f_\gamma \leq f_\beta \leq a1_{T\setminus K} + a\delta \qquad \forall \gamma \geq \beta$$

since $f_\beta \leq a$. Thus we conclude from (3.13.1):

$$\mu(0) = 0 \leq \mu(f_\gamma) \leq \mu_F(a1_{T\setminus K} + a\delta) \leq \varepsilon$$

for all $\gamma \geq \beta$, and so $\mu(f_\gamma) \to \mu(0) = 0$. □

Proposition 3.14. *Let μ be a finite positive measure on (T,B)*

<u>and let</u> ξ <u>be an infinite cardinal, such that</u> μ <u>is</u> ξ<u>-supersmooth at</u> F <u>along</u> $\mathcal{F}$, <u>where</u> $\mathcal{F} \subseteq \mathcal{B}$ <u>is a</u> $(\cap\xi)$<u>-stable paving. Let</u> $A \subseteq T$ <u>and put</u>

$$\mathcal{R} = \{R \subseteq T \mid \exists F \in \mathcal{F}:\quad F \cap A = R \cap A\}$$

<u>Let</u> $f \in \overline{\mathbb{R}}^T$ <u>and let</u> $\{\varphi_\gamma \mid \gamma \in \Gamma\}$ <u>be a net in</u> $U(T,\mathcal{R})$, <u>so that</u>

(3.14.1) $\quad \operatorname{card}(\Gamma) \leq \xi$ and $\varphi_\alpha(t) \leq \varphi_\beta(t)\ \forall \alpha \geq \beta\ \forall t \in A$

(3.14.2) $\quad f^- \varphi_\gamma$ is μ-measurable $\forall \gamma \in \Gamma$

(3.14.3) $\quad f(t) = 0 \quad \forall t \in T \setminus A$

(3.14.4) $\quad \int_* f^- \varphi^- \, d\mu < \infty$ and $\int_* |f| \varphi_\gamma^+ \, d\mu < \infty \quad \forall \gamma \in \Gamma$

<u>where</u> $\varphi = \inf_\gamma \varphi_\gamma$. <u>Then there exist a</u> μ<u>-null set</u> N <u>and an increasing map</u> $\tau: \mathbb{N} \to \Gamma$, <u>such that</u>

(3.14.5) $\quad \varphi(t) = \lim_{n\to\infty} \varphi_{\tau(n)}(t) \quad \forall t \in A \setminus N$

(3.14.6) $\quad f^- \varphi \in L^1(\mu),\quad f^- \varphi_\gamma \in L^1(\mu) \quad \forall \gamma \geq \tau(1)$

(3.14.7) $\quad \int_* f\varphi \, d\mu = \lim_\gamma \int_* f\, \varphi_\gamma \, d\mu = \lim_{n\to\infty} \int_* f\, \varphi_{\tau(n)} d\mu$

<u>Proof</u>. Since $\varphi_\gamma \in U(T,\mathcal{R})$, there exist $F_{jn}(\gamma) \in \mathcal{F}$ such that

(i) $\quad A \cap \{\varphi_\gamma \geq j2^{-n}\} = A \cap F_{jn}(\gamma) \quad \forall \gamma \in \Gamma \quad \forall j \in \mathbb{Z} \quad \forall n \in \mathbb{N}$

And put $F_{jn} = \cap_{\gamma\in\Gamma} F_{jn}(\gamma)$. Then clearly we have

(ii) $\quad A \cap \{\varphi \geq j2^{-n}\} = A \cap F_{jn} \quad \forall j \in \mathbb{Z} \quad \forall n \in \mathbb{N}$.

Let Λ be the set of all finite non-empty subsets of Γ, then $\operatorname{card}(\Lambda) \leq \xi$, and so by ξ-supersmoothness of μ and $(\cap\xi)$-stability of $\mathcal{F}$, there exist $\lambda(n) \in \Lambda$ for $n \in \mathbb{N}$, such that

(iii) $\quad \mu\Big(\bigcap_{\gamma\in\lambda(n)} F_{jn}(\gamma) \setminus F_{jn}\Big) \leq 2^{-n} \quad \forall n \geq 1 \quad \forall |j| \leq 4^n$.

Since Γ is a directed set and $\lambda(n)$ is a finite subset of Γ we can find $\sigma(1), \sigma(2), \ldots$ in Γ such that

(iv) $\qquad \sigma(n) \leq \sigma(n+1) \quad \forall n \geq 1, \ \gamma \leq \sigma(n) \quad \forall n \geq 1 \quad \forall \gamma \in \lambda(n).$

Now put $\psi = \inf_n \varphi_{\sigma(n)}$ and

$$N_{jn} = \{t \in A \mid \varphi(t) < j2^{-n} \leq \psi(t)\} \quad \forall j \in \mathbb{Z} \quad \forall n \in \mathbb{N} .$$

Let $j \in \mathbb{Z}$ and $n \in \mathbb{N}$ be given, choose $p \geq n$ so that $|j| \leq 2^p$. Let $m \geq p$, then $j2^{-n} = k2^{-m}$, where $k = j \cdot 2^{m-n} \in \mathbb{Z}$, hence

$$\begin{aligned} N_{jn} = N_{km} &\subseteq A \cap \{\varphi_{\sigma(m)} \geq k2^{-m}, \quad \varphi < k2^{-m}\} \\ &\subseteq A \cap [\{\varphi_\gamma \geq k2^{-m} \quad \forall \gamma \in \lambda(m)\} \setminus \{\varphi \geq k2^{-m}\}] \\ &= A \cap \bigcup_{\gamma \in \lambda(m)} F_{km}(\gamma) \setminus F_{km} \end{aligned}$$

since $\varphi_\gamma \geq \varphi_{\sigma(m)}$ on A $\forall \gamma \in \lambda(m)$ by (3.14.1) and (iv). Now since $|k| \leq 2^{p+m-n} \leq 4^m$, we have by (iii) that $\mu^*(N_{jn}) \leq 2^{-m}$ for all $m \geq p$. Hence letting $m \to \infty$ we see that N_{jn} is a μ-nullset and so

$$N = \bigcup_{j=-\infty}^{\infty} \bigcup_{n=1}^{\infty} N_{jn}$$

is a μ-nullset; and

(v) $\qquad \varphi_{\sigma(n)}(t) \downarrow \psi(t) \quad \forall t \in A, \quad \psi(t) = \varphi(t) \quad \forall t \in A \setminus N$

Hence we have by (3.14.3) that $f^+\varphi_{\sigma(n)} \downarrow f^+\varphi = f^+\psi$ μ - a.e. and so by (3.14.4) and the decreasing convergence theorem for lower integrals we get

$$\begin{aligned} \int_* f^+\varphi \, d\mu &= \lim_{n\to\infty} \int_* f^+\varphi_{\sigma(n)} \, d\mu \\ &\geq \lim_\gamma \int_* f^+\varphi_\gamma \, d\mu \\ &\geq \int_* f^+\varphi \, d\mu \end{aligned}$$

since $f^+\varphi_\gamma \downarrow f^+\varphi$ by (3.14.1) and (3.14.3). Hence we have

(vi) $$\lim_{\gamma} \int_* f^+ \varphi_\gamma \, d\mu = \lim_{n\to\infty} \int_* f^+\varphi_{\sigma(n)} d\mu = \int_* f^+\varphi \, d\mu$$

And similarly we get

(vii) $$\lim_{\gamma} \int_* f^-\varphi_\gamma \, d\mu = \lim_{n\to\infty} \int_* f^-\varphi_{\sigma(n)} d\mu = \int_* f^-\varphi \, d\mu$$

Moreover since $f^-\varphi = \lim f^-\varphi_{\sigma(n)}$ μ - a.e. it follows from (3.14.2) that $f^-\varphi$ is μ-measurable and since

$$|f^-\varphi| = f^-\varphi^+ + f^-\varphi^- \leq |f|\varphi^+_{\sigma(1)} + f^-\varphi^-$$

we see that $f^-\varphi$ is μ-integrable, and so by (vii) there exist $k \geq 1$ so that $\int_* f^-\varphi_{\sigma(k)} d\mu \neq \pm\infty$, and since

$$|f^-\varphi_\gamma| = f^-\varphi^+_\gamma + f^-\varphi^-_\gamma \leq |f|\varphi^+_{\sigma(k)} + f^-\varphi^-$$

for all $\gamma \geq \sigma(k)$, it follows from (3.14.2) that $f^-\varphi_\gamma \in L^1(\mu)$ for all $\gamma \geq \sigma(k)$. So putting $\tau(n) = \sigma(n+k)$, we see that τ is an increasing from $\mathbb{N}$ in Γ, such that (3.14.5) and (3.14.6) holds. Now let $\gamma \geq \tau(1)$, then we have (see (D.1)):

$$\int_* f^+\varphi_\gamma \, d\mu - \int f^-\varphi_\gamma \, d\mu = \int_* f^+\varphi_\gamma \, d\mu + \int_* (-f^-\varphi_\gamma) d\mu$$

$$\leq \int_* f\varphi_\gamma \, d\mu \leq \int_* f^+\varphi_\gamma \, d\mu + \int^* (-f^-\varphi_\gamma) d\mu$$

$$= \int_* f^+\varphi_\gamma \, d\mu - \int f^-\varphi_\gamma \, d\mu$$

Hence by (vi) and (vii) we have

$$\lim_{n\to\infty} \int_* f\varphi_{\tau(n)} d\mu = \lim_{\gamma} \int_* f\varphi_\gamma \, d\mu$$

$$= \int_* f^+\varphi \, d\mu - \int f^-\varphi \, d\mu$$

and as above we have

$$\int_* f\varphi \, d\mu = \int_* f^+\varphi \, d\mu - \int f^-\varphi \, d\mu$$

and so (3.14.7) holds. □

Definition 3.15. Let $\theta: T \rightsquigarrow S$ be a correspondance, i.e. a map from T into 2^S, if $f \in \bar{\mathbb{R}}^T$, $g \in \bar{\mathbb{R}}^S$, $F \subseteq \bar{\mathbb{R}}^T$ and $G \subseteq \bar{\mathbb{R}}^S$ we define

$$f\theta^{-1} = \{h \in \bar{\mathbb{R}}^S \mid h(s) = f(t)\ \forall (t,s) \in Gr(\theta)\}$$

$$g\theta = \{h \in \bar{\mathbb{R}}^T \mid h(t) = g(s)\ \forall (t,s) \in Gr(\theta)\}$$

$$\theta(F) = \bigcup_{f \in F} f\theta^{-1}, \quad \theta^{-1}(G) = \bigcup_{g \in G} g\theta$$

where $Gr(\theta) = \{(t,s) \in T \times S \mid s \in \theta(t)\}$ is the graph of θ. Note that $f \in g\theta$ if and only if $g \in f\theta^{-1}$, and if θ is an ordinary point map: $T \to S$, then $g\theta$ consists of the single function $g \circ \theta$, and $h \in f\theta^{-1}$ if and only if $f = h \circ \theta$.

Now let φ be a map from $F \subseteq \bar{\mathbb{R}}^T$ into a set X. Then we say that φ is θ-saturated, if φ is constant on $F \cap (g\theta)$ for all $g \in \bar{\mathbb{R}}^S$. If φ is θ-saturated we defined the image functional, $\theta\varphi$, on $\theta(F)$ by

$$(\theta\varphi)(g) = \varphi(f) \quad \text{if} \quad g \in \theta(F) \quad \text{and} \quad f \in F \cap (g\theta).$$

Note that $\theta\varphi$ is well defined since $F \cap (g\theta) \neq \emptyset$ for $g \in \theta(F)$ and φ is constant on $F \cap (g\theta)$. If φ is θ-saturated and $G \subseteq \theta(F)$, then $\theta_G\varphi$ denote the image functional on G, i.e. the restriction of $\theta\varphi$ to G. Note that if $\theta: T \rightsquigarrow S$ is a correspondance and φ is a θ-saturated map from $F \subseteq \bar{\mathbb{R}}^T$ into X; then we have

(3.15.1) $\quad \theta\varphi$ is θ^{-1}-saturated and $\varphi(f) = \theta^{-1}(\theta\varphi)(f)$

for all $f \in F \cap \theta^{-1}(\theta(F))$

Proposition 3.16. Let $\theta: T \rightsquigarrow S$ be a correspondance and let $\alpha: E^Q \to E$ be a map where $E \subseteq \bar{\mathbb{R}}$ and Q is a set. If $F \subseteq \bar{\mathbb{R}}^T$ and (T,A) is an algebraic function space, then we have

(3.16.1) $\quad$ [the α-closure of $\theta(F)$] $\subseteq \theta$(the α-closure of F)

(3.16.2) $\overline{\theta(\bar{A})} \subseteq \theta(\bar{A})$

(3.16.3) $\overline{\theta(\bar{F})} \subseteq \theta(\bar{F})$ if $\theta^{-1}(S)$ is dense in $(T,\tau(F))$

(3.16.4) $B(S,\mathcal{S}) \subseteq \theta(B(T,\mathcal{T}))$

<u>if</u> $\mathcal{T} = \sigma(H)$ <u>and</u> $\mathcal{S} = \sigma(\theta(H))$ <u>for some</u> $H \subseteq \bar{\mathbb{R}}^T$.

<u>Now suppose that</u> $\lambda: F \to \bar{\mathbb{R}}$ <u>is an increasing</u> θ-<u>saturated functional, then we have</u>

(3.16.5) F is $(\wedge f)$-stable $\Rightarrow$ $\theta\lambda$ <u>is increasing</u>

(3.16.6) $(\theta\lambda)_*(h) \leq \lambda_*(h_\theta)$ and $\lambda^*(h^\theta) \leq (\theta\lambda)^*(h)$

<u>for all</u> $h \in \bar{\mathbb{R}}^S$, <u>where</u>

$$h_\theta(t) = \inf_{s\in\theta(t)} h(s), \quad h^\theta(t) = \sup_{s\in\theta(t)} h(s).$$

<u>Moreover if</u> $\theta\lambda$ <u>is increasing and</u> λ <u>is</u> ξ-<u>supersmooth at</u> U <u>along</u> V, <u>where</u> ξ <u>is an infinite cardinal</u>, $U \subseteq F$, $V \subseteq F$ <u>and</u> V <u>is</u> $(\wedge f)$-<u>stable. If</u> L <u>and</u> M <u>are subsets of</u> $\theta(F)$, <u>satisfying</u>

(3.16.7) $M \subseteq \theta(V)$ <u>and</u> M <u>is</u> $(\wedge f)$-<u>stable</u>

(3.16.8) $\theta^{-1}(L) \cap V_\xi \subseteq U$

<u>then</u> $\theta\lambda$ <u>is</u> ξ-<u>supersmooth at</u> L <u>along</u> M.

<u>Proof</u> (3.16.1): Let $\hat{g} = (g_q)$ be a map: $S \to E^Q$, such that $g_q \in \theta(\hat{F})$, where $\hat{F}$ is the α-closure of F. Since $F \subseteq E^T$ and E^T is α-stable (since α maps E^Q into E) we have that $\hat{F} \subseteq E^T$. And there exists $f_q \in \hat{F}$ so that $f_q \in g_q\theta$. Then $\hat{f} = (f_q)$ maps T into E^Q and

$$\hat{f}(t) = \hat{g}(s) \qquad \forall (t,s) \in Gr(\theta)$$

Hence we have that $\alpha \circ \hat{f} \in (\alpha \circ \hat{g})\theta$, and $\alpha \circ \hat{f} \in \hat{F}$ by α-stability of $\hat{F}$.

Thus $\alpha \circ \hat{g} \in \theta(\hat{F})$, and so $\theta(\hat{F})$ is an α-stable set containing $\theta(F)$, and so (3.16.1) follows.

(3.16.2): Let $g \in \overline{\theta(A)}$, then there exist $g_n \in \theta(A)$ and $f_n \in A$, such that $\|g_n - g\| \to 0$ and $f_n \in g_n\theta$. Then clearly we have that $\{f_n\}$ converges uniformly to some function $\hat{f}$ on $\theta^{-1}(S)$. Then $\hat{f}$ is bounded and uniformly continuous on $\theta^{-1}(S)$ with respect to the uniformity induced by A (see Proposition 2.9). Hence by [8; Corollary 8] we know that $\hat{f}$ admits a uniformly continuous extension f to all of T, and by Proposition 2.9 we have that $f \in \bar{A}$. Moreover if $s \in \theta(t)$ then we have that $t \in \theta^{-1}(S)$ and so

$$f(t) = \hat{f}(t) = \lim_{n\to\infty} f_n(t) = \lim_{n\to\infty} g_n(s) = g(s).$$

Thus $f \in g\theta$ and so $g \in \theta(\bar{A})$.

(3.16.3): Let $g \in \overline{\theta(F)}$, and let $g_n \in \theta(F)$ and $f_n \in F$ so that $\|g_n - g\| \to 0$ and $f_n \in g_n\theta$. Then $\{f_n\}$ converges uniformly on $\theta^{-1}(S)$, and since $\theta^{-1}(S)$ is dense in $(T, \tau(F))$ we have that $f_n \rightrightarrows f$ for some $f \in \bar{F}$, and as above we find that $f \in g\theta$, and so $g \in \theta(\bar{F})$.

(3.16.4): Let $B = B(S) \cap \theta(B(T,\mathcal{T}))$ where $\mathcal{T} = \sigma(H)$ and $H \subseteq \bar{\mathbb{R}}^T$. By (3.16.1) we have that B is an algebra containing 1_S, since $1_T \in 1_S\theta$. Now let $g_n \in B$ so that $g_n(s) \to g(s)$ for all $s \in S$ and $\|g_n\| \leq a$ for all $n \geq 1$, for some $a \in \mathbb{R}_+$.

Then there exist $f_n \in B(T,\mathcal{T}) \cap (g_n\theta)$, such that $\|f_n\| \leq a$ for all $n \geq 1$. Now let

$$f(t) = \begin{cases} \lim_{n\to\infty} f_n(t) & \text{if the limit exist} \\ 0 & \text{otherwise} \end{cases}$$

Then $f \in B(T,\mathcal{T})$, and if $t \in T$ and $s \in \theta(t)$ then

$$f(t) = \lim_{n\to\infty} f_n(t) = \lim_{n\to\infty} g_n(s) = g(s).$$

Thus $f \in g\theta$ and so $g \in B$.

Hence B is closed under dominated, pointwise, sequential convergence, and so by [2; Theorem I.21] we have that $B = B(S,\sigma(B))$, and since $\sigma(\theta(H)) \subseteq B$ by (3.16.1), we see that (3.16.4) follows.

(3.16.5): Let $g_1, g_2 \in \theta(F)$ so that $g_1 \leq g_2$, and choose $f_j \in F \cap (g_j\theta)$ for $j = 1,2$. Then $f = f_1 \wedge f_2 \in F$ and $f \in F \cap (g_1\theta)$. Hence

$$\theta\lambda(g_1) = \lambda(f_1) \leq \lambda(f_2) = \theta\lambda(g_2)$$

and so $\theta\lambda$ is increasing.

(3.16.6): Let $g \in \theta(F)$ so that $g \leq h$. Then there exist $f \in F \cap (g\theta)$, and so

$$f(t) = g(s) \leq h(s) \quad \forall s \in \theta(t) \ \forall t \in T$$

Hence $f \leq h_\theta$ and so $\theta\lambda(g) = \lambda(f) \leq \lambda_*(h_\theta)$. Thus the first part of (3.16.5) follows and the second is proved similarly.

Now let us prove the last part of the proposition. So let $\{g_j \mid j \in J\}$ be a decreasing net in M, such that $\operatorname{card}(J) \leq \xi$ and $g_j \downarrow g$ for some $g \in L$. Let Γ be the set of all finite non-empty subsets of J, and put

$$g_\gamma = \min_{j\in\gamma} g_j, \quad f_\gamma = \min_{j\in\gamma} f_j$$

where f_j is chosen so that $f_j \in V \cap (g_j\theta)$ for all $j \in J$. Then $g_\gamma \in M \subseteq \theta(F)$ and $f_\gamma \in V \subseteq F$, since M and V are $(\wedge f)$-stable. Moreover $g_\gamma \downarrow g$ and $f_\gamma \downarrow f$ for some $f \in V_\xi$. Clearly we have that $f \in g\theta$, and so

$$f \in \theta^{-1}(L) \cap V_\xi \subseteq U .$$

Thus by assumption we have that

$$\theta\lambda(g) = \lambda(f) = \lim_\gamma \lambda(f_\gamma) = \lim_\gamma \theta\lambda(g_\gamma)$$

since $f_\gamma \in g_\gamma\theta$. Now let σ be an increasing cofinal map from Γ into J, such that $\sigma(\gamma) = j$ if $\gamma = \{j\}$. Then $g_\gamma \geq g_{\sigma(\gamma)}$ for all $\gamma \in \Gamma$ and since σ is cofinal we have

$$\theta\lambda(g) = \lim_\gamma \theta\lambda(g_\gamma) \geq \lim_\gamma \theta\lambda(g_{\sigma(\gamma)}) = \lim_j \theta\lambda(g_j) \geq \theta\lambda(g)$$

since $\theta\lambda$ is increasing. Hence $\theta\lambda$ is ξ-supersmooth at L along M. □

Proposition 3.17. Let λ be an increasing functional on T and let $\{F_j \mid j \in J\}$ be an increasing net in $\bar{\mathbb{R}}^T$ such that $F = \cup_j F_j \subseteq D(\lambda)$. Let Σ be a set of increasing maps: $\mathbb{N} \to J$ and let $f \in D(\lambda)$ such that

(3.17.1) If $\sigma \in \Sigma$ and $f_n \in F_{\sigma(n)}$ $\forall n \geq 1$ such that $f_n \downarrow f$ then $\lambda(f) = \lim_{n\to\infty} \lambda(f_n)$

(3.17.2) If $I \subseteq J$ is countable and $f \in \cup_{i\in I} F_i$, then there exist $\sigma \in \Sigma$ and $h \in F_{\sigma(1)}$, so that $h \geq f$ and $\forall i \in I$ $\exists\, n \in \mathbb{N}$ with $i \leq \sigma(n)$.

Then λ is σ-supersmooth at f along F.

Proof. Let $\{f_n \mid n \in \mathbb{N}\} \subseteq F$, so that $f_n \downarrow f$, then there exist $\gamma(n) \in J$, such that $f_n \in F_{\gamma(n)}$. Then by (3.17.2) there exist $\sigma \in \Sigma$ and $h \in F_{\sigma(1)}$, satisfying

(i) $h \geq f_1$ and $\forall k \in \mathbb{N}$ $\exists\, n \in \mathbb{N}$ so that $\gamma(k) \leq \sigma(n)$

Then we may find integers $\alpha(1) < \alpha(2) < \dots$ so that $\gamma(k) \leq \sigma(\alpha(k))$ for all $k \geq 1$. Now put

$$\hat{f}_n = \begin{cases} h & \text{if } 1 \leq n < \alpha(1) \\ f_k & \text{if } \alpha(k) \leq n < \alpha(k+1) \end{cases}$$

Then $\hat{f}_n \in F_{\sigma(n)}$ since $\gamma(k) \leq \sigma(n)$ if $\alpha(k) \leq n$, and $h \in F_{\sigma(1)}$. Moreover $\hat{f}_n \downarrow f$ since $h \geq f_1$, and so

$$\lambda(f) = \lim_{n\to\infty} \lambda(\hat{f}_n) = \lim_{k\to\infty} \lambda(f_{\alpha(k)}) = \lim_{k\to\infty} \lambda(f_k)$$

by (3.17.1), since $\hat{f}_{\alpha(k)} = f_k$. □

Proposition 3.18. Let λ be an increasing functional on T and let $\{F_j \mid j \in J\}$ be an increasing net of $(\wedge\xi)$-stable subsets of $D(\lambda)$, where ξ is an infinite cardinal. Now suppose that λ is ξ-supersmooth at F_j along F_j for all $j \in J$ and that $f \in D(\lambda)$ satisfies

(3.18.1) $\qquad \lambda(f) = \lim_j \lambda(f_j)$ if $f_j \downarrow f$ and $\exists\, i \in J\colon f_j \in F_j \;\; \forall\, j \geq i$.

Then λ is ξ-supersmooth at f along $F = \bigcup_{j\in J} F_j$

Proof. Let $\{f_q \mid q \in Q\}$ be a decreasing net in F such that $f_q \downarrow f$. Let Δ be the set of all finite non-empty subset of Q ordered by inclusion, and put

$$f_\delta = \min_{q\in\delta} f_q \quad \forall \delta \in \Delta$$

Since $\{F_j\}$ is increasing and F_j is $(\wedge f)$-stable for all j, we have that F is $(\wedge f)$-stable. Thus $f_\delta \in F$ for all $\delta \in \Delta$ and $f_\delta \downarrow f$. Now put

$$Q_j = \{q \in Q \mid f_q \in F_j\}, \quad \Delta_j = \{\delta \in \Delta \mid \delta \subseteq Q_j\}$$

for $j \in J$. Then evidently $Q_j \uparrow Q$ and $\Delta_j \uparrow \Delta$. Now observe that $\{f_\delta \mid \delta \in \Delta_j\}$ is a decreasing net in F_j and $\mathrm{card}(\Delta_j) \leq \xi$. Hence if

$$h_j = \inf_{\delta\in\Delta_j} f_\delta \qquad \forall\, j \geq i$$

where $i \in I$ is chosen so that $Q_i \neq \emptyset$. Then we have

(i) $\qquad h_j \in F_j$ and $\lambda(h_j) = \inf_{\delta\in\Delta_j} \lambda(f_\delta) \qquad \forall\, j \geq i$

since F_j is $(\wedge\xi)$-stable and λ is ξ-supersmooth on F_j along F_j for all $j \in J$.

Moreover since $\Delta_j \uparrow \Delta$, we have that $h_j \downarrow f$ and so by (3.18.1) and (i) we find that

$$\lambda(f) = \inf_{j\in J} \lambda(h_j) = \inf_{j\in J} \inf_{\delta\in\Delta_j} \lambda(f_\delta)$$
$$= \lim_\delta \lambda(f_\delta)$$

Now let σ be an increasing map from Δ into Q so that (see the proof of Proposition 2.8)

(ii) $\qquad \sigma(\delta) = q \quad$ if $\quad \delta = \{q\} \quad$ for some $\; q \in Q$.

Then $\sigma(\delta) \geq q$ for all $q \in \delta$, and so $f_\delta \geq f_{\sigma(\delta)} \geq f$, and since σ is cofinal we find

$$\lambda(f) = \lim_\delta \lambda(f_\delta) \geq \lim_\delta \lambda(f_{\sigma(\delta)}) = \lim_q \lambda(f_q) \geq \lambda(f)$$

and so λ is ξ-supersmooth at f along F. □

<u>Lemma 3.19</u>. <u>Let</u> λ_n <u>be an increasing functional on</u> T_n, <u>let</u> B_n <u>be a</u> $(\downarrow c)$-<u>stable subset of</u> $D(\lambda_n)$, <u>let</u> $R_n: B_{n+1} \to B_n$ <u>be a map, let</u> $\sigma_n: T_n \rightsquigarrow T_{n+1}$ <u>be a correspondance, and let</u> $\varepsilon_n \in \mathbb{R}_+$ <u>for</u> $n \in \mathbb{N}$, <u>such that</u>

(3.19.1) $\qquad \lambda_1$ <u>is</u> σ-<u>supersmooth at</u> B_1 <u>along</u> B_1

(3.19.2) $\qquad R_n(\cdot|t)$ <u>is</u> σ-<u>supersmooth at</u> B_{n+1} <u>along</u> B_{n+1} $\forall t \in T_n$

(3.19.3) $\qquad \lambda_{n+1}(f) \leq \varepsilon_n + \lambda_n(R_n f) \quad \forall f \in B_{n+1}$

(3.19.4) $\qquad R_n(f|t) \leq \sup\{f(u) \mid u \in \sigma_n(t)\} \quad \forall f \in B_{n+1} \quad \forall t \in \sigma_n^{-1}(T_{n+1})$

(3.19.5) $\qquad \sigma_n(T_n) \subseteq \sigma_{n+1}^{-1}(T_{n+2})$

<u>for all</u> $n \geq 1$, <u>where</u> $R_n(f|t) = R_n(f)(t)$ <u>for</u> $f \in B_{n+1}$ <u>and</u> $t \in T_n$.

Now let $\varphi_n \in B_n$, such that $\varphi_n \geq R_n \varphi_{n+1}$ for all $n \geq 1$, and put

$$\varphi(t) = \sup \left\{ \inf_n \varphi_n(t_n) \;\middle|\; \begin{array}{l} (t_n) \in \Pi T_n \ \ \text{so that} \ \ t_1 = t \\ \text{and} \ \ t_{n+1} \in \sigma_n(t_n) \quad \forall n \geq 1 \end{array} \right\}$$

if $t \in \sigma_1^{-1}(T_2)$ and $\varphi(t) = \varphi_1(t)$ if $t \in T_1 \setminus \sigma_1^{-1}(T_2)$. Then we have

(3.19.6) $$\limsup_{n\to\infty} \lambda_n(\varphi_n) \leq (\lambda_1)_*(\varphi) \dotplus \sum_{n=1}^{\infty} \varepsilon_n$$

Remark. If $t \in \sigma_1^{-1}(T_2)$, then by (3.19.5) there exist (t_n) so that $t_1 = t$, $t_{n+1} \in \sigma_n(t_n)$, and so supremum defining $\varphi(t)$ is a proper (i.e. non empty) supremum.

Proof. Let $R_{kn} = R_k \circ \ldots \circ R_{n-1}$ if $1 \leq k \leq n$, and let R_{nn} be the identity map: $B_n \to B_n$. Then R_{kn} is an increasing map from B_n into B_k for all $1 \leq k \leq n$. Now put

$$\psi_k = \inf_{n \geq k} R_{kn} \varphi_n \qquad \forall k \geq 1$$

Since $R_n \varphi_{n+1} \leq \varphi_n$ and R_{kn} is increasing we have

$$R_{kn+1} \varphi_{n+1} = R_{kn}(R_n \varphi_{n+1}) \leq R_{kn} \varphi_n$$

and so $R_{kn} \varphi_n \downarrow \psi_k$ as $n \uparrow \infty$. Hence $\psi_k \in B_k$ by $(\downarrow c)$-stability of B_k and so by (3.19.2) we have

$$R_k \psi_{k+1} = \lim_{n\to\infty} R_k(R_{k+1n}\varphi_n) = \lim_{n\to\infty} R_{kn} \varphi_n = \psi_k$$

and since $\psi_k \leq R_{kk} \varphi_k = \varphi_k$ we have

(i) $$R_k \psi_{k+1} = \psi_k \leq \varphi_k \quad \forall k \geq 1.$$

Now note that by (3.19.3) we have

$$\lambda_{k+1}(R_{k+1n}\varphi_n) \leq \varepsilon_k + \lambda_k(R_{kn}\varphi_n)$$

for all $1 \leq k \leq n-1$, and since $R_{nn} \varphi_n = \varphi_n$, we find

$$\lambda_n(\varphi_n) \le \lambda_1(R_{1n}\varphi_n) + \sum_{j=1}^{n-1} \varepsilon_j$$

and so letting $n \to \infty$, we find

(ii) $$\limsup_{n\to\infty} \lambda_n(\varphi_n) \le \lambda_1(\psi_1) \dotplus \sum_{j=1}^{\infty} \varepsilon_j$$

by (3.19.1), since $R_{1n}\,\varphi_n \downarrow \psi_1$.

Now let $t_1 \in \sigma_1^{-1}(T_2)$ and let $\alpha < \psi_1(t_1)$. Since $\psi_1 = R_1\,\psi_2$ by (i), we have by (3.19.4) that there exist $t_2 \in \sigma_1(t_1)$, such that $\alpha < \psi_2(t_2)$. But then $t_2 \in \sigma_2^{-1}(T_2)$ by (3.19.5), and so by (3.19.4) and (i) there exist $t_3 \in \sigma_2(t_2)$ such that $\alpha < \psi_3(t_3)$. Continuing like this we obtain a sequence (t_n), satisfying

(iii) $$t_{n+1} \in \sigma_n(t_n) \quad \text{and} \quad \alpha < \psi_n(t_n) \ \forall n \ge 1$$

Now since $\psi_n \le \varphi_n$ we see that $\alpha \le \varphi(t_1)$. Now letting $\alpha \uparrow \psi_1(t_1)$, and noting that $\psi_1 \le \varphi_1$, we see that $\psi_1 \le \varphi$, and so (3.19.6) follows from (ii), since $\psi_1 \in B_1 \subseteq D(\lambda_1)$. □

Lemma 3.20. *Let λ be an increasing functional on T, such that λ is σ-smooth at F_δ along F, for some $F \subseteq \overline{\mathbb{R}}^T$ with $F_\delta \subseteq D(\lambda)$. Let ξ be an infinite cardinal and let $g \in D(\lambda)$ satisfying the following condition: If $\varepsilon > 0$ and $\varphi \in F_\delta$, then there exists $T_0 \subseteq T$, there exists an increasing functional ν on $\overline{\mathbb{R}}^T$, and there exist increasing maps $R_j\colon F_j \to H$ for $j \in J$, where J is a directed set $\operatorname{card}(J) \le \xi$ and $F_j \subseteq \overline{\mathbb{R}}^T$ and $H \subseteq \overline{\mathbb{R}}^T$ such that*

(3.20.1) $\{F_j \mid j \in J\}$ *is increasing and* $F_j \uparrow F$

(3.20.2) ν *is* ξ-*supersmooth at* $\overline{\mathbb{R}}^T$ *along* H

(3.20.3) $R_j f \le R_i f \quad \forall i \le j \quad \forall f \in F_i$

(3.20.4) $(R_j f)(t) \le f(t) \quad \forall t \in T_0 \quad \forall f \in F_j \quad \forall j \in J$

(3.20.5) $h \in H_\xi,\ h(t) \le g(t) \quad \forall t \in T_0 \ \Rightarrow\ \nu(h) \le \varepsilon + \lambda(g)$

(3.20.6) $\lambda(\varphi) \le \varepsilon + \nu(R_j f)$ <u>if</u> $f \in F_j$ <u>and</u> $\lambda_*(f \wedge \varphi) = \lambda(\varphi)$.

<u>Then</u> λ <u>is</u> ξ-<u>supersmooth at</u> g <u>along</u> F

<u>Proof</u>. Let $\{f_q \mid q \in Q\}$ be a net in F, such that $f_q \downarrow g$ and $\mathrm{card}(Q) \le \xi$. Then there exists an increasing map $\alpha\colon \mathbb{N} \to Q$, such that

(i) $$\lim_{n\to\infty} \lambda(f_{\alpha(n)}) = \inf_q \lambda(f_q)$$

Now let $\varepsilon > 0$ be given and put $\varphi_n = f_{\alpha(n)}$. Then $\varphi_n \in F$ and $\varphi_n \downarrow \varphi$ for some φ. Hence by σ-supersmoothness of λ we have

(ii) $$\inf_q \lambda(f_q) = \lim_{n\to\infty} \lambda(f_{\alpha(n)}) = \lambda(\varphi)$$

Now we choose T_0, ν, F_j, H and R_j according to the assumption And we put

$$\Gamma = \{(j,q) \in J \times Q \mid f_q \in F_j\}$$

ordered by the usual product ordering. Then it follows easily from (3.20.1) that Γ is a directed set, and clearly we have that $\mathrm{card}(\Gamma) \le \xi$. If $\gamma = (j,q)$ we put $h_\gamma = R_j\, f_q$, then $\{h_\gamma\}$ is a net in H. If $\alpha = (i,q)$ and $\beta = (j,r)$ belongs to Γ and $\alpha \le \beta$ then by increasingness of R_j and (3.20.3) we have

$$h_\beta = R_j\, f_r \le R_j\, f_q \le R_i\, f_q = h_\alpha$$

since $i \le j$ and $f_r \le f_q$. Hence $\{h_\gamma \mid \gamma \in \Gamma\}$ is a decreasing net in H and so $h_\gamma \downarrow h$ for some $h \in H_\xi$. Since $R_j\, f_q \le f_q$ on T_0 for $(j,q) \in \Gamma$ we see that $h(t) \le g(t)$ for all $t \in T_0$, since $f_q \downarrow g$. Hence by (3.20.5) we have

(iii) $$\nu(h) \le \varepsilon + \lambda(g)$$

If $\lambda(g) = \infty$, then the conclusion is clear. So suppose that $\lambda(g) < \infty$, and let $a \in \mathbb{R}$ so that $a > \lambda(g)$. Then $\nu(h) < \varepsilon + a$, and so by (3.20.2), there exists $j \in J$ and $q \in Q$ so that

(iv) $$f_q \in F_j \quad \text{and} \quad \nu(R_j f_q) < \varepsilon + a.$$

Since Q is directed there exist an increasing map $\beta: \mathbb{N} \to Q$ such that $\beta(n) \geq q$ and $\beta(n) \geq \alpha(n)$ for all $n \in \mathbb{N}$. Then $f_{\beta(n)} \leq f_{\alpha(n)}$, and $f_{\beta(n)} \downarrow \psi$ for some $\psi \in F_\delta$. Hence by σ-supersmoothness of λ and (ii) we find

$$\lambda(\psi) = \lim_{n\to\infty} \lambda(f_{\beta(n)}) \leq \lim_{n\to\infty} \lambda(f_{\alpha(n)})$$
$$= \lambda(\varphi) = \inf_q \lambda(f_q) \leq \lim_{n\to\infty} \lambda(f_{\beta(n)}) = \lambda(\psi).$$

And so $\lambda(\psi) = \lambda(\varphi)$. Now note that $\psi \leq f_q \wedge \varphi$, since $f_{\beta(n)} \leq f_q \wedge f_{\alpha(n)}$, hence we have

$$\lambda(\varphi) = \lambda(\psi) \leq \lambda_*(f_q \wedge \varphi) \leq \lambda(\varphi)$$

Hence by (3.20.6) and (iv) we have

$$\lambda(\varphi) \leq \varepsilon + \nu(R_j f_q) \leq 2\varepsilon + a$$

So by (i) we find

$$\lambda(g) \leq \lim_q \lambda(f_q) \leq 2\varepsilon + a$$

for all $\varepsilon > 0$ and all $a > \lambda(g)$. Thus $\lambda(g) = \lim \lambda(f_q)$ and so λ is ξ-supersmooth at g along F. □

4. Probability contents.

Let (T,A) be an algebraic function space, i.e. T is a set and A is an algebra of real valued functions on T, such that $1_T \in A \subseteq B(T)$. Then a probability content on A is a positive linear map $\lambda: A \to \mathbb{R}$, such that $\lambda(1_T) = 1$, and we put

$$Pr(A) = \{\lambda \mid \lambda \text{ is a probability content on } A\}.$$

We shall in this section study integral representations of probability contents. If $\lambda \in Pr(A)$, then a representing measure for λ is a probability measure on T such that $A \subseteq L^1(\mu)$ and

$$\lambda(f) = \int_S f d\mu \quad \forall f \in A.$$

In order to obtain good integral representation, we shall need some smoothness or tightness of our probability content.

Let (T,A) be an algebraic function space, and let λ be a probability content on A. If ξ is an infinite cardinal and K is a paving on T, we say that λ is ξ-smooth, resp. K-tight, if λ is ξ-supersmooth at 0 along A, resp. if λ is K-tight along A. If λ is $K_\xi(A)$-smooth we say that λ is ξ-compact. We use the terminology: σ-smooth and semicompact if $\xi = \aleph_0$, and we use the terminology: τ-smooth and compact, if λ is ξ-smooth resp. ξ-compact for all cardinals ξ, or equivalently if λ is ξ-smooth resp. ξ-compact for some $\xi \geq \text{weight}(A)$ (see Propositions 2.6 and 2.9). And we put

$$Pr_\xi(A) = \{\lambda \in Pr(A) \mid \lambda \text{ is } \xi\text{-smooth}\}$$

$$Pr_{c\xi}(A) = \{\lambda \in Pr(A) \mid \lambda \text{ is } \xi\text{-compact}\}$$

$$Pr_\sigma(A) = Pr_\xi(A), \quad Pr_{c\sigma} = P_{c\xi}(A) \quad \text{if} \quad \xi = \aleph_0$$

$$Pr_\tau(A) = Pr_\xi(A), \quad Pr_{c\tau} = P_{c\xi}(A) \quad \text{if} \quad \xi \geq \text{weight}(A)$$

Finally we say that $\lambda \in Pr(A)$ is <u>perfect</u> if and only if the restriction of λ to any countably generated subalgebra af A is semicompact And we put

$$Pr_\pi(A) = \{\lambda \in Pr(A) \mid \lambda \text{ is perfect.}\}$$

If $\xi \geq \eta$ are infinite cardinal then clearly we have

(4.1) $$Pr_\tau(A) \subseteq Pr_\xi(A) \subseteq Pr_\eta(A) \subseteq Pr_\sigma(A)$$

(4.2) $$Pr_{c\tau}(A) \subseteq Pr_{c\xi}(A) \subseteq Pr_{c\eta}(A) \subseteq Pr_{c\sigma}(A).$$

And from Propositions 2.10 and 3.13 we find

(4.3) $$Pr_{c\xi}(A) \subseteq Pr_\xi(A) \quad \text{and} \quad Pr_\pi(A) \subseteq Pr_\sigma(A).$$

If $\lambda \in Pr(A)$, and K is a paving on T, then by (F.5) , (3.8) and Proposition 3.12 we have

(4.4) $$\lambda \in Pr_{c\,\xi}(A) \iff \lambda \text{ is } \overline{K}_\xi(A)\text{-tight}$$

(4.5) $$\lambda \text{ is } K\text{-tight} \iff \inf\{\lambda_*(T \setminus K) \mid K \in K\} = 0$$

(4.6) $$\lambda \text{ is } K\text{-tight} \iff \sup\{\lambda^*(K) \mid K \in K\} = 1 .$$

Let $\lambda \in Pr(A)$, then evidently we have that λ is $\|\cdot\|$-continuous on A, and so λ admit a unique $\|\cdot\|$-continuous extension, which we denote $\bar{\lambda}$, to $\bar{A}$, and clearly we have that $\bar{\lambda} \in Pr(\bar{A})$. Moreover if ξ is an infinite cardinal and α denotes one of the three symbols: ξ, $c\xi$, or π, then

(4.7) $$\lambda \in Pr_\alpha(A) \iff \bar{\lambda} \in Pr_\alpha(\bar{A}).$$

In the rest of this section we let (T,A) denote a fixed algebraic function space.

<u>Theorem 4.1</u>. <u>Let</u> $\lambda \in Pr_\xi(A)$ <u>for some infinite cardinal</u> ξ. <u>Then</u>

λ admits a unique representing measure μ on $(T,\sigma(A_\xi))$ such that μ is ξ-supersmooth at $F_\xi(A)$ along $F_\xi(A)$. Moreover the representing measure μ satisfies

(4.1.1) $\int f d\mu = \lambda^*(f) = \inf\{\lambda(\varphi) \mid \varphi \in A,\ \varphi \geq f\}$ $\quad \forall f \in A_\xi$

(4.1.2) $\int_* h d\mu = \sup\{\lambda^*(f) \mid f \in A_\xi,\ f \leq h\}$ $\quad \forall h \in \bar{\mathbb{R}}^T$

(4.1.3) $\int g\, d\mu = \lambda_*(g) = \sup\{\lambda(\varphi) \mid \varphi \in A,\ \varphi \leq g\}$ $\quad \forall g \in A^\xi$

(4.1.4) $\int^* h d\mu = \inf\{\lambda_*(g) \mid g \in A^\xi,\ g \geq h\}$ $\quad \forall h \in \bar{\mathbb{R}}^T$

(4.1.5) $\lambda_*(h) \leq \int_* h d\mu \leq \int^* h d\mu \leq \lambda^*(h)$ $\quad \forall h \in \bar{\mathbb{R}}^T$

(4.1.6) $\mu^*(B) = \inf\{\mu(G) \mid G \in G_\xi(A),\ G \supseteq B\}$ $\quad \forall B \subseteq T$

(4.1.7) $\mu_*(B) = \sup\{\mu(F) \mid F \in F_\xi(A),\ F \subseteq B\}$ $\quad \forall B \subseteq T$

(4.1.8) λ^* is ξ-supersmooth at $\bar{\mathbb{R}}^T$ along A_ξ

(4.1.9) λ_* is ξ- subsmooth at $\bar{\mathbb{R}}^T$ along A^ξ

(4.1.10) $\mu^*(K) = \lambda^*(K) = \inf\{\lambda(\varphi) \mid \varphi \in A,\ \varphi \geq 1_K\}$ $\quad \forall K \in K_\xi(A)$

(4.1.11) $\forall f^* \in R^*(I^*,A_\xi)$ $\exists f \in A_\delta$: $f \geq f^*$ and $f = f^*$ μ - a.s.

(4.1.12) $\forall g^* \in R_*(I^*,A^\xi)$ $\exists g \in A_\sigma$: $g \leq g^*$ and $g = g^*$ μ - a.s.

(4.1.13) A is dense in $(L^p(\mu), \|\cdot\|_p)$ for all $0 \leq p < \infty$

(4.1.14) If $h \in \bar{\mathbb{R}}^T$ is μ-measurable, then $h = h_0$ μ - a.s. for some $\sigma(A)$-measurable function $h_0 \in \bar{\mathbb{R}}^T$

where $I^*(h) = \int^* h d\mu$ in (4.1.11) and (4.1.12). Moreover the following four statements are equivalent:

(4.1.15) λ is ξ-compact

(4.1.16) $\forall \varepsilon > 0$ $\exists K \in K_\xi(A)$ so that $\mu^*(K) \geq 1-\varepsilon$

(4.1.17) $\forall \varepsilon > 0 \ \exists K \in \bar{K}_\xi(A)$ <u>so that</u> $\mu_*(T \setminus K) \leq \varepsilon$

(4.1.18) $\mu(B) = \sup\{\mu^*(K) \mid K \in \bar{K}_\xi(A),\ K \subseteq B\} \qquad \forall B \in M(\mu)$

<u>Proof</u>. Since $\lambda^0 = \lambda$ we have that $\lambda_* = (\lambda^*)^0$ and $\lambda^* = (\lambda_*)^0$ by (3.3). And by Propositions 2.7-2.8 and 3.5-3.7, it follows easily that we have

(i) $\bar{A}$ is $(\wedge f, \vee f)$-stable, and $\lambda^* = \lambda_* = \bar{\lambda}$ on $\bar{A}$

(ii) λ^* is subadditive on $\bar{\mathbb{R}}^T$ and modular on A_ξ

(iii) λ^* is ξ-supersmooth at $\bar{\mathbb{R}}^T$ along A_ξ

(iv) ν is supermodular on $\bar{\mathbb{R}}^T$ and σ-supersmooth at $\bar{\mathbb{R}}^T$ along $L = \{h \mid \nu(h) < \infty\}$

where ν denotes the right hand side of (4.1.2).

Now we apply Theorem 3.11 with $\lambda = \lambda$, $V = A$, $U = \bar{A}^+$ and $F = A_\xi^+$. Then (3.11.1)-(3.11.5) follows easily from (i)-(iii), and if C and $\hat{C}$ are defined as in Theorem 3.11, then $C = \hat{C} = F_\xi(A)$ and $\sigma(C) = \sigma(A_\xi)$ by Proposition 2.9.

Thus by Theorem 3.11 there exist a probability measure μ on $(T, \sigma(A_\xi))$ satisfying (3.11.6)-(3.11.14), since $\bar{A}^+$ is a convex cone and λ^* is additive on $\bar{A}^+$ and positively homogenuous on $\bar{\mathbb{R}}^T$.. By (3.11.12) we have that μ is ξ-supersmooth at $F_\xi(A)$ along $F_\xi(A)$, and by (3.11.13) and (3.11.14) we have that (4.1.2) holds for all $h \in \bar{\mathbb{R}}_+^T$ and thus (by adding a constant) for all $h \in B_*(T)$. But then (4.1.1) holds for all $f \in A_\xi \cap B(T)$, and so μ is a representing measure for λ, and we shall now μ has all the properties of the theorem

(4.1.1): Since (4.1.1) holds for all $f \in A_\xi \cap B(T)$ and since λ^* and $\int d\mu$ are σ-supersmooth at A_ξ along A_ξ we see that (4.1.1) holds.

(4.1.2): We know that (4.1.2) holds for all $h \in B_*(T)$, and since ν and $\int_* d\mu$ are σ-supersmooth on the sets where they are $< \infty$ we have that (4.1.2) holds for all $h \in \bar{\mathbb{R}}^T$ with $\int_* h^+ d\mu < \infty$. But then it follows easily that (4.1.2) holds in general.

(4.1.3-9) follows easily from (4.1.1), (4.1.2) and (iii)

(4.1.10): We shall apply Proposition 3.10 with $\mu = \mu_*$, $\nu = \mu^*$, $F = F_\xi(A)$, $\varphi = 1_K$ and $f = 1_T$, then (3.10.1) and (3.10.2) holds (see (3.8) and note that $\mu_F = \mu_*$). So let us verify condition (3.10.4). Hence let $F_1 \in F_\xi(A)$, so that $F_1 \cap K = \emptyset$. Then by (F.6) there exist $Q \subseteq A$ and $\hat{F}_1 \subseteq \bar{\mathbb{R}}^Q$, so that $F_1 = p_Q^{-1}(\hat{F}_1)$ and $\text{card}(Q) \leq \xi$. Then $p_Q(K)$ is compact and thus closed in $\bar{\mathbb{R}}^Q$ by Proposition 2.10, and $\hat{F}_1 \cap p_Q(K) = \emptyset$. Hence $F_0 = p_Q^{-1}(p_Q(K))$ belongs to $F_\xi(A)$ by (F.6), and $F_0 \supseteq K$ and $F_0 \cap F_1 = \emptyset$. Thus (3.10.4) holds and so by Proposition 3.10 we have

$$\mu^*(K) \geq \inf\{\mu_*(F) \mid F \in F_\xi(A),\ F \supseteq K\}$$

so by (4.1.1) and (4.1.5) we have that $\mu^*(K) = \lambda^*(K)$.

(4.1.11): Since A_ξ is $(\wedge c)$-stable there exist $f_0 \in A_\xi$ so that $f_0 \geq f^*$ and $\int^* (f_0 - f^*) d\mu = 0$. I.e. $f_0 = f^*$ μ-a.s. Moreover by (4.1.1) and (4.1.8) there exist $f \in A_\delta$ so that $f \geq f_0 \geq f^*$ and $f = f_0$ μ-a.s. Thus (4.1.11) follows

(4.1.12): Follows from (4.1.11)

(4.1.13): By (4.1.1+2) and (4.1.8+9) we have that A is dense in $(L^1(\mu), \|\cdot\|_1)$, and so A is dense in $(L^p(\mu), \|\cdot\|_p)$ for $0 \leq p \leq 1$. Now let $1 < p < \infty$, and let $h \in L^p(\mu)$, and let $\varepsilon > 0$ be given. Let $h_0 \in B(T) \cap L^1(\mu)$ so that $\|h - h_0\|_p < \varepsilon/4$. Let $a = 1 + \|h_0\|$, and choose $\varphi_0 \in A$ such that

$$\|\varphi_0 - h_0\|_1 \leq 2^{1-2p} a^{1-p} \varepsilon^p.$$

Put $\psi = (\varphi_0 \vee a) \wedge (-a)$ and choose $\varphi \in A$, such that $\|\varphi - \psi\| \leq \varepsilon/4$ (see Proposition 2.7). Then we have

$$\|h-\varphi\|_p \le \|h-h_0\|_p + \|h_0-\psi\|_p + \|\psi-\varphi\|_p$$
$$\le \tfrac{1}{2}\varepsilon + \|h_0-\varphi_0\|_p$$
$$\le \tfrac{1}{2}\varepsilon + \{(2a)^{p-1}\|\varphi_0-h_0\|_1\}^{1/p}$$
$$\le \varepsilon$$

since $|h_0-\psi|^p \le |h_0-\varphi_0|^p \le (2a)^{p-1}|h_0-\varphi_0|$. Hence A is dense in $(L^p(\mu), \|\cdot\|_p)$ for all $0 \le p < \infty$.

(4.1.14): follows easily from (4.1.13).

Now let us show that (4.1.15)-(4.1.18) are equivalent:

(4.1.15) ⇒ (4.1.16): Follows from (4.1.10) and (4.6)

(4.1.16) ⇒ (4.1.17): Follows from (F.5)

(4.1.17) ⇒ (4.1.18): Follows from Proposition 3.12 with $\mu = \nu = \mu^*$, $F = 2^T$, $G = F_\xi(A)$ and $a = b = 1$.

(4.1.18) ⇒ (4.1.15): Follows from (4.4) and (4.6).

Finally let us prove the uniqueness of μ. So suppose, that ν is a representing measure for λ on $(T,\sigma(A_\xi))$ such that ν is ξ-supersmooth at $F_\xi(A)$ along $F_\xi(A)$. Then by Proposition 3.14 (with $F = F_\xi(A)$, $A = T$, and $f = 1$) we have that $\int d\nu$ is ξ-supersmooth at A_ξ along A_ξ, and so we have

$$\text{(v)} \qquad \int f d\nu = \lambda^*(f) \qquad \forall f \in A_\xi$$

since λ^* is ξ-smooth at A_ξ along A_ξ and $\int d\nu$ and λ^* coincides on A. Now let

$$R = \{B \in \sigma(A_\xi) \mid \sup\{\nu(F) \mid F \in F_\xi(A),\ F \subseteq B\} = \nu(B)\}$$

$$B = \{B \subseteq T \mid B \in R,\ T \setminus T \in R\}$$

Since $F_\xi(A)$ is $(\cap c, \cup f)$-stable if follows easily that R is $(\cap c, \cup c)$-

stable and thus $\mathcal{B}$ is a σ-algebra. Now let $F \in F_\xi(A)$ and $\varepsilon > 0$, then evidently $F \in \mathcal{R}$, and by Propositions 2.8 and 2.9 and ξ-supersmoothness of ν, there exist $\varphi \in A$ such that

$$\varphi \geq 1_F \quad \text{and} \quad \int \varphi d \leq \varepsilon + \nu(F).$$

Now let $H = \{\varphi \leq 1 - \varepsilon\}$, then $H \in F_\xi(A)$ and $H \subseteq T \setminus F$. Moreover since $1 - \varphi \leq 1_H + \varepsilon$ we have

$$\nu(T \setminus F) = 1 - \nu(F) \leq \varepsilon + \int (1-\varphi) d\nu \leq 2\varepsilon + \nu(H)$$

and so $T \setminus F \in \mathcal{R}$. Thus the σ-algebra $\mathcal{B}$ contains $F_\xi(A)$ and so $\mathcal{B} = \sigma(A_\xi)$. But then (v), (4.1.1) and (4.1.7) shows that $\nu = \mu$. Thus μ is unique and the theorem is proved. □

<u>Definition 4.2</u>. Let $\lambda \in \Pr_\xi(A)$ for some infinite cardinal number ξ, then <u>the unique representing probability</u> on $(T, \sigma(A_\xi))$ satisfying (4.1.1)-(4.1.14) is denoted λ_ξ, and as usual we put $\lambda_\sigma = \lambda_\xi$ if $\xi = \aleph_0$, and $\lambda_\tau = \lambda_\xi$ if $\xi \geq$ weight(A).

Note that if $\lambda \in \Pr(A)$ admits a representing measure μ, such that μ is ξ-supersmooth at $\emptyset$ along $F_\sigma(A)$ then $\lambda \in \Pr_\xi(A)$. However, μ need not be equal to λ_ξ on $M(\mu) \cap \sigma(A_\xi)$.

By Theorem 4.1 we have

(4.8) If $\lambda \in \Pr_\xi(A)$, then $\lambda_\eta = \lambda_\xi$ on $\sigma(A_\eta)$ $\forall\, \aleph_0 \leq \eta \leq \xi$

(4.9) If $\lambda \in \Pr_\tau(A)$, then λ_τ is a regular τ-smooth Borel probability on $(T, \tau(A))$

(4.10) λ is compact, if and only if $\lambda \in \Pr_\tau(A)$ and λ_τ is a Radon probability on $(T, \tau(A))$.

Note that by (4.1.14) we have that λ_ξ is a socalled <u>null-extension</u> of λ_σ, but λ_ξ need not coincide with the Lebesque extension of λ_σ.

In our next theorem we elaborate on (4.10).

Theorem 4.3. Let θ be a topology on T, and let $\mathcal{K} \subseteq \mathcal{K}(T,\theta)$ be a paving satisfying

(4.3.1) $K \cap F$ is A-saturated $\forall K \in \mathcal{K}$ $\forall F \in \mathcal{F}(T,\theta)$

(4.3.2) $\varphi | K$ is θ-continuous $\forall K \in \mathcal{K}$ $\forall \varphi \in A$.

Then $K \cap F \in \overline{\mathcal{K}}(A)$ for all $K \in \mathcal{K}$ and all $F \in \mathcal{F}(T,\theta)$. And if $\lambda \in \Pr(A)$ is $\mathcal{K}$-tight, then λ is compact, and there exist a topology $\hat{\theta}$ on T such that

(4.3.3) $\hat{\theta}$ is stronger than both θ and $\tau(A)$

(4.3.4) $\mathcal{B}(T,\hat{\theta}) \subseteq \mathcal{M}(\lambda_\tau)$

(4.3.5) λ_τ is a Radon probability on $(T,\hat{\theta})$.

Remark. The topology $\hat{\theta}$ constructed below, has the property that $(T,\hat{\theta})$ is a normal k-space, and

$$T = D \cup \bigcup_{n=1}^{\infty} C_n$$

where D is discrete and $\hat{\theta}$-clopen in T, D is a λ_τ-nullset, $\{C_n\}$ is an increasing sequence in $\mathcal{K}(T,\hat{\theta}) \cap \mathcal{F}(T,\theta) \cap \mathcal{F}(A)$, such that $C_n \cap D = \emptyset$ $\forall n \geq 1$, and every $\hat{\theta}$-compact set K is contained in $C_n \cup \alpha$ for some $n \geq 1$ and some finite set $\alpha \subseteq D$.

Proof. By (4.3.2) we have that $\theta | K$ is stronger than $\tau(A) | K$ for all $K \in \mathcal{K}$. Hence $K \cap F \in \mathcal{K}(A)$ for all $K \in \mathcal{K}$ and all $F \in \mathcal{F}(T,\theta)$, and so by Proposition 2.10 and (4.3.1) we have that

(i) $F \cap K \in \overline{\mathcal{K}}(A) \subseteq \mathcal{F}(A)$ $\forall F \in \mathcal{F}(T,\theta)$ $\forall K \in \mathcal{K}$

Let $\lambda \in \Pr(A)$ be $\mathcal{K}$-tight. Since $\mathcal{K} \subseteq \overline{\mathcal{K}}(A)$ by (i) we have that λ is compact. Now we choose an increasing sequence $\{C_n\}$ in $\mathcal{K}$ such that

(ii) $\lambda_\tau(T \setminus C_n) \leq 2^{-n} \quad \forall n \geq 1$

(see (4.5) and (4.1.10) and note that $K \subseteq B(A)$).

Then $F = \{F \subseteq T \mid F \cap C_n \in F(A) \quad \forall n \geq 1\}$ is a $(\cap a, \cup f)$-stable paving on T containing $\emptyset$ and T. Hence there exist a unique topology $\hat{\theta}$, such that $F = F(T, \hat{\theta})$, i.e.

(iii) F is $\hat{\theta}$-closed $\Leftrightarrow F \cap C_n \in F(A) \quad \forall n \geq 1$.

Hence by (i) we have that $\hat{\theta}$, θ and $\tau(A)$ all coincides on C_n for $n \geq 1$, and we see that (4.3.3) holds. By (ii) and (iii) we see that $F(T, \hat{\theta}) \subseteq M(\lambda_\tau)$ and so (4.3.4) follows. Moreover by (4.1.7) and Proposition 3.12 we have that

$$C = \{C_n \cap F \mid n \geq 1, \quad F \in F(A)\}$$

is an inner approximating paving for μ, and since $C \subseteq \bar{K}(T, \hat{\theta})$, we see that (4.3.5) holds. □

<u>Theorem 4.4</u>. <u>Let</u> (T, B, μ) <u>be a probability space, and let</u> $T_0 \subseteq T$ <u>and</u> $U \subseteq \mathbb{R}^T$ <u>satisfy the following two conditions</u>

(4.4.1) $\mu^*(T_0) = 1$ and f is B-measurable $\forall f \in U$

(4.4.2) If $f, g \in U$ and $f = g$ μ-a.s., then $f(t) = g(t) \quad \forall t \in T_0$

<u>Now let</u> π <u>be the topology on</u> U <u>of pointwise convergence on</u> T_0, <u>and put</u>

$$B_0 = \{B_0 \subseteq T \mid \exists B \in B \text{ so that } B_0 \cap T_0 = B \cap T_0\}$$

$$\mu_0(B_0) = \mu^*(B_0) \quad \forall B_0 \in B_0$$

$$\rho_K(t', t'') = \sup_{f \in K} |f(t') - f(t'')| \quad \forall t', t'' \in T \quad \forall K \subseteq \mathbb{R}^T.$$

<u>Then</u> ρ_K <u>is a pseudo-metric on</u> T, <u>and</u> (T, B_0, μ_0) <u>is a probability space, such that</u> $B_0 \supseteq B$ <u>and</u> $\mu_0 = \mu$ <u>on</u> B.

Now let $K \subseteq U$, and let $\hat{K} = cl_\pi K$ be the closure of K in (U,π). Then $(\hat{K},\pi)$ is compact and pseudo-metrizable, in either of the following three cases

Case 1°: K is relatively sequentially compact in (U,π)

Case 2°: K is convex and relatively countably compact in (U,π)

Case 3°: μ_0 is a perfect measure and K is relatively countably countably compact in (U,π)

Moreover if $K \subseteq U$ is compact and pseudo-metrizable then there exist a pseudo-metric ρ on T such that

(4.4.3) $\rho \leq \rho_K$, and ρ is $\mathcal{B} \otimes \mathcal{B}$-measurable

(4.4.4) $\rho(t',t") = \rho_K(t',t") < \infty \quad \forall t',t" \in T_0$

(4.4.5) T_0 is ρ_K-separable, and $\mathcal{B}(T,\rho_K) \subseteq \mathcal{B}_0$

(4.4.6) μ_0 is a τ-smooth Borel probability on (T,ρ_K).

Remarks:

(1): Note that (U,π) is Hausdorff, if and only if $f = g$ whenever $f,g \in U$ and $f = g$ μ-a.s.

(2): Note that the $\|\cdot\|_0$-"norm" on $L^0(\mu)$, is a pseudo-metric on U, and we shall actually show that the $\|\cdot\|_0$-topology on $\hat{K}$ equals the π-topology on $\hat{K}$ in case 1^0, 2^0 and 3^0.

(3): If K is relatively compact in (U,π) and if every countable subset of K is stable in the sense of [23; p.97-98], then it follows easily from step 1^0 in the proof below and [23; Theorem 9-5-2, p.110] that K is relatively sequentially compact in (U,π) and so $(\hat{K},\pi)$ is compact and pseudo-metrizable

(4): Suppose that $T_0 = T$ and that K is compact in (U,π). Then in [23; Theorem 12-4-2, p.147-151] it is shown that (K,π) is metrizable if one assumes either of the following two axioms:

Axiom M: The union of less than $2^{\aleph_0}$ Lebesgue-null sets in [0,1] is a Lebesgue-nullset.

Axiom L_1: The union of $\aleph_1$ Lebesgue-nullsets in [0,1] is a Lebesgue nullset.

Both axioms are known to be consistent with the usual axioms of set theory. Note that the continuum hypothesis ($\aleph_1 = 2^{\aleph_0}$) implies Axiom M but contradicts Axiom L_1. Also Martin's axiom implies Axiom M by [6 ; Theorem 5.3, p.36]

(5): Note that if T_0 is μ-measurable, then $\mathcal{B}_0 \subseteq M(\mu)$, and μ_0 equals the Lebesgue extension of μ on $\mathcal{B}_0$. Thus in this case we have that μ itself is a τ-smooth Borel probability on (T,ρ_K) whenever (K,π) is compact and pseudo-metrizable.

(6): If T is a topological space and μ is a τ-smooth Borel probability on $(T,\mathcal{B}(T))$, then $U = C(T)$ and $T_0 = \mathrm{supp}(\mu)$ satisfies (4.4.1) and (4.4.2), and T_0 is μ-measurable (actually closed).

<u>Proof</u>. Let $K \subseteq U$ be a given set which is relatively countably compact in (U,π), let $\hat{K}$ denote the closure of K in (U,π), and let θ denote the $\|\cdot\|_0$-topology on $L^0(\mu)$.

Let Λ denote the set of all non-empty finite subsets of T_0, and put

$$V_\varepsilon^\lambda(f) = \{g \in U \mid |f(t) - g(t)| \leq \varepsilon \quad \forall t \in \lambda\}$$

for all $f \in U$, $\lambda \in \Lambda$ and $\varepsilon > 0$. Then $\{V_\varepsilon^\lambda(f) \mid \lambda \in \Lambda, \varepsilon > 0\}$ is a neighbourhood base at f in (U,π).

We shall divide the proof into five steps:

<u>Step 1°</u>. Let $\{f_n\}$ be a sequence in K, such that $f_n \to f$ in $(L^0(\mu),\theta)$ for some $f \in L^0(\mu)$. Then there exist $h \in \hat{K}$, such that $f = h$ μ-a.s., $f_n \to h$ in (U,π) and $f_n \to h$ μ-a.s.

<u>Proof</u>. Let $\{f_{\alpha(n)}\}$ be a subsequence of $\{f_n\}$, such that $f_{\alpha(n)}(t) \to f(t)$ for all $t \in T \setminus N$, where $N \in \mathcal{B}$ is a μ-nullset. Let $h \in U$ be a limit point of $\{f_{\alpha(n)}\}$ in (U,π), then $h \in \hat{K}$ and $h(t) = f(t)$ for all $t \in T_0 \setminus N$. Since $\{h = f\}$ is μ-measurable and $\mu^*(T_0 \setminus N) = 1$ by (4.4.1), we have that $h = f$ μ-a.s.

Let $t_0 \in T_0$ and suppose that $f_n(t_0) \not\to h(t_0)$, then there exist a subsequence $\{f_{\beta(n)}\}$ of $\{f_n\}$ and $\varepsilon > 0$, such that $f_{\beta(n)} \to h$ μ-a.s., and

$$|f_{\beta(n)}(t_0) - h(t_0)| \geq \varepsilon.$$

As above there exist $h_0 \in U$, such that h_0 is a limit point of $\{f_{\beta(n)}\}$ in (U,π) and $h = h_0$ μ-a.s. But then $|h_0(t_0) - h(t_0)| \geq \varepsilon$ which contradicts (4.4.2), and so $f_n \to h$ in (U,π). Now since $\{f_n \to h\}$ is μ-measurable by (4.4.1) and this set contains T_0, we see that $f_n \to h$ μ-a.s.

<u>Step 2°</u>. Let $\{f_n\}$ be a sequence in K and let $f \in U$, then the following three statements are equivalent

(a) $f_n \to f$ μ-a.s.

(b) $f_n \to f$ in $(L^0(\mu),\theta)$

(c) $f_n \to f$ in (U,π)

<u>Proof</u>. (a) $\Rightarrow$ (b): Evident

(b) $\Rightarrow$ (c): By step 1° there exist $h \in U$, such that $h = f$ μ-a.s. and $f_n \to h$ in (U,π). But then $h(t) = f(t)$ for all $t \in T_0$ by (4.4.2) and so $f_n \to f$ in (U,π)

(c) $\Rightarrow$ (a): Follows from (4.4.1)

Step 3°. Let $K_0 = U \cap \mathrm{cl}_\theta(K)$, where cl_θ denotes the closure in $(L^0(\mu),\theta)$. Then $K_0 \subseteq \hat{K}$, and the identity map: $(K_0,\theta) \to (U,\pi)$ is continuous. Moreover the identity map: $(U,\pi) \to (U,\theta)$ is sequentially continuous.

Proof. Since θ is pseudo-metrizable, it follows from step 2° that $K_0 \subseteq \hat{K}$. Now let $f_0 \in K_0$, $\lambda \in \Lambda$ and $\varepsilon > 0$ be given. Then by step 2° and pseudo-metrizablity of θ there exist an open neighbourhood V of f_0 in $(L^0(\mu),\theta)$, such that $V \cap K \subseteq V_\varepsilon^\lambda(f_0)$. Since V is open in $(L^0(\mu),\theta)$, we have

$$V \cap K_0 = V \cap \mathrm{cl}_\theta(K) \cap U \subseteq U \cap \mathrm{cl}_\theta(V \cap K)$$

$$\subseteq U \cap \mathrm{cl}_\theta(K \cap V_\varepsilon^\lambda(f_0)).$$

If $f \in U \cap \mathrm{cl}_\theta(K \cap V_\varepsilon^\lambda(f_0))$, then there exist $f_n \in K \cap V_\varepsilon^\lambda(f_0)$ such that $f_n \to f$ in $(L^0(\mu),\theta)$. But then $f_n \to f$ in (U,π) by step 2°, and so $f \in V_\varepsilon^\lambda(f_0)$, since $V_\varepsilon^\lambda(f_0)$ in π-closed. Hence $V \cap K_0 \subseteq V_\varepsilon^\lambda(f_0)$, and thus the identity map: $(K_0,\theta) \to (U,\pi)$ is continuous.

If $f \in U$ and $\{f_n\}$ is a sequence in U, so that $f_n \to f$ in (U,π). Then $f_n \to f$ μ-a.s. by (4.4.1) and so $f_n \to f$ in (U,θ). Thus the identity map: $(U,\pi) \to (U,\theta)$ is sequentislly continuous.

Step 4°. If K is relatively sequentially compact in (U,π), then $K_0 = \hat{K}$ and $(\hat{K},\pi)$ is compact, and π and θ coincides on $\hat{K}$.

Proof. Since the identity map: $(U,\pi) \to (U,\theta)$ is sequentially continuous we have that K is relatively sequentially compact in (U,θ), and since θ is pseudometrizable we have that $K_0 = U \cap \mathrm{cl}_\theta(K)$ is compact in the θ-topology. Since the identity map: $(K_0,\theta) \to (U,\pi)$ is continuous by step 3° we have that K_0 is compact in the π-topology.

Now let $f \in \hat{K}$, then $V_\varepsilon^\lambda(f) \cap K_0 \neq \emptyset$, and $\{V_\varepsilon^\lambda(f)\}$ are π-closed and filters downwards. Hence by π-compactness of K_0, there exist $h \in K_0$, such that $h \in V_\varepsilon^\lambda(f)$ for all $\lambda \in \Lambda$ and all $\varepsilon > 0$. But then $h(t) = f(t)$ for all $t \in T_0$, and so $h = f$ μ-a.s. Hence f belongs to the θ-closure of K_0 in U, and so $f \in K_0$ since K_0 is closed relatively in (U,θ). Thus $\hat{K} \subseteq K_0$, and the converse inclusion follows from step 3°. Hence $\hat{K} = K_0$ and $(\hat{K},\pi)$ is compact.

By step 3° we know that the identity map from $(\hat{K},\theta)$ into $(\hat{K},\pi)$ is continuous. I.e. the π-topology on $\hat{K}$ is weaker than the θ-topology on $\hat{K}$. Now let F be closed relatively in $(\hat{K},\theta)$. Then F is compact in θ and thus in π. If $f \in cl_\pi(F)$, then $F \cap V_\varepsilon^\lambda(f) \neq \emptyset$ for all $\lambda \in \Lambda$ and all $\varepsilon > 0$, and so as above there exist $h \in F$ such that $h = f$ μ a.s. Since $cl_\pi(F) \subseteq \hat{K}$ we have that $f \in \hat{K}$, and so f belongs to the θ-closure of F in $\hat{K}$, thus $f \in F$, and so F is π-closed. Hence the π-topology and the θ-topology coincides on $\hat{K}$, and step 4° is proved.

<u>Step 5°</u>. If K is convex, then K is relatively sequentially compact in (U,π).

<u>Proof</u>. By Segal's localisation principle, see [4; Theorem IV, 11.6], there exist $g_n \in K$ such that $|f| \leq g$ μ-a.s. $\forall f \in K$, where $g = \sup_n |g_n|$. Now by relatively countably compactness of K we have that $g(t) < \infty$ for all $t \in T_0$, and so g is finite μ-a.s. Hence there exist a $\mathcal{B}$-measurable φ satisfying

(i) $\quad 1 \leq \varphi(t) < \infty \qquad \forall t \in T$

(ii) $\quad |f| \leq \varphi$ μ-a.s. $\qquad \forall f \in K$

Now let $\{f_n\}$ be a sequence in K, and put $h_n = f_n/\varphi$. Then $|h_n| \leq 1$ μ-a.s., and so by [4; Theorem IV.8.9] there exist a subsequence $\{h_{\alpha(n)}\}$ of $\{h_n\}$ and $h \in L^1(\mu)$ such that $h_{\alpha(n)} \to h$ in the weak topology of $L^1(\mu)$. Now let

$$D = \text{co}\{h_{\alpha(n)} \mid n \geq 1\}.$$

Then h belongs to the closure of D in $(L^1(\mu), \|\cdot\|_1)$, since D is convex. Hence there exists $u_n \in D$ such that $u_n \to h$ in $(L^1(\mu), \|\cdot\|_1)$ and μ-a.s. Now note that $\varphi u_n \in K$ since K is convex and $1 \leq \varphi < \infty$ everywhere. Then $\varphi u_n \to \varphi h$ μ-a.s. and so by step 1° there exist $f \in U$, such that $f = \varphi h$ μ-a.s. and $\varphi u_n \to f$ in (U, π).

Now let $t_0 \in T_0$, and suppose that $f_{\alpha(n)}(t_0) \not\to f(t_0)$. Then there exist $\varepsilon > 0$, such that one of the sets

$$P_1 = \{n \in \mathbb{N} \mid f_{\alpha(n)}(t_0) \geq f(t_0) + \varepsilon\}$$

$$P_2 = \{h \in \mathbb{N} \mid f_{\alpha(n)}(t_0) \leq f(t_0) - \varepsilon\}$$

are infinite. Suppose that P_1 is infinite and put

$$D_1 = \text{co}\{h_{\alpha(n)} \mid n \in P_1\}.$$

Then exactly as above there exist $u_n^* \in D_1$ and $f^* \in U$ such that $f^* = \varphi h$ μ-a.s. and $v_n^* = \varphi u_n^* \to f^*$ in (U, π). Now since $1 \leq \varphi < \infty$ we have that

$$v_n^* \in \text{co}\{f_{\alpha(n)} \mid n \in P_1\}$$

and so $v_n^*(t_0) \geq f(t_0) + \varepsilon$ for all $n \geq 1$ by definition of P_1. Hence $f^*(t_0) \geq f(t_0) + \varepsilon$ since $t_0 \in T_0$, and $f^* = f$ μ-a.s. which contradicts (4.4.2). Similarly one shows that we obtain a contradiction if P_2 is infinite. Thus we have that $f_{\alpha(n)} \to f$ in (U, π) and so K is relatively sequentially compact, and step 5° is proved.

Now let us turn to the proof of pseudo-metrizability of $(\hat{K}, \pi)$. Note that case 1° and case 2° follows immediately from step 4° and step 5°.

<u>Case 3°</u>. Let $\{g_n\} \subseteq K$, and put $f_n = 1_{T_0} g_n$. If τ is the pro-

duct topology on $\mathbb{R}^T$, then every subsequence of $\{f_n\}$ has a τ-limit point f of the form $f = g\, 1_{T_0}$ for some $g \in U$. Hence f is $\mathcal{B}_0$-measurable, and so by Fremlin's subsequence theorem [23; Theorem 8-1, p.93], we have that $\{f_n\}$ has a subsequence $\{f_{\alpha(n)}\}$, such that $f_{\alpha(n)} \to f$ μ_0-a.s. for some f. But then $\{g_{\alpha(n)}\}$ is μ-a.s. convergent by (4.4.1), and so $\{g_{\alpha(n)}\}$ converges in (U,π) by step 1°. Thus K is relatively sequentially compact in (U,π) and so $(\hat{K},\pi)$ is pseudo-metrizable and compact by step 4°.

Now let $K \subseteq U$ such that K is compact and pseudometrizable in π. Then (K,π) is separable and so there exist a countable dense set $L \subseteq K$. Now put $\rho = \rho_L$, then clearly (4.4.3) and (4.4.4) holds. Moreover $(C(K), \|\cdot\|)$ is separable and $\varepsilon_t(f) = f(t)$ belongs to $C(K)$ for all $t \in T_0$. Hence there exist a countable set $Q \subseteq T_0$, such that $\{\varepsilon_t \mid t \in Q\}$ is $\|\cdot\|$-dense in $\{\varepsilon_t \mid t \in T_0\}$, and since

$$\|\varepsilon_u - \varepsilon_v\| = \rho_K(u,v) \quad \forall u, v \in T$$

we see that Q is dense in (T_0,ρ_K). Now let $G \in \mathcal{G}(T,\rho_K)$, then there exist $\{t_j \mid j \in J\} \subseteq T$ and $\{r_j \mid j \in J\} \subseteq \mathbb{R}_+$ such that

$$G = \bigcup_{j \in J} b_K(t_j, r_j)$$

where $b_K(t,r) = \{u \mid \rho_K(t,u) < r\}$ if $t \in T$, $r \in \mathbb{R}_+$ and $K \subseteq \mathbb{R}^T$. Moreover since T_0 is separable there exist a countable set $I \subseteq J$ so that

$$G \cap T_0 = \bigcup_{j \in I} T_0 \cap b_K(t_j, r_j) = \bigcup_{j \in I} T_0 \cap b_L(t_j, r_j)$$

by (4.4.4). Now put

$$G_0 = \bigcup_{j \in I} b_L(t_j, r_j).$$

Then $G_0 \in \mathcal{B}$ by (4.4.3) and countability of I, and since $G \cap T_0 = G_0 \cap T_0$ we see that $G \in \mathcal{B}_0$. Since this holds for all $G \in \mathcal{G}(T,\rho_K)$, we see that (4.4.5) holds, and (4.4.6) follows easily from (4.4.5). □

Corollary 4.5. *Let* T *be a pseudo-metrizable locally convex linear space, and let* μ *be a probability measure on* T *satisfying*

$$\sigma(T') \subseteq M(\mu) \quad \text{and} \quad \mu^*(T_0) = 1 \tag{4.5.1}$$

where T' *is the topological dual of* T *and*

$$N = \{t' \in T' \mid t' = 0 \quad \mu\text{-a.s.}\}$$

$$T_0 = \{t \in T \mid t'(t) = 0 \quad \forall t \in N\}.$$

Then T_0 *is separable, and there exist a* τ*-smooth Borel probability* μ_0 *on* $(T, \mathcal{B}(T))$, *so that* μ_0 *is an extension of* μ *and* $\mu_0(T_0) = 1$. *Moreover we have*

(1): *If* T_0 *is* μ*-measurable, then* $\mathcal{B}(T) \subseteq M(\mu)$, *and* μ *is a* τ*-smooth Borel probability on* T.

(2): *If* T_0 *is complete (e.g. if* T *is so), then* $\hat{\mu}$ *is a Radon measure on* T.

(3): *If* $\lambda \in Pr_\tau(A)$, *where* (T,A) *is an algebraic function space, so that* $\{t \in T \mid t'(t) = 0\} \in F(A)$ *and* t' *is* $\mathcal{B}(A)$*-measurable for all* $t' \in T'$, *then* $\mu = \lambda_\tau$ *satisfies (4.5.1) and* $T_0 \in F(A)$. *Hence* $\mathcal{B}(T) \subseteq M(\lambda_\tau)$ *and* λ_τ *is a* τ*-smooth Borel probability on* T.

(4): *Every* τ*-smooth Borel probability on* (T,w), *where* $w = \tau(T')$ *is the weak topology on* T, *is a* τ*-smooth Borel probability on* T.

Proof. Let $\{q_n \mid n \geq 1\}$ be a sequence of seminorms on T, such that $\{q_n\}$ induces the topology on T, and put

$$K_n = \{t' \in T' \mid |t'(t)| \leq 1 \quad \text{if} \quad q_n(t) \leq 1\}.$$

Now note that (4.4.1) holds with $U = T'$, $\mathcal{B} = M(\mu)$ and $T_0 = T_0$, and

$$q_n(u-v) = \rho_{K_n}(u,v) \qquad \forall u, v \in T$$

By Alauglo's theorem, [11; Theorem 20.9.(4)] we have that K_n is convex and π-compact, so by Theorem 4.4 we have that T_0 is q_n-separable for all $n \geq 1$. But then T_0 is separable, and so if we define μ_0 as in Theorem 4.4 we find that μ_0 is an extension of μ, $\mu_0(T_0) = 1$ and μ_0 is a τ-smooth Borel probability on T. Moreover (1)-(4) follows easily from this and the definition of μ_0. □

Definition 4.6. Let S and T be sets, then a kernel on $T|S$ is a map ρ from a subset, usually denoted $D(\rho)$, of $\bar{\mathbb{R}}^T$ into $\bar{\mathbb{R}}^S$, such that $\rho(.|s)$ is an increasing functional for all $s \in S$, where

$$\rho(f|s) = \rho(f)(s) \qquad \forall f \in D(\rho) \quad \forall s \in S$$

If $\rho(\cdot|s)$ ξ-subadditive on $F \subseteq D(\rho)$ for all $s \in S$, we say that is ξ-subadditive on F, and similar for the other properties introduced in section 3. And we put

$$\Pr(A|S) = \{\rho: A \to \mathbb{R}^S \mid \rho(\cdot|s) \in \Pr(A) \quad \forall s \in S\}.$$

And we define $\Pr_\xi(A|S)$, $\Pr_{c\xi}(A|S)$ and $\Pr_\pi(A|S)$ similarly.

If $\mathcal{B}$ is a σ-algebra on T, then a Markov kernel on $(T,\mathcal{B})|S$ is a kernel ρ, such that $D(\rho) = \mathcal{B}$ and $\rho(\cdot|s)$ is a probability measure on $(T,\mathcal{B})$ for all $s \in S$. Let ρ be a Markov kernel on $(T,\mathcal{B})|S$, and let $(S,\mathcal{S},\mu)$ be a probability space, then we say that ρ is $\mathcal{S}$-measurable resp. μ-measurable if $\rho(B|\cdot)$ is so for all $B \in \mathcal{B}$. And if $\mathcal{C}$ is a σ-algebra on $S \times T$ then we say that ξ is $(\mathcal{S},\mathcal{C})$-admissible if

(4.6.1) $f(s,\cdot)$ is $\mathcal{B}$-measurable $\forall s \in S$, $\forall f \in B(S \times T,\mathcal{C})$

(4.6.2) $s \sim \int_T f(s,t)\rho(dt|s)$ is $\mathcal{S}$-measurable $\forall f \in B(S \times T,\mathcal{C})$,

and we say ρ is $(\mu,\mathcal{C})$-admissible if

(4.6.3) $f(s,\cdot)$ is $\rho(\cdot|s)$-measurable for μ-a.a. $s \in S$

(4.6.4) $\quad s \rightsquigarrow \int_T^* f(s,t)\rho(dt|s)$ is μ-measurable

for all $f \in B(S \times T, C)$. Let $F(\rho|S)$ resp. $F(\rho|\mu)$ be the set of all $f \in B(S \times T)$ satisfying (4.6.1)-(4.6.2) resp. (4.6.3)-(4.6.4). Then clearly we have that $F(\rho|S)$ and $F(\rho|\mu)$ are linear subspaces of $B(S \times T)$, which are stable under dominated, pointwise, sequential convergence. Hence if $F_0 \subseteq B(S \times T)$ is a <u>semigroup</u> (i.e. $f \cdot g \in F_0$ $\forall f, g \in F_0$), and $C = \sigma(F_0)$, then by [2 ; Theorem I.21] we have

(4.6.5) $\quad \rho$ is (S,C)-admissible $\iff F_0 \in F(\rho|S)$

(4.6.6) $\quad \rho$ is (μ,C)-admissible $\iff F_0 \subseteq F(\rho|\mu)$

(4.6.7) $\quad \rho$ is $(S\ S \otimes B)$-admissible $\iff \rho$ is S-measurable

(4.6.8) $\quad \rho$ is $(\mu, S \otimes B)$-admissible $\iff \rho$ is μ-measurable

Now suppose that ρ is (μ,C)-admissible, then we may define <u>the Fubini product of</u> μ <u>and</u> ρ <u>on</u> C by

$$\nu(C) = \int_S \rho(C(s)|s)\mu(ds) \qquad \forall C \in C$$

Then ν is a probability measure on $(S \times T, C)$, and we write $\nu = \mu \overset{\rightarrow}{\otimes} \rho$ on C (or $\nu = \rho \overset{\leftarrow}{\otimes} \mu$ on C if we consider ν as a measure on $T \times S$). If $\rho(\cdot|s) = \lambda$ for all $s \in S$ for some probability measure λ, then we write $\nu = \mu \overset{\rightarrow}{\otimes} \lambda$ on C (or $\nu = \lambda \overset{\leftarrow}{\otimes} \mu$ on C).

If ρ is μ-measurable, then the Fubini product of μ and ρ exists on $S \otimes B$, and the Fubini product on $S \otimes B$ is denoted $\mu \otimes \rho$ (or $\rho \otimes \mu$). And if λ is a probability measure on (T,B), then $\mu \otimes \lambda$ denote the usual <u>product measure</u> on $(S \times T, S \otimes B)$.

Suppose that $\nu = \mu \overset{\rightarrow}{\otimes} \rho$ on C, and let f be a ν-measurable function in $B^*(S \times T) \cup B_*(S \times T)$, then it is easily checked that we have

(4.6.9) $$\int_{S \times T} f d\nu = \int_S \mu(ds) \int_T f(s,t)\rho(dt|s)$$

If (T,A) is an algebraic function space, and $\rho \in Pr_\xi(A|S)$ for some infinite cardinal ξ, then we let ρ_ξ denote the Markov kernel on $(T,\sigma(A_\xi))|S$ given by

$$\rho_\xi(\cdot|s) = \rho(\cdot|s)_\xi \quad \forall s \in S.$$

If $H = \{\rho(f|\cdot) \mid f \in A\}$ and $\mathcal{S} = \sigma(H_\xi)$, then clearly we have that ρ_ξ is $\mathcal{S}$-measurable.

Theorem 4.7. Let $(T_0,\mathcal{B}_0,\mu)$ be a probability space, let ρ be a Markov kernel on $(T_1,\mathcal{B}_1)|T_0$, and let $\mathcal{F}_j \subseteq \mathcal{B}_j$ be $(\cup f,\cap\xi)$-stable pavings containing $\emptyset$ and T_j for $j = 0,1$, where ξ is an infinite cardinal. Suppose that we have

(4.7.1) $\quad \rho(\cdot|t_0)$ is ξ-supersmooth at $\mathcal{F}_1$ along $\mathcal{F}_1$ $\forall t_0 \in T_0$

(4.7.2) $\quad \rho(F_1|\cdot) \in U(T_0,\mathcal{F}_0) \quad \forall F_1 \in \mathcal{F}_1$

Let $\mathcal{F}$ be the $(\cup f,\cap\xi)$-closure of $\{F_0 \times F_1 \mid F_0 \in \mathcal{F}_0,\ F_1 \in \mathcal{F}_1\}$, and let $\mathcal{B} = \sigma(\mathcal{F})$. Then ρ is $(\mathcal{B}_0,\mathcal{B})$-admissible, and if $\nu = \mu \vec{\otimes} \rho$ on $\mathcal{B}$ and $f \in B^*(T_0 \times T_1) \cap U(T_0 \times T_1,\mathcal{F})$, then we have

(4.7.3) $\quad f(t_0,\cdot) \in U(T_1,\mathcal{F}_1) \quad \forall t_0 \in T_0$

(4.7.4) $\quad f(\cdot,t_1) \in U(T_0,\mathcal{F}_0) \quad \forall t_1 \in T_1$

(4.7.5) $\quad t_0 \curvearrowright \int_{T_1} f(t_0,t_1)\rho(dt_1|t_0)$ belongs to $U(T_0,\mathcal{F}_0)$

Moreover if μ is ξ-supersmooth at $\mathcal{F}_0$ along $\mathcal{F}_0$, then $\nu = \mu \otimes \rho$ is ξ-supersmooth at $\mathcal{F}$ along $\mathcal{F}$.

Proof. Let $T = T_0 \times T_1$ and $U_j = B^*(T_j) \cap U(T_j,\mathcal{F}_j)$ for $j = 0,1$. Then by Proposition 2.1 and $(\cup f,\cap\xi)$-stability of $\mathcal{F}_j$ we have

(i) $\quad U_j$ is $(\wedge\xi,\vee f)$-stable, and $U_j \cap \mathbb{R}^{T_j}$ is a convex cone containing $\pm 1_{T_j}$ for $j = 0,1$.

Hence if U is the set of all $f \in B^*(T)$ satisfying (4.7.3)-(4.7.5), then by (4.7.1) and Proposition 3.14 we have

(ii) U is $(\downarrow\xi)$-stable and $U \cap B(T)$ is a convex cone containing $\pm 1_T$.

Thus if $\hat{F} = U \cap 2^T$, then by Proposition 2.2 we have

(iii) $U(T,\hat{F}) \cap B^*(T) \subseteq U$

since U is $(\downarrow c)$-stable.

Now let $F^* = \{F_0 \times F_1 \mid F_0 \in F_0,\ F_1 \in F_1\}$, and let F^{**} be the $(\cup f)$-closure of F^*. Let $f = 1_F$ where $F \in F^{**}$, then there exist $F_1,\dots,F_n \in F_0$ and $H_1,\dots,H_n \in F_1$ such that $F = \cup(F_j \times H_j)$. Let $t_0 \in T_0$ be given and put

$$\pi = \{1 \leq j \leq n \mid t_0 \in F_j\}, \quad H = \bigcup_{j\in\pi} H_j$$

then $f(t_0,\cdot) = 1_H$ and so (4.7.3) holds by $(\cup f)$-stability of F_1. Similarly we have that (4.7.4) holds. Let

$$\psi(x_1,\dots,x_n,t_1) = \max_{1\leq j\leq n} \{\varepsilon(x_j) 1_{H_j}(t_1)\}$$

$$\varphi(x_1,\dots,x_n,t_0) = \int_{T_1} \psi(x_1,\dots,x_n,t_1)\rho(dt_1|t_0)$$

for $x = (x_1,\dots,x_n) \in \mathbb{R}^n$ and $(t_0,t_1) \in T_0 \times T_1$, where $\varepsilon = 1_{[1,\infty[}$. Then $\psi(\cdot,t_1)$ is increasing and uppersemicontinuous for all $t_1 \in T_1$, hence by Fatou's lemma we have that $\varphi(\cdot,t_0)$ is increasing and upper semicontinuous, since $0 \leq \psi \leq 1$. Let $x = (x_1,\dots,x_n) \in \mathbb{R}^n$ be given and put

$$\pi = \{1 \leq j \leq n \mid x_j \geq 1\}, \quad H = \bigcup_{j\in\pi} H_j$$

Then $\psi(x,\cdot) = 1_H$ and $\varphi(x,\cdot) = \rho(H|\cdot)$. Hence $\varphi(x,\cdot) \in U(T_0,F_0)$ for all $x \in \mathbb{R}^n$ by (4.7.2) and $(\cup f)$-stability of F_1. Hence by Proposition 2.11 we have that

$$\varphi(1_{F_1}(t_0),\ldots,1_{F_n}(t_0),t_0) = \int_T f(t_0,t_1)\rho(dt_1|t_0)$$

belongs to $U(T_0,F_0)$. Thus f satisfies (4.7.3)-(4.7.5) and so we conclude that $F^{**} \subseteq \hat{F}$. Now since F^{**} is $(\cup f,\cap f)$-stable, we have that F is the $(\downarrow\xi)$-closure of F^{**}, and since $\hat{F}$ is $(\downarrow\xi)$-stable by (ii), we have that $F \subseteq \hat{F}$, and so by (iii) we find

(iv) $$U(T,F) \cap B^*(T) \subseteq U.$$

Thus (4.7.3)-(4.7.5) holds for all $f \in U(T,F) \cap B^*(T)$. In particular we have that $F \subseteq F(\rho|B_0)$, and so by (4.6.5) there exist a Fubini product of μ and ρ on $B = \sigma(F)$, and ρ is (B_0,B)-admissible.

Finally suppose that μ is ξ-supersmooth at F_0 along F_0, and let $\{F_\gamma \mid \gamma \in \Gamma\}$ be a decreasing net in F with $\mathrm{card}(\Gamma) \leq \xi$ and $F_\gamma \downarrow F$. Then $F \in F$, and $\rho(F_\gamma|t_0) \downarrow \rho(F|t_0)$ for all $t_0 \in T_0$ by (4.7.1). Hence by ξ-supersmoothnes of μ, (4.7.5) and Proposition 3.14 we have that $\mu \vec{\otimes} \rho(F_\gamma) \downarrow \mu \vec{\otimes} \rho(F)$. Thus $\mu \vec{\otimes} \rho$ is ξ-supersmooth at F along F. □

<u>Corollary 4.8</u>. <u>Let</u> T_0 <u>and</u> T_1 <u>be topological spaces, let</u> μ <u>be a Borel probability on</u> T_0 <u>and let</u> ρ <u>be a Markov kernel on</u> $(T_1,B(T_1)) \mid T_0$, <u>such that</u>

(4.8.1) $\rho(\cdot|t_0)$ <u>is a</u> τ-<u>smooth Borel measure on</u> T_1 $\forall t_0 \in T_0$

(4.8.2) $\rho(G_1|\cdot) \in Lsc(T_0)$ $\forall G_1 \in G_1$

<u>where</u> G_1 <u>is a</u> $(\cup f)$-<u>stable open base for</u> T_1. <u>Then</u> ρ <u>is</u> $(B(T_0), B(T_0 \times T_1))$-<u>admissible and we have</u>

(4.8.3) $t_0 \sim \int_{T_1} f(t_0,t_1)\rho(dt_1|t_0)$ <u>is upper semicontinuous</u>

<u>for all</u> $f \in B^*(T_0 \times T_1) \cap Usc(T_0 \times T_1)$. <u>If moreover</u> μ <u>is a</u> τ-<u>smooth probability on</u> T_0, <u>then</u> $\mu \vec{\otimes} \rho$ <u>is a</u> τ-<u>smooth Borel probability on</u>

$(T_0 \times T_1, \mathcal{B}(T_0 \times T_1))$. In particular we have

(1): If μ is a probability on $(T_0, \mathcal{B}(T_0))$, and λ is a τ-smooth Borel probability on $(T_1, \mathcal{B}(T_1))$. Then there exist a Fubini product $\nu = \mu \vec{\otimes} \lambda$ on $\mathcal{B}(T_0 \times T_1)$ and λ is $(\mathcal{B}(T_0), \mathcal{B}(T_0 \times T_1))$-admissible

(2): If μ_j is a τ-smooth Borel probability on $(T_j, \mathcal{B}(T_j))$ for $j = 0,1$, then the product measure $\mu_0 \otimes \mu_1$ admits a unique τ-smooth extension $\mu_0 \hat{\otimes} \mu_1$ to $\mathcal{B}(T_0 \times T_1)$, and we have $\mu_0 \hat{\otimes} \mu_1 = \mu_o \overleftarrow{\otimes} \mu_1 = \mu_0 \vec{\otimes} \mu_1$ on $\mathcal{B}(T_0 \times T_1)$

Remarks. (1): A probability measure λ on $(T_1, \mathcal{B}_1)$, is said to be $(\mathcal{B}_0, \mathcal{B})$-admissible, where $\mathcal{B}_0$ is a σ-algebra on T_0 and $\mathcal{B}$ is a σ-algebra on $T_0 \times T_1$, if and only if $\rho(\cdot|t_0) = \lambda \quad \forall t_0 \in T_0$ is so

(2): A Markov kernel satisfying (4.8.1) and (4.8.2) for some $(\cup f)$-stable open base $\mathcal{G}_1$ for T_1, is said to be continuous. Hence if ρ is continuous, then

(4.8.4) $\quad t_0 \sim \int_{T_1} f(t_0,t_1)\, \rho(dt_1|t_0)$ is uppersemicontinuous

(4.8.5) $\quad t_0 \sim \int_{T_1} g(t_0,t_1)\rho(dt_1|t_0)$ is lowersemicontinuous

(4.8.6) $\quad t_0 \sim \int_{T_1} \varphi(t_0,t_1)\rho(dt_1|t_0)$ is continuous

if $f \in B^*(T) \cap Usc(T)$, $g \in B_*(T) \cap Lsc(T)$ and $h \in C(T)$, where $T = T_0 \times T_1$ with its product topology.

(3): If ρ satisfies (4.8.1) and $\Phi \subseteq B_*(T_1) \cap Lsc(T_1)$ satisfies

(4.8.7) $\quad \forall G \in \mathcal{G}(T_1) \quad \exists \{\varphi_\gamma\} \subseteq \Phi$ so that $\varphi_\gamma \uparrow 1_G$

(4.8.8) $\quad \int_{T_1} \varphi(\cdot,t_1)\rho(dt_1|\cdot) \in Lsc(T_0) \qquad \forall \varphi \in \Phi$

then evidently ρ is continuous.

(4): Let (T,A) be an algebraic function space, and let $\rho \in Pr_\tau(A|S)$. Put $H = \{\rho(f|\cdot) \mid f \in A\}$, if T has the $\tau(A)$-topology and S has the $\tau(H)$-topology, then the Markov kernel ρ_τ on $(T,\mathcal{B}(A))|S$ is continuous.

Proof. Easy consequence of Theorem 4.7. □

Theorem 4.9. *Let* (T,A) *be an algebraic function space, and let* $(S,\mathcal{B},\mu)$ *and* (Ω,F,P) *be probability spaces. Let* $\varphi: \Omega \to (S,\mathcal{B})$ *and* $\psi: \Omega \to (T,\sigma(A))$ *be P-measurable functions, such that* $\mu = \varphi P$ *and put*

$$\lambda(f) = \int_\Omega f(\psi(\omega))P(d\omega) \qquad \forall f \in A.$$

If λ $Pr_{c\xi}(A)$ *for some infinite cardinal* ξ, *then there exist a kernel* $\rho \in Pr_{c\xi}(A|S)$ *satisfying*

(4.9.1) $\qquad \rho^*(h|\cdot)$ *is* μ-*measurable* $\quad \forall h \in \bar{\mathbb{R}}^T$

(4.9.2) $\qquad \int_S \rho^*(h|s)\mu(ds) = \lambda^*(h) \qquad \forall h \in \bar{\mathbb{R}}^T$

(4.9.3) $\qquad \exists\{K_n \mid n \in \mathbb{N}\} \subseteq \bar{K}_\xi(A): \lim_{n\to\infty} \rho_\xi^*(K_n|s) = 1 \qquad \forall s \in S$

(4.9.4) $\qquad \rho_\xi$ *is* μ-*measurable*

(4.9.5) $\qquad \int_\Omega g(\varphi(\omega),\psi(\omega))P(d\omega) = \int_S \mu(ds) \int_T g(s,t)\rho_\xi(dt|s)$

for all $g \in B(S \times T, \mathcal{B} \otimes \sigma(A))$.

Moreover if $H = \{\psi^{-1}(F) \mid F \in F_\xi(A)\} \subseteq F$ *and* P *is* ξ-*supersmooth at* H *along* H, *then (4.9.5) holds for all* $g \in B(S \times T, \mathcal{B} \otimes \sigma(A_\xi))$.

Remark. Note that (4.9.5) states that ρ_ξ is a regular conditional P-distribution of ψ given φ.

In [16; Theorem 3.5] a similar result is proved in the case where $\xi = \aleph_0$

<u>**Proof**</u>. Let $\mathcal{B}_0 = M(\mu)$ and let $\theta: L^\infty(\mu) \to B(S,\mathcal{B}_0)$ be a lifting, see [7 , p.258]. Let $L = \theta(L^\infty(\mu))$, then L is a $\|\cdot\|$-closed algebra, such that $1 \in L \subseteq B(S,\mathcal{B}_0)$ and by [7 ; Lemma 1 and 2] and Proposition 3.5 we have

(i) $f,g \in L,\ f \geq g \quad \mu\text{-a.s.} \Rightarrow f(s) \geq g(s) \quad \forall s \in S$

(ii) $L_\tau \subseteq \overline{L}\ (\mu)$ and $\int d\mu$ is ξ-supersmooth at L_τ along L_τ.

Now let $f \in \bar{A}$, then by (i) and Radon-Nikodym's theorem [4;Theorem III.10.2] there exist a unique $\rho_0(f|\cdot)$ such that

(iii) $\rho_0(f|\cdot) \in L$

(iv) $\int_B \rho_0(f|s)\mu(ds) = \int_{\varphi^{-1}(B)} f(\psi(\omega))P(d\omega) \quad \forall B \in \mathcal{B}$.

Moreover by (i), (iii) and (iv) it follows easily that $\rho_0 \in \Pr(\bar{A}|S)$. Now let $h \in \bar{\mathbb{R}}^T$, since $\bar{A}$ is $(\wedge f)$-stable we have that $\{\rho_0(f) \mid f \in \bar{A}, f \geq h\}$ filters downwards to $\rho_0^*(h)$, and since

$$\int_S \rho_0(f|s)\mu(ds) = \int_\Omega (f \circ \psi)\,dP = \bar{\lambda}(f) \quad \forall f \in \bar{A}$$

by (iv) and definition of λ, we have by (ii) that

(v) $\rho_0^*(h|\cdot)$ is μ-measurable $\quad \forall h \in \bar{\mathbb{R}}^T$

(vi) $\int_S \rho_0^*(h|s)\mu(ds) = \lambda^*(h) \quad \forall h \in \bar{\mathbb{R}}^T$

since $\lambda^* = \bar{\lambda}^*$.

By assumption we have that $\lambda \in \Pr_{c\xi}(A)$, so by Theorem 4.1 there exist an increasing sequence $\{K_n\}$ in $\bar{K}_\xi(A)$, such that

(vii) $\lambda^*(K_n) = \lambda_\xi^*(K_n) \geq 1-2^{-n} \quad \forall n \geq 1$

Thus by (vi) we have that

$$\int_S \rho_0^*(K_n|s)\mu(ds) \geq 1-2^{-n}$$

and so there exist a μ-nullset $N \in \mathcal{B}$, such that

$$\text{(viii)} \qquad \lim_{n\to\infty} \rho_0^*(K_n|s) = 1 \quad \forall s \in S \setminus N.$$

Now let t_0 be an arbitrary but fixed point in K_1, and put

$$\rho(f|s) = \begin{cases} \rho_0(f|s) & \text{if } f \in A \text{ and } s \in S \setminus N \\ f(t_0) & \text{if } f \in A \text{ and } s \in N. \end{cases}$$

Then $\rho^*(K_n|s) = 1$ for all $s \in N$ and all $n \geq 1$, and $\rho^*(\cdot|s) = \rho_0^*(\cdot|s)$ for all $s \in S \setminus N$. Hence we have

$$\text{(ix)} \qquad \lim_{n\to\infty} \rho^*(K_n|s) = 1 \quad \forall s \in S$$

and so $\rho \in Pr_{c\xi}(A|S)$, and (4.9.1), (4.9.2) and (4.9.3) holds (see (4.1.10)), and since

$$\int_T f_0(t)\rho_\xi(dt|s) = \rho^*(f_0|s) \qquad \forall f_0 \in A_\xi$$

by (4.1.1), it follows easily from (4.9.1) and [2 ; Theorem I.21] that (4.9.4) holds. By (iv) we have that (4.9.5) holds if $g(s,t) = 1_B(s)f(s)$ where $B \in \mathcal{B}$ and $f \in A$, hence as above we conclude that (4.9.5) holds for all bounded $\mathcal{B} \otimes \sigma(A)$-measurable functions g.

Now suppose that P is ξ-supersmooth at H along H, and let $F \in F_\xi(A)$ and $B \in \mathcal{B}$. Then there exist a decreasing net $\{F_\gamma \mid \gamma \in \Gamma\}$ in $F_0(A)$, such that $F_\gamma \downarrow F$ and $\operatorname{card}(\Gamma) \leq \xi$. Since $F_0(A) \subseteq \sigma(A)$ we have

$$\begin{aligned} P(\varphi^{-1}(B) \cap \psi^{-1}(F_\gamma)) &= \int_B \rho_\xi(F_\gamma|s)\mu(ds) \\ &= \int_B \rho^*(F_\gamma|s)\mu(ds) = \int_B \rho_0^*(F_\gamma|s)\mu(ds) \end{aligned}$$

by (4.1.1). Hence by (ii) and Proposition 3.14 we have

$$\begin{aligned} P(\varphi^{-1}(B) \cap \psi^{-1}(F)) &= \lim_\gamma P(\varphi^{-1}(B) \cap \psi^{-1}(F_\gamma)) \\ &= \int_B \lim_\gamma \rho_0^*(F_\gamma|s)\mu(ds) = \int_B \lim_\gamma \rho^*(F_\gamma|s)\mu(ds) \end{aligned}$$

$$= \int_B \lim_\gamma \rho_\xi(F_\gamma|s)\mu(ds) = \int_B \rho_\xi(F|s)\mu(ds)$$

since $\psi^{-1}(F_\gamma) \in \mathcal{H}$ and $\psi^{-1}(F_\gamma) \downarrow \psi^{-1}(F)$, and since $\rho_\xi(\cdot|s)$ is ξ-supersmooth at $\mathcal{F}_\xi(A)$ along $\mathcal{F}_\xi(A)$ by (4.1.8). Hence (4.9.5) holds if $g = 1_{B\times F}$ where $B \in \mathcal{B}$ and $F \in \mathcal{F}_\xi(A)$. But then as above we have that (4.9.5) holds for all bounded $\mathcal{B} \otimes \sigma(A_\xi)$-measurable functions g, since $\sigma(\mathcal{F}_\xi(A)) = \sigma(A_\xi)$ by Proposition 2.9. □

Theorem 4.10. *Let $\lambda \in \Pr_\pi(A)$, then λ_σ is a perfect probability measure. And if $\lambda \in \Pr_{c\xi}(A)$ for some infinite cardinal ξ, then λ_ξ is a ξ-compact probability measure, more precisely there exists a ξ-compact $(\cup f,\cap\xi)$-stable paving $\mathcal{K} \subseteq \mathcal{F}_\xi(A)$ satisfying*

(4.10.1) $$(\lambda_\xi)_*(E) = \sup\{\lambda_\xi(K) \mid K \in \mathcal{K},\ K \subseteq E\} \quad \forall E \subseteq T.$$

Proof. Supoose that $\lambda \in \Pr_{c\xi}(A)$, and put

$$S = \{s \in \mathbb{R}^A \mid |s(f)| \le \|f\| \quad \forall f \in A\}$$

$$\varepsilon_f(s) = s(f) \quad \forall s \in S \quad \forall f \in A$$

$$B = \text{the algebra spanned by } \{\varepsilon_f \mid f \in A\}$$

$$\mathcal{B} = \sigma(B_\xi), \quad \mathcal{A} = \sigma(A_\xi)$$

$$\varphi(t) = (f(t))_{f\in A}: \quad T \to \mathbb{R}^A$$

Then φ maps T into S, and since $f = \varepsilon_f \circ \varphi$ for all $f \in A$, we have that $\hat{f} \circ \varphi \in A$ if $\hat{f} \in B$, and so we have $\hat{f} \circ \varphi \in A_\xi$ for all $\hat{f} \in B_\xi$. Hence φ is measurable from $(T,\mathcal{A})$ into $(S,\mathcal{B})$, so let μ be the image measure on $(S,\mathcal{B})$ of λ_ξ under the map φ.

Now we apply Theorem 4.9 with $(\Omega,\mathcal{F},P) = (T,\mathcal{A},\lambda_\xi)$ and $\psi(t) = t$ for all $t \in T$. Then there exist a kernel $\rho \in \Pr_{c\xi}(A|S)$, such that

(i) $$\lambda_\xi(D \cap \varphi^{-1}(E)) = \int_E \rho_\xi(D|s)\mu(ds) \quad \forall D \in \mathcal{A} \quad \forall E \in \mathcal{B}$$

Now let

$$\mathcal{K} = \left\{ K \subseteq T \;\middle|\; \begin{array}{l} \exists C \in F_\xi(B) \quad \text{so that} \quad K = \varphi^{-1}(C) \\ \text{and} \quad \rho_\xi(K|s) = 1 \qquad \forall s \in C \end{array} \right\}$$

By the argument above we have that $\mathcal{K} \subseteq F_\xi(A)$. Let $K_1, K_2 \in \mathcal{K}$, and let $C_j \in F_\xi(B)$ so that $K_j = \varphi^{-1}(C_j)$ and

$$\rho_\xi(K_j|s) = 1 \quad \forall s \in C_j .$$

Then $K_1 \cup K_2 = \varphi^{-1}(C_1 \cup C_2)$ and

$$\rho_\xi(K_1 \cup K_2|s) = 1 \quad \forall s \in C_1 \cup C_2$$

and $K_1 \cap K_2 = \varphi^{-1}(C_1 \cap C_2)$ and

$$\rho_\xi(K_1 \cap K_2|s) = 1 \quad \forall s \in C_1 \cap C_2$$

Now since $F_\xi(B)$ is $(\cup f, \cap f)$-stable we find that $\mathcal{K}$ is $(\cup f, \cap f)$-stable. And clearly we have that $\emptyset \in \mathcal{K}$ and $T \in \mathcal{K}$. Now let $\{K_\gamma\}$ be a decreasing net in $\mathcal{K}$, such that $\text{card}(\Gamma) \leq \xi$ and $K_\gamma \downarrow K$. Let $C_\gamma \in F_\xi(B)$ be chosen, such that

$$K_\gamma = \varphi^{-1}(C_\gamma) \quad \text{and} \quad \rho_\xi(K_\gamma|s) = 1 \quad \forall s \in C_\gamma$$

and put $C = \cap C_\gamma$. Then $C \in F_\xi(A)$ and $K = \varphi^{-1}(C)$. Moreover if $s \in C$, then $\rho_\xi(K_\gamma|s) = 1$ for all γ. Now $\mathcal{K} \subseteq F_\xi(A)$ and so $\rho_\xi(\cdot|s)$ is ξ-supersmooth at K along $\mathcal{K}$ by (4.1.1) and (4.1.8). Hence we have that $\rho_\xi(K|s) = 1$ for all $s \in C$, and so $K \in \mathcal{K}$. Moreover if $K = \emptyset$, we see that $C = \emptyset$, and since $\tau(B)$ equals the product topology on S, we have that $(S, \tau(B))$ is a compact Hausdorff space and so

$$C_\gamma \in F_\xi(B) \subseteq F(B) = \tilde{K}(B) .$$

Hence there exist a finite set $\pi \subseteq \Gamma$ such that

$$\bigcap_{\gamma \in \pi} C_\gamma = \emptyset$$

and there exist $\alpha \in \Gamma$ so that $\alpha \geq \gamma$ for all $\gamma \in \pi$. Thus we find

$$K_\alpha \subseteq \bigcap_{\gamma\in\pi} K_\gamma = \varphi^{-1}(\bigcap_{\gamma\in\pi} C_\gamma) = \emptyset$$

and so we conclude

(ii) $\quad \mathcal{K}$ is $(\cup f, \cap\xi)$-stable and $\mathcal{K}$ is ξ-compact.

Now let us show that (4.10.1) holds. So let $E \subseteq T$ and $a < (\lambda_\xi)_*(E)$ be given. Then by (4.1.7) there exist $F_0 \in \mathcal{F}_\xi(A)$, such that $F_0 \subseteq E$ and $\lambda_\xi(F_0) > a$. By (F.6) we have that $F_0 = \varphi^{-1}(H_0)$ for some $H_0 \in \mathcal{F}_\xi(B)$, so by (i) we have

$$\lambda_\xi(F_0) = \mu(H_0) = \lambda_\xi(F_0 \cap \varphi^{-1}(H_0)) = \int_{H_0} \rho_\xi(F_0|s)\mu(ds)$$

and since $0 \le \rho_\xi(F_0|\cdot) \le 1$, we see that if

$$G_0 = \{s \in H_0 \mid \rho_\xi(F_0|s) = 1\}$$

then $\mu(G_0) = \mu(H_0) > a$. Now note that μ is ξ-supersmooth at $\mathcal{F}_\xi(B)$ along $\mathcal{F}_\xi(B)$, since $\hat{f} \circ \varphi \in A_\xi$ for all $\hat{f} \in B_\xi$, so if

$$\nu(f) = \int \hat{f}\, d\mu = \int (f \circ \varphi)\, d\lambda_\xi = \lambda(\hat{f} \circ \varphi) \qquad \forall \hat{f} \in B$$

then $\nu_\xi = \mu$ by Theorem 4.1. Hence there exist $H_1 \in \mathcal{F}_\xi(B)$ so that $H_1 \subseteq G_0$ and $\mu(H_1) = \lambda_\xi(F_1) > a$, where $F_1 = \varphi^{-1}(H_1)$, and as above we have that $\mu(H_1) = \mu(G_1)$ where

$$G_1 = \{s \in H_1 \mid \rho_\xi(F_1|s) = 1\}.$$

Continuing like this we can find $H_n \in \mathcal{F}_\xi(A)$, such that

(iii) $\quad H_{n+1} \subseteq G_n \subseteq H_n \qquad \forall n \ge 0$

(iv) $\quad \mu(H_n) = \lambda_\xi(F_n) = \mu(G_n) > a \qquad \forall n \ge 0$

where $F_n = \varphi^{-1}(H_n)$ and

$$G_n = \{s \in H_n \mid \rho_\xi(F_n|s) = 1\}.$$

Now put $H = \cap H_n$ and $F = \cap F_n$. Then $F = \varphi^{-1}(H)$ and $H \in \mathcal{F}_\xi(B)$, moreover if $s \in H$, then $s \in G_n$ for all n and so

$$\rho_\xi(F_n | s) = 1 \quad \forall n \geq 1$$

So by (iii) and (iv) we have that $\lambda_\xi(F) \geq a$ and $\rho_\xi(F|s) = 1$ for all $s \in H$. Thus $F \in K$, $F \subseteq E$ and $\lambda_\xi(F) \geq a$, and so we find

$$(\lambda_\xi)_*(E) \leq \sup\{\lambda_\xi(K) \mid K \in K,\ K \subseteq E\}$$

but the converse inequality is evident and so (4.10.1) holds and λ_ξ is a ξ-compact measure.

Now suppose that $\lambda \in Pr_\pi(A)$, and let $B \subseteq \sigma(A)$ be a countably generated σ-algebra. Then there exist a countably generated algebra $A_0 \subseteq A$ such that $B \subseteq \sigma(A_0)$. Let $\nu = \lambda | A_0$, then ν is semicompact, and by the argument above we have that $\nu_\sigma = \lambda_\sigma | \sigma(A_0)$ is a semicompact measure. Hence $\lambda_\sigma | B$ is semicompact by [16; Proposition 4.1], and so λ_σ is a perfect measure. □

Theorem 4.11. *Let $\lambda \in Pr_\pi(A)$ and let Q be a subset of A of countable weight, then λ is $\bar{K}(Q)$-tight. Hence if A is of countable weight then*

$$\text{(4.11.1)} \qquad Pr_\pi(A) = Pr_{c\xi}(A) = Pr_{c\tau}(A)$$

for all infinite cardinals ξ.

Proof. Clearly we may assume that Q is countable. Now consider the map

$$p_Q(t) = (f(t))_{f \in Q} : T \to \mathbb{R}^Q .$$

Then p_Q is measurable: $(T, \sigma(A)) \to (\mathbb{R}^Q, B(\mathbb{R}^Q))$, and λ_σ is a perfect measure by Theorem 4.10. Hence by [17; Theorem 3] there exist compact sets $C_n \subseteq \mathbb{R}^Q$ such that $C_n \subseteq p_Q(T)$ and

$$\lambda_\sigma(p_Q^{-1}(C_n)) \geq 1-2^{-n}$$

Let $K_n = p_Q^{-1}(C_n)$, then $p_Q(K_n) = C_n$, since $C_n \subseteq p_Q(T)$, and so $K_n \in K(Q) \cap F(Q)$. Hence the theorem follows. □

<u>Theorem 4.12</u>. <u>Let</u> μ <u>be a probability measure on</u> T, <u>and let</u> $A \subseteq B(T, M(\mu))$ <u>be an algebra containing</u> 1_T. <u>Let us put</u>

$$\lambda(f) = \int_T f d\mu \quad \forall f \in A,$$

<u>Then</u> $\lambda \in Pr_\sigma(A)$, <u>and if</u> μ_0 <u>is the restriction of</u> μ <u>to</u> $\sigma(A)$, <u>then we have</u> $\mu_0 = \lambda_\sigma$ <u>and</u>

(4.12.1) $\quad \lambda \in Pr_\pi(A) \iff \mu_0$ <u>is a perfect measure</u>

(4.12.2) $\quad$ <u>If</u> $F_\xi(A) \subseteq M(\mu)$ <u>and</u> μ <u>is</u> ξ-<u>supersmooth at</u> $F_\xi(A)$ <u>along</u> $F_\xi(A)$, <u>then</u> $\mu(B) = \lambda_\xi(B) \quad \forall B \in \sigma(A_\xi)$

(4.12.3) $\quad \sup_{K \in K} \mu^*(K) = 1 \Rightarrow \lambda$ <u>is</u> K-<u>tight</u>

<u>where</u> ξ <u>is a given infinite cardinal, and</u> K <u>is a given paving on</u> T.

<u>Proof</u>. Clearly $\lambda \in Pr_\sigma(A)$ and by the uniqueness of λ_σ we have that $\lambda_\sigma = \mu_0$.

(4.12.1): The implication "⇒" follows from Theorem 4.10. So suppose that $\mu_0 = \lambda_\sigma$ is perfect and let A_0 be a countably generated subalgebra of A. Let $Q \subseteq A$ be countable set generating A_0, and consider the map

$$p_Q(t) = (f(t))_{f \in Q}: T \to \mathbb{R}^Q .$$

Then p_Q is continuous: $(T, \tau(A_0)) \to \mathbb{R}^Q$, and by [17; Theorem 3] and perfectness of μ_0, there exist compact sets $K_n \subseteq p_Q(T)$, so that

(i) $$\mu_0(p_Q^{-1}(K_n)) \geq 1-2^{-n} .$$

Let $C_n = p_Q^{-1}(K_n)$, then $K_n = p_Q(C_n)$, since $C_n \subseteq p_Q(T)$. Hence $C_n \in K(Q) = K(A_0)$ and so $\lambda|A_0$ is compact by (i), (4.1.10), (4.6) and (4.4). Thus $\lambda \in Pr_\pi(A)$

(4.12.2): Easy consequence of the uniqueness of λ_ξ

(4.12.3): Follows from (4.6) since $\mu^* \leq \lambda^*$. □

We shall now study the image functional $\theta\lambda$, where $\theta: T \rightsquigarrow S$ is a correspondance and λ is a θ-saturated probability functional on (T,A). Usually θ is an ordinary map from T into S, or the inverse of an ordinary map from S into T. If φ is a map from S into T, then $\varphi^{-1}\lambda$ is called <u>the co-image functional of</u> λ <u>under</u> φ. The main advantage of using correspondances instead of ordinary maps is that, in this way we may treat the image and the co-image functional simultaneously.

Let (T,A) and (S,B) be algebraic function spaces and let $\theta: T \rightsquigarrow S$ be a correspondance, such that $B \subseteq \theta(A)$. Then it is easily checked that we have (see (H.7) and Proposition 3.16)

(4.11) θ is upper and lower continuous on $\theta^{-1}(S)$

(4.12) $\theta(K) \in K_\xi(B) \quad \forall K \in K_\xi(A)$ with $K \subseteq \theta^{-1}(S)$

(4.13) $\bar{B} \subseteq \theta(\bar{A})$, $B_\xi \subseteq \theta(A_\xi)$, $B^\xi \subseteq \theta(A_\xi)$

(4.14) $B(S,\sigma(B_\xi)) \subseteq \theta(B(T,\sigma(A_\xi))$

(4.15) $(S, B(S) \cap \theta(A))$ is an algebraic function space

whenever ξ is an infinite cardinal. And if $\lambda \in Pr(A)$ then we have

(4.16) λ is θ-saturated, if and only of $\lambda(f) = 0$, whenever $f \in A$ and $f(t) = 0 \quad \forall t \in \theta^{-1}(S)$

(4.17) If λ is θ-saturated, then $\theta_B\lambda$ is a linear functional on B.

However λ may be θ-saturated and $\theta_B\lambda \notin Pr(B)$. To see this put

$$S = T = [0,1], \quad \theta(t) = \{t\} \cap [0,\tfrac{1}{2}] \quad \forall t \in T$$

$$A = \text{all polynomials on } T, \quad \lambda(f) = f(1) \quad \forall f \in A$$

$$B = \{g \in C[0,1] \mid \exists f \in A: \ f = g \text{ on } [0,\tfrac{1}{2}]\}.$$

Then $B \subseteq \theta(A)$ and λ is θ-saturated by (4.16). But if $g(x) = (\frac{1}{2} - x)^+$, then $g \in B^+$ and $(\theta_B\lambda)(g) = -\frac{1}{2}$. Thus $\theta_B\lambda$ is not a positive functional.

<u>Theorem 4.13</u>. <u>Let</u> (T,A) <u>and</u> (S,B) <u>be algebraic function spaces, and let</u> $\theta: T \rightsquigarrow S$ <u>be a correspondance, such that</u> $B \subseteq \theta(A)$, <u>and let</u> $\lambda \in Pr(A)$. <u>Then we have</u>

(4.13.1) $\qquad \bar{\lambda}$ <u>is</u> θ-<u>saturated</u> $\iff \lambda^*(D) = 1$

<u>where</u> $D = \theta^{-1}(S)$. <u>Now suppose that</u> $\bar{\lambda}$ <u>is θ-saturated, and that</u> ξ <u>is an infinite cardinal, then we have</u>

(4.13.2) $\qquad \theta_B \lambda \in Pr(B)$ <u>and</u> $(\overline{\theta_B\lambda}) = \theta_{\bar{B}} \bar{\lambda}$

(4.13.3) $\qquad \theta_B \lambda \in Pr_\xi(B)$ <u>if</u> $\lambda \in Pr_\xi(A)$ and $\lambda_\xi^*(D) = 1$

(4.13.4) $\qquad \theta_B \lambda \in Pr_{c\xi}(B)$ <u>if</u> $\lambda \in Pr_{c\xi}(A)$ and $(\lambda_\xi)_*(D) = 1$

(4.13.5) $\qquad \theta_B \lambda \in Pr_\pi(B)$ <u>if</u> $\lambda \in Pr_\pi(A)$ and $(\lambda_\sigma)_*(D) = 1$

<u>Remark</u>. Note that $D = \{t \mid \theta(t) \neq \emptyset\}$ is <u>the domain</u> of θ. Hence if θ is a <u>proper correspondance</u>, i.e. if $D = T$, then $\bar{\lambda}$ is θ-saturated, $\theta_B\lambda \in Pr(B)$, and we have

(4.13.6) $\qquad \lambda \in Pr_\alpha(A) \Rightarrow \theta_B\lambda \in Pr_\alpha(B)$

whenever α denotes one of the three symbols: ξ, $c\xi$ or π, where ξ is an infinite cardinal.

<u>Proof</u> (4.13.1): Suppose that $\bar{\lambda}$ is θ-saturated, and that $f \in A$ so that $f \geq 1_D$. Then $f \wedge 1$ be longs to $\bar{A}$ by Proposition 2.7 and $f \wedge 1 = 1$ on D. Hence $f \wedge 1 \in \theta 1_S$ and so

$$\lambda(f) \geq \bar{\lambda}(f \wedge 1) = \lambda(1_T) = 1$$

Hence $\lambda^*(D) = 1$.

Conversely of $\lambda^*(D) = 1$ and $f \in \bar{A}$ so that $f(t) = 0$ for all $t \in D$. Now let $a = 1 + \|f\|$, then $f + a \geq 0$ and $a - f \leq 0$, and so

$$a^{-1}(f+a) \geq 1_D, \quad a^{-1}(a-f) \geq 1_D$$

and so

$$1 = \lambda^*(D) = \bar{\lambda}^*(D) \leq a^{-1}(\bar{\lambda}(f) + a)$$

$$1 = \lambda^*(D) = \bar{\lambda}^*(D) \leq a^{-1}(a - \bar{\lambda}(f)).$$

Hence $\bar{\lambda}(f) = 0$, and so $\bar{\lambda}$ is θ-saturated by (4.16).

(4.13.2): Since $\bar{A}$ is $(\wedge f)$-stable we have by (4.17) and Proposition 3.16 that $\theta_{\bar{B}}\bar{\lambda} \in Pr(\bar{B})$. And since $\theta_B\lambda$ is the restriction of $\theta_{\bar{B}}\bar{\lambda}$ to B we see that (4.13.2) holds.

(4.13.3): Let $\{g_\gamma \mid \gamma \in \Gamma\}$ be a decreasing net in B, so that $g_\gamma \downarrow 0$ and card $\Gamma \leq \xi$. Then there exist $f_\gamma \in A \cap (\theta g_\gamma)$ for all $\gamma \in \Gamma$. Now put

$$E = \{t \in T \mid f_\alpha(t) \leq f_\beta(t) \quad \forall \alpha \geq \beta\}.$$

Then $E \in F_\xi(A)$ and $E \supseteq D$, since $g_\alpha \leq g_\beta$ for all $\alpha \geq \beta$. Hence $\lambda_\xi(E) = 1$ by assumption and

$$(\theta_B\lambda)(g_\gamma) = \lambda(f_\gamma) = \int_T 1_E \, f_\gamma \, d\lambda_\xi \,.$$

Putting $\mu = \lambda_\xi$, $F = F_\xi(A)$, $A = E$, $\varphi_\gamma = f_\gamma$ and $f = 1_E$ in Proposition 3.14 we see that (3.14.1)-(3.14.4) holds. Hence if $\varphi = \inf f_\gamma$, then by Proposition 3.14 we have

$$\lim_{\gamma}(\theta_B\lambda)(g_\gamma) = \int_E \varphi \, d \lambda_\xi .$$

Since $g_\gamma \downarrow 0$ we have that $\varphi(t) = 0$ for all $t \in D$, and so $\varphi = 0$ λ_ξ - a.s., since φ is λ_ξ-mesuarable and $\lambda_\xi^*(D) = 1$. Hence

$$\lim_{\gamma}(\theta_B\lambda)(g_\gamma) = 0$$

an so $\theta_B\lambda \in Pr_\xi(B)$.

(4.13.4): Let $\varepsilon > 0$ be given. Since D is λ_ξ-measurable and $\lambda_\xi(D) = 1$, we have by (4.1.17) that there exist $K \in \bar{K}_\xi(A)$, so that

(i) $$K \subseteq D \quad \text{and} \quad \lambda_\xi^*(K) \geq 1 - \varepsilon .$$

Now let $C = \theta(K)$, then $C \in K_\xi(B)$ by (4.12), and if $g \in B$ and $g \geq 1_C$, then there exist $f \in A \cap (\theta g)$, and we have

$$f(t) = g(s) \qquad \forall s \in \theta(t) \quad \forall t \in D .$$

Hence $f(t) \geq 1$ for all $t \in K$ and $f(t) \geq 0$ for all $t \in D$. Hence $f \geq 1_K$ λ_ξ - a.s. and so

$$(\theta_B\lambda)(g) = \lambda(f) \geq \lambda_\xi^*(K) \geq 1 - \varepsilon$$

and so by $\theta_B\lambda \in Pr_{c\xi}(B)$ by (4.6).

(4.13.5): Let B_0 be a countably generated subalgebra of B containing 1_S. Let H be a countable set of generators for B_0, then there exist a countable set $Q \subseteq A$ so that $H \subseteq \theta(Q)$, and by assumption there exist $D_0 \in \sigma(A)$ so that

(ii) $$D_0 \subseteq D \quad \text{and} \quad \lambda_\sigma(D_0) = 1.$$

Then there exist a countably generated algebra $A_0 \subseteq A$, so that

(iii) $$Q \cup \{1_T\} \subseteq A_0 \quad \text{and} \quad D_0 \in \sigma(A_0).$$

Since $H \subseteq \theta(Q)$ we have that $B_0 \subseteq \theta(A_0)$, and so we may put

$$\mu = \theta_B\lambda, \quad \lambda_0 = \lambda|A_0 \quad \text{and} \quad \mu_0 = \theta_{B_0}\lambda_0.$$

Then clearly μ_0 is the restriction of μ to B_0, and since $\lambda_0 \in Pr_c(A_0)$ it follows from (ii), (iii) and (4.13.4) that $\mu_0 \in Pr_c(B_0)$. Thus μ is perfect and the theorem is proved. □

Theorem 4.14. *Let* (T,A) *and* (S,B) *be algebraic function spaces, and let* $\theta: T \rightsquigarrow S$ *be a correspondance so that* $B \subseteq \theta(A)$. *Let* $\lambda \in Pr_\xi(A)$, *where* ξ *is an infinite cardinal, and suppose that* $\lambda_\xi^*(\theta^{-1}(S)) = 1$. *Then* $\mu = \theta_B\lambda \in Pr_\xi(B)$ *and*

(4.14.1) $$\int_T f d\lambda_\xi = \int_S g d\mu_\xi \quad \forall g \in L^1(\mu_\xi) \quad \forall f \in L^1(\lambda_\xi) \cap (g\theta)$$

(4.14.2) $$\int_S g d\mu_\xi = \int^* g^* d\lambda_\xi = \int_* g_* \, d\lambda_\xi \quad \forall g \in L^1(\mu_\xi)$$

where

$$g^*(t) = \begin{cases} \sup\{g(s) \mid s \in \theta(t)\} & \forall t \in \theta^{-1}(S) \\ \inf\{g(s) \mid s \in S\} & \forall t \in T \setminus \theta^{-1}(S) \end{cases}$$

$$g_*(t) = \begin{cases} \inf\{g(s) \mid s \in \theta(t)\} & \forall t \in \theta^{-1}(S) \\ \sup\{g(s) \mid s \in S\} & \forall t \in T \setminus \theta^{-1}(S) \end{cases}$$

And if $g: S \to \bar{\mathbb{R}}$ *is* μ_ξ*-measurable, there exist* $\sigma(B_\xi)$*-measurable maps* g_0 *and* g_1, *and there exist* $\sigma(A_\xi)$*-measurable maps* $f_j \in g_j\theta$ *for* $j = 0,1$, *so that*

(4.14.3) $$g_0 \le g \le g_1, \quad f_0 \le f_1 \wedge g_*, \quad f_1 \ge f_0 \vee g^*$$

(4.14.4) $$f_0 = f_1 \quad \lambda_\xi\text{-a.s., and} \quad g_0 = g = g_1, \quad \mu_\xi\text{-a.s.}$$

Remark. Note that (4.14.1) simply states that

(4.14.5) $$\theta_L \lambda_\xi = (\theta_B\lambda)_\xi$$

where $L = \sigma(B_\xi)$.

•

<u>Proof</u>. Let $\hat{A} = B(T,\sigma(A_\xi))$ and $\hat{B} = B(S,\sigma(B_\xi))$, and put

$$\hat{\lambda}(f) = \int_T f d\lambda_\xi \qquad \forall f \in \hat{A}$$

$$\hat{\mu}(g) = \int_S g d\mu_\xi \qquad \forall g \in \hat{B}$$

(note that $\mu \in Pr_\xi(B)$ by Theorem 4.13). Then $\hat{B} \subseteq \theta(\hat{A})$ and

$$\hat{\lambda}^*(\theta^{-1}(S)) = \lambda_\xi^*(\theta^{-1}(S)) = 1.$$

Hence by Theorem 4.13 we have that $\hat{\lambda}$ is θ-saturated and

$$\nu = \theta_{\hat{B}} \lambda \in Pr_\sigma(\hat{B}) .$$

Putting $\lambda = \hat{\lambda}$, $U = V = A_\xi \cap B(T)$ and $L = M = B_\xi \cap B(S)$ in Proposition 3.16 we see that ν is ξ-supersmooth at $B_\xi \cap B(S)$ along $B_\xi \cap B(S)$. And since $\nu = \mu$ on B, we see that $\nu = \hat{\mu}$ by uniqueness of μ_ξ. Hence

$$\int_T f d\lambda_\xi = \int_S g d\mu_\xi \quad \forall g \in B \quad \forall f \in \hat{A} \cap (g\theta).$$

From which (4.14.1) follows easily.

Now let $g\colon S \to \bar{\mathbb{R}}$ be μ_ξ-measurable, and put $a = m(g)$ and $b = M(g)$. Then there exist $\sigma(B_\xi)$-measurable maps g_0 and g_1 so that

(i) $\qquad a \le g_0 \le g \le g_1 \le b, \quad g_0 = g = g_1 \quad \mu_\xi\text{-a.s.}$

And by (4.14) there exist $\sigma(A_\xi)$-measurable functions $f_0 \in g_0\theta$ and $f_1 \in g_1\theta$, and clearly we may assume that

(ii) $\qquad a \le f_0 \le f_1 \le b$

Since $f_0(t) \le g(s) \le f_1(t)$ for all $s \in \theta(t)$, and $f_0 \le b$, and $f_1 \ge a$, we see that (4.14.3) holds. Moreover

$$\text{Arctg } f_j \in (\text{Arctg } g_j)\theta$$

and so

$$\int_T (\text{Arctg } f_0)d\lambda_\xi = \int_S g d\mu_\xi = \int_T (\text{Arctg } f_1)d\lambda_\xi$$

by (4.14.1). Since $f_0 \leq f_1$, this implies that $f_0 = f_1$ λ_ξ-a.s. and so (4.14.4) holds.

Now suppose that $g \in L^1(\mu_\xi)$, then it follows easily that $f_j \in L^1(\lambda_\xi)$ and

$$\int_S g d\mu_\xi = \int_T f_0 d\lambda_\xi = \int_T f_1 d\lambda_\xi$$

Since $f_0 \leq g_*$ and $f_1 \geq g^*$ we find

(iii) $$\int^* g^* d\lambda_\xi \leq \int_S g d\mu_\xi \leq \int_* g_* d\lambda_\xi .$$

Now let $h \in L^1(\lambda_\xi)$ so that $h \geq g^*$. Then

$$h(t) \geq g(s) \geq f_0(t) \quad \forall s \in \theta(t).$$

Hence $h \geq f_0$ on $\theta^{-1}(S)$ and so $h \geq f_0$ λ_ξ-a.s. Thus we have

$$\int_S h d\lambda_\xi \geq \int_S f_0 d\lambda_\xi = \int_T g d\mu_\xi \ .$$

Taking infimum over h we see that

(iv) $$\int^* g^* d\lambda_\xi \geq \int_S g d\mu_\xi$$

And similarly we find that

(v) $$\int_* g_* d\lambda_\xi \leq \int_S g d\mu_\xi$$

Thus (4.14.2) follows from (iii)-(v). □

5. Outer probability content

Let T be a set, then an outer (resp. inner) probability content on T is a positive functional χ from $\overline{\mathbb{R}}^T$ into $\overline{\mathbb{R}}$, such that χ is subadditive on $\overline{\mathbb{R}}^T$ (resp. superadditive on $\overline{\mathbb{R}}^T$), χ is positively homogenuous, i.e. $\chi(af) = a\,\chi(f)$ $\forall f \in \overline{\mathbb{R}}^T$ $\forall a \in \mathbb{R}_+$, and $\chi(\pm 1_T) = \pm 1$. Clearly we have

(5.1) χ is an outer (inner) probability content, if and only if χ° is an inner (outer) probability content.

And if χ is an outer probability content on T, then we have (see (3.8) and (3.9))

(5.2) $\chi^\circ(f) \,\dot{+}\, \chi^\circ(g) \le \chi^\circ(f \dot{+} g) \le \chi^\circ(f \dot{+} g) \le \chi^\circ(f) \,\dot{+}\, \chi(g)$

(5.3) $\chi^\circ(f) \,\dot{+}\, \chi(g) \le \chi(f \dot{+} g) \le \chi(f \dot{+} g) \le \chi(f) \,\dot{+}\, \chi(g)$

(5.4) $\chi^\circ(f) \le \chi(f)$

for all $f,g \in \overline{\mathbb{R}}^T$.

If χ is an outer probability content on T, then we define

$$L^1(\chi) = \{f \in \mathbb{R}^T \mid -\infty < \chi^\circ(f) = \chi(f) < \infty\}$$

$$L(\chi) = \{f \in \overline{\mathbb{R}}^T \mid \chi^\circ(f) = \chi(f)\}$$

$$\|f\|_\chi = \chi(|f|) \qquad \forall f \in \overline{\mathbb{R}}^T$$

Then $\|\cdot\|_\chi$ is a seminorm on $\overline{\mathbb{R}}^T$, and the reader easily verifies that we have

(5.5) $\|f\|_\chi \le \|f\|$ $\quad \forall f \in \overline{\mathbb{R}}^T$

(5.6) $L^1(\chi)$ is a $\|\cdot\|_\chi$-closed linear subspace of $\mathbb{R}^T$

(5.7) χ is positive linear functional on $L^1(\chi)$

Putting $\mu = \chi$ and $\nu = \chi^\circ$ in Proposition 3.4 gives

(5.8) $L(\chi) = R^*(\chi, L(\chi)) = R_*(\chi, L(\chi))$

(5.9) $G \cap F^{\xi} \subseteq L(\chi)$ if $F \subseteq L(\chi)$ is $(\wedge f)$-stable and χ is ξ-subsmooth at G along F

(5.10) $f \in L(\chi)$ if $\forall \varepsilon > 0 \ \exists f_0, f_1 \in \overline{\mathbb{R}}^T$ so that $f_0 \leq f \leq f_1$ and $\chi(f_1) \leq \varepsilon + \chi^{\circ}(f_0)$

(5.11) $G \subseteq L(\chi)$ if χ is σ-subsmooth at G along G, and G satisfies (3.4.9) for some $(\wedge f, \vee f)$-stable set F, such that $F \subseteq L(\chi) \cap (-G)$

We let $Pr^*(T)$ denote the set of all outer probability contents on T. If (T,A) is an algebraic function space and $\chi \in Pr^*(T)$ we define

$$Pr(A|\chi) = \{\lambda \in Pr(A) \mid \lambda(\varphi) \leq \chi(\varphi) \quad \forall \varphi \in A\}$$

and if ξ is an infinite cardinal and α denotes any of the five symbols: $\xi, c\xi, \pi, \sigma$ or τ then we put

$$Pr_{\alpha}(A|\chi) = Pr(A|\chi) \cap Pr_{\alpha}(A)$$

<u>Theorem 5.1</u>. <u>Let</u> $\chi \in Pr^*(T)$, <u>let</u> $H \subseteq B(T)$ <u>and let</u> θ <u>be a map from</u> H <u>in</u> $\mathbb{R}$ <u>satisfying</u>

(5.1.1) $$\sum_{j=1}^{n} \theta(h_j') - \sum_{i=1}^{m} \theta(h_j'') \leq \chi\left(\sum_{j=1}^{n} h_j' - \sum_{i=1}^{m} h_j''\right)$$

<u>whenever</u> $n, m \in \mathbb{N}_0$ <u>and</u> $h_1', \ldots, h_n', h_1'', \ldots, h_m'' \in H$. <u>Then there exist</u> $\lambda \in Pr(B(T)|\chi)$ <u>such that</u> $\lambda(h) = \theta(h)$ <u>for all</u> $h \in H$.

<u>Proof</u>. Let $h_1,\ldots,h_n \in H$ be given and define

$$F(x) = \sum_{j=1}^{n} x_j\theta(h_j), \qquad G(x) = \chi(\sum_{j=1}^{n} x_j h_j)$$

for $x = (x_1,\ldots,x_n) \in \mathbb{R}^n$. Then F is linear and G is positively homegenuous, finite and subadditive on $\mathbb{R}^n$. Hence G is convex and so F and G are continuous. By repeating h_j a finite number of times in (5.1.1) we see that $F(x) \le G(x)$ whenever $x \in \mathbb{Z}^n$. By positive homogenuity of F and G we have that $F(x) \le G(x)$ for all x with rational coordinates. Hence by continuity of F and G we have

$$\text{(i)} \qquad \sum_{j=1}^{n} x_j\theta(h_j) \le \chi(\sum_{j=1}^{n} x_j h_j) \qquad \forall\,(x_1,\ldots,x_n) \in \mathbb{R}^n$$

Replacing x by $-x$ we find

$$\text{(ii)} \qquad \sum_{j=1}^{n} x_j\theta(h_j) \ge \chi^o(\sum_{j=1}^{n} x_j h_j)$$

Hence if we put

$$\lambda_0(f) = \sum_{j=1}^{n} x_j\theta(h_j) \quad \text{for} \quad f = \sum_{j=1}^{n} x_j h_j, \quad h_1\ldots h_n \in H$$

then λ_0 is a well defined linear functional on the linear span of H and

$$\text{(iii)} \qquad \chi^o(f) \le \lambda_0(f) \le \chi(f) \qquad \forall f \in \operatorname{span} H$$

And by Hahn-Banach's theorem we have that λ_0 admits an extension λ to $B(T)$, such that

$$\text{(iv)} \qquad \chi^o(f) \le \lambda(f) \le \chi(f) \qquad \forall f \in B(T)$$

But then clearly $\lambda \in \Pr(B(T)|\chi)$ and by definition of λ_0 we have $\lambda(h) = \lambda_0(h) = \theta(h)$ for all $h \in H$. □

Corollary 5.2. *Let* $\chi \in Pr^*(T)$ *and let* H *be a subset of* $B(T)$ *satisfying*

$$(5.2.1) \qquad \chi\left(\sum_{j=1}^{n} h_j\right) = \sum_{j=1}^{n} \chi(h_j) \qquad h_1,\ldots h_n \in H$$

Then there exist $\lambda \in Pr(B(T)|\chi)$ *so that* $\lambda(h) = \chi(h)$ *for all* $h \in H$.

And if $f \in B(T)$ *and* $\chi^{\circ}(f) \leq a \leq \chi(f)$, *then there exist* $\nu \in Pr(B(T)|\chi)$ *so that* $\nu(f) = a$.

Remark. If $H \subseteq B(T) \cap L^1(\chi)$ then (5.2.1) holds, since χ is linear on $L^1(\chi)$. We shall later see that (5.2.1) also holds if the order structure of H is sufficently regular.

Proof. Suppose that (5.2.1) holds and let $f_j \in H$ for $j = 1,\ldots,n+m$ Then by (5.3) and (5.2.1) we have

$$\sum_{j=1}^{n} \chi(f_j) - \sum_{j=n+1}^{n+m} \chi(f_j) = \chi\left(\sum_{j=1}^{n} f_j\right) - \chi\left(\sum_{j=n+1}^{n+m} f_j\right)$$

$$= \chi\left(\sum_{j=1}^{n} f_j\right) + \chi^{\circ}\left(-\sum_{j=n+1}^{n+m} f_j\right)$$

$$\leq \chi\left(\sum_{j=1}^{n} f_j - \sum_{j=n+1}^{n+m} f_j\right)$$

Thus the first part of the corollary follows from Theorem 5.1.

Putting $H = \{f\}$ and $\theta(f) = a$, we see that the second part of the corollary follows from Theorem 5.1. □

Theorem 5.3. *Let* (T,A) *be an algebraic function space let* $\chi \in Pr^*(T)$ *and let* $M \subseteq Pr(A)$ *such that*

$$(5.3.1) \qquad \chi(\varphi) = \sup\{\lambda(\varphi) \mid \lambda \in M\} \qquad \forall \varphi \in A$$

Let $0 \in H \subseteq A_\xi$, where ξ is an infinite cardinal number, and let F be the set of all $f \in \overline{\mathbb{R}}^T$ satisfying

(5.3.2) $$\chi(f) = \inf\{\chi(g) + \chi(h) \mid g \in A^\xi, h \in H, f \leq g+h\}$$

Then $F \supseteq A^\xi \cup H$ and the following four statements are equivalent:

(5.3.3) χ is ξ-supersmooth at H along A

(5.3.4) χ is ξ-supersmooth at F along A_ξ

(5.3.5) $\tilde{M} \subseteq Pr_\xi(A)$ and $\lambda^*(h) \leq \chi(h) \quad \forall \lambda \in \tilde{M} \ \forall h \in H$

(5.3.6) $\lambda \in Pr_\xi(A)$ and $\lambda^*(f) \leq \chi(f) \quad \forall \lambda \in Pr(A|\chi) \quad \forall f \in F \cap A_\xi$ and $\forall f \in A_\xi \cap F \ \exists \lambda \in \tilde{M}$ so that $\lambda^*(f) = \chi(f)$.

where $\tilde{M}$ is the closure of M in the product topology on $\mathbb{R}^A$. In particular if M satisfies (5.3.1), then

(5.3.7) $\tilde{M} \subseteq Pr_\xi(A) \Leftrightarrow \chi$ is ξ-supersmooth at 0 along A

(5.3.8) χ is ξ-supersmooth at A_ξ along A_ξ, if and only if $\tilde{M} \subseteq Pr_\xi(A)$ and $\lambda^*(f) \leq \chi(f) \quad \forall \lambda \in \tilde{M} \ \forall f \in A_\xi$

Remarks (1): Note that $Pr(A)$ and $Pr(A|\chi)$ are closed compact subsets of $\mathbb{R}^A$, and notice that $M = Pr(A|\chi)$ satisfies (5.3.1).

(2): Suppose that $\lambda \in Pr_\xi(A|\chi)$ and $\lambda^*(f) \leq \chi(f)$ for all $f \in A_\xi$. Then by Theorem 4.1 we have

(5.3.9) $$\int_* h d\lambda_\xi \leq \chi(h), \quad \chi^\circ(h) \leq \int^* h d\lambda_\xi \qquad \forall h \in \overline{\mathbb{R}}^T$$

(3): Let M be a subset of $Pr(A)$ satisfying (5.3.1), then by Hahn-Banach's theorem we have

(5.3.10) $Pr(A|\chi) =$ the closed convex hull of M in $\mathbb{R}^A$.

Proof. (5.3.3) ⇒ (5.3.4): Let $\{\varphi_\gamma | \gamma \in \Gamma\}$ be a decreasing net in A_ξ, such that $\mathrm{card}(\Gamma) \leq \xi$ and $\varphi_\gamma \downarrow f$ for some $f \in F$. Let $g \in A^\xi$ and $h \in H$ be chosen so that $f \leq g + h$. Then $f - g \leq h$ and if $\psi_\gamma = (\varphi_\gamma - g) \vee h$, then $\psi_\gamma \downarrow h$ and $\psi_\gamma \in A_\xi$. By Proposition 3.5. (2) and (5.3.3) we have that

$$\lim_\gamma \chi(\psi_\gamma) = \chi(h)$$

And since $\varphi_\gamma \leq g + \psi_\gamma$ we find

$$\chi(f) \leq \lim_\gamma \chi(\varphi_\gamma) \leq \lim_\gamma (\chi(g) + \chi(\psi_\gamma))$$

$$\leq \chi(g) + \chi(h)$$

And since f satisfies (5.3.2) we have that

$$\chi(f) = \lim \chi(\varphi_\gamma)$$

and so (5.3.4) holds.

(5.3.4) ⇒ (5.3.6): Let $\lambda \in \Pr(A|\chi)$. Since $0 \in F$ we see that $\lambda \in \Pr_\xi(A)$. Now let $f \in F \cap A_\xi$, then there exist $\{\varphi_\gamma | \gamma \in \Gamma\} \subseteq A$, so that $\mathrm{card}(\Gamma) \leq \xi$, and $\varphi_\gamma \downarrow f$. Hence we have by (5.3.4) and Theorem 4.

$$\lambda^*(f) = \lim_\gamma \lambda(\varphi_\gamma) \leq \lim_\gamma \chi(\varphi_\gamma) = \chi(f)$$

for all $\lambda \in \Pr(A|\chi)$, since $\Pr(A|\chi) \subseteq \Pr_\xi(A)$. And by (5.3.1) there exist $\lambda_{n\gamma} \in M$ so that

(i) $$\lambda_{n\gamma}(\varphi_\gamma) \geq \chi(\varphi_\gamma) - 2^{-n} \quad \forall \gamma \in \Gamma \quad \forall n \geq 1$$

Let $\mathbb{N} \times \Gamma$ have its product ordering, then $\{\lambda_{n\gamma}\}$ is a net in M, and since $M \subseteq \Pr(A|\chi)$ and $\Pr(A|\chi)$ is compact in $\mathbb{R}^A$, there exist a limit point $\lambda \in \widetilde{M} \subseteq \Pr(A|\chi)$ of $\{\lambda_{n\gamma}\}$ in $\mathbb{R}^A$. If (n,γ) is given then we have

$$\lambda(\varphi_\gamma) \geq \inf\{\lambda_{m\beta}(\varphi_\gamma) \mid m \geq n, \beta \geq \gamma\}$$

$$\geq \inf\{\lambda_{m\beta}(\varphi_\beta) \mid m \geq n, \beta \geq \gamma\}$$

$$\geq \inf_\beta \chi(\varphi_\beta) - 2^{-n} \geq \chi(f) - 2^{-n}$$

So by Theorem 4.1 we have that $\lambda^*(f) \geq \chi(f)$ and the converse inequality has just been proved above. Thus $\lambda \in \widetilde{M}$ and $\lambda^*(f) = \chi(f)$.

(5.3.6) ⇒ (5.3.5): Since $M \subseteq Pr(A|\chi)$ by (5.3.1) and $H \subseteq F$ we see that (5.3.6) implies (5.3.5).

(5.3.5) ⇒ (5.3.1): Let $\{\varphi_\gamma \mid \gamma \in \Gamma\}$ be a net in A so that $\text{card}(\Gamma) \leq \xi$, and $\varphi_\gamma \downarrow h$ for some $h \in H$. Then by (5.3.1) there exist $\lambda_{n\gamma} \in M$ satisfying (i), and as above there exist a limit point $\lambda \in \widetilde{M}$ of $\{\lambda_{n\gamma}\}$ in $\mathbb{R}^A$. Since $\widetilde{M} \subseteq Pr_\xi(A)$ by assumption we find as above that

$$\chi(h) \leq \lim_\gamma \chi(\varphi_\gamma) \leq \lim_\gamma \lambda(\varphi_\gamma) = \lambda^*(h)$$

and since $\lambda^*(h) \leq \chi(h)$ by assumption we see that χ is ξ-supersmooth at H along A. □

If M is a subset of $Pr(A)$, and ξ is an infinite cardinal, then we say that M is <u>uniformly</u> ξ-<u>compact</u> if

(5.11) $\quad \forall \varepsilon > 0 \ \exists K \in K_\xi(A)$ so that $\lambda^*(K) \geq 1 - \varepsilon \quad \forall \lambda \in M$

If $\xi =$, then we say that M is <u>uniformly semicompact</u>, and if $\xi \geq \text{weight}(A)$, then we say that M is <u>uniformly compact</u>.

Theorem 5.4. *Let* (T,A) *be an algebraic function space, let* $\chi \in Pr^*(T)$ *and let* ξ *be an infinite cardinal. Let* F *be the set of all* $f \in \overline{\mathbb{R}}^T$ *satisfying*

$$(5.4.1) \qquad \chi(f) = \sup\{\chi(\varphi) \mid \varphi \in A_\xi, \varphi \leq f\}$$

and let M *be a subset of* Pr(A) *satisfying*

$$(5.4.2) \qquad \chi(\varphi) = \sup\{\lambda(\varphi) \mid \lambda \in M\} \qquad \forall \varphi \in A$$

Then $F \supseteq A_\xi$ *and the following three statements are equivalent*

(5.4.3) χ *is* $\bar{K}_\xi(A)$*-tight along* F

(5.4.4) χ *is* $K_\xi(A)$*-tight along* A

(5.4.5) M *is uniformly* ξ*-compact*

Proof. Since F evidently contains A we see that (5.4.3) implies (5.4.4), and obviously we have that (5.4.4) and (5.4.2) implies (5.4.5).

(5.4.5) ⇒ (5.4.3): Let $0 < \varepsilon < 1$ be given, then there exist $K \in \bar{K}_\xi(A)$, so that $\lambda_*(T \setminus K) < \varepsilon$ for all $\lambda \in M$. Now let $f \in F$ so that

$$\text{(i)} \qquad f \leq \varepsilon + 1_{T \setminus K}$$

and choose $\varphi \in A_\xi$ such that $\varphi \leq f$. Then there exist $\{\varphi_\gamma \mid \gamma \in \Gamma\} \subseteq A$, so that $card(\Gamma) \leq \xi$, $\varphi_\gamma \downarrow \varphi$ and $\varphi_\gamma \leq 2$ for all γ. Since $\varphi(t) \leq \varepsilon$ for all $t \in K$, we have

$$\{\varphi_\gamma \geq 2\varepsilon\} \cap K \downarrow \emptyset$$

And so there exist $\beta \in \Gamma$, such that $K \cap \{\varphi_\beta \geq 2\varepsilon\} = \emptyset$, hence $\varphi_\beta \leq 2\varepsilon + 2\, 1_{T \setminus K}$, and so

$$\lambda(\varphi_\beta) \leq 2\varepsilon + 2\lambda_*(T \setminus K) \leq 4\varepsilon \qquad \forall \lambda \in M$$

So by (5.4.2) we have $\chi(\varphi) \leq \chi(\varphi_\beta) \leq 4\varepsilon$ for all $\varphi \in A_\xi$ with $\varphi \leq f$. And since f satisfies (5.4.1) we have that $\chi(f) \leq 4\varepsilon$, and thus χ is $\bar{K}_\xi(A)$-tight along F. □

Theorem 5.5. Let $\chi \in Pr^*(T)$ and let M be a set of increasing functionals on T. Suppose that F and G are subsets of $\overline{\mathbb{R}}^T$ satisfying

(5.5.1) F is $(\wedge f, \vee f)$-stable

(5.5.2) χ is σ-supersmooth at $G \cap F_\delta$ along F

(5.5.3) χ is σ-subsmooth at G along G

(5.5.4) $\forall g \in G\ \exists \varphi \in F_\delta : \varphi \leq g$ and $G \supseteq \{f \in S(F) \mid \varphi \leq f \leq g\}$

(5.5.5) $\chi(f) = \sup\{\lambda^*(f) \mid \lambda \in M\} \quad \forall f \in F \cup (G \cap F_\delta)$

Let $H = G \cap F_\delta$ and let $g \in G \cap S(F)$, then we have

(5.5.6) $\chi(g) = \sup\{\lambda^*(h) \mid h \in H, h \leq g, \lambda \in M\}$

(5.5.7) $\chi(g) = \sup\{\chi(h) \mid h \in H, h \leq g\}$

Moreover if $F \subseteq A_\xi$ and $M \subseteq Pr_\xi(A)$ for some algebra $A \subseteq B(T)$ containing 1_T, and some infinite cardinal ξ, then

(5.5.8) $\chi(g) = \sup\{\int_T g\, d\lambda_\xi \mid \lambda \in M\}$

for all $g \in G \cap S(F)$ satisfying

(5.5.9) $\sup\{\lambda^*(\varphi) \mid \varphi \in F_\delta, \varphi \leq g\} > -\infty \qquad \forall \lambda \in M$

<u>Remark</u>. Since $F \subseteq A_\xi$, it follows that g is λ_ξ-measurable for all $g \in S(F)$ and all $\lambda \in Pr_\xi(A)$. Thus the integrals in (5.5.8) exist by (5.5.9).

<u>Proof</u>. We shall apply the last part of Proposition 3.4 with $F = F$, $G = G$, $\mu = \chi$ and

$$\nu(f) = \sup\{\lambda^*(f) \mid \lambda \in M\} \qquad \forall f \in \overline{\mathbb{R}}^T$$

Then by (5.5.5) we have that $\nu(f) = \chi(f)$ for all $f \in H \cup F$, so $F \subseteq \{f \mid \chi(f) \leq \nu(f)\}$. Clearly G satisfies (3.4.9) by (5.5.4), and χ is σ-subsmooth at G along G, by (5.5.3). Since $\chi = \nu$ on $F \cup H$ it follows from (5.5.2) that ν is σ-supersmooth at H along F. Hence by Proposition 3.4 we have

$$\begin{aligned}\chi(g) &\leq \nu_H(g) = \sup\{\lambda^*(h) \mid h \in H, h \leq g, \lambda \in M\}\\ &= \sup\{\nu(h) \mid h \in H, h \leq g\}\\ &= \sup\{\chi(h) \mid h \in H, h \leq g\}\\ &\leq \chi(g)\end{aligned}$$

for all $g \in G \cap S(F)$, since $\nu = \chi$ on H. Hence (5.5.6) and (5.5.7) holds.

Now suppose that $M \subseteq Pr_\xi(A)$ and $F \subseteq A_\xi$ for some algebra $A \subseteq B(T)$ containing 1_T. Then by (5.5.6) and Theorem 4.1 we have

$$\text{(i)} \qquad \chi(g) \leq \sup\{\textstyle\int_* g d\lambda_\xi \mid \lambda \in M\} \qquad \forall g \in G \cap S(F)$$

since $H \subseteq F_\delta \subseteq A_\xi$. Now let $\lambda \in M$ be given and let G_λ be the set of all $g \in G$ satisfying

(ii) $\sup\{\lambda^*(\varphi) \mid \varphi \in F_\delta, \varphi \leq g\} > -\infty$

If $g \in G_\lambda$, then there exist $\varphi_0, \varphi_1 \in F_\delta$, such that $\varphi_j \leq g$ for $j = 0,1$ and

(iii) $\lambda^*(\varphi_0) > -\infty$ and $G \supseteq \{f \in S(F) \mid \varphi_1 \leq f \leq g\}$

Then $\varphi = \varphi_0 \vee \varphi_1 \in F_\delta$ by $(\wedge f, \vee f)$-stability of F, and $\varphi \leq f$. Hence

$$G_\lambda \supseteq \{f \in S(F) \mid \varphi \leq f \leq g\}$$

since $\lambda^*(\varphi) \geq \lambda^*(\varphi_0) > -\infty$. Thus G_λ satisfies (3.4.9), and so applying the last part of Proposition 3.4 with $F = F$, $G = G_0$, $\nu = \lambda^*$ and

$$\mu(f) = \int^* f d\lambda_\xi$$

we see that

(iv) $\int_T g d\lambda_\xi \leq \sup\{\lambda^*(h) \mid h \in H, h \leq g\}$

for all $g \in G_\lambda \cap S(F)$.

If $g \in G \cap S(F)$ satisfies (5.5.9), then $g \in G_\lambda$ for all $\lambda \in M$, and so (5.5.8) follows from (i), (iv) and (5.5.6). □

<u>Theorem 5.6</u>. <u>Let</u> $\chi \in Pr^*(T)$, <u>let</u> L <u>be a linear subspace of</u> $\overline{\mathbb{R}}^T$, <u>and let</u> M <u>be a convex set of linear functionals on</u> L, <u>such that</u>

(5.6.1) M <u>is closed in product topology of</u> $\mathbb{R}^L$

(5.6.2) $\inf\{\mu(\varphi) \mid \mu \in M\} \leq \chi(\varphi) < \infty \quad \forall \varphi \in L$

Then there exist $\mu \in M$ satisfying

(5.6.3) $\chi^{o}(\varphi) \leq \mu(\varphi) \leq \chi(\varphi) \quad \forall \varphi \in L.$

Proof. Let L be the set of all linear functionals μ on L satisfying (5.6.3). Let $\varphi \in L$ then

$$\infty > \chi(\varphi) \geq \chi^{o}(\varphi) = -\chi(-\varphi) > -\infty$$

by (5.6.2), since $(-\varphi) \in L$. Hence by Tychonov's theorem we have that L is a compact convex subset of $\mathbb{R}^L$.

Let L* be the set of all linear functionals on L, equipped with its $\sigma(L^*,L)$-topology. Then M is a convex closed subset of L* and L is a convex compact subset L*.

Now suppose that $M \cap L = \emptyset$, then by [11, 20.7.(1) p. 243] there exist $\varphi_0 \in L$, such that

(i) $$\sup_{\lambda \in L} \lambda(\varphi_0) < \inf_{\mu \in M} \mu(\varphi_0)$$

And by Hahn-Banach's theorem, there exist $\lambda_0 \in L^*$ such that $\lambda_0(\varphi_0) = \chi(\varphi_0)$, and $\lambda_0(\varphi) \leq \chi(\varphi)$ for all $\varphi \in L$. But then

$$\chi(\varphi) \geq \lambda_0(\varphi) = -\lambda_0(-\varphi) \geq -\chi(-\varphi) = \chi^{o}(\varphi)$$

for all $\varphi \in L$, and so $\lambda_0 \in L$. Hence by (i) we have that

$$\chi(\varphi_0) = \lambda_0(\varphi_0) < \inf_{\mu \in M} \mu(\varphi_0)$$

But this contradicts (5.6.2) and so $M \cap L \neq \emptyset$. Thus there exist $\mu \in M$ satisfying (5.6.3). □

<u>Proposition 5.7.</u> <u>Let</u> $\chi \in Pr^*(T)$ <u>and let</u> $M \subseteq Pr^*(T)$. <u>Now put</u>

$$\zeta(f) = \sup\{\mu(f) \mid \mu \in M\} \qquad \forall f \in \overline{\mathbb{R}}^T$$

$$E = \{f \in \overline{\mathbb{R}}^T \mid \chi(f) = \zeta(f)\}$$

$$E_0 = \{f \in \overline{\mathbb{R}}^T \mid \exists \mu \in M \text{ so that } \mu(f) = \chi(f)\}$$

<u>Then</u> $\zeta \in Pr^*(T)$, <u>and if</u> $\zeta \leq \chi$, <u>then we have</u>

(5.7.1) E <u>is closed in</u> $(\overline{\mathbb{R}}^T, \|\cdot\|_\chi)$

(5.7.2) <u>If</u> $f_0 \in E_0$ <u>and</u> $\|f-f_0\|_\chi = 0$, <u>then</u> $f \in E_0$

<u>Proof</u>. I shall leave the simple proof to the reader. □

<u>Definition 5.8</u>. Let $(X,\leq)$ be a <u>preordered set</u>, i.e. $\leq$ is reflexive and transitive), then $I(X)$ denotes the set of all <u>increasing functions</u> from X into $\overline{\mathbb{R}}$, and if $\xi \geq 1$ is a cardinal number then we put

$$I^\xi(x) = \left\{\varphi \in I(x) \;\middle|\; \begin{array}{l} \forall Y \subseteq X:\ 0 < \text{card } Y \leq \xi \quad \exists x \in X \quad \text{so that} \\ x \geq y \quad \forall y \in Y \quad \text{and} \quad \varphi(x) = \sup\limits_{y \in Y} \varphi(y) \end{array}\right\}$$

$$I_\xi(X) = \left\{\varphi \in I(X) \;\middle|\; \begin{array}{l} \forall Y \subseteq X:\ 0 < \text{card } Y \leq \xi \quad \exists x \in X \quad \text{so that} \\ x \leq y \quad \forall y \in Y \quad \text{and} \quad \varphi(x) = \inf\limits_{y \in Y} \varphi(y) \end{array}\right\}$$

As usual we write $I^\sigma(X)$ and $I_\sigma(X)$ if $\xi = \aleph_0$, and we write $I^\tau(X)$ and $I_\tau(X)$ if $\xi \geq \text{card } X$. It is easily checked that we have

(5.8.1) $I^\sigma(\overline{\mathbb{R}}) = I^\tau(\mathbb{R}) = I(\overline{\mathbb{R}}) \cap Lsc(\overline{\mathbb{R}}) = I(\overline{\mathbb{R}}) \cap C_\ell(\overline{\mathbb{R}})$

(5.8.2) $I_\sigma(\overline{\mathbb{R}}) = I_\tau(\mathbb{R}) = I(\overline{\mathbb{R}}) \cap Usc(\overline{\mathbb{R}}) = I(\overline{\mathbb{R}}) \cap C_r(\overline{\mathbb{R}})$

where $C_\ell(\overline{\mathbb{R}})$ and $C_r(\overline{\mathbb{R}})$ denote the set of all left resp. right continuous functions from $\overline{\mathbb{R}}$ into $\overline{\mathbb{R}}$.

Let $\{f_\gamma | \gamma \in \Gamma\}$ be a net in $\overline{\mathbb{R}}^T$, and let $f \in \overline{\mathbb{R}}^T$, then we write $\lim_\gamma \sup f_\gamma << f$, if and only if:

$$\text{(5.8.3)} \qquad \lim \sup f_\gamma(t_\gamma) \leq \sup_\gamma f(t_\gamma) \quad \forall (t_\gamma) \in T^\Gamma$$

and we write $f << \lim_\gamma \inf f_\gamma$, if $\lim \sup(-f_\gamma) << (-f)$. Now let ψ be a map: $S \to T$, and let $\varphi : \overline{\mathbb{R}} \to \overline{\mathbb{R}}$ be increasing and continuous, then clearly we have

$$\text{(5.8.4)} \qquad \lim_\gamma \sup f_\gamma << f \Rightarrow \lim_\gamma \sup f_\gamma \circ \psi << f \circ \psi$$

$$\text{(5.8.5)} \qquad \lim_\gamma \sup f_\gamma << f \Rightarrow \lim_\gamma \sup \varphi \circ f_\gamma << \varphi \circ f$$

$$\text{(5.8.6)} \qquad \lim_\gamma \sup f_\gamma << f \Rightarrow \lim_\gamma \sup f_\gamma \leq f$$

and the converse implication in (5.8.6) holds if $\{f_\gamma\}$ is an increasing net (or more generally if $f_\gamma \leq f$ for all $\gamma \in \Gamma$).

<u>Theorem 5.9</u>. <u>Let</u> $f \in \overline{\mathbb{R}}^T$ <u>and let</u> $\{f_\gamma | \gamma \in \Gamma\}$ <u>be a net in</u> $\overline{\mathbb{R}}^T$. <u>Let</u> ξ <u>be an infinite cardinal, such that</u> Γ <u>admits a cofinal set of cardinality at most</u> ξ. <u>Then the following seven statements are equivalent</u>

$$\text{(5.9.1)} \qquad \lim_\gamma \sup f_\gamma << f$$

$$\text{(5.9.2)} \qquad \lim_\gamma \sup M_S(f_\gamma) \leq M_S(f) \quad \forall S \subseteq T$$

$$\text{(5.9.3)} \qquad \lim_\gamma \sup M_S(f_\gamma) \leq M_S(f) \quad \forall S \subseteq T: \text{card}(S) \leq \xi$$

(5.9.4) $\exists \varphi_\gamma \in I_\tau(\overline{\mathbb{R}}) : f_\gamma \leq \varphi_\gamma \circ f$ <u>and</u> $\lim_\gamma \sup \varphi_\gamma(x) \leq x \quad \forall x \in \overline{\mathbb{R}}$

(5.9.5) $\exists (X, \leq)$ <u>a preordered set</u>, $\exists \varphi_\gamma \in I(X)$ $\exists \varphi \in I^\xi(X)$ $\exists \hat{f} : T \to X$, <u>so that</u> $f_\gamma \leq \varphi_\gamma \circ \hat{f}$, $\varphi \circ \hat{f} \leq f$ <u>and</u> $\lim_\gamma \sup \varphi_\gamma \leq \varphi$

(5.9.6) $\quad \lim_\gamma \sup f_\gamma(t_\gamma) \le \lim_\gamma \sup f(t_\gamma) \quad \forall (t_\gamma) \in T^\Gamma$

(5.9.7) $\quad \forall \varepsilon > 0 \ \forall a,b \in \mathbb{R} \ \exists \beta \in \Gamma : f_\gamma(t) \le a \vee (f(t)+\varepsilon)$

<u>for all</u> $\gamma \ge \beta$ <u>and all</u> $t \in \{f \le b\}$

<u>And if</u> $\{f_\gamma\}$ <u>is a decreasing net, then (5.9.1-7) is equivalent to</u>

(5.9.8) $\quad \inf_\gamma f_\gamma(t_\gamma) \le \sup_\gamma f(t_\gamma) \quad \forall (t_\gamma) \in T^\Gamma$

<u>Remark</u> (1): Note that (5.9.7) shows that (5.9.1) is a sort of uniform lim sup.

(2): The smallest cardinal ξ we may use in the lemma equals $\mathrm{cof}(\Gamma)$, where

$$\mathrm{cof}(\Gamma) = \min\{\mathrm{card}\,\Delta \mid \Delta \text{ is cofinal in } \Gamma\}$$

whenever Γ is a upwards directed set. If $\Delta \subseteq \Gamma$ is cofinal in Γ, then obviously we have that Δ is upwards directed, and the reader easily verifies that

(5.9.9) $\quad \mathrm{cof}(\Delta) = \mathrm{cof}(\Gamma) \quad \text{if} \quad \Delta \text{ is cofinal in } \Gamma$

<u>Proof</u>. (5.9.1) ⇒ (5.9.2): If S is empty then $M_S(f) = M_S(f_\gamma) = -\infty$ and so (5.9.2) holds. So suppose that $S \neq \emptyset$, and let $a < \lim\sup M_S(f_\gamma)$. Then

$$\Delta = \{\gamma \in \Gamma \mid M_S(f_\gamma) > a\}$$

is cofinal in Γ, and for all $\delta \in \Delta$, there exist $t_\delta \in S$ so that $f_\delta(t_\delta) > a$. Now choose $t_0 \in S$ and put $t_\gamma = t_0$ if $\gamma \in \Gamma \setminus \Delta$, then

$$M_S(f) \geq \sup_{\gamma\in\Gamma} f(t_\gamma) \geq \limsup_{\gamma\in\Gamma} f_\gamma(t_\gamma)$$

$$\geq \inf_{\delta\in\Delta} f_\delta(t_\delta) \geq a$$

since Δ is cofinal in Γ. Thus (5.9.2) holds

(5.9.2) ⇒ (5.9.3): Evident!

(5.9.3) ⇒ (5.9.4): Let $E_x = \{f \leq x\}$ for $x \in \overline{\mathbb{R}}$ and define

$$\psi_\gamma(x) = \sup\{f_\gamma(t) \mid t \in E_x\} \qquad (\sup \emptyset = -\infty)$$

$$\varphi_\gamma(x) = \inf\{\psi_\gamma(y) \mid y > x\} \qquad (\inf \emptyset = +\infty)$$

for $x \in \overline{\mathbb{R}}$. Then φ_γ is increasing and right continuous and so $\varphi_\gamma \in I_\tau(\overline{\mathbb{R}})$ by (5.8.2). By definition we have

(i) $\qquad f_\gamma \leq \psi_\gamma \circ f \leq \varphi_\gamma \circ f \qquad \forall \gamma \in \Gamma$

So let us show that

(ii) $\qquad \varphi(x) = \limsup_\gamma \varphi_\gamma(x) \leq x \qquad \forall x \in \overline{\mathbb{R}}$

If $\varphi(x) = -\infty$ or $x = \infty$ then (ii) is obvious. So suppose that $\varphi(x) > -\infty$ and $x < \infty$, and choose $a,b \in \mathbb{R}$, such that $a < \varphi(x)$ and $b > x$. Then by (5.9.9) there exist a cofinal set $\Delta \subseteq \Gamma$, so that card $\Delta \leq \xi$ and

$$\psi_\gamma(b) \geq \varphi_\gamma(x) > a \qquad \forall \gamma \in \Delta$$

By definition of $\psi_\gamma(b)$ there exist $t_\gamma \in E_b$, such that $f_\gamma(t_\gamma) > a$ for all $\gamma \in \Delta$. Let $S = \{t_\gamma \mid \gamma \in \Delta\}$ then $\text{card}(S) \leq \xi$ and by (5.9.3) we have

$$a \leq \inf_{\delta\in\Delta} M_S(f_\delta) \leq \limsup_\gamma M_S(f_\gamma) \leq M_S(f) \leq b$$

since Δ is cofinal, $S \subseteq E_b$ and $M_S(f_\delta) \geq f_\delta(t_\delta) \geq a$. Letting $a \uparrow \varphi(x)$ and $b \downarrow x$ we see that (ii) and thus (5.9.4) holds.

(5.9.4) ⇒ (5.9.5): Evident!

(5.9.5) ⇒ (5.9.6): Let $(t_\gamma) \in T^\Gamma$ and $a < \limsup f_\gamma(t_\gamma)$. Then by (5.9.9) there exist a cofinal set $\Delta \subseteq \Gamma$, so that $\operatorname{card} \Delta \leq \xi$ and $f_\delta(t_\delta) > a$ for all $\delta \in \Delta$. Now choose $(X,\leq)$, φ_δ, φ and $\hat{f}$ according to (5.9.5), and let $\Delta(\alpha) = \{\delta \in \Delta \mid \delta \geq \alpha\}$. Then $\operatorname{card} \Delta(\alpha) \leq \xi$ and $\varphi \in I^\xi(x)$, hence there exist $x_\alpha \in X$ so that

$$\text{(iii)} \qquad x_\alpha \geq \hat{f}(t_\delta) \quad \forall \delta \in \Delta(\alpha), \quad \varphi(x_\alpha) = \sup_{\delta\in\Delta(\alpha)} \varphi(\hat{f}(t_\delta))$$

for all $\alpha \in \Gamma$. Hence by (5.9.5) we have

$$\begin{aligned} a &\leq \inf_{\delta\in\Delta(\alpha)} f_\delta(t_\delta) \leq \inf_{\delta\in\Delta(\alpha)} \varphi_\delta(\hat{f}(t_\delta)) \\ &\leq \inf_{\delta\in\Delta(\alpha)} \varphi_\delta(x_\alpha) \leq \lim_\gamma \sup \varphi_\gamma(x_\alpha) \\ &\leq \varphi(x_\alpha) \leq \sup_{\gamma\geq\alpha} \varphi(\hat{f}(t_\gamma)) \\ &\leq \sup_{\gamma\geq\alpha} f(t_\gamma) \end{aligned}$$

for all $\alpha \in \Gamma$, since φ_δ is increasing and $\Delta(\alpha)$ is cofinal in Γ for all α. Hence (5.9.6) follows by letting $\alpha\to\infty$ and $a \uparrow \limsup f_\gamma(t_\gamma)$.

(5.9.6) ⇒ (5.9.7): Suppose that (5.9.7) does not hold. Then there exist $\varepsilon > 0$, $a,b \in \mathbb{R}$, a map $\sigma: \Gamma \to \Gamma$, and $(t_\gamma) \in T^\Gamma$, such that

$$\text{(iv)} \qquad \sigma(\gamma) \geq \gamma, \text{ and } f(t_\gamma) \leq b \qquad \forall \gamma \in \Gamma$$

$$\text{(v)} \qquad f_{\sigma(\gamma)}(t_\gamma) \geq a \vee (\varepsilon + f(t_\gamma)) \qquad \forall \gamma \in \Gamma$$

Let $m = \limsup f(t_\gamma)$ and choose $c \in \mathbb{R}$ so that $c < a$. Now put

$$p = m - \varepsilon/4, \qquad q = c \vee (m+\varepsilon/4)$$

Then $p < m$ if $m > -\infty$, and $p = -\infty$ if $m = -\infty$, moreover since $m \leq b < \infty$ by (iv) we have that $m < q$. Hence there exist a cofinal set $\Delta \subseteq \Gamma$ and $\alpha \in \Delta$ so that

$$\text{(vi)} \qquad p \leq f(t_\gamma) \quad \forall \gamma \in \Delta, \qquad f(t_\gamma) \leq q \quad \forall \gamma \geq \alpha$$

Let $\Lambda = \{\sigma(\gamma) \mid \gamma \in \Delta, \gamma \geq \alpha\}$, then Λ is cofinal in Γ by (iv) and there exist a map $\tau : \Gamma \to \Gamma$, such that $\tau(\gamma) = \alpha$ if $\alpha \in \Gamma \setminus \Lambda$ and

$$\text{(vii)} \qquad \tau(\gamma) \in \sigma^{-1}(\gamma) \cap \{\beta \in \Gamma \mid \beta \in \Delta, \beta \geq \alpha\} \qquad \forall \gamma \in \Lambda$$

Now put $u_\gamma = t_{\tau(\gamma)}$ for $\gamma \in \Gamma$. If $\gamma \in \Lambda$ and $\beta = \tau(\gamma)$ then $\sigma(\beta) = \gamma$, and $\beta \in \Delta$, hence by (v) and (vi) we have

$$f_\gamma(u_\gamma) = f_{\sigma(\beta)}(t_\beta) \geq a \vee (\varepsilon + f(t_\beta)) \geq a \vee (\varepsilon + p)$$

Since Λ is cofinal we have by (5.9.6) and (vi)

$$\begin{aligned} a \vee (\varepsilon + p) &\leq \limsup_{\lambda \in \Lambda} f_\lambda(u_\lambda) \leq \lim_\gamma \sup f_\gamma(u_\gamma) \\ &\leq \lim_\gamma \sup f(u_\gamma) \leq \sup_{\gamma \in \Gamma} f(t_{\tau(\gamma)}) \\ &\leq q \end{aligned}$$

since $\tau(\gamma) \geq \alpha$ for all γ. But $q < a \vee (\varepsilon + p)$, and so we have derive a contradiction. Thus (5.9.6) implies (5.9.7).

(5.9.7) ⇒ (5.9.1): Let $(t_\gamma) \in T^\Gamma$, and let $m = \sup f(t_\gamma)$ and $r = \lim \sup f_\gamma(t_\gamma)$. If $m = \infty$ or $r = -\infty$, then clearly we have that $r \leq m$. So suppose that $m < \infty$ and $r > -\infty$, and let $a, b \in \mathbb{R}$ be chosen so that $b \geq m$ and $a < r$. Then there exist $\varepsilon > 0$ such that $a + \varepsilon < r$, and by (5.9.7) there exist $\beta \in \Gamma$, such that

$$f_\gamma(t) \leq a \vee (\varepsilon + f(t)) \qquad \forall \gamma \geq \beta \quad \forall t \in \{f \leq b\}$$

Now since $f(t_\gamma) \leq m \leq b$ we have

$$f_\gamma(t_\gamma) \leq a \vee (\varepsilon + f(t_\gamma)) \leq a \vee (\varepsilon + m) \qquad \forall \gamma \geq \beta$$

and so we find

$$a + \varepsilon \leq \limsup f_\gamma(t_\gamma) \leq a \vee (\varepsilon + m)$$

Hence $a \leq m$ for all $a < r$, and so $r \leq m$ and (5.9.1) holds.

(5.9.1) $\Rightarrow$ (5.9.8) is evident.

(5.9.8) $\Rightarrow$ (5.9.2) (if $\{f_\gamma\}$ is increasing): Let $S \subseteq T$ and let $a < \limsup M_S(f_\gamma)$. Since $M_S(f_\gamma)$ is decreasing in γ, we have that $a < M_S(f_\gamma)$ for all $\gamma \in \Gamma$, and so there exist $t_\gamma \in S$ with $f_\gamma(t_\gamma) \geq a$ for all $\gamma \in \Gamma$. Hence by (5.9.8) we have

$$a \leq \inf_\gamma f_\gamma(t_\gamma) \leq \sup f(t_\gamma) \leq M_S(f)$$

since $t_\gamma \in S$ for all γ. Thus (5.9.2) follows by letting $a \uparrow \limsup M_S(f_\gamma)$, and so the theorem is proved. □

<u>Theorem 5.10</u>. *Let $\{\mu_\gamma | \gamma \in \Gamma\}$ be a net of every where defined increasing functionals on T, let F and G be subsets of $\overline{\mathbb{R}}^T$, and let $\mu: G \to \overline{\mathbb{R}}$ be an increasing functional satisfying*

(5.10.1) $\qquad \limsup_\gamma \mu_\gamma(f \wedge (\psi_\gamma \circ g)) \leq \mu(g)$

if $f \in F$, $g \in G$ and $\{\psi_\gamma\}$ is a decreasing net in $I_\tau(\overline{\mathbb{R}})$, such that $\lim \psi_\gamma(x) \leq x$ for all $x \in \overline{\mathbb{R}}$

If $\{h_\gamma | \gamma \in \Gamma\}$ is a net in $\overline{\mathbb{R}}^T$, so that $\sup_\gamma h_\gamma \leq f$ for some $f \in F$, then we have

(5.10.2) $\qquad \limsup_\gamma \mu_\gamma(h_\gamma) \leq \mu^G(h)$ *if* $h \gg \limsup_\gamma h_\gamma$

__Proof__. Let $h \gg \limsup h_\gamma$ and let $f \in F$, such that $h_\gamma \leq f$ for all $\gamma \in \Gamma$. By Theorem 5.9, there exist $\{\varphi_\gamma | \gamma \in \Gamma\} \subseteq I_\tau(\overline{\mathbb{R}})$, so that

(i) $$h_\gamma \leq \varphi_\gamma \circ h \quad \forall \gamma \in \Gamma, \quad \lim_\gamma \sup \varphi_\gamma(x) \leq x$$

Now let $\hat{\psi}_\gamma = \sup_{\beta \geq \gamma} \varphi_\beta$ and $\psi_\gamma(x) = \inf\{\hat{\psi}_\gamma(y) | y > x\}$. Then $\{\psi_\gamma\}$ is a decreasing net in $I_\tau(\overline{\mathbb{R}})$, and

$$\lim_\gamma \psi_\gamma(x) \leq \lim_\gamma \hat{\psi}_\gamma(y) = \lim_\gamma \sup \varphi_\gamma(y) \leq y$$

for all $y > x$. Hence $\lim \psi_\gamma(x) \leq x$ for all $x \in \overline{\mathbb{R}}$, and so by (5.10.1) we have

(ii) $$\limsup \mu_\gamma(f \wedge (\psi_\gamma \circ g)) \leq \mu(g) \qquad \forall g \in G$$

Let $g \in G$, so that $g \geq h$. Since $\varphi_\gamma \leq \hat{\psi}_\gamma \leq \psi_\gamma$ and φ_γ is increasing we have by (i)

$$h_\gamma \leq f \wedge (\varphi_\gamma \circ h) \leq f \wedge (\varphi_\gamma \circ g) \leq f \wedge (\psi_\gamma \circ g)$$

and so (5.10.2) follows from (ii). □

__Corollary 5.11__. __Let__ (T, B, λ) __be a positive measure space, and let__ $\{h_n | n \geq 1\}$ __be a sequence in__ $\overline{\mathbb{R}}^T$ __satisfying__

(5.11.1) $$\int^* (\sup_n h_n) d\lambda < \infty$$

__Then we have__

(5.11.2) $$\limsup_{n \to \infty} \int^* h_n d\lambda \leq \int^* h d\lambda \quad \underline{\text{if}} \quad h \gg \lim_n \sup h_n$$

__Proof__. Put $F = L^1(\lambda)$, $G = L(\lambda)$, $\mu = \mu_n = \int^* d\lambda$ in Theorem 5.10. □

It is wellknown that the upper integral satisfies _the increasing convergence theorem_ and _the lower Fatou lemma_ i.e.

$$\int^* h d\mu = \lim_{n\to\infty} \int^* h_n d\mu \quad \text{if} \quad h_n \uparrow h \quad \text{and} \quad \int^* h_1 d\mu > -\infty \tag{5.11}$$

$$\int^* (\liminf_{n\to\infty} h_n) d\mu \le \liminf_{n\to\infty} \int^* h_n d\mu \quad \text{if} \quad \int^* (\inf_n h_n) d\mu > -\infty \tag{5.12}$$

And it is also wellknown that the upper integral does __not__ satisfy _the decreasing convergence theorem_ and _the upper Fatou lemma_ in general. However Corollary 5.11 shows that the upper Fatou lemma holds for outer integral if we replace $\limsup h_n$ by a function h, such that $h \gg \limsup h_n$.

It is wellknown that the upper integral is subadditive on $\overline{\mathbb{R}}^T$, but not superadditive in general. However, we shall now see that if we replace Σh_j by a larger function in a similar facon as above, then we get a certain form of superadditivity of outer integrals and of outer probability contents. To do this we need some technical lemmas concerning real valued setfunctions.

__Lemma 5.12__. _Let_ J _be a finite set, and let_ ξ _be a map from_ 2^J _into_ $[-\infty,\infty[$ _satisfying_

$$\xi(\alpha) + \xi(\beta) \le \xi(\alpha \cap \beta) + \xi(\alpha \cup \beta) \qquad \forall\, \alpha, \beta \in 2^J \tag{5.12.1}$$

Let $p, q, r \in \mathbb{R}$, _such that_

$$p \le -\xi(\emptyset), \quad q \ge \xi(J), \quad \xi(J) + p \le r \le q - \xi(\emptyset) \tag{5.12.2}$$

Then there exist $(a_j) \in \mathbb{R}^J$, satisfying

(5.12.3) $$\xi(\alpha) + p \leq \sum_{j\in\alpha} a_j \leq q - \xi(J\setminus\alpha) \qquad \forall \alpha \subseteq J$$

(5.12.4) $$\sum_{j\in J} a_j = r$$

Proof. Of course we may assume that $J = \{1,\ldots,n\}$, and we shall prove the lemma by induction. If $n = 1$ then $a_1 = r$ clearly satisfies (5.12.3) and (5.12.4).

So suppose that the lemma holds for some n, and consider the case where $J = \{1,\ldots,n+1\}$. Now put $f(j) = j \wedge n$ for $j \in J$, then f maps J into $I = \{1,\ldots,n\}$, and if

$$\hat{\xi}(\alpha) = \xi(f^{-1}(\alpha)) \qquad \forall \alpha \subseteq I$$

Then $\hat{\xi}$ satisfies (5.12.1) with J replaced by I. Moreover $\hat{\xi}(\emptyset) = \xi(\emptyset)$ and $\hat{\xi}(I) = \xi(J)$, so by induction hypothesis there exist $b_1,\ldots,b_n \in \mathbb{R}$ satisfying

(i) $$\hat{\xi}(\alpha) + p \leq \sum_{j\in\alpha} b_j \leq q - \hat{\xi}(I\setminus\alpha) \qquad \forall \alpha \subseteq I$$

(ii) $$\sum_{j=1}^{n} b_j = r$$

Now put $\hat{p} = p \vee (r-q)$ and $\hat{q} = (p-b_n) \vee (r-b_n-q)$. Since

$$\sum_{j\in\alpha} b_j = r - \sum_{j\notin\alpha} b_j \geq r - q + \hat{\xi}(\alpha)$$

we see that

(iii) $$\hat{\xi}(\alpha) + \hat{p} \leq \sum_{j\in\alpha} b_j \qquad \forall \alpha \subseteq I$$

Let $L = \{1,\dots,n-1\}$, and let $\alpha \subseteq L$. Then

$$\hat{\xi}(\alpha \cup \{n\}) + (p-b_n) \le \sum_{j\in\alpha} b_j$$

$$\sum_{j\in\alpha} b_j = r - b_n - \sum_{j\in L\setminus\alpha} b_j$$

$$\ge r - b_n - q + \hat{\xi}(I\setminus(L\setminus\alpha))$$

and since $I\setminus(L\setminus\alpha) = \alpha \cup \{n\}$, we find

(iv) $$\hat{\xi}(\alpha \cup \{n\}) + \hat{q} \le \sum_{j\in\alpha} b_j \quad \forall \alpha \subseteq L$$

Now let $\alpha,\beta \subseteq L$, then by definition of f we find that

$$(\alpha \cup \{n\}) \cup (\beta \cup \{n+1\}) = f^{-1}(\alpha \cup \beta \cup \{n\})$$

$$(\alpha \cup \{n\}) \cap (\beta \cap \{n+1\}) = f^{-1}(\alpha \cap \beta)$$

Hence by (iii), (iv) and (5.12.1) we have

$$\xi(\alpha \cup \{n\}) + \xi(\beta \cup \{n+1\}) \le \hat{\xi}(\alpha \cup \beta \cup \{n\}) + \hat{\xi}(\alpha \cap \beta)$$

$$\le \sum_{j\in\alpha} b_j + \sum_{j\in\beta} b_j - \hat{p} - \hat{q}$$

Thus if

$$a = \sup\{\xi(\alpha \cup \{n\}) - \sum_{j\in\alpha} b_j + \hat{p} \mid \alpha \subseteq L\}$$

$$b = \inf\{\sum_{j\in\beta} b_j - \xi(\beta \cup \{n+1\}) - \hat{q} \mid \beta \subseteq L\}$$

Then $a \le b$, and since $\xi(\alpha) < \infty$ for all $\alpha \subseteq J$, Then $a < \infty$ and $b > -\infty$. Thus there exist $a_n \in \mathbb{R}$ such that $a \le a_n \le b$, and we define

$$a_j = \begin{cases} b_j & \text{if} \quad 1 \leq j \leq n-1 \\ a_n & \text{if} \quad j = n \\ b_n - a_n & \text{if} \quad j = n+1 \end{cases}$$

Then clearly (5.12.4) follows from (ii). To prove (5.12.3) we divide the discussion in four cases.

<u>Case 1°</u>. $\alpha \subseteq L$: In this case $a_j = b_j$ for $j \in \alpha$, $\alpha = f^{-1}(\alpha)$ and $J \setminus \alpha = f^{-1}(I \setminus \alpha)$, and so (5.12.3) follows from (i).

<u>Case 2°</u>. $n, n+1 \in \alpha$: Let $\beta = \alpha \cap I$, then $\alpha = f^{-1}(\beta)$ and $J \setminus \alpha = f^{-1}(I \setminus \beta)$, and since $b_n = a_n + a_{n+1}$ we have that

$$\sum_{j \in \beta} b_j = \sum_{j \in \alpha} a_j$$

Thus (5.12.3) follows from (i).

<u>Case 3°</u>. $n \in \alpha$, $n+1 \notin \alpha$: Let $\beta = \alpha \cap L = \alpha \setminus \{n\}$, since $a_n \geq a$ and $\beta \subseteq L$ we find

$$\sum_{j \in \alpha} a_j = a_n + \sum_{j \in \beta} a_j \geq a + \sum_{j \in \beta} a_j$$

$$\geq \xi(\beta \cup \{n\}) - \sum_{j \in \beta} b_j + \hat{p} - \sum_{j \in \beta} a_j$$

$$\geq \xi(\alpha) + p$$

since $\hat{p} \geq p$ and $b_j = a_j$ for $j \in \beta$. Similarly since $a_n \leq b$ and $\gamma = L \setminus \beta \subseteq L$ we have

$$\sum_{j \in \alpha} a_j \leq b + \sum_{j \in \beta} a_j$$

$$\leq \sum_{j \in \gamma} b_j + \sum_{j \in \beta} a_j - \xi(\gamma \cup \{n+1\}) - \hat{q}$$

$$= \sum_{j \in L} b_j - \hat{q} - \xi(J - \alpha)$$

$$= r - b_n - \hat{q} - \xi(J \setminus \alpha)$$

$$\leq q - \xi(J \setminus \alpha)$$

since $\hat{q} \geq r - b_n - q$, and $J \setminus \beta = \gamma \cup \{n+1\}$. Thus (5.12.3) holds.

Case 4°. $n \notin \alpha$, $n+1 \in \alpha$: Put $\beta = L \cap \alpha = \alpha \setminus \{n+1\}$, since $a_n \geq a$ and $\gamma = L \setminus \beta \subseteq L$ we have

$$\begin{aligned}
\sum_{j\in\alpha} a_j &= a_{n+1} + \sum_{j\in\beta} a_j = b_n - a_n + \sum_{j\in\beta} b_j \\
&\leq b_n - a + \sum_{j\in\beta} b_j \\
&\leq b_n - \xi(\gamma \cup \{n\}) + \sum_{j\in\gamma} b_j + \sum_{j\in\beta} b_j - \hat{p} \\
&= \sum_{j=1}^{n} b_j - \xi(\gamma \cup \{n\}) - \hat{p} \\
&= r - \hat{p} - \xi(J \setminus \alpha) \\
&\leq q - \xi(J \setminus \alpha)
\end{aligned}$$

since $a_j = b_j$ for $j \in \beta$, $\gamma \cup \{n\} = J \setminus \alpha$ and $\hat{p} \geq r - q$. Similarly since $a_n \leq b$ and $\beta \subseteq L$ we have

$$\begin{aligned}
\sum_{j\in\alpha} a_j &\geq b_n - b + \sum_{j\in\beta} b_j \\
&\geq b_n - \sum_{j\in\beta} b_j + \xi(\beta \cup \{n+1\}) + \hat{q} + \sum_{j\in\beta} b_j \\
&\geq b_n + \hat{q} + \xi(\alpha) \\
&\geq p + \xi(\alpha)
\end{aligned}$$

since $\hat{q} \geq p - b_n$. Then (5.12.3) holds.

Since case 1°-4° exhaust all possible cases, we see that (5.12.3) holds, and thus the induction is completed and the lemma is proved. □

Lemma 5.13. *Let* Ω *be a set, let* $\mathcal{F}$ *be an algebra of subsets of* Ω, *and let* $\xi:\mathcal{F}\to[-\infty,\infty[$ *be a map, so that*

$$(5.13.1)\qquad \xi(F_1)+\xi(F_2)\le\xi(F_1\cap F_2)+\xi(F_1\cup F_2)\qquad \forall F_1,F_2\in\mathcal{F}$$

Let $p,q,r\in\overline{\mathbb{R}}$, *such that*

$$(5.13.2)\qquad p\le-\xi(\emptyset),\ q\ge\xi(\Omega),\ \xi(\Omega)+p\le r\le q-\xi(\emptyset)$$

Then there exists a map $\eta:\mathcal{F}\to\overline{\mathbb{R}}$ *satisfying*

$$(5.13.3)\qquad \xi(F)\,\dot{+}\,p\le\eta(F)\le q\,\dot{-}\,\xi(\Omega\setminus F)\qquad \forall F\in\mathcal{F}$$

$$(5.13.4)\qquad \eta(\emptyset)=0\quad\text{and}\quad\eta(\Omega)=r$$

$$(5.13.5)\qquad {\sum_{j=1}^{n}}_{*}\eta(F_j)\le\eta\Big(\bigcup_{j=1}^{n}F_j\Big)\le\sum_{j=1}^{n}{}^{*}\eta(F_j)$$

whenever $n\ge1$ *and* $F_1,\dots,F_n\in\mathcal{F}$ *are disjoint. Moreover if* $\xi(\emptyset)$ *and* $\xi(\Omega)$ *both are finite, then we may choose* η *satisfying (5.13.3-5), such that*

$$(5.13.6)\qquad 0=\lim_{\gamma}\eta(F_\gamma),\qquad r=\lim_{\gamma}\eta(\Omega\setminus F_\gamma)$$

for every net $\{F_\gamma\mid\gamma\in\Gamma\}$ *in* $\mathcal{F}$ *satisfying*

$$(5.13.7)\qquad \xi(\emptyset)\le\liminf_{\gamma}\xi(F_\gamma),\quad \xi(\Omega)\le\liminf_{\gamma}\xi(\Omega\setminus F_\gamma)$$

$$(5.13.8)\qquad \lim_{\gamma}1_{F_\gamma}(\omega)=0\qquad\forall\omega\in\Omega$$

Proof. Let Σ be the set of all finite subalgebras of $\mathcal{F}$, and let M be the set of all $\eta\in\overline{\mathbb{R}}^{\mathcal{F}}$ which are finitely additive in the sense of (5.13.5). By (A.5) and (A.9) we have that M is compact in the product topology of $\overline{\mathbb{R}}^{\mathcal{F}}$.

Let us first assume that $p,q,r \in \mathbb{R}$. If $\mathcal{G} \in \Sigma$ then we define

$$M(\mathcal{G}) = \left\{ \eta \in M \,\middle|\, \begin{array}{l} \xi(G) + p \leq \eta(G) \leq q - \xi(\Omega \setminus G) \quad \forall G \in \mathcal{G} \\ \eta(\emptyset) = 0 \quad \text{and} \quad \eta(\Omega) = r \end{array} \right\}$$

Then $M(\mathcal{G})$ is a compact subset of M. Let $\mathcal{G} \in \Sigma$, then $\mathcal{G}$ is a finite algebra, and so there exist a partition $G_1, \ldots, G_k$ of Ω, such that

$$\mathcal{G} = \{ \bigcup_{j \in \alpha} G_j \mid \alpha \subseteq \{1, \ldots, k\} \}$$

Now let $J = \{1, \ldots, k\}$ and $f = \sum_{j=1}^{k} j \, 1_{G_j}$, then f maps Ω into J and

(i) $$\mathcal{G} = \{ f^{-1}(\alpha) \mid \alpha \subseteq J \}$$

Let $\hat{\xi}(\alpha) = \xi(f^{-1}(\alpha))$ for all $\alpha \subseteq J$. Then $\hat{\xi}$ satisfies (5.12.1) and $\hat{\xi}(\emptyset) = \xi(\emptyset)$, $\hat{\xi}(J) = \xi(\Omega)$. Hence by Lemma 5.12 there exist $a_1, \ldots, a_k \in \mathbb{R}$ satisfying

(ii) $$\xi(f^{-1}(\alpha)) + p \leq \sum_{j \in \alpha} a_j \leq q - \xi(\Omega \setminus f^{-1}(\alpha))$$

(iii) $$\sum_{j=1}^{k} a_j = r$$

Now let us choose points $\omega_1 \in G_1, \ldots, \omega_k \in G_k$ and define

$$\eta(F) = \sum_{j=1}^{k} a_j 1_F(\omega_j) \qquad \forall F \in \mathcal{F}$$

Then by (i), (ii) and (iii) it follows easily that $\eta \in M(\mathcal{G})$.

Thus $M(\mathcal{G})$ is compact and non-empty in $\overline{\mathbb{R}}^{\mathcal{F}}$, and since $M(\mathcal{G}_0) \subseteq M(\mathcal{G}_1)$ if $\mathcal{G}_0 \supseteq \mathcal{G}_1$, then we see that $\{M(\mathcal{G}) \mid \mathcal{G} \in \Sigma\}$ is filtering downwards. Thus there exist $\eta \in M$ such that

$$\eta \in M(\mathcal{G}) \qquad \forall \mathcal{G} \in \Sigma$$

But then η evidently satisfies (5.13.3)-(5.13.5).

If $p,q,r \in \overline{\mathbb{R}}$ satisfies (5.13.2), then since $\xi(\emptyset) < \infty$ and $\xi(\Omega) < \infty$, there exist $p_n, q_n, r_n \in \mathbb{R}$ satisfying (5.13.2), such that $p_n \to p$, $q_n \to q$ and $r_n \to r$. Thus there exist $\eta_n \in M$ satisfying (5.13.3) (5.13.5) with (p,q,r) replaced by (p_n, q_n, r_n). Since M is compact we know that $\{\eta_n\}$ admits a limit point $\eta \in M$, and evidently η satisfies (5.13.3)-(5.13.5).

Now suppose that $\xi(\emptyset)$ and $\xi(\Omega)$ are finite, then putting $p_0 = -\xi(\emptyset)$, $q_0 = \xi(\Omega)$, $r_0 = \xi(\Omega) - \xi(\emptyset)$ we see that there exist $\mu \in M$ such that

(iv) $$\xi(F) - \xi(\emptyset) \leq \mu(F) \leq \xi(\Omega) - \xi(\Omega \setminus F) \qquad \forall F \in F$$

(v) $$\mu(\Omega) = \xi(\Omega) - \xi(\emptyset)$$

Hence if $\{F_\gamma\}$ satisfies (5.13.7), then

$$0 \leq \lim_\gamma \inf(\xi(F_\gamma) - \xi(\emptyset)) \leq \lim_\gamma \inf \mu(F_\gamma)$$

$$\leq \lim_\gamma \sup \mu(F_\gamma) \leq \lim_\gamma \sup(\xi(\Omega) - \xi(\Omega \setminus F))$$

$$\leq 0$$

And similarly when we replace F_γ by $\Omega \setminus F_\gamma$. Thus

(vi) $$\lim_\gamma \mu(F_\gamma) = 0, \qquad \lim_\gamma \mu(F_\gamma) = r_0 = \xi(\Omega) - \xi(\emptyset)$$

Now suppose that p,q,r satisfy (5.13.2) and let us choose a point $\omega_0 \in \Omega$. Then we put

$$\alpha = r - r_0 = r + \xi(\emptyset) - \xi(\Omega)$$

$$\eta(F) = \mu(F) \dotplus \alpha 1_F(\omega_0) \qquad \forall F \in F$$

By (A.5) and (A.9) we have that $\eta \in M$. Moreover by (5.13.2) we have that $q \geq \xi(\Omega) \vee (\xi(\Omega)+\alpha)$, and so

$$\eta(F) \leq \alpha 1_F(\omega_0) \dot{+} (\xi(\Omega)-\xi(\Omega\setminus F)$$

$$\leq (\alpha 1_F(\omega_0) \dot{+} \xi(\Omega)) \dot{-} \xi(\Omega\setminus F)$$

$$\leq q \dot{-} \xi(\Omega\setminus F)$$

And similarly, since $p \leq (-\xi(\emptyset)) \vee (\alpha-\xi(\emptyset))$ we have

$$\eta(F) \geq \alpha 1_F(\omega_0) \dot{+} (\xi(F)-\xi(\emptyset))$$

$$= (\alpha 1_F(\omega_0)-\xi(\emptyset)) \dot{+} \xi(F)$$

$$\geq p \dot{+} \xi(F)$$

Thus η satisfies (5.13.3-5), since $\eta(\Omega) = r_0 + \alpha = r$. Moreover if $\{F_\gamma\}$ satisfies (5.13.7) and (5.13.8) then there exist $\alpha \in \Gamma$ so that

$$1_{F_\gamma}(\omega_0) = 0, \qquad 1_{\Omega\setminus F_\gamma}(\omega_0) = 1 \qquad \forall \gamma \geq \alpha$$

and so we see that (5.13.6) follows from (vi), and thus the lemma is proved. □

<u>Definition 5.14</u>. A <u>linearly preordered</u> set $(X,\leq)$ is a set X with a preordering $\leq$, such that for all $x,y \in X$ we have that either $x \leq y$ or $y \leq x$. Note that if X is a linearly preordered set, then every non-empty finite set $\sigma \subseteq X$ admits a maximum and a minimum both belonging to σ.

Let $F \subseteq \overline{\mathbb{R}}^T$, the F induces a preordering on T in the following way:

$$t' \leq t'' \Longleftrightarrow f(t') \leq f(t'') \qquad \forall f \in F$$

If this preordering is a linear preordering we say that F is rectilinear. Evidently we have that F is rectilinear if and only if

(5.14.1) $\quad f_0(t') < f_0(t'') \Rightarrow f_1(t') \leq f_1(t'')$

whenever $f_0, f_1 \in F$ and $t', t'' \in T$. In particular we find

(5.14.2) $\quad F$ is rectilinear $\Longleftrightarrow$ $\{f,g\}$ is rectilinear $\forall f,g \in F$

(5.14.3) $\quad$ Every subset of a rectilinear set is rectilinear.

(5.14.4) $\quad F$ is rectilinear if and only if $F \subseteq I(T,\leq)$ for some linear preordering $\leq$ on T.

(5.14.5) $\quad \{\varphi \circ f \mid \varphi \in I(\overline{\mathbb{R}})\}$ is rectilinear $\forall f \in \overline{\mathbb{R}}^T$.

Moreover if $\mathcal{E}$ is a paving on T, then the reader easily verifies that

(5.14.6) $\quad \{1_E \mid E \in \mathcal{E}\}$ is rectilinear, if and only if $\mathcal{E}$ is linearly ordered by inclusion.

Let $\{f_j \mid j \in J\} \subseteq \overline{\mathbb{R}}^T$, and let $f \in \overline{\mathbb{R}}$, then we write: $f \gg \sum_j f_j$ if (cf. (5.9.2)):

(5.14.7) $\quad \sum_{j}{}_{*}\, M_S(f_j) \leq M_S(f) \quad \forall S \subseteq T$

and we write: $f \ll \sum_j f_j$ if $(-f) \gg \sum_j (-f_j)$, i.e. if

(5.14.8) $\quad \sum_j{}^{*}\, m_S(f_j) \geq m_S(f) \quad \forall S \subseteq T$

If $f \in \overline{\mathbb{R}}^T$, $\{f_j \mid j \in J\} \leq \overline{\mathbb{R}}^T$ and $\psi: S \to T$ is a map, then clearly we have

(5.14.9) $\quad \sum_j f_j << f \Rightarrow \sum_{j*} f_j \le f$

(5.14.10) $\quad \sum_j f_j << f \Rightarrow \sum_j f_j o\psi << f o\psi$

If $S \subseteq T$ we write: $\sum_j f_j << f$ on S, if we have that $\sum_j \hat{f}_j << \hat{f}$ where $\hat{f}$ and $\hat{f}_j$ are the restrictions of f and f_j to S.

Lemma 5.15. Let $(X,\le)$ be a linearly preorderet set and let $\{\varphi_j \mid j \in J\} \subseteq I(X)$. Put

$$\alpha = \sum_{j*} \varphi_j \ , \qquad \beta = \sum_j^* \varphi_j$$

Then there exist a countable set $N \subseteq J$, such that

(5.15.1) $\quad \varphi_j(x) \le 0 \quad \forall j \in J \setminus N \quad \forall x \in \{\beta < \infty\}$

(5.15.2) $\quad \varphi_j(x) \ge 0 \quad \forall j \in J \setminus N \quad \forall x \in \{\alpha > -\infty\}$

Moreover if $\{\varphi_j\}$ satisfies

(5.15.3) $\quad \exists a \in X$ so that $\sum_{j \in J} |\varphi_j(a)| < \infty$

then $\alpha(x) = \beta(x)$ for all $x \in X$.

Remark. Let $\{\varphi_j\}$ be any family in $\overline{\mathbb{R}}^J$ and define α and β as above. Then by definition of Σ_* and Σ_* we have

(5.15.4) $\quad \{\alpha > -\infty\} = \{x \mid \sum_j \varphi_j^+(x) < \infty\}$

(5.15.5) $\quad \{\beta < \infty\} = \{x \mid \sum_j \varphi_j^+(x) < \infty\}$

(5.15.6) $\quad \alpha^{-1}(\mathbb{R}) = \beta^{-1}(\mathbb{R}) = \{x \mid \sum_j |\varphi_j(x)| < \infty\}$

(5.15.7) $\quad \alpha(x) = \beta(x) \quad \forall x \in \{\alpha > -\infty\} \cup \{\beta < \infty\}$

Proof. Let $\hat{\beta} = \sum \varphi_j^+$ and $\hat{\alpha} = \sum \varphi_j^-$, and put

$$J_n = \{j \mid \exists\, x \in X \text{ so that } \hat{\beta}(x) \leq n,\ \varphi_j(x) \geq 1/n\}$$

$$J^n = \{j \mid \exists\, x \in X \text{ so that } \hat{\alpha}(x) \leq n,\ \varphi_j(x) \leq -1/n\}$$

Let σ be a finite subset of J_n and let $x_j \in X$ for $j \in \sigma$ be chosen, so that $\hat{\beta}(x_j) \leq n$ and $\varphi_j(x_j) \geq 1/n$ for all $j \in \sigma$. Let $k \in \sigma$ be chosen so that $x_k \geq x_j$ for all $j \in \sigma$. Then

$$n \geq \hat{\beta}(x_k) \geq \sum_{j\in\sigma} {}_j(x_k) \geq \sum_{j\in\sigma} \varphi_j(x_j) \geq n^{-1}\text{card}(\sigma)$$

Hence $\text{card}(\sigma) \leq n^2$, and so $\text{card}\, J_n \leq n^2$. Similarly we have that $\text{card}\, J^n \leq n^2$. Thus

$$N = \bigcup_{n=1}^{\infty} J_n \cup \bigcup_{n=1}^{\infty} J^n$$

is countable. And from (5.15.4) and (5.15.5) we see that (5.15.1) and (5.15.2) holds.

Now suppose that (5.15.3) holds. If $x \leq a$, then $\varphi_j^+(x) \leq \varphi_j^+(a)$, and so $\beta(x) < \infty$ by (5.15.3) and (5.15.5). If $x \geq a$ then $\varphi_j^-(x) \leq \varphi_j^-(a)$ and so $\alpha(x) > -\infty$ by (5.15.3) and (5.15.4). Thus $\alpha = \beta$ by (5.15.7), since for all $x \in X$ we either have $x \leq a$ or $x \geq a$. □

Lemma 5.16. *Let* $(X, \leq)$ *be a linearly ordered set, and let* φ *and* $\{\varphi_j \mid j \in J\}$ *be increasing functions from* X *into* $\overline{\mathbb{R}}$ *satisfying*

(5.16.1) $$\sum_{j}{}_{*}\ \varphi_j(x) \leq \varphi(x) \quad \forall x \in X$$

and suppose that there exist $a \in X$ *satisfying*

(5.16.2) $\sum_{j\in J} |\varphi_j(a)| < \infty$

(5.16.3) $\forall \{x_\gamma | \gamma \in \Gamma\}$ <u>an increasing net in</u> $L(a)$ <u>with</u> $\mathrm{card}(\Gamma) \leq \mathrm{card}(J)$, $\exists\, u \in X$, <u>so that</u> $u \geq x_\gamma \quad \forall \gamma \in \Gamma$ <u>and</u> $\varphi(u) = \sup_\gamma \varphi(x_\gamma)$

<u>where</u> $L(a) = \{x \in \varphi^{-1}(\mathbb{R}) | x \leq a\}$. <u>Then there exist a family</u> $\{\psi_j | j \in J\}$ <u>of increasing functions from</u> X <u>into</u> $\overline{\mathbb{R}}$ <u>satisfying</u>

(5.16.4) $\sum_j{}_* \psi_j(x) = \sum_j{}^* \psi_j(x) = \varphi(x)$

(5.16.5) $\psi_j(x) \geq \varphi_j(x) \quad \forall j \in J \quad \forall x \in X$

<u>Remarks</u> (1): If J is finite then (5.16.3) holds trivially. And if $\xi = \mathrm{card}\, J$, and $\mathcal{L}_a$ is the set of all intervals: $L = \{x | c \leq x \leq a\}$, where $c \leq a$ and $c \in \varphi^{-1}(\mathbb{R})$, then the reader easily virifies that (5.16.3) is equivalent to

(5.16.6) $\varphi|_L \in I^\xi(L) \quad \forall L \in \mathcal{L}_a$

(2): A family $\{\psi_j\}$ of increasing functions from X into $\overline{\mathbb{R}}$ satisfying (5.16.4) is called an <u>increasing partition of</u> φ, and if $X \subseteq \overline{\mathbb{R}}$ and $\varphi(x) = x$, then $\{\psi_j\}$ is called an <u>increasing partition of unity on</u> X. Note that if $\{\psi_j\}$ is an increasing partition of φ, and if X is linearly ordered then

(5.16.7) $\sum_{j\in J} |\psi_j(x)| < \infty \iff \varphi(x) \in \mathbb{R}$

(5.16.8) $\sum_{j\in J} |\psi_j(x) - \psi_j(y)| \leq |\varphi(x) - \varphi(y)| \quad \forall x,y \in \varphi^{-1}(\mathbb{R})$

Thus if φ is continuous on $\varphi^{-1}(\mathbb{R})$ for some topology on X, then so is ψ_j for all $j \in J$.

Proof. Let $\gamma = \Sigma_* \varphi_j$, then by (5.16.2) and Lemma 5.15 we have

(i) $$\gamma(x) = \sum_{*j} \varphi_j(x) = \sum_j^* \varphi_j(x) \leq \varphi(x) \qquad \forall x \in X$$

First we shall prove the lemma under the following assumption:

(a) $$\varphi(x) \in \mathbb{R} \qquad \forall x \in X$$

and so by (a) we have

(ii) $$\sum_j \varphi_j^+(x) < \infty \qquad \forall x \in X$$

Moreover since $\varphi_j^-(x) \leq \varphi_j^-(a)$ for $x \geq a$, then by (ii) and (5.16.2) we have

(iii) $$\sum_j |\varphi_j(x)| < \infty \qquad \forall x \geq a$$

Let $\alpha \subseteq J$, then we define

$$\varphi_\alpha(x) = \sum_{j \in \alpha} \varphi_j(x) \qquad \forall x \in X$$

$$\xi(\alpha) = \sup\{\varphi_\alpha(x) - \varphi(x) \mid x \geq a\}$$

Note that φ_α is unambigously defined by (ii). Then ξ maps 2^J into $\overline{\mathbb{R}}$, and since $\alpha \sim \varphi_\alpha(x)$ is continuous on $2^J = \{0,1\}^J$, with its product topology, for all $x \geq a$ by (iii) we have that ξ is lower semicontinuous on 2^J. Moreover we have

$$-\varphi(a) - \sum_{j\in J} |\varphi_j(a)| \le -\varphi(a) + \varphi_\alpha(a) \le \xi(\alpha)$$

$$= \sup_{x\ge a}\{\varphi_\alpha(x) - \varphi(x)\} = \sup_{x\ge a}\{\gamma(\alpha) - \varphi(x) - \varphi_{J\setminus\alpha}(x)\}$$

$$\le -\varphi_{J\setminus\alpha}(a) \le \sum_{j\in J} |\varphi_j(a)|$$

since $\gamma \le \varphi$ and $\varphi_{J\setminus\alpha}$ is increasing. And since φ is increasing, $\varphi_\emptyset = 0$ and $\varphi_J = \gamma \le \varphi$ we have

(iv) ξ is bounded and lower semicontinuous on 2^J

(v) $\xi(\emptyset) = -\varphi(a)$, $\xi(J) \le 0$

Let $x \ge a$, $y \ge a$ and $\alpha, \beta \in 2^J$. If $x \ge y$ then we have

$$\varphi_\alpha(x) + \varphi_\beta(y) = \varphi_{\alpha\cup\beta}(x) + \varphi_{\alpha\cap\beta}(y) + \varphi_{\beta\setminus\alpha}(y) - \varphi_{\beta\setminus\alpha}(x)$$

$$\le \varphi_{\alpha\cup\beta}(x) + \varphi_{\alpha\cap\beta}(y)$$

since $\varphi_{\beta\setminus\alpha}$ is increasing. And if $y \ge x$ then similarly

$$\varphi_\alpha(x) + \varphi_\beta(y) \le \varphi_{\alpha\cap\beta}(x) + \varphi_{\alpha\cup\beta}(y)$$

Hence we find

(vi) $\xi(\alpha) + \xi(\beta) \le \xi(\alpha\cup\beta) + \xi(\alpha\cap\beta) \quad \forall\, \alpha, \beta \in 2^J$

Now put $p = r = \varphi(a)$ and $q = 0$, then (5.13.2) holds by (v) and so by Lemma 5.13 there exist a map η from 2^J into $\overline{\mathbb{R}}$ satisfying (5.13.3-6) for all nets $\{F_\gamma\} \subseteq 2^J$ satisfying (5.13.7-8). Now since ξ is bounded we have that η is bounded, and so η is a bounded signed measure on 2^J. And since ξ is lower semicontinuous on 2^J, then by (5.13.6) we have $\eta(F_\gamma) \to 0$ for every net $\{F_\gamma\}$ in 2^J with

$F_\gamma \neq \emptyset$. Hence if $a_j = \eta(\{j\})$, then

(vii) $\sum_{j\in J} |a_j| < \infty$ and $\sum_{j\in J} a_j = \eta(J) = \varphi(a)$

(viii) $\xi(\alpha) + \varphi(a) \leq \sum_{j\in\alpha} a_j = \eta(\alpha) \leq -\xi(J\setminus\alpha) \quad \forall\, \alpha \subseteq J$

Now we put

$$\theta_j(x) = \begin{cases} a_j \vee \varphi_j(x) & \text{if } x \geq a \\ \varphi_j(x) & \text{if } x < a \end{cases}$$

Then θ_j is increasing, $\theta_j \geq \varphi_j$ and I claim that

(ix) $\theta_j(a) = a_j \quad \forall\, j, \quad \sum_j \theta_j(a) = \varphi(a)$

(x) $\theta(x) = \theta_j(x) = \sum_j{}^{*} \theta_j(x) \leq \varphi(x) \quad \forall\, x \in X$

By (viii) with $\alpha = \{j\}$ we have that $a_j \geq \varphi_j(a)$, and so (ix) follows from (vii).

By (vii), (ix) and Lemma 5.15 we have that $\Sigma_* \theta_j = \Sigma^* \theta_j$. If $x < a$ then $\theta(x) = \gamma(x) \leq \varphi(x)$. If $x \geq a$ and $\alpha = \{j \in J \mid a_j < \varphi_j(x)\}$ then

$$\theta(x) = \varphi_\alpha(x) + \sum_{j\notin\alpha} a_j \leq \varphi_\alpha(x) - \xi(\alpha) \leq \varphi(x)$$

by (viii) and definition of ξ. Thus (x) holds.

Since $\theta(x) \leq \varphi(x) < \infty$ for all $x \in X$, then by Lemma 5.15 there exist a countable set N_0 such that

(xi) $\theta_j(x) \leq 0 \quad \forall\, j \in J\setminus N_0 \quad \forall\, x \in X$

(xii) $\theta_j(x) = 0 \quad \forall\, j \in J\setminus N_0 \quad \forall\, x \in \{\theta > -\infty\}$

And I claim that there exist a countable set $N \supseteq N_0$, such that

(xiii) $$\sum_{j \in N} \theta_j(x) \leq \varphi(x) \qquad \forall x \in X$$

If $\theta(x) > -\infty$ for all $x \in X$, then $N = N_0$ satisfies (xiii) by (xii) and (x). And if J is countable, then $N = J$ satisfies (xiii). So suppose that J is uncountable and $\theta(c) = -\infty$ for some c. Then $c < a$, since $\theta(a) \in \mathbb{R}$, and there exist a countable $C \subseteq J$, such that $C \supseteq N_0$ and

$$\sum_{j \in C} \theta_j(c) = -\infty$$

Now let $L = \{x \in X \mid c \leq x \leq a\}$, and let Γ be the set of all finite subsets of $J \setminus C$ ordered by inclusion. Put

$$\hat{\theta}_\gamma(x) = \sum_{j \in \gamma \cup C} \theta_j(x) \qquad \forall x \in L \qquad \forall \gamma \in \Gamma$$

$$\hat{\varphi} = \text{the restriction of } \varphi \text{ to } L$$

Since $N_0 \subseteq C$ we have by (xi) that $\hat{\theta}_\gamma(x) \downarrow \theta(x)$ for all $x \in L$, and so $\lim\sup \hat{\theta}_\gamma \leq \hat{\varphi}$. Moreover by (5.16.6) we have that $\hat{\varphi} \in I^\xi(L)$, where $\xi = \text{card}(J) = \text{card}(\Gamma)$. Hence by (5.9.5) we have

$$\lim_\gamma \sup \hat{\theta}_\gamma << \hat{\varphi}$$

Now $\hat{\varphi}$ is bounded on L, and so by (5.9.7) there exist $\alpha_n \in \Gamma$, such that

$$\hat{\theta}_\gamma(x) \leq 2^{-n} + \varphi(x) \qquad \forall \gamma \geq \alpha_n \qquad \forall x \in L$$

Now let

$$N = C \cup \bigcup_{n=1}^{\infty} \alpha_n$$

Then N is countable and $N_0 \subseteq N \subseteq J$. By (xii) and (x) we have that (xiii) holds for all $x \in \{\theta > -\infty\}$. And since $\theta_j(x) \leq 0$ for $j \in N \setminus C$, we have

$$\sum_{j\in N} \theta_j(x) \leq \sum_{j\in C} \theta_j(x) \leq \sum_{j\in C} \theta_j(c) = -\infty$$

for $x \leq c$. So suppose that $\theta(x) = -\infty$ and $x \geq c$ then $x \in L$, and since $\theta_j(x) \leq 0$ for $j \in N \setminus C$, we have

$$\sum_{j\in N} \theta_j(x) \leq \sum_{j\in N\cup\alpha_n} \theta_j(x) \leq 2^{-n} + \varphi(x)$$

for all $n \geq 1$. Hence (xiii) holds for all $x \in X$.

Of course we may assume that $N = \{1,\dots,k\}$ if $\mathrm{card}(N) = k < \infty$ and that $N = \mathbb{N}$ otherwise. If $n \in N$, we put

$$f_0 = 0, \quad f_n = \sum_{j=1}^{n} \theta_j, \qquad f^n = \sum_{j\in N, j>n} \theta_j$$

$$\hat{f}_n(x) = \sup\{f_n(u) - \varphi(u) + \varphi(x) \mid u \geq x\} \qquad \forall x \in X$$

$$\hat{\theta}_n = \hat{f}_n - \hat{f}_{n-1}$$

Let $x \in X$, and put $v = x$ if $x \geq a$ and $v = a$ if $x < a$. Then

$$\hat{f}_n(x) \geq f_n(v) - \varphi(v) + \varphi(x) \geq -\sum_{j\in J} |\theta_j(v)| - \varphi(v) + \varphi(x)$$

and since $\gamma \leq \theta \leq \varphi$, then $\Sigma|\theta_j(v)| < \infty$. If $u \geq a$ then

$$f_n(u) - \varphi(u) + \varphi(x) \leq -f^n(u) + \varphi(x) \leq \varphi(x) + \sum_j |\theta_j(a)|$$

since $f_n + f^n \leq \varphi$ and f^n is increasing. If $x \leq u \leq a$ then

$$f_n(u) - \varphi(u) + \varphi(x) \leq f_n(a) \leq \sum_j |\theta_j(a)|$$

since f_n and φ are increasing. Thus we have

(xiv) $$\beta(x) = \sup_{n \in N} |\hat{f}_n(x)| < \infty \qquad \forall x \in X$$

and $\hat{f}_0 = 0$ since φ is increasing. Next we show

(xv) $$\hat{\theta}_n \text{ is increasing and } \hat{\theta}_n \geq \theta_n \geq \varphi_n \qquad \forall n \in N$$

So let $x \leq y$ and let $u \geq x$ and $v \geq y$. Then if $u \geq y$ we have

$$f_n(u) - \varphi(u) + \varphi(x) + f_{n-1}(v) - \varphi(v) + \varphi(y) \leq \hat{f}_n(y) + \hat{f}_{n-1}(x)$$

since $v \geq y \geq x$. And if $x \leq u \leq y$, then

$$f_n(u) - \varphi(u) + \varphi(x) + f_{n-1}(v) - \varphi(v) + \varphi(y)$$

$$= f_{n-1}(u) - \varphi(u) + \varphi(x) + \theta_n(u) + f_{n-1}(v) - \varphi(v) + \varphi(y)$$

$$\leq f_{n-1}(u) - \varphi(u) + \varphi(x) + f_n(v) - \varphi(v) + \varphi(y)$$

$$\leq \hat{f}_{n-1}(x) + \hat{f}_n(y)$$

since $u \leq v$ and $f_n = f_{n-1} + \theta_n$. Thus we have

$$\hat{f}_n(x) + \hat{f}_{n-1}(y) \leq \hat{f}_{n-1}(x) + \hat{f}_n(y)$$

and so $\hat{\theta}_n = \hat{f}_n - \hat{f}_{n-1}$ is increasing. Moreover since $f_n = \theta_n + f_{n-1}$ we have

$$\theta_n(x) + f_{n-1}(u) - \varphi(u) + \varphi(x) \leq \hat{f}_n(x) \qquad \forall u \geq x$$

and so $\theta_n \leq \hat{\theta}_n$. Thus (xv) is proved. Next we show

(xvi) $$\hat{f}_n(a) = f_n(a) \text{ and } \hat{\theta}_n(a) = \theta_n(a) \qquad \forall n \in N$$

Clearly we have that $\hat{f}_n(a) \geq f_n(a)$, so let $u \geq a$. Then by (xiii) we have $f_n + f^n \leq \varphi$, and by (xii) and (ix) we have $f_n(a) + f^n(a) = \varphi(a)$. Thus we find

$$f_n(u) - \varphi(u) + \varphi(a) \leq -f^n(u) + f^n(a) + f_n(a) \leq f_n(a)$$

since f^n is increasing. Thus $\hat{f}_n(a) = f_n(a)$, and so $\hat{\theta}_n(a) = \theta_n(a)$. Next we show

$$\text{(xvii)} \qquad \sum_{n \in N} |\hat{\theta}_n(x)| < \infty \qquad \forall x \in X$$

So let $x \in X$. Since $\hat{f}_n = \hat{\theta}_1 + \ldots + \hat{\theta}_n$ we have by (xiv) and (xvi)

$$\sum_{j=1}^{n} |\theta_j(x)| = 2 \sum_{j=1}^{n} \hat{\theta}_j^+(x) - \hat{f}_n(x)$$

$$\leq 2 \sum_{j \in J} |\theta_j(a)| + \beta(x) < \infty$$

for $x \leq a$, since $\hat{\theta}_j^+$ is increasing. And for $x \geq a$, we have by (xvi) and (xiv)

$$\sum_{j=1}^{n} |\hat{\theta}_j(x)| \leq \hat{f}_n(x) + 2 \sum_{j=1}^{n} \hat{\theta}_j^-(x)$$

$$\leq \beta(x) + 2 \sum_{j \in J} |\theta_j(a)| < \infty$$

since $\hat{\theta}_j^-$ is decreasing. Hence (xvii) is proved. Next we show

$$\text{(xviii)} \qquad \hat{f}_n(x) - \hat{f}_n(y) \leq \varphi(x) - \varphi(y) \qquad \forall n \in N \quad \forall x \geq y$$

Let $n \geq 1$ be given and let $u \geq x \geq y$. Then we have

$$f_n(u) - \varphi(u) + \varphi(x) \leq f_n(u) - \varphi(u) + \varphi(y) + (\varphi(x) - \varphi(y))$$

$$\leq \hat{f}_n(y) + (\varphi(x) - \varphi(y))$$

and so (xviii) follows.

By (xvii) we have that

$$\hat{f}(x) = \sum_{n\in N} \hat{\theta}_j(x)$$

exist and is finite for all, and I claim that

$$\text{(xix)} \qquad \hat{f}(x) = \varphi(x) \quad \text{if} \quad x \leq a, \quad \hat{f}(x) \leq \varphi(x) \quad \text{if} \quad x \geq a.$$

By (xvi) and (xviii) we have

$$\hat{f}_n(x) \leq \varphi(x) + f_n(a) - \varphi(a) \qquad \text{if} \qquad x \geq a$$

$$\hat{f}_n(x) \geq \varphi(x) + f_n(a) - \varphi(a) \qquad \text{if} \qquad x \leq a$$

if $k = \text{card}(N) < \infty$, then $\hat{f} = \hat{f}_k$, and $f_k(a) = \varphi(a)$ by (ix) and (xii). Thus $\hat{f}(x) \leq \varphi(x)$ for $x \geq a$ and $\hat{f}(x) \geq \varphi(x)$ for $x \leq a$, moreover $f_k(u) \leq \varphi(u)$ by (xiii) and so

$$\hat{f}(x) = \sup_{u\geq x}(f_k(u) - \varphi(u) + \varphi(x)) \leq \varphi(x)$$

Thus (xix) holds if N is finite. If N is infinite then so is J and $f = \lim \hat{f}_n$. And since $f_n(a) \to \varphi(a)$ by (ix) and (xii) we have as above that $\hat{f}(x) \leq \varphi(x)$ for $x \geq a$ and $\hat{f}(x) \geq \varphi(x)$ for $x \leq a$. Now let $x \leq a$ and let $L = \{u \mid x \leq u \leq a\}$ then by (5.16.6) we have that φ restricted to L belongs to $I^\sigma(L)$ (note that J is infinite), moreover $\limsup f_n \leq \varphi$ by (xiii). And since φ is bounded on L, then by Theorem 5.9 we have that for all $\varepsilon > 0$ there exist $m \geq 1$ so that

$$f_n(u) \leq \varepsilon + \varphi(u) \qquad \forall n \geq m \ \forall u \in L$$

Hence if $x \leq u \leq a$, then

$$f_n(u) - \varphi(u) + \varphi(x) \leq \varepsilon + \varphi(x)$$

And if $u \geq a \geq x$, then

$$f_n(u) - \varphi(u) + \varphi(x) \leq \varphi(x) + \hat{f}_n(a) - \varphi(a)$$

$$= \varphi(x) + f_n(a) - \varphi(a)$$

and $f_n(a) \to \varphi(a)$. So if $q \geq m$ is chosen so that $f_n(a) - \varphi(a) < \varepsilon$ for $n \geq q$, then we have

$$f_n(u) - \varphi(u) + \varphi(x) \leq \varepsilon + \varphi(x) \quad \forall n \geq q \quad \forall u \geq x$$

Thus $\hat{f}_n(x) \leq \varepsilon + \varphi(x)$ for all $n \geq q$. And so we have that $\hat{f}(x) = \varphi(x)$ for $x \leq a$. Thus (xix) is proved.

Now let $h(x) = \varphi(x) - \hat{f}(x)$, then $h \geq 0$ by (xix) and h is increasing by (xviii). So if $\lambda_n \in \mathbb{R}_+$ for $n \in N$ is chosen so that $\Sigma\lambda_n = 1$, and

$$\psi_j = \begin{cases} 0 & \text{if} \quad j \in J \setminus N \\ \hat{\theta}_j + \lambda_j h & \text{if} \quad j \in N \end{cases}$$

then ψ_j is increasing, $\sum_j |\psi_j| < \infty$ by (xviii) and

$$\sum_j \psi_j = \sum_{j \in N} \hat{\theta}_j + h = \varphi$$

Moreover since $h \geq 0$, then $\psi_j \geq \varphi_j$ by (xv) and (xi), and so $\{\psi_j\}$ satisfies (5.16.4) and (5.16.5).

Thus the lemma has been proved under the assumption (a). And we shall now prove the lemma under the following assumption:

(b) $\qquad \varphi(a) \in \mathbb{R}$

In this case we let $\hat{X} = \varphi^{-1}(\mathbb{R})$, and we let $\hat{\varphi}_j$ and $\hat{\varphi}$ denote the restrictions of φ_j and φ to $\hat{X}$. Since $a \in X$ we see that $(\hat{X}, \hat{\varphi}_j, \hat{\varphi})$ satisfies (5.16.1-3) and $\hat{\varphi}$ is finite. Hence by case (a) there exist $\hat{\psi}_j \in I(\hat{X})$, satisfying (5.16.4-5) on $\hat{X}$. Then we put

$$\psi_j(x) = \begin{cases} \infty & \text{if } \varphi(x) = \infty \\ \hat{\psi}_j(x) & \text{if } \varphi(x) \in \mathbb{R} \\ \varphi_j(x) & \text{if } \varphi(x) = -\infty \end{cases}$$

Then clearly $\psi_j \geq \varphi_j$ and (5.16.4) holds by (i) and the choice of $\{\hat{\psi}_j\}$. Now let $x \leq y$, if $\varphi(y) = \infty$ then $\psi_j(x) \leq \psi_j(y) = \infty$. If $\varphi(y) \in \mathbb{R}$ then $\varphi(x) < \infty$ and

$$\psi_j(x) = \begin{cases} \varphi_j(x) \leq \varphi_j(y) \leq \psi_j(y) & \text{if } \varphi(x) = -\infty \\ \hat{\psi}_j(x) \leq \psi_j(y) & \text{if } \varphi(x) > -\infty \end{cases}$$

If $\varphi(y) = -\infty$, then $\varphi(x) = -\infty$, and

$$\psi_j(x) = \varphi_j(x) \leq \varphi_j(y) = \psi_j(y)$$

Thus ψ_j is increasing, and the lemma is proved under the assumption (b).

Finally we prove the lemma under the following assumption

(c) $\qquad \varphi(a) = \infty$

In this case we put $m_j = \varphi_j(a)$ and $m = \gamma(a)$. Then by (5.16.2) we have that $\Sigma |m_j| < \infty$ and so $m \in \mathbb{R}$. Now let $\hat{\varphi} = \varphi \wedge m$ and $\hat{\varphi}_j = \varphi_j \wedge m_j$. Then we have

$$\sum_j{}_* \hat{\varphi}_j(x) = \sum_j{}_* \varphi_j(x) = \gamma(x) \leq \hat{\varphi}(x) \qquad \forall x \leq a$$

since $\varphi_j(x) \le m_j$ and $\gamma(x) \le m$ for $x \le a$. And since $\varphi_j(x) \ge m_j$ and $\varphi(x) = \infty$ for $x \ge a$ by (c) we have

$$\sum_{j}{}_{*} \hat{\varphi}_j(x) = \sum_j m_j = m = \hat{\varphi}(x) \qquad \forall x \ge a$$

Hence $(X, \hat{\varphi}, \hat{\varphi}_j)$ satisfies (5.16.1) and (5.16.2). Now let $\{x_\gamma \mid \gamma \in \Gamma\}$ be an increasing net in $\hat{\varphi}^{-1}(\mathbb{R})$, so that $\operatorname{card}(\Gamma) \le \operatorname{card}(J)$ and $x_\gamma \le a$ for all $\gamma \in \Gamma$, and let $q = \sup \hat{\varphi}(x_\gamma)$. Then $q \le m$, and if $q = m$ then $\hat{\varphi}(a) = m = q$ and $a \ge x_\gamma$ for all γ, and if $q < m$, then $\hat{\varphi}(x_\gamma) = \varphi(x_\gamma) \in \mathbb{R}$, and so by (5.16.3) there exist $u \in X$ with $u \ge x_\gamma$ for all $\gamma \in \Gamma$, and such that $\varphi(u) = q < m$. Hence $\hat{\varphi}(u) = \varphi(u) = q$ and we see that $\hat{\varphi}$ satisfies (5.16.3). Moreover $\hat{\varphi}(a) = m \in \mathbb{R}$ so by case (b) there exist increasing functions $\{\hat{\psi}_j \mid j \in J\}$ such that $\hat{\psi}_j \ge \hat{\varphi}_j$ and

$$\sum_{j}{}_{*} \hat{\psi}_j = \sum_{j}{}^{*} \hat{\psi}_j = \hat{\varphi}$$

Now $\varphi - \hat{\varphi} = (\varphi - m)^+$ is increasing and non-negative so if we choose $\lambda_j \in \mathbb{R}_+$ with $\Sigma \lambda_j = 1$ and put

$$\psi_j(x) = \begin{cases} \infty & \text{if } \varphi(x) = \infty \\ \hat{\psi}_j(x) + \lambda_j(\varphi(x) - m)^+ & \text{if } \varphi(x) < \infty \end{cases}$$

Then ψ_j is increasing, $\psi_j \ge \varphi_j$ and since

$$\sum_j \lambda_j (\varphi(x) - m)^+ = (\varphi(x) - m)^+ < \infty$$

for all $x \in \{\varphi < \infty\}$, we see that

$$\sum{}_{*} \psi_j(x) = \sum{}^{*} \psi_j(x) = \hat{\varphi}(x) + (\varphi(x) - m)^+ = \varphi(x)$$

if $\varphi(x) < \infty$, and if $\varphi(x) = \infty$, then evidently we have

$$\Sigma_* \psi_j(x) = \Sigma^* \psi_j(x) = \infty = \varphi(x)$$

Thus the lemma is proved under the assumption (c).

Now observe that $\varphi(a) \geq \gamma(a) > -\infty$ by (5.16.1) and (5.16.2). Hence either $\varphi(a) \in \mathbb{R}$ or $\varphi(a) = \infty$, and so the lemma follows either from case (b) or from case (c). □

<u>Theorem 5.17</u>. <u>Let</u> $f \in \overline{\mathbb{R}}^T$ <u>and</u> $f_j \in \overline{\mathbb{R}}^T$ <u>for all</u> $j \in J$, <u>and put</u> $\xi = \text{card } J$ <u>and</u>

$$S = \{S \subseteq T \; 0 < \text{card}(S) \leq \xi\}$$

$$S_0 = \{S \in S \mid S \text{ is finite and } \sum_j{}_* M_S(f_j) > -\infty\}$$

$$S_1 = \{S \in S \mid S \nsupseteq S_0 \;\; \forall S_0 \in S_0\}$$

<u>Then the following four statements are equivalent</u>

(5.17.1) $\sum_j{}_* M_S(f_j) \leq M_S(f) \quad \forall S \in S_0 \cup S_1$

(5.17.2) $\sum_j f_j \ll f$

(5.17.3) $\exists \{\psi_j \mid j \in J\}$ <u>an increasing partition of unity on</u> $\overline{\mathbb{R}}$, <u>such that</u> $f_j \leq \psi_j \circ f \quad \forall j \in J$

(5.17.4) $\exists (X, \leq)$ <u>a preorderet set</u>, $\exists \psi_j \in I(X)$, $\exists \psi \in I^{\xi}(X)$, $\exists \hat{f}: T \to X$ <u>so that</u> $f_j \leq \psi_j \circ \hat{f}$, $\psi \circ f \leq \hat{f}$, <u>and</u> $\sum_j{}_* \psi_j \leq \psi$

<u>Remarks</u> (1): Let $g = \Sigma_* f_j$, then clearly S_0 contains any finite set S, such that $\text{card}(S) \leq \xi$ and $S \cap \{g > -\infty\} \neq \emptyset$. Thus if $g(t) > -\infty$ for all $t \in T$, then $S_1 = \emptyset$.

(2): If ξ is finite and X is a linearly preordered space then $I^{\xi}(X) = I(X)$.

Proof. (5.17.1) ⇒ (5.17.2): Let S be a subset of T and let $a < \Sigma_* M_S(f_j)$. Then there exist $(a_j) \in \mathbb{R}^J$, such that

(i) $$\sum_j |a_j| < \infty \quad \text{and} \quad \sum_j a_j > a$$

(ii) $$a_j \leq M_S(f_j) \quad \forall j \in J$$

Hence there exist a countable set $N \subseteq J$, such that $a_j = 0$ for $j \in J \setminus N$, and if J is atmost countable we may assume that $N = J$. Then by (i) there exist $(c_j) \in \mathbb{R}^J$, so that

(iii) $$c_j = a_j = 0 \quad \forall j \in J \setminus N \quad \text{and} \quad c_j < a_j \quad \forall j \in N$$

(iv) $$\sum_j |c_j| < \infty \quad \text{and} \quad \sum_j c_j > a$$

Then for each $j \in N$ there exist $t_j \in S$, so that $f_j(t_j) > c_j$, and for each $j \in J \setminus N$, there exist a countable set $S_j \subseteq S$ so that

(v) $$M_{S_j}(f_j) = M_S(f_j) \geq 0 = c_j \quad \forall j \in J \setminus N$$

Now put

$$V = \{t_j \mid j \in N\} \cup \bigcup_{j \in J \setminus N} S_j$$

Since $J \setminus N = \emptyset$ if J is countable we have that $V \in \mathcal{S}$, and by construction we have

(vi) $$M_V(f_j) \geq c_j \quad \forall j \in J \quad \text{and} \quad V \subseteq S$$

Thus if $W \in S_1$, for some $W \subseteq V$, then by (5.17.1) we have

$$a < \sum_j c_j \leq \sum_j{}_* M_W(f_j) \leq M_W(f) \leq M_S(f)$$

If this is not the case, let Γ be the set of all $\alpha \in S_0$ such that $\alpha \subseteq V$. Then Γ is non-empty, and Γ filters upwards to V. Hence we have $\{M_\alpha(f_j) \mid \alpha \in \Gamma\}$ increases upwards to $M_V(f_j)$ for all $j \in J$ and

(vii) $$-\infty < \sum_j{}_* M_\alpha(f_j) \leq M_S(f) \qquad \forall \alpha \in \Gamma$$

so by (A.5+10) we have that

$$a < \sum_j c_j \leq \sum_j{}_* M_V(f_j) \leq \lim_{\alpha\in\Gamma} \sum_j{}_* M_\alpha(f_j) \leq M_S(f)$$

Thus in any case we find that $a < M_S(f)$, and so (5.17.2) holds.

(5.17.2) ⇒ (5.17.3): Let $S(x) = \{f \leq x\}$ for $x \in \overline{\mathbb{R}}$ and put

$$\varphi_j(x) = M_{S(x)}(f_j) \qquad \forall x \in \overline{\mathbb{R}}$$

Then φ_j is an increasing map from $\overline{\mathbb{R}}$ into $\overline{\mathbb{R}}$. Moreover if $t \in T$ and $x = f(t)$, then $t \in S(x)$ and so $f_j(t) \leq \varphi_j(x)$. Hence we have

(viii) $$f_j \leq \varphi_j \circ f \qquad \forall j \in J$$

By (5.17.2) we have

$$\sum_j{}_* \varphi_j(x) \leq M_{S(x)}(f) \leq x \qquad \forall x \in \overline{\mathbb{R}}$$

and so by Lemma 5.16 there exist an increasing partition $\{\psi_j \mid j \in J\}$ of unity on $\overline{\mathbb{R}}$, such that $\psi_j \geq \varphi_j$. Hence $f_j \leq \psi_j \circ f$ for all $j \in J$

by (viii).

(5.17.3) ⇒ (5.17.4): Evident!

(5.17.4) ⇒ (5.17.1): Let $S \in S_0 \cup S_1$, then S is non-empty and $\mathrm{card}(S) \le \xi$. Now let us choose $(X,\le)$, ψ_j, ψ and f according to (5.17.4). Since $\mathrm{card}\,\hat{f}(S) \le \xi$, there exists $u \in X$, such that $u \ge f(t)$ for all $t \in S$ and

$$\psi(u) = \sup_{t \in S} \psi(\hat{f}(t)) \le M_S(f)$$

since $f \ge \psi \circ \hat{f}$. Moreover since ψ_j is increasing and $f_j \le \psi_j \circ \hat{f}$, we have that $M_S(f_j) \le \psi_j(u)$. Hence we conclude that

$$\sum_j{}_* M_S(f_j) \le \sum_j{}_* \psi_j(u) \le \psi(u) \le M_S(f)$$

Thus (5.17.1) holds, and the theorem is proved. □

<u>Theorem 5.18</u>. *Let* $\{f_j \mid j \in J\}$ *be a rectilinear subset of* $\overline{\mathbb{R}}^T$ *and let* $g \ge f$, *where* $f = \sum_j{}_* f_j$. *Put*

$$m_j = \sup\{f_j(t) \mid t \in \{f = -\infty\}\}, \qquad m = \sum_j{}_* m_j$$

$$\mathcal{M} = \{S \subseteq T \mid \sum_j{}_* M_S(f_j) \le M_S(g)\}$$

$$\mathcal{L} = \{S \subseteq T \mid \sum_j f_j \ll g \text{ on } S\} = \{S \subseteq T \mid 2^S \subseteq \mathcal{M}\}$$

Then $\mathcal{L}$ *and* $\mathcal{M}$ *are* (vf)-*stable pavings containing all finite subsets of* T, *and*

(5.18.1) $\quad f(t) \ge m \quad \forall t \in \{f > -\infty\}$

(5.18.2) $\quad m = -\infty$ *if* J *is finite*

(5.18.3) $\quad$ *If* $M_S(g) \le m$, *then* $S \in \mathcal{M}$

(5.18.4) $\{g \geq m\} \in L$

(5.18.5) $\bigcup_{j\in\alpha} \{f_j = -\infty\} \in L \quad \forall\alpha \quad \text{finite} \subseteq J$

(5.18.6) $S \in M$ <u>if:</u> $\forall c < m \; \exists \alpha$ <u>finite</u> $\subseteq J$, <u>such that</u>

$\{g \leq c\} \cap S \subseteq \bigcup_{j\in\alpha} \{f_j = -\infty\}$

<u>In particular we have that</u> $g >> \sum_j f_j$ <u>if</u> $\{g < m\}$ <u>is finite (e.g. if</u> $m = -\infty$)

<u>Proof</u>. Let $U,V \in M$ and let $S = U \cup V$, then we have $M_S(f_j) = M_U(f_j) \vee M_V(f_j)$ for all j. Hence if $M_U(f_j) \geq M_V(f_j)$ for all $j \in J$, then

$$\sum_j{}^* M_S(f_j) = \sum_j{}^* M_U(f_j) \leq M_U(g) \leq M_S(g)$$

And if $M_U(f_i) < M_V(f_i)$ for some $i \in J$, then there exist $t_0 \in V$, so that $f_i(t_0) > f_i(u)$ for all $u \in U$. Let $\leq$ be the linear pre-ordering on T induced by $\{f_j\}$, then $t_0 \geq u$ for all $u \in U$, and so we have that $f_j(u) \leq f_j(t_0)$ for all $j \in J$. Thus $M_U(f_j) \leq M_V(f_j)$ for all $j \in J$ and so as above we have

$$\sum_j{}^* M_S(f_j) \leq M_S(g)$$

Hence $S \in M$ and so M is (vf)-stable.

Let $U,V \in L$ and let $S = U \cup V$. If $W \subseteq S$, then $W \cap U$ and $W \cap V$ both belong to M, and so $W \in M$ by (vf)-stability of M. Thus L is (vf)-stable.

Clearly $\{t\} \in L$ for all $t \in T$, since $g \geq f$, hence all finite sets belongs to L and thus to M.

(5.18.1): Let $t \in \{f > -\infty\}$ and let $u \in \{f = -\infty\}$, then $t \geq u$ and so $f_j(t) \geq f_j(u)$ for all j. But then we have that $f_j(t) \geq m_j$

for all $j \in J$ and all $t \in \{f > -\infty\}$. Thus (5.18.1) holds.

(5.18.2): Suppose that $m > -\infty$, then $m_j > -\infty$ for all $j \in J$, and so there exist $t_j \in \{f = -\infty\}$, such that $f_j(t_j) > -\infty$. If J is finite then there exist $k \in J$, sich that $t_k \geq t_j$ for all $j \in J$, and so

$$-\infty = f(t_k) = \sum_j{}_* f_j(t_j) > -\infty$$

which is impossible. Thus (5.18.2) holds.

(5.18.3): Suppose that $M_S(g) \geq m$. If $S \subseteq \{f = -\infty\}$ then we have

$$\sum_j{}_* M_S(f_j) \leq \sum_j{}_* m_j = m \leq M_S(g)$$

and so $S \in M$. So suppose that $S \not\subseteq \{f = -\infty\}$, i.e. if $\hat{T} = \{f > -\infty\}$, then $U = S \cap \hat{T} \neq \emptyset$. Since M contains all finite sets we have

$$-\infty < \sum_j{}_* M_F(f_j) \leq M_F(g)$$

for all non-empty finite sets $F \subseteq \hat{T}$. Hence by Theorem 5.17 we have, that $g >> \Sigma f_j$ on $\hat{T}$. Thus $U \in M$, and since $u \geq t$ for all $u \in U$ and all $t \in S \setminus \hat{T}$, we see that $M_U(f_j) = M_S(f_j)$ for all $j \in J$, since $U \neq \emptyset$. Hence

$$\sum_j{}_* M_S(f_j) = \sum_j{}_* M_U(f_j) \leq M_U(g) \leq M_S(g)$$

and so $S \in M$.

(5.18.4): Follows trivially from (5.18.3).

(5.18.5): Let α be a finite subset of L, and suppose that

(i) $$S \subseteq \bigcup_{j \in \alpha} \{f_j = -\infty\}$$

Then I claim that $M_S(f_j) = -\infty$ for at least one $j \in \alpha$. So suppose that $M_S(f_j) > -\infty$ for all $j \in \alpha$, then there exist $t_j \in S$ such that

$f_j(t_j) > -\infty$ for all $j \in \alpha$, and since α is finite then by rectilinearity there exist $k \in \alpha$ with $t_k \geq t_j$ for all $j \in \alpha$. But then

$$f_j(t_k) \geq f_j(t_j) > -\infty \qquad \forall j \in \alpha$$

which contradicts (i), since $t_k \in S$, and so we must have that $M_S(f_j) = -\infty$ for at least one $j \in \alpha$. Hence we conclude that

$$\sum_j{}^* M_S(f_j) = -\infty \leq M_S(g)$$

and so $S \in M$. Thus (5.18.5) follows.

(5.18.6): If $M_S(g) \geq m$, then $S \in M$ by (5.18.3). If $c = M_S(g) < m$, then $S \subseteq \{g \leq c\}$, and so under the assumption of (5.18.6) we see that $S \in M$ by (5.18.5). □

<u>Theorem 5.19</u>. <u>Let</u> $\{\nu_j \mid j \in J\}$ <u>be a family of increasing everywhere defined functionals on</u> T, <u>let</u> $G \subseteq \overline{\mathbb{R}}^T$ <u>and let</u> $\nu : G \to \overline{\mathbb{R}}$ <u>be an increasing functional, such that</u>

(5.19.1) $$\sum_{j \in J}{}^* \nu_j(\psi_j \circ g) \leq \nu(g)$$

<u>whenever</u> $g \in G$ <u>and</u> $\{\psi_j \mid j \in J\}$ <u>is an increasing partitition of unity on</u> $\overline{\mathbb{R}}$ <u>Then we have</u>

(5.19.2) $$\sum_{j \in J}{}^* \nu_j(f_j) \leq \nu^G(f) \quad \text{if} \quad f \gg \sum_j{}^* f_j$$

<u>Proof</u>. By Theorem 5.17 there exist an increasing partition $\{\psi_j \mid j \in J\}$ of unity on $\overline{\mathbb{R}}$, such that $f_j \leq \psi_j \circ f$. Now let $g \in G$ so that $g \geq f$. Then $f_j \leq \psi_j \circ g$, since ψ_j is increasing, hence by (5.19.1) we have

$$\sum_j{}_* \nu_j(f_j) \le \sum_j{}_* \nu_j(\psi_j \circ g) \le \nu(g)$$

and so (5.19.2) follows by taking infimum over $g \in G$ with $g \geq f$. □

We are now ready for the promised version of superadditivity for outer integrals.

<u>Theorem 5.20</u>. <u>Let</u> $(T,\mathcal{B},\mu)$ <u>be a positive measure space, and let</u> $f \in \overline{\mathbb{R}}^T$ <u>and</u> $\{f_j \mid j \in J\} \subseteq \overline{\mathbb{R}}^T$, <u>such that for every</u> $a > \int^* f d\mu$ <u>there exist</u> $h \in \overline{\mathbb{R}}^T$ <u>satisfying</u>

(5.20.1) $\quad \sum_j f_j \ll h$, <u>and</u> $\int^* h d\mu \le a$

<u>Then we have</u>

(5.20.2) $\quad \sum_j{}_* \int^* f_j f\mu \le \int^* f d\mu$

<u>Remark</u>. If $\sum_j f_j \ll f$, then (5.20.1) clearly holds with $h = f$. Suppose that $\sum_j f_j \ll f$ on $T\setminus S$, for some $S \subseteq T$ and put

$$c = \sum_j{}_* M_S(f_j) \ , \qquad h = 1_{T\setminus S} f + c 1_S$$

Then the reader easily verifies that

(5.20.3) $\quad \sum_j f_j \ll h$ and $\int^* h d\mu \le \int^* f d\mu + \int_S^* (c-f) d\mu$

Thus if S or $(c-f)$ is sufficiently small we see that (5.20.1) holds.

<u>Proof</u>. Let $G = \{g \in L(\mu) \mid g(t) > -\infty \ \ \forall t \in T\}$. Then we have that

$$\text{(i)} \qquad \int^* \varphi d\mu = \inf\{\int g d\mu \mid g \in G, g \geq \varphi\}$$

for all $\varphi \in \overline{\mathbb{R}}^T$.

First we show that (5.19.1) holds if $\nu_j = \nu = \int^* d\mu$. So let $\{\psi_j \mid j \in J\}$ be an increasing partition of unity on $\overline{\mathbb{R}}$, and let $g \in G$. We shall then show that

$$\text{(ii)} \qquad \sum_{j}{}_* \int^* (\psi_j \circ g) d\mu \leq \int g d\mu$$

So let $m_j = \int^* (\psi_j \circ g) d\mu$ and let $m = \Sigma_* m_j$. If $m = -\infty$ or if $\int g d\mu = \infty$, then (ii) holds trivially. So suppose that $m > -\infty$ and $\int g d\mu < \infty$. Then we have

$$\text{(iii)} \qquad \sum_j m_j^- < \infty \quad \text{and} \quad g^+ \in L^1(\mu)$$

Let $g_j = \psi_j \circ g$, and if $\alpha \subseteq J$, then we put

$$\psi_\alpha = \sum_{j \in \alpha} \psi_j \quad , \quad g_\alpha = \psi_\alpha \circ g = \sum_{j \in \alpha} g_j$$

By Lemma 5.15 there exist a countable set $N \subseteq J$, such that $\psi_j(x) = 0$ for all $x \in \mathbb{R}$ and all $j \in J \setminus N$, and $\psi_j(\infty) \geq 0$ for all $j \in J \setminus N$. Now since $g^+ \in L^1(\mu)$ and $g > -\infty$ we have that

$$\text{(iv)} \qquad g_j = 0 \quad \mu\text{-a.e.} \quad \forall j \in J \setminus N$$

$$\text{(v)} \qquad g_N = g \quad \mu\text{-a.e.}$$

By (5.16.8) we have that $\psi_\alpha(x) \leq \psi_\alpha(0) + x$ for all $x \geq 0$, and so if $\psi_\alpha(0) \leq 0$, then $\psi_\alpha(x) \leq x^+$ for all $x \in \overline{\mathbb{R}}$. Let $\pi = \{j \in N \mid \psi_j(0) \leq 0\}$, then $\pi \neq \emptyset$ and $\psi_\alpha(0) \leq 0$ for all $\alpha \subseteq \pi$, so we have that $g_\alpha \leq g^+$ for all $\alpha \subseteq \pi$; moreover if α is a finite subset of π, then we have

$$\int g_\alpha d\mu = \sum_{j\in\alpha} \int g_j d\mu = \sum_{j\in\alpha} m_j \geq \sum_{j\in J} m_j^- > -\infty$$

Hence $g_\alpha \in L^1(\mu)$ for all finite subsets α of π, and since π is atmost countable (recall that $\pi \subseteq N$) then by the Fatou lemma we have

$$\int g_\pi d\mu \geq \limsup_{n\to\infty} \sum_{j\in\alpha(n)} m_j \geq - \sum_{j\in J} m_j^- > -\infty$$

where $\{\alpha(n) | n \geq 1\}$ is an increasing sequence of finite subsets of π, such that $\alpha(n) \uparrow \pi$. And since $g_\pi \leq g^+$ we see that

(vi) $\qquad g_\pi \in L^1(\mu)$

Now let α be any subset of N. then evidently we have that $\psi_{\alpha\cap\pi}(0) \leq 0$ and $\psi_{\alpha\cup\pi}(0) \leq 0$, and so by the argument above we have

$$g_\alpha = g_{\alpha\cap\pi} + g_{\alpha\cup\pi} - g_\pi \leq 2g^+ + |g_\pi|$$

Let $\{\beta(n) | n \geq 1\}$ be finite sets so that $\beta(n) \uparrow N$. Since $2g^+ + |g_\pi| \in L^1(\mu)$ by (iii) and (vi) we have

$$\int g_N d\mu \geq \limsup_{n\to\infty} \sum_{j\in\beta(n)} m_j$$

$$\geq \liminf_{n\to\infty} \sum_{j\in\beta(n)} m_j$$

$$\geq \sum_{j\in N} m_j$$

since $m_j \geq -m_j^-$ and $\Sigma m_j^- < \infty$ by (iii). Moreover by (iv) we have $m_j = 0$ for $j \in J\setminus N$, and so by (v) we see that

$$\int g d\mu = \int g_N d\mu \geq \sum_{j\in N} m_j = \sum_{j\in J} m_j$$

and thus (ii) holds.

Now let $a > \int^* f d\mu$, and choose $h \in \overline{\mathbb{R}}^T$ satisfying (5.20.1). Then by (i), (ii) and Theorem 5.19 we have

$$\sum_{j}{}_* \int^* f_j d\mu \leq \int^* h d\mu \leq a$$

and so (5.20.2) holds. □

<u>Theorem 5.21</u>. <u>Let</u> $\chi \in Pr^*(T)$ <u>and let</u> G <u>be a subset of</u> $\overline{\mathbb{R}}^T$, <u>such that</u> $\psi \circ g \in L(\chi)$ <u>for all</u> $g \in G$ <u>and all increasing functions</u> $\psi : \overline{\mathbb{R}} \to \overline{\mathbb{R}}$ <u>satisfying that</u> $\psi(\mathbb{R}) \subseteq \mathbb{R}$ <u>and</u>

(5.21.1) $$|\psi(x) - \psi(y)| \leq |x-y| \qquad x,y \in \mathbb{R}$$

<u>If</u> $H \subseteq B(T)$ <u>is rectilinear, then we have</u>

(5.21.2) $$\chi\left(\sum_{j=1}^n h_j\right) \leq \sum_{j=1}^n (h_j) \leq \chi^G\left(\sum_{j=1}^n h_j\right)$$

<u>for all</u> $h_1, \ldots, h_n \in H$.

<u>Proof</u>. Let $\{\psi_1, \ldots, \psi_n\}$ be an increasing partition of unity on $\overline{\mathbb{R}}$. Then $\psi_j(\mathbb{R}) \subseteq \mathbb{R}$ and ψ_j satisfies (5.21.1) for all $j = 1, \ldots, n$. Hence $\psi_j \circ g = g_j$ belongs to $L(\chi)$ and $g \in L(\chi)$, thus we have

$$\sum_{j=1}^n{}_* \chi(g_j) = \sum_{j=1}^n{}_* \chi^o(g_j) \leq \chi^o\left(\sum_{j=1}^n g_j\right) = \chi^o(g) = \chi(g)$$

by (5.2).

Let $h_1,\dots,h_n$, and let $h = h_1 + \dots + h_n$. Then by Theorem 5.18 we have that $h >> \Sigma h_j$. Thus the last inequality in (5.21.2) follows from Theorem 5.19, and the first inequality follows from (5.3). □

6. Marginal and projective systems

A marginal system of probability constants, is a collection of the form

$$\mathcal{E} = \{T \xrightarrow{q_\gamma} (T_\gamma, A_\gamma, \lambda_\gamma) \mid \gamma \in \Gamma\}$$

where T and Γ are sets (the target space and the index set), (T_γ, A_γ) for $\gamma \in \Gamma$ are algebraic function spaces (the marginal spaces), λ_γ for $\gamma \in \Gamma$ are probability contents on (T_γ, A_γ) (the marginals), and q_γ for $\gamma \in \Gamma$ are functions from T into T_γ (the projections). Let $\mathcal{E}$ be a marginal system, if $\gamma \in \Gamma$, $\Sigma \subseteq \Gamma$ and θ is a map from a set Δ into Γ, then we put

$$A^\gamma = \{\varphi \circ q_\gamma \mid \varphi \in A_\gamma\} = q_\gamma^{-1}(A_\gamma)$$

$$L^\Sigma = \text{the linear span of } \bigcup_{\gamma \in \Sigma} A_\gamma, \quad L^\theta = L^{\theta(\Delta)}$$

$$A^\Sigma = \text{the algebraic generated by } L^\Sigma, \quad A^\theta = A^{\theta(\Delta)}$$

$$\mathcal{E}_\theta = \{T \xrightarrow{q_{\theta(\delta)}} (T_{\theta(\delta)}, A_{\theta(\delta)}, \lambda_{\theta(\delta)}) \mid \delta \in \Delta\}$$

Let ξ be an infinite cardinal, and let α denote one of the following seven symbols: ξ, $c\xi$, τ, $c\tau$, σ, $c\sigma$ or π, then we define

$$M(\mathcal{E}) = \{\lambda \in Pr(A^\Gamma) \mid \lambda(\varphi \circ q_\gamma) = \lambda_\gamma(\varphi) \quad \forall \gamma \in \Gamma \quad \forall \varphi \in A_\gamma\}$$

$$M_\alpha(\mathcal{E}) = M(\mathcal{E}) \cap Pr_\alpha(A^\Gamma)$$

A functional $\lambda \in M(\mathcal{E})$ is said to have marginals $\{\lambda_\gamma\}$.

If $M(\mathcal{E}) \neq \emptyset$, then we say that $\mathcal{E}$ is consistent, and we say that $\mathcal{E}$ is α-consistent (resp. fully α-consistent) if $M_\alpha(\mathcal{E}) \neq \emptyset$ (resp. if $M(\mathcal{E}) = M_\alpha(\mathcal{E}) \neq \emptyset$), whenever α denotes one of the symbols

ξ, $c\xi$, τ, $c\tau$, σ, $c\sigma$ or π.

If $\mathcal{E}$ is consistent, then by Theorem 4.1.3, we have that $\mathcal{E}$ satisfies the socalled weak consistency conditions:

$$(6.1) \qquad \overline{\lambda_\gamma}(\varphi) = \lambda_\beta(\psi) \quad \text{if} \quad \varphi o q_\gamma = \psi o q_\beta \quad \text{and} \quad \varphi \in \bar{A}_\gamma,\ \psi \in \bar{A}_\beta$$

A marginal system satisfying (6.1) is said to be weakly consistent. It is also clear that a consistent system $\mathcal{E}$ satisfies the socalled strong consistency conditions:

$$(6.2) \qquad \sum_{\gamma\in\sigma} \lambda_\gamma(\varphi_\gamma) \geq 0 \quad \text{if} \quad \sigma \in 2^{(\Gamma)},\ \varphi_\gamma \in A_\gamma \ \forall \gamma \in \sigma,\ \sum_{\gamma\in\sigma} \varphi_\gamma o q_\gamma \geq 0$$

where $2^{(\Gamma)}$ is the set of all non-empty finite subsets of Γ, and below we shall see that $\mathcal{E}$ is consistent if and only if $\mathcal{E}$ satisfies the strong consistency conditions (6.2).

The general marginal problem is the problem of fininding necessary and/or sufficient conditions for consistency, α-consistency or fully α-consistency for marginal systems.

The weak consistency conditions do not imply consistency. However, note that if $\mathcal{E}$ is weakly consistent, then by Theorem 4.13 there exists $\lambda^\gamma \in Pr(A^\gamma)$ such that $\lambda_\gamma = q_\gamma\lambda^\gamma$ for all $\gamma \in \Gamma$, and λ^γ and λ^β coincides on $A^\gamma \cap A^\beta$ for all $\gamma, \beta \in \Gamma$, and $\mathcal{E}$ is consistent if and only if $\{\lambda^\gamma \mid \gamma \in \Gamma\}$ admits a simultaneous extension belonging to $Pr(A^\Gamma)$. If the algebras $\{A^\gamma\}$ are filtering upwards then $A^\Gamma = \cup A^\gamma$ and so this extension is possible and unique. A marginal system $\mathcal{E}$ is said to be projective if $\mathcal{E}$ is weakly consistent and if the algebras $\{A^\gamma \mid \gamma \in \Gamma\}$ are filtering upwards. By the argument we have that a projective system is consistent and $M(\mathcal{E})$ contains exactly one element, denoted $\varprojlim \mathcal{E}$ and called the projective limit of $\mathcal{E}$.

Let Θ be a family of maps θ from a set $\Delta(\theta)$ into Γ, then we say that Θ is <u>finitely exhausting for $\mathcal{E}$</u>, if

(6.3) $$\forall F \text{ finite} \subseteq \bigcup_{\gamma\in\Gamma} A^{\gamma},\ \exists\theta\in\Theta:\ F \subseteq \bigcup_{\delta\in\Delta(\theta)} A^{\theta(\delta)}$$

And if ξ is an infinite cardinal, then we say that Θ is <u>ξ-exhausting for</u> $\mathcal{E}$ if

(6.4) $$\forall F \subseteq \bigcup_{\gamma\in\Gamma} A^{\gamma} \text{ with } \operatorname{card}(F) \leq \xi,\ \exists\theta\in\Theta:\ F \subseteq A^{\theta}$$

And we define σ-<u>exhausting</u> and τ-<u>exhausting</u> similarly, i.e. Θ is τ-exhausting, if and only if $A^{\theta} = A^{\Gamma}$ for some θ.

Let $\mathcal{E}$ be a marginal system, we shall then define a functional (outer probability content, if $\mathcal{E}$ is consistent) associated to $\mathcal{E}$, which will play a central role in all that follows

$$\mathcal{E}^*(f) = \inf\left\{ \sum_{\gamma\in\sigma} \lambda_{\gamma}(\varphi_{\gamma}) \;\middle|\; \begin{array}{l} \sigma\in 2^{(\Gamma)}, \varphi_{\gamma}\in A_{\gamma}\ \forall\gamma\in\sigma \\ \text{and } \sum_{\gamma\in\sigma} \varphi_{\gamma}\circ q_{\gamma} \geq f \end{array} \right\}$$

for all $f \in \overline{\mathbb{R}}^{T}$. Evidently we have that $\mathcal{E}^*$ is increasing and subadditive, and its dual functional, denoted $\mathcal{E}_* = (\mathcal{E}^*)^0$, is given by

$$\mathcal{E}_*(f) = \sup\left\{ \sum_{\gamma\in\sigma} \lambda_{\gamma}(\varphi_{\gamma}) \;\middle|\; \begin{array}{l} \sigma\in 2^{(\Gamma)}, \varphi_{\gamma}\in A_{\gamma}\ \forall\gamma\in\sigma \\ \text{and } \sum_{\gamma\in\sigma} \varphi_{\gamma}\circ q_{\gamma} \leq f \end{array} \right\}$$

for all $f \in \overline{\mathbb{R}}^{T}$.

<u>Theorem 6.1. Let</u> $\mathcal{E} = \{T \xrightarrow{q_\gamma} (T_\gamma, A_\gamma, \lambda_\gamma) \mid \gamma \in \Gamma\}$ <u>be a marginal system, and let</u> Θ <u>be a finitely exhausting family of maps for</u> $\mathcal{E}$. <u>Then the following five statements are equivalent</u>

(6.1.1) $\mathcal{E}$ <u>is consistent.</u>

(6.1.2) $\mathcal{E}$ <u>is weakly consistent, and</u> $\mathcal{E}_\theta$ <u>is consistent</u> $\forall\, \theta \in \Theta$

(6.1.3) $\mathcal{E}$ <u>satisfies the strong consistency conditions (6.2)</u>

(6.1.4) $\exists\, f \in B(T)$ <u>such that</u> $\mathcal{E}^*(f) \neq -\infty$

(6.1.5) $\mathcal{E}^*$ <u>is an outer probability content on</u> T

<u>Proof</u>. (6.1.1) $\Rightarrow$ (6.1.2): Evident.

(6.1.2) $\Rightarrow$ (6.1.3): Let $\sigma \in 2^{(\Gamma)}$ and let $\varphi_\gamma \in A_\gamma$ for $\gamma \in \sigma$, such that $\Sigma\, \varphi_\gamma \circ q_\gamma \geq 0$. By (6.3) there exist $\theta \in \Theta$, so that

$$\varphi_\gamma \circ q_\gamma \in \bigcup_{\delta \in \Delta(\theta)} A^{\theta(\delta)}$$

Now let $\lambda \in M(\mathcal{E}_\theta)$, and choose $\delta(\gamma) \in \Delta(\theta)$ for $\gamma \in \sigma$, such that $\varphi_\gamma \circ q_\gamma \in A^{\alpha(\gamma)}$, where $\alpha(\gamma) = \theta(\delta(\gamma))$. Then there exist $\psi_\gamma \in A_{\alpha(\gamma)}$, such that $\varphi_\gamma \circ q_\gamma = \psi_\gamma \circ q_{\alpha(\gamma)}$, and so

$$0 \leq \lambda\Big(\sum_{\gamma\in\sigma} \varphi_\gamma \circ q_\gamma\Big) = \sum_{\gamma\in\sigma} \lambda(\psi_\gamma \circ q_{\alpha(\gamma)})$$

$$= \sum_{\gamma\in\sigma} \lambda_{\alpha(\gamma)}(\psi_\gamma) = \sum_{\gamma\in\sigma} \lambda_\gamma(\varphi_\gamma)$$

Hence $\mathcal{E}$ satisfies the strong consistency conditions

(6.1.3) $\Rightarrow$ (6.1.4): Evident.

(6.1.4) $\Rightarrow$ (6.1.5): Let $a = \|f\|_T$, then $f \leq a1_T$, and so $-\infty < \mathcal{E}^*(f) \leq \mathcal{E}^*(a1_T) \leq a$. Hence $\alpha = \mathcal{E}^*(1_T)$ is finite and $\alpha \leq 1$. By definition of $\mathcal{E}^*$ we clearly have that $\mathcal{E}^*$ is subadditive and

and $\mathcal{E}^*(rg) = r\mathcal{E}^*(g)$ for all $g \in \overline{\mathbb{R}}^T$ and all $0 < r < \infty$. Hence, if $c \in \mathbb{R}$ and $b > c^+$ then we have

$$c\alpha = b\alpha + (c-b)\alpha = \mathcal{E}^*(b1_T) + (c-b)\alpha$$

$$\leq \mathcal{E}^*(c1_T) + \mathcal{E}^*((b-c)1_T) - (b-c)\alpha$$

$$= \mathcal{E}^*(c1_T) \leq c$$

Thus we find that $\alpha = 1$ and $\mathcal{E}^*(c1_T) = c$ for all $c \in \mathbb{R}$. Thus $\mathcal{E}^*$ is an outer probability content.

(6.1.5) ⇒ (6.1.1): By Theorem 5.1 there eixst $\lambda \in \Pr(A^\Gamma | \mathcal{E}^*)$. If $\gamma \in \Gamma$ and $\varphi \in A_\gamma$, then

$$\lambda_\gamma(\varphi) \leq \mathcal{E}_*(\varphi) \leq \lambda(\varphi \circ q_\gamma) \leq \mathcal{E}^*(\varphi) \leq \lambda_\gamma(\varphi)$$

and so $\lambda \in M(\mathcal{E})$. Thus $\mathcal{E}$ is consistent. □

<u>Proposition 6.2</u>. <u>Let</u> $\mathcal{E} = \{T \xrightarrow{q_\gamma} (T_\gamma, A_\gamma, \lambda_\gamma) \mid \gamma \in \Gamma\}$ <u>be a consistent marginal system, let</u> $\lambda \in M(\mathcal{E})$ <u>and let</u> $f \in \overline{\mathbb{R}}^T$. <u>Thus we have</u>

(6.2.1) $\quad L^\Gamma \subseteq L^1(\mathcal{E}^*) \subseteq B(T)$

(6.2.2) $\quad \mathcal{E}_* \leq \lambda_* \leq \lambda^* \leq \mathcal{E}^*$

(6.2.3) $\quad M(\mathcal{E}) = \Pr(A^\Gamma | \mathcal{E}^*)$

(6.2.4) $\quad \mathcal{E}^*(f) = \inf\{\mathcal{E}^*(g) \mid g \in L^\Gamma, g \geq f\}$

(6.2.5) $\quad \mathcal{E}^*(f) = \sup\{\lambda^*(f) \mid \lambda \in M(\mathcal{E})\}$

(6.2.6) $\quad \mathcal{E}_*(f) = \sup\{\mathcal{E}_*(g) \mid g \in L^\Gamma, g \leq f\}$

(6.2.7) $\quad \mathcal{E}_*(f) = \inf\{\lambda_*(f) \mid \lambda \in M(\mathcal{E})\}$

_If_ $H \subseteq B(T)$ _and_ $\mu: H \to \mathbb{R}$ _is a map satisfying:_

(6.2.8) $$\sum_{j=1}^{n} \mu(h_j') - \sum_{j=1}^{m} \mu(h_j'') \leq \mathcal{E}^*\left(\sum_{j=1}^{n} h_j' - \sum_{j=1}^{m} h_j''\right)$$

for all $n,m \geq 0$ _and all_ $h_1',\ldots,h_n',h_1'',\ldots,h_m'' \in H$, _then there exist_ $\lambda \in M(\mathcal{E})$, _so that_ $\lambda_*(h) \leq \mu(h) \leq \lambda^*(h) \quad \forall h \in H$.

_If_ $H \subseteq B(T)$ _satisfies_

(6.2.9) $$\sum_{j=1}^{n} \mathcal{E}^*(h_j) = \mathcal{E}^*\left(\sum_{j=1}^{n} h_j\right) \quad \forall h_1,\ldots,h_n \in \mathcal{E}^* \quad \forall n \geq 1$$

then there exist $\lambda \in M(\mathcal{E})$, _so that_ $\lambda^*(h) = \mathcal{E}^*(h) \quad \forall h \in H$.

_If_ $f \in B(T)$ _and_ $\mathcal{E}_*(f) \leq a \leq \mathcal{E}^*(f)$, _then there exist_ $\lambda \in M(\mathcal{E})$, _so that_ $\lambda_*(f) \leq a \leq \lambda^*(f)$.

Proof (6.2.1): Let $h \in L^\Gamma$, then $h = \Sigma_{\gamma\in\sigma} \varphi_\gamma \circ q_\gamma$ for some $\sigma \in 2^{(\Gamma)}$ and some $\varphi_\gamma \in A_\gamma$. Hence

$$\sum_{\gamma\in\sigma} \lambda_\gamma(\varphi_\gamma) \leq \mathcal{E}_*(h) \leq \mathcal{E}^*(h) \leq \sum_{\gamma\in\sigma} \lambda_\gamma(\varphi_\gamma)$$

and so $h \in L^1(\mathcal{E}^*)$. If $g \in \overline{\mathbb{R}}^T$ and $M_T(g) = \infty$, then evidently $\mathcal{E}^*(g) = \infty$, and if $m_T(g) = -\infty$, then similarly $\mathcal{E}_*(g) = -\infty$, thus $L^1(\mathcal{E}^*) \subseteq B(T)$.

(6.2.2): Let $\sigma \in 2^{(\Gamma)}$ and let $\varphi_\gamma \in A_\gamma$ for $\gamma \in \sigma$, so that $f \leq \sum_{\gamma\,\sigma} \varphi_\gamma \circ q_\gamma$, then we have

$$\lambda^*(f) \leq \lambda\left(\sum_{\gamma\in\sigma} \varphi_\gamma \circ q_\gamma\right) = \sum_{\gamma\in\sigma} \lambda_\gamma(\varphi_\gamma)$$

and so $\lambda^* \leq \mathcal{E}^*$, and similarly $\mathcal{E}_* \leq \lambda_*$.

(6.2.3): Easy consequence of (6.2.1) and (6.2.2).

(6.2.4): Easy consequence of (6.2.1) and (6.2.2).

(6.2.5): If $M_T(f) = \infty$, then both sides equals $+\infty$ and thus we have equality in (6.2.5). So suppose that f is bounded above, and put $f_n = f \vee (-n)$ for $n \in \mathbb{N}$. By Corollary 5.2 there exist $\nu_n \in Pr(B(T) \mid \mathcal{E}^*)$, such that $\nu_n(f_n) = \mathcal{E}^*(f_n)$ for all $n \geq 1$. Since $Pr(B(T) \mid \mathcal{E}^*)$ is compact in $\mathbb{R}^{B(T)}$, we have that $\{\nu_n\}$ admits a limit point $\nu \in Pr(B(T) \mid \mathcal{E}^*)$. Since $\nu_n(f_k) \geq \nu_n(f_n) \geq \mathcal{E}^*(f)$ for all $n \geq k$ we have

$$\nu(f_k) \geq \liminf_{n\to\infty} \nu_n(f_k) \geq \mathcal{E}^*(f) \qquad \forall k \geq 1$$

Now let $\lambda = \nu | A^\Gamma$, then $\lambda \in M(\mathcal{E})$ by (6.2.3) and

$$\lambda^*(f_k) \geq \nu(f_k) \geq \mathcal{E}^*(f) \qquad \forall k \geq 1$$

Let $\varphi \in A^\Gamma$ so that $\varphi \geq f$, then $\varphi \geq f_k$ for all $k \geq \|\varphi\|$, and so $\lambda(\varphi) \geq \mathcal{E}^*(f)$. Thus $\lambda^*(f) \geq \mathcal{E}^*(f)$ and so by (6.2.2) we see that we have equality in (6.2.5).

(6.2.6) and (6.2.7) are easy consequences of (6.2.4) and (6.2.5). And the last three statements of the proposition follows directly from Theorem 5.1 and Corollary 5.2. □

<u>Proposition 6.3</u>. <u>Let</u> $\mathcal{E} = \{T \xrightarrow{q_\gamma} (T_\gamma, A_\gamma, \lambda_\gamma) \mid \gamma \in \Gamma\}$ <u>be a consistent marginal system, and let</u> $\theta: \Delta \to \Gamma$ <u>be a map. Let</u> ξ <u>be an infinite cardinal and let</u> α <u>denote one of the seven symbols:</u> $\xi, c\xi, \tau, c\tau, \sigma, c\sigma$ or π, <u>then we have</u>

(6.3.1) $\quad \lambda | A^\theta \in M(\mathcal{E}_\theta) \qquad \forall \lambda \in M(\mathcal{E})$

(6.3.2) <u>If</u> $\mathcal{E}$ <u>is</u> α-<u>consistent, then so is</u> $\mathcal{E}_\theta$

(6.3.3) <u>If</u> $\mathcal{E}$ <u>is</u> α-<u>consistent, then</u> $\lambda_\gamma \in Pr_\alpha(A_\gamma) \quad \forall \gamma \in \Gamma$

(6.3.4) $\mathcal{E}^* \leq \mathcal{E}^*_\theta$ <u>with equality on</u> L^θ .

<u>And if $\mathcal{E}_\theta$ is projective then we have</u>

$$(6.3.5) \quad \mu(\varphi) = \lambda(\varphi) = \mathcal{E}_\theta^*(\varphi) = \mathcal{E}^*(\varphi) \qquad \forall \varphi \in A^\theta \ \forall \lambda \in M(\mathcal{E}) \ \forall \mu \in M(\mathcal{E}_\theta)$$

<u>Remark</u>. There exists a projective $c\tau$-consistent system $\mathcal{E}$ and a map $\theta: \Delta \to \Gamma$, such that $\mathcal{E}_\theta$ is not fully σ-consistent. And so there exists $\mu \in M(\mathcal{E}_\theta)$ which admits no extensions belonging to $M(\mathcal{E})$. To see this let I be the unit interval and put

$$T_{2n} = I^n, \quad A_{2n} = C(I^n), \quad \lambda_{2n}(f) = \int_0^1 f(t,\cdots,t)\,dt$$

$$T_{2n-1} = I, \quad A_{2n-1} = C(I), \quad \lambda_{2n-1}(f) = \int_0^1 f(t)\,dt$$

$$T = \{(t_j) \in I^{\mathbb{N}} \mid \lim_{n\to\infty} (\min_{2\le j\le n} 4^j|t_1-t_j|) = 0\}$$

$$\varphi_{2n}(t) = (t_1,\cdots,t_n), \quad \varphi_{2n-1}(t) = t_n \qquad \forall t = (t_j) \in T$$

$$\theta(n) = 2n-1 : \mathbb{N} \longrightarrow \mathbb{N}$$

Then we have that

$$\mathcal{E} = \{T \xrightarrow{q_n} (T_n, A_n, {}_n) \mid n \in \mathbb{N}\}$$

is projective and $c\tau$-consistent, but $\mathcal{E}_\theta$ is not fully σ-consistent.

<u>Proof.</u> (6.3.1) is evident and (6.3.2) follows easily from (6.3.1). Moreover (6.3.3) follows from Theorem 4.13.

(6.3.4): Evidently we have that $\mathcal{E}^* \le \mathcal{E}_\theta^*$, and if $\lambda \in M(\mathcal{E})$, then by (6.2.1) and (6.2.2) we have

$$\lambda(\varphi) = \mathcal{E}^*(\varphi) = \mathcal{E}_\theta^*(\varphi) \qquad \forall \varphi \in L^\theta$$

and so 6.3.4 holds.

(6.3.5): If $\mathcal{E}_\theta$ is projective then $A^\theta = L^\theta$, and $M(\mathcal{E}_\theta)$ consists of exactly one element. Thus (6.3.5) holds.

□

Theorem 6.4. Let $\mathcal{E} = \{T \xrightarrow{q_\gamma} (T_\sigma, A_\gamma, \lambda_\gamma) \mid \gamma \in \Gamma\}$ be a marginal system and let ξ be an infinite cardinal. If $\mu \in M_\xi(\mathcal{E})$ then we have

(6.4.1) $\qquad \lambda_\gamma \in \Pr_\xi(A_\gamma)$ and $\lambda^*_{\gamma\xi}(q_\gamma(T)) = 1 \;\; \forall \gamma \in \Gamma$

(6.4.2) $\qquad \int_T (f \circ q_\gamma)\, d\mu_\xi = \int_{T_\gamma} f\, d\lambda_{\gamma\xi} \quad \forall f \in L^1(\lambda_{\gamma\xi}) \quad \forall \gamma \in \Gamma$

(6.4.3) $\qquad \mathcal{E}_*(\mu) \le \int_* h\, d\mu_\xi \le \int^* h\, d\mu_\xi \le \mathcal{E}^*(h) \qquad \forall h \in \overline{\mathbb{R}}^T$

where μ_ξ and $\lambda_{\gamma\xi}$ are the representing ξ-smooth measures for μ and λ_γ respectively. Moreover if $\mu \in M_{c\xi}(\mathcal{E})$, then we have

(6.4.4) $\qquad \lambda_\gamma \in \Pr_{c\xi}(A_\gamma) \qquad \forall \gamma \in \Gamma$

(6.4.5) $\qquad (\lambda_{\gamma\xi})_*(q_\gamma(T)) = 1 \quad \forall \gamma \in \Gamma$ so that weight $(A_\gamma) \le \xi$ and $q_\gamma(T)$ is A_γ-saturated.

Proof (6.4.1+2): By Theorem 4.14 with $\theta = q_\gamma$ we have that $\lambda_\gamma \in \Pr_\xi(A_\gamma)$ and (6.4.2) holdes, but then (6.4.1) follows.

(6.4.3): By Theorem 4.1 and (6.2.2) we have

$$\mathcal{E}_*(h) \le \mu_*(h) \le \int_* h\, d\mu_\xi \le \int^* h\, d\mu_\xi \le \mu^*(h) \le \mathcal{E}^*(h)$$

and so (6.4.3) holds.

(6.4.4) follows from Theorem 4.13.

(6.4.5): Let $\varepsilon > 0$ be given, then by (4.1.16) there exists $K \in K_\xi(A^\Gamma)$ so that $\mu_\xi^*(K) \geq 1-\varepsilon$. Let $C_\gamma = q_\gamma(K)$ and put

$$\bar{C}_\gamma = \{u \in T_\gamma \mid \exists\, t \in C_\gamma \text{ so that } u \equiv t (\mathrm{mod}\, A_\gamma)\}$$

$$K_\gamma = \text{the } \tau(A_\gamma)\text{-closure of } C_\gamma$$

By (4.12) we have that $C_\gamma \in K_\xi(A_\gamma)$, so if $\xi \geq \mathrm{weight}(A_\gamma)$, then $C_\gamma \in K(A_\gamma)$, and evidently we have that

$$C_\gamma \subseteq \bar{C}_\gamma \subseteq K_\gamma$$

Now since $\tau(A_\alpha)$ is regular we have that $\bar{C}_\gamma \in K(A_\gamma)$, and since $\bar{C}_\gamma$ is A_γ-saturated, we have that $\bar{C}_\gamma = K_\gamma$ by Proposition 2.10. Moreover if $q_\gamma(T)$ is A_γ-saturated, then $K = \bar{C}_\gamma \subseteq q_\gamma(T)$, and so

$$(\lambda_{\gamma\xi})_*(q_\gamma(T)) \geq \lambda_{\gamma\xi}(C_\gamma) = \mu_\xi(q_\gamma^{-1}(C_\gamma))$$

$$\geq \mu_\xi^*(K) \geq 1-\varepsilon$$

since $K \subseteq q^{-1}(C_\gamma)$. □

Theorem 6.1 solves the general marginal consistency problem. However, this result, which is only a simple consequence of the Hahn-Banach theorem, is mainly of interest as a starting point for the search of smooth or compact functionals with the given marginals. Our next two theorems give necessary and sufficient conditions for fully $c\xi$-consistency resp. ξ-consistency of marginal systems.

<u>Theorem 6.5. Let</u> $\mathcal{E} = \{T \xrightarrow{q_\gamma} (T_\gamma, A_\gamma, \lambda_\gamma) \mid \gamma \in \Gamma\}$ <u>be a consistent marginal system, and let</u> ξ <u>be an infinite cardinal. Then the following three statements are equivalent</u>

(6.5.1) $\mathcal{M}(\mathcal{E})$ <u>is uniformly compact.</u>

(6.5.2) $\mathcal{E}^*$ <u>is</u> $K_\xi(A^\Gamma)$<u>-tight along</u> A^Γ

(6.5.3) $\mathcal{E}^*$ <u>is</u> $\bar{K}_\xi(A^\gamma)$<u>-tight along</u> $(A^\Gamma)_\xi$

<u>Moreover if</u> $(T,\tau(A^\Gamma))$ <u>is a Prokorov space (see</u> [9]), <u>and if</u> $\xi \geq \mathrm{weight}(A^\Gamma)$, <u>then (6.5.1)-(6.5.3) is equivalent to</u>

(6.5.4) $\mathcal{E}$ <u>is fully</u> $c\tau$<u>-consistent.</u>

<u>Proof</u>. Since $\mathcal{E}^*$ is an outer probability content satisfying (6.2.5) we have that (6.5.1)-(6.5.3) are equivalent by Theorem 5.4.

Now suppose that $\xi \geq \mathrm{weight}(A^\Gamma)$ and that $(T,\tau(A^\Gamma))$ is a Prokorov space. Then bu Proposition 2.9 we have that

(i) $$C(T,\tau(A^\Gamma)) = (A^\Gamma)_\xi \cap (A^\Gamma)^\xi$$

Let $M^+(T)$ be the set of all positive finite Radon measures on $(T,\tau(A^\Gamma))$. Then $\lambda \rightsquigarrow \lambda_\tau$ maps $Pr_{c\tau}(A^\Gamma)$ into $M(T)$, and since $\lambda \rightsquigarrow \lambda(\varphi)$ is continuous in the product topology on $Pr_{c\tau}(A^\Gamma)$ for all $\varphi \in A^\Gamma$, we see that $\lambda \rightsquigarrow \lambda^*(h)$ and $\lambda \rightsquigarrow \lambda_*(h)$ are upper resp. lower semi-continuous on $Pr_{c\tau}(A^\Gamma)$ for all $h \in \overline{\mathbb{R}}^T$. Hence by (i) we see that $i(\lambda) = \lambda_\tau$ is a continuous map from $Pr_{c\tau}(A^\Gamma)$ with its product topology into $M^+(T)$ with its w^*-topology (see [9]). Hence, if (6.5.4) holds, then $i(\mathcal{M}(\mathcal{E}))$ is a compact subset of $M^+(T)$ and thus $\mathcal{M}(\mathcal{E})$ is uniformly compact by the Prokorov property of T. The converse implication is evident.

□

<u>Remark</u>. Note that the category of Prokorov includes all Čech complete spaces, all complete metric spaces and all locally compact spaces. □

Applying Theorem 5.3 with $\chi = \mathcal{E}^*$, $A = A^\Gamma$ and $M = M(\mathcal{E})$ we obtain the following result by (6.2.2) and (6.2.5):

<u>Theorem 6.6</u>. <u>Let</u> $\mathcal{E} = \{T \xrightarrow{q_\gamma} (T_\gamma, A_\gamma, \lambda_\gamma) \mid \gamma \in \Gamma\}$ <u>be a consistent marginal system, and let</u> ξ <u>be an infinite cardinal, then the following three statements are equivalent:</u>

(6.6.1) $\mathcal{E}$ <u>is fully</u> ξ-<u>consistent</u>

(6.6.2) $\mathcal{E}^*$ <u>is</u> ξ-<u>supersmooth at</u> 0 <u>along</u> A^Γ

(6.6.3) $\mathcal{E}^*$ <u>is</u> ξ-<u>supersmooth at</u> $\overline{\mathbb{R}}^T$ <u>along</u> $(A^\Gamma)_\xi$ □

Motivated by Theorem 6.6 we shall now study ξ-smoothness of $\mathcal{E}^*$ or more generally of an outer probability content χ more closely.

Let χ be an outer probability content on T and let ξ be an infinite cardinal. If $h \in \overline{\mathbb{R}}^T$ we define

$$S_\xi(\chi|h) = \{F \subseteq \overline{\mathbb{R}}^T \mid \chi \text{ is } \xi\text{-supersmooth at } h \text{ along } F\}$$
$$S_\xi(\chi) = \{F \subseteq \overline{\mathbb{R}}^T \mid \chi \text{ is } \xi\text{-supersmooth at } \overline{\mathbb{R}}^T \text{ along } F\}$$

and we define $S_\sigma(\chi,h), S_\sigma(\chi), S_\tau(\chi|h)$ and $S_\tau(\chi)$ similarly.

<u>Lemma 6.7</u>. <u>Let</u> χ <u>be an outer probability content on</u> T, <u>let</u> $\varphi \in B^+(T)$, $g \in \mathbb{R}^T$ <u>and</u> $h \in \overline{\mathbb{R}}^T$ <u>so that</u> $\chi(g+h) = \chi(g) \dot{+} \chi(h)$, <u>and let</u> V <u>be a paving on</u> T. <u>Then we have</u>

(6.7.1) $F \in S_\xi(\chi|h)$, F <u>is</u> <u>$(\wedge f)$-stable</u> $\Rightarrow F_\xi \in S_\xi(\chi|h)$

(6.7.2) $F \in S_\xi(\chi|h)$, $f + a1_T \in \overline{F}$ $\forall f \in F$ $\forall a \in \mathbb{R}_+ \Rightarrow \overline{F} \in S_\xi(\chi|h)$

(6.7.3) $F \in S_\xi(\chi|h) \Rightarrow \{f+g \mid f \in F\} \in S_\xi(\chi|g+h)$

(6.7.4) $F_\xi \in S_\xi(\chi)$, $F_\xi \subseteq L^1(\chi) \Rightarrow \{\varphi f \mid f \in F_\xi\} \in S_\xi(\chi)$

(6.7.5) $V \in S_\xi(\chi)$, $V_\xi \subseteq L^1(\chi) \Rightarrow B(T) \cap U(T,V) \subseteq S_\xi(\chi)$

(6.7.6) $S_\xi(\chi|h)$ <u>is</u> <u>$(\vee f)$-stable.</u>

<u>Remark</u>. If either $g \in L^1(\chi)$ or $h \in L^1(\chi)$ then we have that $\chi(g+h) = \chi(g) \dotplus \chi(h)$.

<u>Proof</u> (6.7.1) and (6.7.2) follow directly from Proposition 3.5.

(6.7.3): Let $\hat{F} = \{f+g \mid f \in F\}$ and let $\{\hat{f}_q | q \in Q\}$ be a decreasing net in $\hat{F}$ so that $\text{card}(Q) \le \xi$ and $\hat{f}_q \downarrow h+g$. Then $f_q = \hat{f}_q - g \in F$, and $f_q \downarrow h$. Hence we have

$$\chi(g+h) \le \lim_q \chi(\hat{f}_q) \le \chi(g) \dotplus \lim_q \chi(f_q)$$

$$= \chi(g) \dotplus \chi(h) = \chi(g+h)$$

and so $\hat{F} \in S_\xi(\chi|g+h)$.

(6.7.4): Let $\hat{F} = \{\varphi f | f \in F_\xi\}$ and let $\{\hat{f}_q \mid q \in Q\}$ be a decreasing net in $\hat{F}$ with $\text{card}(Q) \le \xi$ and $\hat{f} = \lim \hat{f}_q$. Then there exists $f_q \in F$ so that $\hat{f}_q = \varphi f_q$ for all $q \in Q$. Now let $\varepsilon > 0$ be given and let $a > \|\varphi\|$. Let $2^{(Q)}$ be ordered by inclusion and put

$$f_\pi = \min_{q \in \pi} f_q, \quad f = \inf_{q \in Q} f_q \qquad \forall \pi \in 2^{(Q)}$$

Then $f_\pi \in F_\xi \in S_\xi(\chi)$, and $f \in F_\xi \subseteq L^1(\chi)$, and $f_\pi \downarrow f$. Hence there exists $\pi \in 2^{(Q)}$ such that

$$0 \leq \chi(f_\pi - f) = \chi(f_\pi) - \chi(f) < \varepsilon/a$$

Now let $q \in Q$ so that $q \geq r$ for all $r \in \pi$. Then $\hat{f} = f\varphi$ and $0 \leq \hat{f}_q - \hat{f}_\pi \leq a(f_\pi - f)$, hence we find

$$\chi(\hat{f}) \leq \chi(\hat{f}_q) \leq \chi(\hat{f}) \dotplus \chi(\hat{f}_q - \hat{f}) \leq \chi(\hat{f}) \dotplus a\chi(f_\pi - f)$$

$$\leq \chi(\hat{f}) + \varepsilon$$

and so $\chi(\hat{f}_q) \longrightarrow \chi(\hat{f})$. Thus $\hat{F} \in S_\xi(\chi)$.

(6.7.5): By (6.7.3) and (6.7.4) it suffices to show that $\hat{F} \in S_\xi(\chi)$ where

$$\hat{F} = \{f \in U(T,V) \mid 0 \leq f \leq 1\}$$

So let $\{f_q \mid q \in Q\}$ be a decreasing net in $\hat{F}$ with $f = \lim_q f_q$ and $\operatorname{card}(Q) \leq \xi$. Now let $\varepsilon > 0$ be given and choose $m \in \mathbb{N}$ so that $m \geq 2/\varepsilon$. Let

$$D_j(q) = \{f_q \geq j/m\} \quad , \ D_j = \{f \geq j/m\}$$

for $q \in Q$ and $j = 0,1,\dots,m$. Then $D_j(q) \in V$ and $D_j(q) \downarrow D_j$ for all $0 \leq j \leq m$. Hence there exists $r \in Q$, such that

(i) $$\chi(D_j(q)) \leq \chi(D_j) + \varepsilon/2 \quad \forall\, 0 \leq j \leq m \quad \forall q \geq r$$

since $V \in S_\xi(\chi)$. Moreover, since

$$f_q \leq \frac{1}{m} + \frac{1}{m}\sum_{j=1}^{m} 1_{D_j(q)} \;, \qquad\qquad f \geq \frac{1}{m}\sum_{j=1}^{m} 1_{D_j}$$

then for every $q \geq r$ we have by (i) that

$$\begin{aligned}\chi(f) \leq \chi(f_q) &\leq \frac{1}{m} + \frac{1}{m}\sum_{j=1}^{m}\chi(D_j(q))\\ &\leq \frac{1}{2}\varepsilon + \frac{1}{m}\sum_{j=1}^{m}(\chi(D_j) + \varepsilon/2)\\ &= \varepsilon + \chi(\frac{1}{m}\sum_{j=1}^{m} 1_{D_j})\\ &\leq \varepsilon + \chi(f)\end{aligned}$$

since $D_j \in V_\xi \subseteq L^1(\chi)$ for all $o \leq j \leq m$. Hence $\chi(f_q) \to \chi(f)$ and so χ is ξ-supersmooth at $\overline{\mathbb{R}}^T$ along $\hat{F}$, and thus (6.7.5) is proved.

(6.7.6): Let $F_1, F_2 \in S_\xi(\chi|h)$, and let $\{f_q \mid q \in Q\}$ be a net in $F_1 \cup F_2$, so that $f_q \downarrow h$ and $\operatorname{card}(Q) \leq \xi$. Put $Q_j = \{q \in Q \mid f_q \in F_j\}$ for $j = 1,2$, then $Q_1 \cup Q_2 = Q$, and so either Q_1 or Q_2 is cofinal in Q. If Q_1 is cofinal in Q then $\{f_q \mid q \in Q_1\}$ is a decreasing net in F_1, which decreases to h, and so

$$\chi(h) = \lim_{q \in Q_1} \chi(f_q) = \lim_{q \in Q} \chi(f_q)$$

since $\{\chi(f_q) \mid q \in Q\}$ decreases and Q_1 is cofinal in Q. Thus $F_1 \cup F_2 \in S_\xi(\chi|h)$. □

<u>Lemma 6.8</u>. <u>Let</u> χ_0 <u>and</u> χ_1 <u>be outer probability contents on</u> T, <u>let</u> $F \in S_\xi(\chi_0)$ <u>and let</u> ξ <u>be an infinite cardinal. Then</u> $F \in S_\xi(\chi_0 \mid g)$ <u>for every</u> $g \in \overline{\mathbb{R}}^T$ <u>satisfying:</u>

<u>For all</u> $\varepsilon > 0$ <u>and all</u> $\varphi \in F_\delta$ <u>there exists</u> $T_0 \subseteq T$, <u>there exists</u> $H \in S_\xi(\chi_1)$ <u>and there exists an increasing map</u> R <u>from</u> F <u>into</u> H <u>such that</u>

(6.8.1) $\quad Rf \leq f$ <u>on</u> $T_0 \quad \forall f \in F$

(6.8.2) $\quad \chi_1(h) \leq \varepsilon + \chi_0(g)$ <u>if</u> $h \in H_\xi$ <u>and</u> $h \leq g$ <u>on</u> T.

(6.8.3) $\quad \chi_0(\varphi) \leq \varepsilon + \chi_1(Rf)$ <u>if</u> $F \in F$ <u>and</u> $\chi_0(\varphi) = \chi_0(f \wedge \varphi)$

<u>Remark</u>. In applications we often take

(6.8.4) $\quad R(f|t) = \inf\{f(u) \mid u \in \sigma(t)\} \qquad \forall t \in T$

(6.8.5) $\quad T_0 = \{t \in T \mid t \in \sigma(t)\}$

where $\sigma: T \rightsquigarrow T$ is a suitably chosen correspondence. Then R is an increasing map: $\overline{\mathbb{R}}^T \to \overline{\mathbb{R}}^T$, satisfying (6.8.1) for all $f \in \overline{\mathbb{R}}^T$. Moreover we have

(6.8.6) $\quad \{Rf < a\} = \sigma^{-1}(\{f < a\}) \qquad \forall f \in \overline{\mathbb{R}}^T \qquad \forall a \in \mathbb{R}$

(6.8.7) $\quad Rf \in U(T,\mathcal{F}) \quad \forall f \in F$

if $F \subseteq \overline{\mathbb{R}}^T$ and $\mathcal{F}$ is a paving on T, such that

(6.8.8) $\quad T \setminus \sigma^{-1}(\{f < a\}) \in \mathcal{F} \qquad \forall f \in F \qquad \forall a \in \mathbb{R}$

Hence the set H in the lemma may be chosen to be a suitable subset of $U(T,\mathcal{F})$.

<u>Proof</u>. Immediate consequence of Lemma 3.20. □

In (6.7.6) we saw that $S_\xi(\chi|h)$ is $(\vee f)$-stable, in the context of marginal systems (in particular projective systems), we often need stability under infinite (increasing) unions. Our next lemma collects a series of results in this direction.

Lemma 6.9. Let χ be an outer probability content on T, let $\{F_j \mid j \in J\}$ be an increasing net of subsets of $\overline{\mathbb{R}}^T$, and let $h \in \overline{\mathbb{R}}^T$, such that J is infinite, and

(6.9.1) $\lim_j \chi(f_j) = \chi(h)$ if $f_j \downarrow h$ and $\exists i: f_j \in F_j \;\; \forall j \geq i$

Let ξ be an infinite cardinal and let $F = \cup_{j \in J} F_j$, then we have

(1): If $F_j \in S_\xi(\chi|h) \;\; \forall j \in J$, and if every subset of F of cardinality $\leq \xi$ is contained in some F_j, then we have that $F \in S_\xi(\chi \mid h)$.

(2): If $F_j \in S_\sigma(\chi \mid h) \;\; \forall j \in J$, then $F \in S_\sigma(\chi \mid h)$.

(3): If $F_j \in S_\xi(\chi)$ and F_j is $(\wedge\xi)$-stable $\forall j \in J$, then $F_\xi \in S_\xi(\chi \mid h)$ and $F_\xi \subseteq F_\eta$, where $\eta = \mathrm{cof}(J)$.

(4): If $F_j \in S_\xi(\chi)$ and F_j is $(\wedge\xi_j)$-stable $\forall j \in J$, where $\{\xi_j\}$ is an increasing net of cardinals (possibly finite), such that $\xi_j \geq \mathrm{card}\{i \in J \mid i \leq j\}$ for all $j \in J$, then $F_\eta \in S_\eta(\chi|h)$ where $\eta = \xi \wedge \Sigma_{j \in J} \xi_j$.

(5): If $F_j \in S_\xi(\chi)$ and F_j is $(\wedge f)$-stable $\forall j \in J$, and if J is finitely founded (i.e. $\{i \in J \mid i \leq j\}$ is finite for all $j \in J$), then $F_\eta \in S_\eta(\chi \mid h)$ where $\eta = \xi \wedge \mathrm{card}(J)$.

(6): <u>if</u> $F_j \in S_\xi(\chi)$, <u>if</u> $a1_T + f$ <u>and</u> $f \wedge g$ <u>belongs to</u> $\bar{F}_j$ $\forall j \in J$, $\forall a \in [0,1]$, $\forall f,g \in F_j$, <u>and if</u> J <u>is finitely founded, then</u> $\bar{F}_\eta \in S_\eta(\chi|h)$ <u>where</u> $\eta = \xi$ card J.

Remark. Recall that cof(J) is the minimal cardinal of a cofinal subset of J. Note that $\mathbb{N}$ is finitely founded in its natural ordering, and so is any set of finite sets ordered by inclusion.

Proof (1): Evident.

(2): Let $\{f_n \mid n \in \mathbb{N}\}$ be a decreasing sequence in F, such that $f_n \downarrow h$. We shall then show:

(i) $$\lim_{n\to\infty} \chi(f_n) = \chi(h)$$

Since J is upwards directed and $\{F_j\}$ is increasing then there exists an increasing map $\beta: \mathbb{N} \to J$, such that $f_n \in F_{\beta(n)}$ for all $n \geq 1$. If β is bounded above, then $f_n \in F_j$ for all $n \geq 1$ for some $j \in J$, and so (i) follows, since $F_j \in S_\sigma(\chi|h)$ by assumption. So suppose that β is unbounded, then

$$\gamma(j) = \min\{n \geq 0 \mid \beta(n+1) \not\leq j\} \quad \forall j \in J$$

is a well defined increasing map from J into $\mathbb{N}_0$, such that $\gamma(\beta(n)) \geq n-1$ for all $n \geq 1$, since β is increasing. Let $h_j = f_{\gamma(j)}$, where $f_0 = \infty$, then $h_j \downarrow h$, and since $\beta(\gamma(j)) \leq j$ if $\gamma(j) \geq 1$, we see that $h_j \in F_j$ for all $j \geq \beta(2)$. Hence by (6.9.1) we have

$$\chi(h) = \lim_j \chi(h_j) = \lim_j \chi(f_{\gamma(j)}) = \lim_{n\to\infty} \chi(f_n)$$

since γ is increasing and cofinal. This proves (i) and thus (2) is proved.

(3): The first part follows from Proposition 3.18 and (6.9.1) since F is $(\wedge f)$-stable. Let $f \in F_\xi$, and let J_0 be a cofinal subset of J with $\mathrm{card}(J_0) = \mathrm{cof}(J)$. Then there exists $\{f_q \mid q \in Q\} \subseteq F$, so that $\mathrm{card}(Q) \leq \xi$ and $f = \inf_q f_q$. Now put

$$Q_j = \{q \in Q \mid f_q \in F_j\}, \quad I = \{i \in J_0 \mid Q_i \neq \emptyset\}$$

$$\hat{f}_i = \inf\{f_q \mid q \in Q_i\} \qquad \forall i \in I$$

Then $\hat{f}_i \in F_i$ by $(\wedge\xi)$-stability of F_i, and

$$f = \inf_{i \in I} \hat{f}_i$$

since $Q_j \uparrow Q$ and J_0 is cofinal in J. Thus $f \in F_\eta$.

(4): Let $\eta_j = \mathrm{card}\{i \in J \mid i \leq j\}$, and let $\{\hat{f}_j \mid j \in J\}$ be a decreasing net, so that

$$\text{(ii)} \qquad \hat{f}_j \downarrow h \quad \exists m \in J: \quad \hat{f}_j \in \hat{F}_j = (F_j)_\eta \quad \forall j \geq m$$

we shall then show that

$$\text{(iii)} \qquad \lim_j \chi(\hat{f}_j) = \chi(h)$$

Let $\{f_{jq} \mid q \in Q\} \subseteq F_j$ for $j \geq m$, so that $\mathrm{card}(Q) \leq \eta$, and

$$\hat{f}_j = \inf_{q \in Q} f_{jq} \qquad \forall j \geq m$$

Since $\eta \leq \Sigma \xi_j$, there exist $Q_j \subseteq Q$, so that

$$Q = \bigcup_{j \in J} Q_j \quad \text{and} \quad \operatorname{card}(Q_j) \leq \xi_j \quad \forall j$$

Now let $I = \{j \in J \mid j \geq m\}$ and let

$$R_j = \{(i,q) \in I \times Q \mid i \leq j, \ q \in \bigcup_{k \leq j} Q_k\}$$

$$f_j = \inf\{f_{iq} \mid (i,q) \in R_j\}$$

for $j \in J$. Then we have

$$\operatorname{card}(R_j) \leq \eta_j \cdot \sum_{k \leq j} \xi_k \leq \xi_j^{\,3}$$

Since $\xi_j^{\,3} = \xi_j$ if $\xi_j = 1$ or ξ_j is infinite, and $2 \leq \xi_j^{\,3} < \infty$ otherwise we have that F_j is $(\wedge \xi_j^{\,3})$-stable, and so $f_j \in F_j$ for $j \geq m$, and

$$f_j \downarrow \inf_{i \in I} \inf_{q \in Q} f_{iq} = \inf_{i \in I} \hat{f}_i = h$$

Hence we have by (6.9.1) that

$$\chi(h) = \lim_j \chi(f_j) \geq \lim_j \chi(\hat{f}_j) \geq \chi(h)$$

since $f_j \geq \hat{f}_j \geq h$ for $j \geq m$. Thus (iii) holds, and so by (3) we have that $F_\eta \in S_\eta(\chi \mid h)$.

(5): Follows from (4) with $\xi_j = \operatorname{card}\{i \in J \mid i \leq j\}$.

(6): I claim that (6.9.1) holds with F_j replaced by $\bar{F}_j$.

So let $f_j \downarrow h$ such that $f_j \in \bar{F}_j \quad \forall j \geq m$ for some $m \in J$. Consider the equivalence relation on J:

$$i \equiv j \Leftrightarrow i \leq j \text{ and } j \leq i$$

Then we have that $f_i = f_j$ if $i \equiv j$, hence for $j \geq m$ there exist $\varphi_j \in F_j$ such that

$$\text{(iv)} \qquad \|\varphi_j - (f_j + 3\cdot 2^{-\eta(j)})\| \leq 2^{-\eta(j)} \qquad \forall j \geq m$$

$$\text{(v)} \qquad \varphi_i = \varphi_j \quad \text{if } i \equiv j, \; j \geq m$$

where $\eta(j) = \text{card}\{i \in J \mid i \leq j\}$. Now note that η is increasing and $\eta(j) \geq 1 + \eta(i)$ if $i \leq j$ and $i \not\equiv j$. Hence exactly as in the proof of Proposition 2.8 we find that $\varphi_j \downarrow h$ and $\varphi_j \geq f_j$ for $j \geq m$ (put $\varphi_j = \infty$ if $j \not\geq m$). Hence by (6.9.1) we have

$$\chi(h) = \lim_j \chi(\varphi_j) \geq \lim_j \chi(f_j) \geq \chi(h)$$

and so (6.9.1) holds with F_j replaced by $\bar{F}_j$. Moreover by assumption we have that $\bar{F}_j$ is $(\wedge f)$-stable and so (6) follows from (5). □

Let me illustrate the use of the lemmas above to prove an extension theorem for measures, which is useful to prove ξ-consistency of marginal systems.

Theorem 6.10. Let $(T,\mathcal{B},\mu)$ be a probability space, and let $\{\mathcal{F}_j \mid j \in J\}$ be an increasing net of $(\cap f)$-stable pavings on T, such that $\mathcal{F}_j \uparrow \mathcal{F}$ and

(6.10.1) $\hat{F} \subseteq B$, where $\hat{F}_j = (F_j)_\xi$ and $\hat{F} = \bigcup_{j \in J} \hat{F}_j$

(6.10.2) μ is ξ-supersmooth at B along $\hat{F}_j$ $\forall j \in J$

And suppose that for all $\varepsilon > 0$ and all $D \in F_\delta$, there exists $T_0 \subseteq T$, there exists I countable $\subseteq J$, and there exists an increasing map ρ from F into $\hat{F}_I = (\bigcup_{i \in I} F_i)_\xi$ satisfying

(6.10.3) $T_0 \cap \rho(F) \subseteq F \quad \forall F \in F$

(6.10.4) $\mu(H) < \varepsilon$ if $H \in \hat{F}_I$ and $H \cap T_0 = \emptyset$

(6.10.5) $\mu(D) < \varepsilon + \mu(\rho(F))$ if $F \in F$ and $\mu(D \setminus F) = 0$

Then μ^* is ξ-supersmooth at $\emptyset$ along H_ξ, where $H = F \cup \{\emptyset, T\}$.

Moreover if F is $(\cup f)$-stable, and if μ satisfies

(6.10.6) $\mu(B) = \sup\{\mu(F) \mid F \in B \cap H_\xi, \ F \subseteq B\} \quad \forall B \in B$

Then μ admits a σ-additive extension $\hat{\mu}$ to a σ-algebra $\hat{B}$, such that $\hat{B} \supseteq B \cup H_\xi$ and

(6.10.7) $\hat{\mu}$ is ξ-supersmooth at H_ξ along H_ξ

(6.10.8) $\hat{\mu}(F) = \inf\{\mu(F_0) \mid F_0 \in F, \ F_0 \supseteq F\} \quad \forall F \in F_\xi$

(6.10.9) $\hat{\mu}(B) = \sup\{\hat{\mu}(F) \mid F \in H_\xi, \ F \subseteq B\} \quad \forall B \in \hat{B}$

Remarks (1): Let $F_I = \bigcup_{i \in I} F_i$ and $\hat{F}_I = (F_I)_\xi$ whenever $I \subseteq J$. If $I \subseteq J$ is countable then

(6.10.10) $\hat{F}_I \subseteq F_\delta \subseteq B$

(6.10.11) μ is ξ-supersmooth at $\hat{F}_I$ along $\hat{F}_I$.

by (6.10.1+2) and Lemma 6.9(3). Thus $\mu(H)$ and $\mu(\rho(F))$ in (6.10.4) and (6.10.5) are well defined.

(2): From the proof below it follows that if S is any given subset of T, such that there exists $T_0 \subseteq T$, there exists I countable $\subseteq J$, and there exists an increasing map $\rho: F \to \hat{F}_I$, satisfying (6.10.3), (6.10.5) and

$$(6.10.4)^* \qquad \mu(H) < \varepsilon + \mu(S) \quad \text{if} \quad H \in \hat{F}_I \quad \text{and} \quad H \cap T_0 \subseteq S$$

Then we have

$$(6.10.12) \qquad \mu^* \text{ is } \xi\text{-supersmooth at } S \text{ along } F_\xi$$

(3): Note if F is $(\cup f)$-stable and (6.10.6) holds, then $\hat{\mu}(F) = \mu^*(F)$ for $F \in F_\xi$ by (6.10.8), and so μ^* is ξ-supersmooth at F_ξ along F_ξ by (6.10.7).

Proof. As noted above we have by Lemma 6.9.(3) that (6.10.10) and (6.10.11) holds. Hence by Lemma 6.8 with $\chi_0 = \chi_1 = \mu^*$, $F = F$ and $g = 0$ we have that μ^* is ξ-supersmooth at $\emptyset$ along F, since $\hat{F}_I \in S_\xi(\mu^*)$ by (6.10.11). Then evidently we have that μ^* is ξ-supersmooth at $\emptyset$ along H where $H = F \cup \{\emptyset, T\}$. And since H evidently is $(\cap f)$-stable we have that μ^* is ξ-supersmooth at $\emptyset$ along H_ξ by (6.7.1). Now put

$$U = B \cap H_\xi , \qquad V = \{G \subseteq T \mid T \setminus G \in U\}$$
$$\lambda = \mu \text{ restricted to } V$$

and suppose that F is $(\cup f)$-stable and (6.10.6) holds. Then $\lambda^* = \mu^*$, $\emptyset \in V \cap U$ and U is $(\cap f, \cup f)$-stable. And it is easily seen that (3.11.1)-(3.11.5) holds with $F = H_\xi$, moreover if C, $\hat{C}$ and R are the pavings defined in Theorem 3.11, we have $C = \hat{C} = H_\xi$

and:

(i) $\qquad B \cap C \in \mathcal{R} \quad \forall B \in \mathcal{B}\ \forall C \in \mathcal{H}_\xi$

Hence by Theorem 3.11 there exists a σ-algebra $\hat{\mathcal{B}} \supseteq \mathcal{B} \cup \mathcal{H}_\xi$ a probability measure $\hat{\mu}$ on $(T,\hat{\mathcal{B}})$, satisfying (6.10.7) and such that

(ii) $\qquad \hat{\mu}(H) = \mu^*(H) \quad \forall H \in \mathcal{H}_\xi$

(iii) $\qquad \hat{\mu}(B) = \sup\{\mu^*(H) \mid H \in \mathcal{H}_\xi, H \subseteq B\} \quad \forall B \in \hat{\mathcal{B}}$

Thus by (6.10.7) and (ii) we have that (6.10.8) holds and by (ii) and (iii) we have that (6.10.9) holds. By (iii) we have that $\hat{\mu}(B) \leq \mu(B)$ for all $B \in \mathcal{B}$, and since $\hat{\mu}$ and μ both are probability measures, we see that $\hat{\mu}$ is an extension of μ. Thus the theorem is proved. □

Now let us return to the fully ξ-consistency problem for marginal systems. By Theorem 6.6 and Lemmas 6.7-6.9 we obtain the following corollaries

Corollary 6.11. Let $\mathcal{E} = \{T \xrightarrow{q_\gamma} (T_\gamma,A_\gamma,\lambda_\gamma) \mid \gamma \in \Gamma\}$ be a consistent marginal system and let Θ be a ξ-exhausting family of maps for $\mathcal{E}$, such that $\mathcal{E}_\theta$ is fully ξ-consistent for all $\theta \in \Theta$. Then $\mathcal{E}$ is fully ξ-consistent.

Proof. Since $\mathcal{E}^* \leq \mathcal{E}_\theta^*$ we see that the corollary follows from Theorem 6.6 and Lemma 6.9.(1).

Corollary 6.12. Let $\mathcal{E} = \{T \xrightarrow{q_\gamma} (T_\gamma,A_\gamma,\lambda_\gamma) \mid \gamma \in \Gamma\}$ be a consistent marginal system, and let Θ be an upwards directed set of maps $\theta\colon \Delta(\theta) \to \Gamma$, satisfying

(6.12.1) $L^{\theta'} \subseteq L^{\theta''}$ <u>if</u> $\theta' \leq \theta''$ <u>and</u> $L^{\theta} \uparrow L^{\Gamma}$

(6.12.2) $\lim_{\theta} \mathcal{E}^*(\varphi_\theta) = 0$ <u>if</u> $\varphi_\theta \downarrow 0$ <u>and</u> $\varphi_\theta \in A^\theta \ \forall \theta$

<u>Then we have:</u>

(1): <u>If</u> $\mathcal{E}_\theta$ <u>is fully</u> σ-<u>consistent</u> $\forall \theta \in \Theta$, <u>then</u> $\mathcal{E}$ <u>is fully</u> σ-<u>consistent.</u>

(2): <u>If</u> Θ <u>is finitely founded and</u> $\mathcal{E}_\theta$ <u>is fully</u> η-<u>consistent</u> $\forall \theta \in \Theta$, <u>where</u> $\eta = \text{card}(\Theta)$, <u>then</u> $\mathcal{E}$ <u>is fully</u> η-<u>consistent.</u>

(3): <u>if</u> $\mathcal{E}_\theta$ <u>is fully</u> ξ-<u>consistent</u> $\forall \theta \in \Theta$, <u>and if (6.12.2) holds with</u> A^θ <u>replaced by</u> $(A^\theta)_\xi$, <u>where</u> ξ <u>is a given infinite cardinal, then</u> $\mathcal{E}$ <u>is fully</u> ξ-<u>consistent.</u>

<u>Remark</u> (1): Let $\Delta \subseteq \Gamma$, then put θ_Δ equal to the identity map from Δ into Γ, and if $\Delta = \{\gamma\}$ is a singleton then we just write θ_γ. Hence if $\mathcal{D}$ is a paving on Γ, then $\mathcal{D}$ induces a family

$$\tilde{\mathcal{D}} = \{\theta_\Delta \mid \Delta \in \mathcal{D}\}$$

of maps into Γ. If we order $\tilde{\mathcal{D}}$ by inclusion, then (6.12.1) holds if $\mathcal{D}$ filters up to Γ, and $\tilde{\mathcal{D}}$ is finitely founded if $\mathcal{D} \subseteq 2^{(\Gamma)}$. Similarly if $\Gamma_0 \subseteq \Gamma$ then Γ_0 induces a family

$$\Gamma_0 = \{\theta_\gamma \mid \gamma \in \Gamma_0\}$$

of maps into Γ.

(2): Let $(T,\mathcal{B},\mu)$ be the unit interval with Lebesgue measure, and let $\{S_n\}$ be a sequence of subsets of T, such that $S_n \downarrow \emptyset$ and $\mu^*(S_n) = 1 \ \forall n \geq 1$. Put

$$A_n = \{f \in B(T,\mathcal{B}) \mid f(t) = f(s) \quad \forall t,s \in S_n\} \quad \forall n \geq 1$$

$$\lambda_n(f) = \int_0^1 f(t)\,dt \qquad \forall f \in A_n \qquad \forall n \geq 1$$

$$q_n = \text{the identity map: } T \to T$$

Then λ_n is τ-smooth, $A_1 \subseteq A_2 \subseteq \cdots$, and

$$\mathcal{E} = \{T \xrightarrow{q_n} (T,A_n,\lambda_n) \mid n \in \mathbb{N}\}$$

is a fully σ-consistent projective system. However $\mathcal{E}$ is not τ-consistent, since

$$1_{S_n} \in (A_n)_\tau, \quad 1_{S_n} \downarrow 0 \quad \text{and} \quad \mathcal{E}^*(1_{S_n}) = 1$$

Thus (6.12.1+2) and full τ-consistency of $\mathcal{E}_\theta$ do <u>not</u> imply τ-consistency of $\mathcal{E}$.

<u>Proof</u>. Let us first show that

(i) $$\mathcal{E}^*(f) = \inf_\theta \mathcal{E}_\theta^*(f) = \lim_\theta \mathcal{E}_\theta^*(f) \quad \forall f \in \overline{\mathbb{R}}^T$$

So let $a > \mathcal{E}^*(f)$, then by (6.2.4) there exists $g \in L^\Gamma$ such that $g \geq f$ and $L^*(g) < a$. Then by (6.12.1) there exists θ_0 so that $g \in L^\theta$ for all $\theta \geq \theta_0$, and so by (6.3.4) we have

$$\mathcal{E}_\theta^*(f) \leq \mathcal{E}_\theta^*(g) = \mathcal{E}^*(g) \leq a \quad \forall \theta \geq \theta_0$$

and so we have by (6.3.4)

$$\limsup_\theta \mathcal{E}_\theta^*(f) \leq \mathcal{E}^*(f) \leq \inf_\theta \mathcal{E}_\theta^*(f)$$

and thus (i) holds. Next we show

(ii) $$F \in S_\xi(\mathcal{E}_\theta^*) \quad \forall\, \theta \geq \pi \Rightarrow F \in S_\xi(\mathcal{E}^*)$$

So let $\{f_q \mid q \in Q\}$ be a net in F, such that $f_q \downarrow f$ and $\mathrm{card}(Q) \leq \xi$. Then by (i) we have

$$\mathcal{E}^*(f) = \inf_{\theta \geq \pi} \mathcal{E}_\theta^*(f) = \inf_{\theta \geq \pi} \inf_q \mathcal{E}_\theta^*(f_q)$$

$$= \inf_q \inf_{\theta \geq \pi} \mathcal{E}_\theta^*(f_q) = \inf_q \mathcal{E}^*(f_q)$$

since $\mathcal{E}_\theta^*$ is ξ-supersmooth at $\overline{\mathbb{R}}^T$ along F. Thus (ii) holds.

But then the corollary follows from (ii), Theorem 6.6 and Lemma 6.9. □

Corollary 6.13. *Let $\mathcal{E} = \{T \xrightarrow{q_\gamma} (T_\gamma, A_\gamma, \lambda_\gamma) \mid \gamma \in \Gamma\}$ be a fully σ-consistent marginal system, and let ξ be a given infinite cardinal, such that for all $\varepsilon > 0$ and all $\varphi \in (A^\Gamma)_\delta$, there exists a map $\theta: \Delta \to \Gamma$, there exists $T_0 \subseteq T$, and there exists an increasing map R from A^Γ into $(A^\theta)_\xi$, satisfying*

(6.13.1) $Rf \leq f$ on T_0

(6.13.2) $\mathcal{E}_\theta^*(h) \leq \varepsilon$ *if* $h \in (A^\theta)_\xi$ *and* $h \leq 0$ on T_0

(6.13.3) $\mathcal{E}^*(\varphi) \leq \varepsilon + \mathcal{E}_\theta^*(Rf)$ *if* $f \in A^\Gamma$ *and* $\mathcal{E}^*(\varphi) = \mathcal{E}^*(\varphi \wedge f)$

(6.13.4) $\mathcal{E}_\theta$ *is fully ξ-consistent.*

Then $\mathcal{E}$ is fully ξ-consistent.

Proof. Easy consequence of Lemma 6.8.

Corollary 6.14. *Let* $\mathcal{E} = \{T \xrightarrow{q_\gamma} (T_\gamma, A_\gamma, \lambda_\gamma) \mid \gamma \in \Gamma\}$ *be a projective system and let* Θ *be a set of functions* θ *from* $\mathbb{N}$ *into* Γ, *satisfying*

(6.14.1) $A^{\theta(n)} \subseteq A^{\theta(n+1)} \quad \forall n \geq 1$

(6.14.2) Θ *is* σ*-exhausting for* $\mathcal{E}$

(6.14.3) $\lambda_{\theta(n)}(\varphi_n) \to 0$ *if* $\varphi_n \circ q_{\theta(n)} \downarrow 0$ *and* $\varphi_n \in E_{\theta(n)} \; \forall n$

where $E_\gamma = \{\varphi \in A_\gamma \mid 0 \leq \varphi \leq 1\}$. *Then* $\mathcal{E}$ *is fully* σ*-consistent.*

Moreover if ξ *is an infinite cardinal, such that for all* $\theta \in \Theta$ *we have*

(6.14.4) $\lambda_\gamma \in Pr_\xi(A_\gamma)$ *and* $\lambda^*_{\gamma\xi}(q_\gamma(T)) = 1 \quad \forall \gamma \in \theta(\mathbb{N})$

(6.14.5) $\lambda^*_{\theta(n)}(f_n) \to 0$ *if* $f_n \circ q_{\theta(n)} \downarrow 0$ *and* $f_n \in F_{\theta(n)} \forall n$

where $F_\gamma = \{f \in (A_\gamma)_\xi \mid 0 \leq f \leq 1\}$. *Then* $\mathcal{E}_\theta$ *is fully* ξ*-consistent for all* $\theta \in \Theta$. *And so if for all* $\varepsilon > 0$ *and all* $\varphi \in (A^\Gamma)_\delta$, *there exist* $\theta \in \Theta$, *there exists* $T_0 \subseteq T$, *and there exists an increasing map* $R: A^\Gamma \to (A^\theta)_\xi$ *satisfying (6.13.1)-(6.13.3), then* $\mathcal{E}$ *is fully* ξ*-consistent.*

Remark (1): Every projective system admits a set $\Theta \subseteq \Gamma^{\mathbb{N}}$ satisfying (6.14.1) and (6.14.2), e.g. the set of all $\Theta: \mathbb{N} \to \Gamma$ satisfying (6.14.1) will satisfy (6.14.1) and (6.14.2).

(2): For projective systems we obviously have that α-consistency and fully α-consistency coincide, whenever α is one of the symbols: $\xi, c\xi, \tau, c\tau, \sigma, c\sigma$ or π.

(3): Condition (6.14.3) is evidently also necessary for σ-consistency of $\mathcal{E}$, and similarly condition (6.14.4) are necessary for ξ-consistency of $\mathcal{E}$.

Proof. Let $\lambda = \underleftarrow{\lim}\, \mathcal{E}$, and let $\theta \in \Theta$. Then I claim that we have

(i) $$\lambda(\psi_n) \longrightarrow 0 \quad \text{if } \psi_n \downarrow 0 \quad \text{and } \psi_n \in A^{\theta(n)} \quad \forall n \geq 1$$

To see this we choose $f_n \in A_{\theta(n)}$, so that $\psi_n = f_n \circ q_{\theta(n)}$ for all $n \geq 1$. Let $a = \|\psi_1\|$ and $h_n = f_n^+ \wedge a$, then we have that $\psi_n = h_n \circ q_{\theta(n)}$, since $0 \leq \psi_n \leq a$. And by Proposition 2.7 there exists $\varphi_n \in A_{\theta(n)}$, so that

$$h_n + 2^{-n-1} \leq \varphi_n \leq h_n + 2^{-n}$$

Then $0 \leq \varphi_n < 1+a$ and it is easily seen that $\varphi_n \circ q_{\theta(n)} \downarrow 0$. Hence by (6.14.3) we have

$$\lim_{n\to\infty} \lambda_{\theta(n)}(\varphi_n) = (1+a) \lim_{n\to\infty} \lambda_{\theta(n)}(\varphi_n/(1+a)) = 0$$

And since $\varphi_n \circ q_{\theta(n)} \geq h_n \circ q_{\theta(n)} = \psi_n$ we have

$$0 \leq \lim_{n\to\infty} \lambda(\psi_n) \leq \lim_{n\to\infty} \lambda_{\theta(n)}(\varphi_n) = 0$$

Thus (i) is proved.

By (i) and (6.14.1) we have that λ is σ-supersmooth at 0 along $A^{\theta(n)}$, and so by Lemma 6.9.(2) we have that λ is σ-supersmooth at 0 along A^{θ}. Thus by (6.14.2) and Lemma 6.9.(1) we conclude that $\lambda \in Pr_{\sigma}(A^{\Gamma})$ and so $\mathcal{E}$ is fully σ-consistent.

Now suppose that (6.14.4) and (6.14.5) hold. By (6.14.4) and Theorem 4.13 we have that λ is ξ-smooth on $A^{\theta(n)}$, and so by ξ-smoothness of $\lambda_{\theta(n)}$ and λ we have

$$\text{(ii)} \qquad \lambda^*(f \circ q_{\theta(n)}) = \lambda^*_{\theta(n)}(f) \qquad \forall f \in (A_{\theta(n)})_{\xi}$$

Butthen one shows exactly as above (actually a bit easier since $h_n \in (A_{\theta(n)})_{\xi}$) that we have

$$\text{(iii)} \qquad \lambda^*(f_n) \to 0 \ \text{ if } \ f_n \downarrow 0 \ \text{ and } \ f_n \in (A^{\theta(n)})_{\xi} \qquad \forall n \geq 1$$

But then λ is ξ-supersmooth at 0 along A^{θ} by Lemma 6.9.(3), and then $\mathcal{E}_{\theta}$ is fully ξ-consistent by Theorem 6.6, since $\mathcal{E}^* = \lambda$ on A^{θ} by (6.3.5). □

Corollary 6.14 reduces the ξ-consistency problem for projective systems to the problem of verifying (6.14.3) or (6.14.5) for projective systems with index set $\mathbb{N}$, such that $A^1 \subseteq A^2 \subseteq \cdots$. Lemma 3.19 is a very useful tool for doing this, and the lemma generates a variety of ξ-consistency result for projective systems, depending on the choice of (R_n) and (σ_n). We shall here restrict ourselves to the case where (R_n) is a disintegration and (σ_n) is an atomic maximal sequence.

Definition 6.15. Let $\mathcal{E} = \{T \xrightarrow{q_n} (T_n, A_n, \lambda_n) \mid n \in \mathbb{N}\}$ be a projective system with index set $\mathbb{N}$, and let ξ be an infinite cardinal, then we say that $\mathcal{E}$ is of type $(\mathbb{N},\xi)$ if $\mathcal{E}$ satisfies

(6.15.1) $$A^1 \subseteq A^2 \subseteq \cdots \subseteq A^n \subseteq \cdots$$

(6.15.2) $$\lambda_n \in \Pr_\xi(A_n) \quad \text{and} \quad \lambda^*_{n\xi}(q_n(T)) = 1 \quad \forall n \geq 1$$

where $\lambda_{n\xi}$ is the representing measure of λ_n on $(T_n, \sigma(A_{n\xi}))$. We define projective systems of type $(\mathbb{N},\sigma)$ or type $(\mathbb{N},\tau)$ similarly. Let $\mathcal{E}$ be of type $(\mathbb{N},\xi)$ then a ξ-disintegration of $\mathcal{E}$ is a sequence (ρ_n) of $\lambda_{n\xi}$-measurable Markov kernels ρ_n on $(T_{n+1}, \sigma((A_{n+1})_\xi)) \mid T_n$ satisfying

(6.15.3) $$\lambda^*_{n+1}(f) = \int_{T_n} \lambda_{n\xi}(du) \int_T f(v)\rho_n(dv|u) \quad \forall f \in (A_{n+1})_\xi$$

By Theorem 4.1 and [2 , Theorem I.21] it follows easily that (6.15.3) implies:

(6.15.4) f is $\rho_n(\cdot|u)$-measurable for $\lambda_{n\xi}$- a.a. $u \in T_n$

(6.15.5) $$\int_{T_{n+1}} f\, d\lambda_{(n+1)\xi} = \int_{T_n} \lambda_{n\xi}(du) \int_{T_{n+1}} f(v)\rho_n(dv|u)$$

for all $f \in L^1(\lambda_{(n+1)\xi})$.

Let $\mathcal{E}$ be a marginal system with index set $\mathbb{N}$, then an atomic maximal sequence for $\mathcal{E}$ is a sequence (σ_n) of correspondences $\sigma_n: T_n \rightsquigarrow T_{n+1}$ satisfying

(6.15.6) $$\sigma_n^{-1}(T_n) \leq \sigma_{n+1}^{-1}(T_{n+2}) \quad \forall n \geq 1$$

(6.15.7) If $k \in \mathbb{N}$ and $u_{n+1} \in \sigma_n(u_n)$ $\forall n \geq k$, then $\exists\, t \in T$, so that $q_n(t) \equiv u_n \pmod{A_n}$ $\forall n \geq k$

(Recall that $u \equiv v \pmod{A}$ means that $\varphi(u) = \varphi(v)$ for all $\varphi \in A$). Note that if $u_{n+1} \in \sigma_n(u_n)$ for some $n \geq 1$, then by (6.15.6) there exists $u_j \in T_j$ for $j \geq n+2$, such that $u_{j+1} \in \sigma_j(u_j)$ for all $j \geq n$, hence by (6.15.7) we have

(6.15.8) If $k \leq n$ and $u_{j+1} \in \sigma_j(u_j)$ for $k \leq j \leq n$, then $\exists t \in T$ so that $q_j(t) \equiv u_j \pmod{A}$ $\forall k \leq j \leq n+1$.

Let me give two typical examples of atomic maximal sequences:

(1): Let $S_1, S_2, \ldots$ be a given sequence of sets, and let us put

$$T_n = S_1 \times \ldots \times S_n \quad , T = \prod_{j=1}^{\infty} S_j$$

$$q_n = \text{the natural projection: } T \to T_n$$

$$\sigma_n(u) = \{u\} \times S_{n+1} \qquad \forall u \in T_n$$

Then $\{\sigma_n\}$ is an atomic maximal sequence no matter how we choose the algebras A_n.

(2): Let T be a given and let $\{A_n\}$ be an increasing sequence of function algebras on T. Now put

$$T_n = T \text{ and } q_n(t) = t \quad \forall t \in T \quad \forall n \geq 1$$

$$\sigma_n(u) = \{t \in T \mid t \equiv u \pmod{A_n}\} \quad \forall u \in T \quad \forall n \geq 1$$

i.e. $\sigma_n(u)$ is the A_n-atom containing u. Then (σ_n) satisfies (6.15.6) and (6.15.8) and it is easily checked that (6.15.7) is equivalent to the usual <u>atomic maximality condition</u>:

(6.15.9) $\bigcap_1^\infty \alpha_n \neq \emptyset$ whenever α_n is an A_n-atom for all $n \geq 1$, and $\alpha_1 \supseteq \alpha_2 \supseteq \cdots \supseteq \cdots$

Theorem 6.16. Let $\mathcal{E} = \{T \xrightarrow{q_n} (T_n, A_n, \lambda_n) \mid n \in \mathbb{N}\}$ be a projective system of type $(\mathbb{N}, \xi)$ where ξ is a given infinite cardinal. Suppose that for all $\varepsilon > 0$, there exists a ξ-disintegration (ρ_n) of $\mathcal{E}$ and an atomic maximal sequence (σ_n) for $\mathcal{E}$ satisfying

(6.16.1) $$\sum_{n=1}^\infty \lambda_{n\xi}\{v \in T_n \mid \rho_n^*(\sigma_n(v) \mid v) < 1\} < \varepsilon$$

where $\lambda_{n\xi}$ is the representing measure of λ_n on $(T_n, \sigma(A_{n\xi}))$. Then $\mathcal{E}$ is fully ξ-consistent

Proof. Let $\varphi_n \in A_n$, so that $0 \leq \varphi_n \leq 1$ and $\varphi_n \circ q_n \downarrow 0$, then by Corollary 6.14 it suffices to show

(*) $$\lim_{n\to\infty} \lambda_n^*(\varphi_n) = 0$$

So let $\varepsilon > 0$ be given and choose (ρ_n) and (σ_n) according to the assumption of the theorem. Let $\mathcal{B}_n = \sigma(A_{n\xi})$ then ρ_n is a $\lambda_{n\xi}$-measurable Markov kernel on $(T_{n+1}, \mathcal{B}_{n+1}) \mid T_n$ satisfying (6.15.3)-(6.15.5). Now put

$$\mu_n(f) = \int_{T_n} f d\lambda_{n\xi} \quad \text{for} \quad f \in B(T_n, \mathcal{B}_n)$$

$$\varepsilon_n = \lambda_{n\xi}^*(v \in T_n \mid \rho_n^*(\sigma_n(v) \mid v) < 1)$$

and choose $D_n \in \mathcal{B}_n$ such that

(i) $$\lambda_{n\xi}(D_n) = 1-\varepsilon_n \quad \text{and} \quad \rho_n^*(\sigma_n(v) \mid v) = 1 \quad \forall v \in D_n$$

We shall now construct sequences $\{\mathcal{B}_{nk} \mid n\geq 1,\ k\geq 0\}$ and $\{T_{nk} \mid n\geq 1,\ k\geq 0\}$ satisfying

(ii) $\mathcal{B}_{no} = \sigma(\{\varphi_n, D_n\}),\quad T_{no} = D_n$

(iii) $\mathcal{B}_{nk}$ is a countably generated σ-algebra on T_n.

(iv) $\mathcal{B}_{nk-1} \subseteq \mathcal{B}_{nk} \subseteq \mathcal{B}_n$ and $T_{nk} \in \mathcal{B}_{nk}$

(v) $T_{nk-1} \supseteq T_{nk}$ and $\lambda_{n\xi}(T_{nk}) = 1-\varepsilon_n$

(vi) $1_{T_{nk}} \rho_n(B|\cdot)$ is $\mathcal{B}_{nk}$-measurable $\forall B \in \mathcal{B}_{n+1,k-1}$

for all $n \geq 1$ and all $k \geq 1$. This will be done by induction in k.

For $k = 0$ we define $\mathcal{B}_{no}$ and T_{no} by (ii). Then $\mathcal{B}_{no}$ is countably generated, $T_{no} \in \mathcal{B}_{no}$ and $\lambda_{n\xi}(T_{no}) = 1-\varepsilon_n$.

Now suppose that $\{\mathcal{B}_{nk}, T_{nk}\}$ has been defined for $n \geq 1$ and $0 \leq k \leq m$ satisfying (ii)-(vi). Then we choose a countable algebra $\mathcal{C}_{nm}$, such that $\mathcal{B}_{nm} = \sigma(\mathcal{C}_{nm})$. Since ρ_n is $\lambda_{n\xi}$-measurable and $\mathcal{C}_{nm}$ is countable there exists a set $S_n \in \mathcal{B}_n$, such that $\lambda_{n\xi}(S_n) = 1$ and

$$1_{S_n} \rho_n(C|\cdot) \text{ is } \mathcal{B}_n\text{-measurable } \forall C \in \mathcal{C}_{n+1m}$$

Now put $T_{n\,m+1} = S_n \cap T_{nm}$ and let $\mathcal{B}_{nm+1}$ be the σ-algebra generated by

$$\mathcal{C}_{nm} \cup \{T_{nm+1}\} \cup \{1_{S_n} \rho_n(C|\cdot) \mid C \in \mathcal{C}_{n+1m}\}$$

Then $\mathcal{B}_{nm+1}$ is countably generated, $\mathcal{B}_{nm} \subseteq \mathcal{B}_{nm+1} \subseteq \mathcal{B}_n$, $T_{nm+1} \in \mathcal{B}_{nm+1}$, and since $\lambda_{n\xi}(S_n) = 1$ we have

$$T_{nm+1} \subseteq T_{nm} \quad \text{and} \quad \lambda_{n\xi}(T_{nm+1}) = \lambda_{n\xi}(T_{nm}) = 1-\varepsilon_n)$$

Moreover if

$$\mathcal{V}_n = \{B \in \mathcal{B}_{n+1} \mid 1_{T_{nm+1}} \rho_n(B|\cdot) \text{ is } \mathcal{B}_{nm+1}\text{-measurable}\}$$

Then $\mathcal{V}_n$ is a $(\downarrow c, \uparrow c)$-stable paving containing the algebra $\mathcal{C}_{n+1,m}$. Thus $\mathcal{V} \supseteq \mathcal{B}_{n+1,m}$, and so we see that (ii)-(vi) holds.

This completes the induction and so the existence of $(\mathcal{B}_{nk})$ and (T_{nk}) satisfying (ii)-(vi) is established. Now put

$$S_n = \bigcap_{k=0}^{\infty} T_{nk} \qquad \mathcal{B}^n = \sigma(\bigcup_{k=0}^{\infty} \mathcal{B}_{nk})$$

Then by (i)-(vi) we have

(vii) $\quad \mathcal{B}^n \subseteq \mathcal{B}_n$ and $\varphi_n \in B(T_n, \mathcal{B}^n)$

(viii) $\quad S_n \in \mathcal{B}^n$ and $\lambda_{n\xi}(S_n) = 1-\varepsilon_n$

(ix) $\quad \rho_n^*(\sigma_n(v) \mid v) = 1 \qquad \forall v \in S_n$

(x) $\quad 1_{S_n} \rho_n(B|\cdot) \in B(T_n, \mathcal{B}^n) \qquad \forall B \in \mathcal{B}^{n+1}$

Now let us put

$$B_n = \{f \in B(T_n, \mathcal{B}^n) \mid 0 \le f \le 1\}$$

$$R_n(f|v) = 1_{S_n}(v) \int_{T_{n+1}} f(u)\rho_n(du|v) \quad \forall f \in B_{n+1}$$

Then B_n is $(\downarrow c)$-stable, μ_1 satisfies (3.19.1), and by (x) we have that R_n is an increasing map from B_{n+1} into B_n satisfying (3.19.2). By (viii) and (6.15.5) we have

$$\mu_{n+1}(f) \leq \varepsilon_n + \mu_n(R_n f) \qquad \forall f \in B_{n+1} \quad \forall n \geq 1$$

and so (3.19.3) holds. And (3.19.4) follows from (ix) and definition of R_n. Finally (3.19.5) holds by (6.15.6). Now note that $\varphi_n \in B_n$ by (vii), and if $v \in S_n$ and $u \in \sigma_n(v)$, then by (6.15.8) there exists $t \in T$ so that

$$\varphi_{n+1}(u) = \varphi_{n+1}(q_{n+1}(t)) \leq \varphi_n(q_n(t)) = \varphi_n(v)$$

since $\varphi_{n+1} \circ q_{n+1} \leq \varphi_n \circ q_n$. Hence by (ix) we have that $R_n \varphi_{n+1} \leq \varphi_n$. Then by (6.16.1) and Lemma 3.19 we have

(xi) $$\lim_{n\to\infty} \mu_n(\varphi_n) \leq \mu_1(\varphi) + \varepsilon$$

where $\varphi(u) = \varphi_1(u)$ for $u \notin \sigma_1^{-1}(T_2)$ and

$$\varphi(u) = \sup\{\inf_n \varphi_n(t_n) \mid t_1 = u, \quad t_{j+1} \in \sigma_j(t_j) \quad \forall j \geq 1\}$$

for $u \in \sigma_1^{-1}(T_2)$. Now if $t_{j+1} \in \sigma_j(t_j)$ for all $j \geq 1$, then by (6.15.7) there exists $t \in T$ such that

$$\varphi_j(t_j) = \varphi_j(q_j(t)) \quad \forall j \geq 1$$

and since $\varphi_j \circ q_j \downarrow 0$, we see that $\varphi(u) = 0$ for $u \in \sigma_1{}^{-1}(T_2)$. Moreover by (ix) we have that $S_1 \subseteq \sigma_1{}^{-1}(T_2)$, and so

$$\mu_1(\varphi) < \lambda_{1\xi}(T \setminus S_1) = \varepsilon_1 \leq \varepsilon$$

Thus by (4.1.1) and (xi) we have $\lim \lambda_n^*(\varphi_n) \leq 2\varepsilon$ for all $\varepsilon > 0$. Thus (*) and the theorem is proved. □

Let (T,A) be an algebraic function space, let $\lambda \in Pr(A)$ and let ξ be an infinite cardinal, then we say that λ is *ξ-perfect*, if $\lambda|B$ is ξ-compact for all algebras $B \subseteq A$ which are generated by a set of cardinality atmost ξ (or equivalently for all algebras $B \subseteq A$ with weight$(B) \leq \xi$, see Proposition 2.9 and Theorem 4.1). Clearly we have

(6.5) ξ-compact $\Rightarrow$ ξ-perfect $\Rightarrow$ ξ-smooth

(6.6) σ-perfect $\Leftrightarrow$ perfect

(6.7) τ-perfect $\Leftrightarrow$ compact

where σ-perfect and τ-perfect are defined using the usual conventions: σ = countable and τ = arbitrary large cardinal. I.e. if we put

$$Pr_{\pi\xi}(A) = \{\lambda \in Pr(A) \mid \lambda \text{ is } \xi\text{-perfect}\}$$

Then we have

(6.8) $Pr_{c\xi}(A) \subseteq Pr_{\pi\xi}(A) \subseteq Pr_{\xi}(A)$

(6.9) $Pr_{\pi\sigma}(A) = Pr_{\pi}(A)$

(6.10) $Pr_{\pi\tau}(A) = Pr_{c\tau}(A)$

Note that if B is a subalgebra of A with weight atmost ξ, then by Proposition 2.9 we have that

$$\overline{K}_{\xi}(B) = \overline{K}(B) \subseteq F(B) = F_{\xi}(B) \subseteq F_{\xi}(A)$$

So by Theorem 4.1 we find (cf. Theorem 4.11):

(6.11) If $\lambda \in Pr_{\pi\xi}(A)$ and $Q \subseteq A$ so that weight$(Q) \leq \xi$, then λ is $\overline{K}(Q)$-tight along A.

<u>Corollary 6.17. Let</u> $\mathcal{E} = \{T \xrightarrow{q_n} (T_n, A_n, \lambda_n) \mid n \in \mathbb{N}\}$ <u>be a projective system of type</u> $(\mathbb{N}, \xi)$, <u>where</u> ξ <u>is an infinite cardinal, and suppose that</u> λ_n <u>is</u> ξ<u>-perfect for all</u> $n \geq 1$, <u>and that</u> $\mathcal{E}$ <u>admits an atomic maximal sequence</u> (τ_n) <u>satisfying</u>

(6.17.1) $$q_{n+1}(t) \in \tau_n(q_n(t)) \quad \forall\, t \in T \quad \forall\, n \geq 1$$

(6.17.2) $$\Sigma_n = \{(u,v) \in T_n \times T_{n+1} \mid v \in \tau_n(u)\} \in \mathcal{B}_n \otimes \mathcal{B}_{n+1} \quad \forall\, n \geq 1$$

<u>where</u> $\mathcal{B}_n = \sigma(A_{n\xi})$. <u>Then</u> $\mathcal{E}$ <u>is fully</u> ξ<u>-consistent.</u>

<u>Proof</u>. Let $\varphi_n \in A_{n\xi}$, so that $0 \leq \varphi_n \leq 1$ and $\varphi_n \circ q_n \downarrow 0$, then by Corollary 6.14 it suffices to show

(*) $$\lim_{n\to\infty} \lambda_n^*(\varphi_n) = 0$$

By (6.17.2) there exists $Q_n \subseteq A_n$, so that $\operatorname{card}(Q_n) \leq \xi$, $\varphi_n \in Q_{n\xi}$ and $\Sigma_n \in \mathcal{D}_n \otimes \mathcal{D}_{n+1}$ where $\mathcal{D}_n = \sigma(Q_{n\xi})$. Let $\varphi \in Q_k$, then $\varphi \circ q_k \in A^k \subseteq A^n$ for all $n \geq k$, hence there exist $\{\varphi_{nk} \mid k \leq n\} \subseteq A_n$, such that

(i) $$\varphi_{nn} = \varphi \qquad \varphi_{nk} \circ q_n = \varphi \circ q_k \qquad \forall\, k \leq n \quad \forall \varphi \in Q_k$$

Now let $\hat{A}_n$ be the algebra generated by

$$\hat{Q}_n = \{\varphi_{nk} \mid k \in \{1,\dots,n\} \quad \varphi \in Q_k\}$$

Then $\operatorname{card}(\hat{Q}_n) \leq \xi$, and if $\hat{\mathcal{B}}_n = \sigma(\hat{A}_{n\xi})$, then we have that $\Sigma_n \in \hat{\mathcal{B}}_n \otimes \hat{\mathcal{B}}_{n+1}$. Moreover if $\hat{A}^n = q_n^{-1}(\hat{A}_n)$ then by construction we have that $\hat{A}^n \subseteq \hat{A}^{n+1}$. Hence if $\hat{\lambda}_n$ is the restriction of λ_n to $\hat{A}_n$, then

$$\hat{\mathcal{E}} = \{T \xrightarrow{q_n} (T_n, \hat{A}_n, \hat{\lambda}_n) \mid n \in \mathbb{N}\}$$

is a projective system of type $(\mathbb{N},\xi)$, since

$$\hat{\lambda}^*_{n\xi}(q_n(T)) \geq \lambda^*_{n\xi}(q_n(T)) = 1$$

And since $A_n \supseteq \hat{A}_n$ we have that (τ_n) is an atomic maximal sequence for $\hat{\mathcal{E}}$.

By Theorem 4.14 there exists a probability measure P_n on (T, F_n), where $F_n = \sigma((\hat{A}^n)_\xi)$, such that

(ii) $$\int_T (f \circ q_n) dP_n = \int_{T_n} f d\hat{\lambda}_{n\xi} \qquad \forall\, f \in L^1(\hat{\lambda}_{n\xi})$$

and since $\hat{\lambda}_{n\xi}$ is ξ-compact by assumption, then by Theorem 4.9, there exists a $\hat{\lambda}_{n\xi}$-measurable Markov kernel ρ_n on $(T_{n+1}, B_{n+1}) \mid T_n$ satisfying

(iii) $$\int_T f(q_n(t), q_{n+1}(t))\ P_{n+1}(dt)$$

$$= \int_{T_n} \hat{\lambda}_{n\xi}(du) \int_{T_{n+1}} f(u,v)\rho_n(dv|u)$$

for all bounded $B_n \otimes B_{n+1}$-measurable functions f. In particular we see that (ρ_n) is a ξ-disintegration of $\hat{\mathcal{E}}$, and putting $f = 1_{\Sigma_n}$ in (iii) gives

$$\int_{T_n} \rho_n(\tau_n(u)|u)\ \hat{\lambda}_{n\xi}(du) = 1$$

by (6.17.1) and (6.17.2). Hence we have

$$\hat{\lambda}_{n\xi}\{u|\rho_n(\tau_n(u)|u) < 1\} = 0 \qquad \forall\, n \geq 1$$

and so by Theorem 6.16 we have, that $\hat{\mathcal{E}}$ is fully ξ-consistent. And since $\varphi_n \in \hat{A}_{n\xi}$ we see that (*) and thus the theorem holds. □

Theorem 6.16 and its Corollary 6.17 gives a fairly satisfactory solution to the ξ-consistency problem for projective systems of type $(\mathbb{N},\xi)$, and combining these with Corollaries 6.13 and 6.14 we get a solution to the ξ-consistency problem for arbitrary projective system, which covers most cases. The cξ-consistency is in some sense much easier, much also much more restrictive. Let me just give one result in this direction.

__Theorem 6.18.__ _Let_ $\mathcal{E} = \{T \xrightarrow{q_\gamma} (T_\gamma, A_\gamma, \lambda_\gamma)|\gamma \in \Gamma\}$ _be a marginal system, such that_ $\lambda_\gamma \in Pr_\xi(A_\gamma)$ _for all_ $\gamma \in \Gamma$, _where_ ξ _is a given infinite cardinal. Let_ $\mathcal{K}_\gamma$ _be a paving on_ T_γ, _such that_

(6.18.1) $$\forall\, \gamma \in \Gamma \quad \forall\, \varepsilon > 0 \quad \exists\, K \in \mathcal{K}_\gamma : \lambda^*_{\gamma\xi}(T_\gamma \setminus K_\gamma) < \varepsilon$$

And let us define

$$\mathcal{K}_\Delta = \{\bigcap_{\gamma\in\Delta} q_\gamma^{-1}(K_\gamma)\,|\,K_\gamma \in \mathcal{K}_\gamma \quad \forall\, \gamma \in \Delta\} \qquad \forall\, \Delta \subseteq \Gamma$$

_If_ Δ _is a countable subset of_ Γ _and_ $\mu \in M_\xi(\mathcal{E})$, _then_

(6.18.2) $$\forall\, \varepsilon > 0 \quad \exists\, K \in \mathcal{K}_\Delta \quad \text{so that} \quad \mu^*_\xi(T\setminus K) < \varepsilon$$

<u>Remark</u>. It is wellknown that of an infinite product measure is Radon, then all but a countable number of the factors must have compact support. Thus the countably of Δ is indispensable.

<u>Proof</u>. Since Δ is countable, there exist $\varepsilon_\gamma > 0$ for $\gamma \in \Delta$ so that $\Sigma_{\gamma\in\Delta}\, \varepsilon_\gamma < \varepsilon$. Hence by (6.18.1) there exists $K_\gamma \in \mathcal{K}_\gamma$ for $\gamma \in \Delta$ so that

$$\lambda^*_{\gamma\xi}(T\setminus K_\gamma) < \varepsilon_\gamma$$

Let $C = \cap_{\gamma\in\Delta}\, q_\gamma^{-1}(K_\gamma)$, then $C \in \mathcal{K}_\Delta$ and

$$\mu^*_\xi(T\setminus C) \leq \sum_{\gamma\in\Delta} \mu^*_\xi(q_\gamma^{-1}(T_\gamma\setminus K_\gamma))$$

$$\leq \sum_{\gamma\in\Delta} \lambda^*_{\gamma\xi}(T_\gamma\setminus K_\gamma) < \varepsilon$$

by Theorem 6.4. Thus (6.18.2) holds. □

In our last two theorems of this section we shall elaborate on (6.2.5) and the last part of Proposition 6.2.

<u>Theorem 6.19</u>. <u>Let</u> $\mathcal{E} = \{T \xrightarrow{q_\gamma} (T_\gamma, A_\gamma, \lambda_\gamma) \mid \gamma \in \Gamma\}$ <u>be a fully ξ-consistent marginal system, where</u> ξ <u>is a given infinite cardinal, and let</u> $F = (A^\Gamma)_\xi$. <u>Then for all</u> $h \in R_*(\mathcal{E}^*, F)$ <u>we have</u>

(6.19.1) $\quad h \in B^*(T)$ <u>and</u> h <u>is</u> $\mu\xi$-<u>measurable</u> $\forall\, \mu \in M_\xi(\mathcal{E})$

(6.19.2) $\quad \exists\, \nu \in M_\xi(\mathcal{E})$ <u>so that</u> $\mathcal{E}^*(h) = \int_T h\,d\nu_\xi$

<u>Moreover if</u> $\mathcal{E}^*$ <u>is σ-subsmooth at</u> $B(T)$ <u>along</u> $B^+(T)\cap S(F)$ <u>then for every</u> $g \in B(T)\cap S(F)$ <u>we have</u>

$$(6.19.3) \qquad \mathcal{E}^*(g) = \sup\{\int_T g d\mu_\xi \mid \mu \in M_\xi(\mathcal{E})\}$$

$$(6.19.4) \qquad \mathcal{E}^*(g) = \sup\{\mathcal{E}^*(f) \mid f \in F, \quad f \le g\}$$

<u>Remark</u>. Recall that $S(F)$ is the set of all Souslin F-functions, and that every $g \in S(F)$ is an upper $S(F_\xi(A^\Gamma))$-function. Hence by [20, Corollary 2.9.3, p. 42] we have that every $g \in S(F)$ is μ_ξ-measurable for every $\mu \in M_\xi(\mathcal{E})$, and so the integrals in (6.19.3) are welldefined.

<u>Proof</u> (6.19.1): Since $h \in R_*(\mathcal{E}^*, F)$, there exist $f_n \in F$ such that $f_n \le h$ and $\mathcal{E}^*(h-f_n) \le 2^{-n}$. Hence by definition of $\mathcal{E}^*$ we have that $h-f_n \in B^*(T)$ and since $h \le f_n \dot{+} (h-f_n)$ and $f_n \in B^*(T)$ we have that h is bounded above, i.e. $h \in B^*(T)$. Moreover by Theorem 6.4 we have for all $\mu \in M_\xi(\mathcal{E})$, that

$$\int^* (h-f_n) d\mu_\xi \le \mathcal{E}^*(h-f_n) \le 2^{-n}$$

and so $h-f_n \to 0$ μ_ξ - a.s. Thus h is μ_ξ-measurable since f_n is so for all $n \ge 1$.

(6.19.2): By Proposition 6.2 there exist $\mu \in M(\mathcal{E})$ such that $\mu^*(h) = \mathcal{E}^*(h)$. Now let $f_n \in F$ be chosen as above, then by (6.2.2), (4.1.1) and (6.4.3) we have

$$\mathcal{E}^*(h) = \mu^*(h) \leq \mathcal{E}^*(h-f_n) \dot{+} \mu^*(f_n)$$

$$\leq 2^{-n} + \int_T f_n \, d\mu_\xi \leq 2^{-n} + \int_T h d\mu_\xi$$

$$\leq 2^{-n} + \mathcal{E}^*(h)$$

Thus letting $n \to \infty$ we see that (6.19.2) holds.

(6.19.3+4): Now suppose that $\mathcal{E}^*$ is ξ-subsmooth at $B(T)$ along $B^+(T) \cap S(F)$. Let $G = B(T) \cap S(F)$, since

$$\mathcal{E}^*(a1_T+g) = a + \mathcal{E}^*(g)$$

for all $a \in \mathbb{R}$ and all $g \in \bar{\mathbb{R}}^T$, it follows easily that $\mathcal{E}^*$ is σ-subsmooth at G along G. Now we shall apply Theorem 5.5 with $\chi = \mathcal{E}^*$, $F = (A^\Gamma)_\xi$ and $M = M(\mathcal{E})$. Then (5.5.1) holds by Proposition 2.9, (5.5.2) holds by Theorem 6.6 since $\mathcal{E}$ is fully ξ-consistent, (5.5.3) has just been verified. Let $g \in G$, then $c = m(g) \in \mathbb{R}$ and so $c1_T \in F$ and

$$\{f \in S(F) \mid c1_T \leq f \leq g\} \subseteq G$$

Thus (5.5.4) holds, and (5.5.5) holds by (6.2.5). But then (6.19.3) and (6.19.4) follows from (5.5.7) and (5.5.8). □

The σ-smoothness of $\mathcal{E}^*$ is indispensable for the validity of (6.19.3) and (6.19.4), to see this let $(I,\mathcal{B})$ be the unit interval with its Borel τ-algebra, and put

$$T_n = I, \quad A_n = \mathcal{B}(I,\mathcal{B}(I)), \quad \lambda_n(f) = \int_0^1 f(t)dt$$

$$\mathcal{E} = \{I^{\mathbb{N}} \xrightarrow{q_n} (T_n, A_n, \lambda_n) | n \in \mathbb{N}\}$$

where q_n is the n-th coordinate function on $I^{\mathbb{N}}$. In our next section we shall see that $\mathcal{E}$ is fully π-consistent. Now let

$$f_n(t) = \inf_{k \geq n} t_k, \quad f(t) = \lim_{n \to \infty} \inf t_n \quad \forall\ t \in (t_n) \in I^{\mathbb{N}}$$

Then $0 \leq f_n \leq f \leq 1$ and $f_n \uparrow f$, however the reader easily verifies, that

$$\mathcal{E}^*(f) = 1, \quad \mathcal{E}^*(f_n) = \tfrac{1}{2} \quad \forall\ n \geq 1$$

$$\sup\{\int f d\mu_\tau | \mu \in M_\tau(\mathcal{E})\} = \tfrac{1}{2}$$

Thus $\mathcal{E}^*$ is not τ-subsmooth, and (6.19.3) and (6.19.4) fails.

Now we give one case in which $\mathcal{E}^*$ becomes σ-subsmooth at $B(T)$ along $B(T)$.

<u>Theorem 6.20</u>. <u>Let</u> $\mathcal{E} = \{T \xrightarrow{q_\gamma} (T_\gamma, A_\gamma, \lambda_\gamma) | \gamma \in \Gamma\}$ <u>be a consistent marginal system satisfying</u>

(6.20.1) Γ <u>is finite.</u>

(6.20.2) $q(T) = \prod_{\gamma \in \Gamma} T_\gamma$ <u>where</u> $q(t) = (q_\gamma(t))_{\gamma \in \Gamma} \quad \forall\ t \in T$

(6.20.3) $A_\gamma = B(T_\gamma, \sigma(A_\gamma))$ <u>and</u> $\lambda_\gamma \in Pr_\sigma(a_\gamma) \quad \forall\ \gamma \in \Gamma$

<u>Then</u> $\mathcal{E}^*$ <u>is</u> τ-<u>subsmooth at</u> $B(T)$ <u>along</u> $B(T)$.

<u>Remark</u>. The example above shows that even in very nice cases we can not surpress (6.20.1). Moreover let $(I,\mathcal{B})$ be the uniterval with its Borel τ-algebra and put

$$T_n = I, \quad A_n = B(I,\mathcal{B}(I)), \quad \lambda_n(f) = \int_0^1 f(t)dt, \quad n = 1,2$$

$$T = \{(x,y) \mid 0 \le x \le y \le 1\}$$

$$\mathcal{E} = \{T \xrightarrow{q_n} (T_n, A_n, \lambda_n) \mid n = 1,2\}$$

where $q_1(x,y) = x$ and $q_2(x,y) = y$. Then $\mathcal{M}(\mathcal{E})$ contains exactly one element, viz. the Lebesgue measure on the diagonal, and so $\mathcal{E}$ is fully $c\tau$-consisttent. And if

$$f = 1_E \ , \quad f_n(x,y) = 1 \wedge (ny-nx)$$

where $E = \{(x,y) \in T \mid x < y\}$, then $f_n \uparrow f$, and the reader easily verifies that

$$\mathcal{E}^*(f_n) = 0 \quad \forall\, n \ge 1, \quad \mathcal{E}^*(f) = 1$$

$$\sup\{\int_T f d\mu_\tau \mid \mu \in \mathcal{M}_\tau(\mathcal{E})\} = 0$$

Thus $\mathcal{E}^*$ is not σ-subsmooth, and (6.19.3) and (6.19.4) fails. Thus condition (6.20.2) can not be surpressed in general, it is however possible to weaken (6.20.2) somewhat, but I shall leave this as an exercise for the reader.

<u>Proof</u>. Let $f \in \bar{\mathbb{R}}_+^T$, then I claim that

(i) $$\mathcal{E}^*(f) = \inf\{ \sum_{\gamma\in\Gamma} \lambda_\gamma(\varphi_\gamma) \mid \varphi_\gamma \in A_\gamma^+ \ \forall \gamma \in \Gamma, \sum_{\gamma\in\Gamma} \varphi_\gamma \circ q_\gamma \geq f\}$$

So let $\mathcal{E}^*(f) < a$, then there exist $\psi_\gamma \in A_\gamma$ for $\gamma \in \Gamma$ (recall that Γ is finite), so that

$$f \leq \sum_{\gamma\in\Gamma} \psi_\gamma \circ q_\gamma \quad \text{and} \quad \sum_{\gamma\in\Gamma} \lambda_\gamma(\psi_\gamma) < a$$

By (6.20.2) and non-negativity of f we have

$$\sum_{\gamma\in\Gamma} \psi_\gamma(t_\gamma) \geq 0 \quad \forall\ (t_\gamma) \in \prod_{\gamma\in\Gamma} T_\gamma$$

Hence if $m_\gamma = m(\psi_\gamma)$, then $m = \sum_{\gamma\in\Gamma} m_\gamma \geq 0$. Now put

$$\varphi_\gamma = \psi_\gamma + k^{-1}(m-m_\gamma)1_{T_\gamma}$$

where $k = \text{card}(\Gamma)$, then $\varphi_\gamma \in A_\gamma^+$ and

$$\sum_\gamma \psi_\gamma \circ q_\gamma = \sum_\gamma \varphi_\gamma \circ q_\gamma \geq f$$

$$a > \sum_\gamma \lambda_\gamma(\psi_\gamma) = \sum_\gamma \lambda_\gamma(\varphi_\gamma)$$

since $\Sigma k^{-1}(m-m_\gamma) = 0$. Hence we see that $\mathcal{E}^*(f)$ is $\geq$ the right hand side of (i), and the converse inequality is evident. Thus (i) is proved.

By (i) and Corollary 3.9 we have that $\mathcal{E}^*$ is σ-subsmooth at $B^+(T)$ along $B^+(T)$, and since

$$\mathcal{E}^*(a1_T+g) = a + \mathcal{E}^*(g) \quad \forall\, a \in \mathbb{R} \quad \forall\, g \in \bar{\mathbb{R}}^T$$

it follows easily that $\mathcal{E}^*$ is σ-subsmooth at $B(T)$ along $B(T)$, and thus the theorem is proved. □

7. Functionals on product spaces

We shall in this section study a particular kind of marginal systems, viz. the socalled product systems. But first we study functionals on product spaces.

Let $\{(T_j, A_j) \mid j \in J\}$ be an indexed family of algebraic function spaces, then we put

$$T^\gamma = \prod_{j\in\gamma} T_j, \quad p_{\beta\gamma} = \text{the projection}: T^\gamma \to T^{\beta\cap\gamma}$$

$$p_\gamma = p_{\gamma J}, \quad p_{j\gamma} = p_{\{j\}\gamma}, \quad p_j = p_{\{j\}}$$

for all $\beta, \gamma \subseteq J$ and all $j \in J$. Let $\gamma \in 2^{(J)}$ and let $f_j : T_j \to \mathbb{R}$ be maps for $j \in \gamma$, then we define <u>the tensor product</u> and <u>the direct sum</u> of $\{f_j\}$ by

$$(\bigotimes_{j\in\gamma} f_j)(t) = \prod_{j\in\gamma} f_j(t_j) \qquad \forall\, t = (t_j) \in T^J$$

$$(\bigoplus_{j\in\gamma} f_j)(t) = \sum_{j\in\gamma} f_j(t_j) \qquad \forall\, t = (t_j) \in T^J$$

And we define the tensor product and the direct sum of $\{A_j\}$ by

$$\bigotimes_{j\in J} A_j = \left\{ \sum_{k=1}^{n} \bigotimes_{j\in\gamma} \varphi_{jk} \;\middle|\; \begin{array}{l} \gamma \in 2^{(J)}, \quad n \in \mathbb{N} \quad \text{and} \\ \varphi_{jk} \in A_j \quad \forall\, j \in \gamma \quad \forall\, 1 \le k \le n \end{array} \right\}$$

$$\bigoplus_{j\in J} A_j = \{ \bigoplus_{j\in\gamma} \varphi_j \mid \gamma \in 2^{(J)}, \ \varphi_j \in A_j \quad \forall\, j \in \gamma \}$$

Then $\otimes A_j$ is an algebra and $\oplus A_j$ is a vector space.

A <u>product system</u> is a marginal system of probability contents ot the form

$$\mathcal{E} = \{ T^J \xrightarrow{p_\gamma} (T^\gamma, \bigotimes_{j\in\gamma} A_j, \lambda_\gamma) \mid \gamma \in \Gamma \} \tag{7.1}$$

where $\{(T_j, A_j) \mid j \in J\}$ is a family of algebraic function spaces, Γ is a paving on J, and T^γ and p_γ are defined as above. If Γ is a covering on J, Then we say that $\mathcal{E}$ is a <u>proper product system</u>, and if Γ is the paving of all singletons, i.e. if

$$\mathcal{E} = \{T^J \xrightarrow{p_j} (T_j, A_j, \lambda_j) \mid j \in J\}$$

then we say that $\mathcal{E}$ is a <u>pure product system</u>.

Let $\mathcal{E}$ be a product system of the form (7.1). If Δ is a subset of Γ and $I = \cup_{\gamma \in \Delta} \gamma$, thus with the notation of section 6 we have

$$(7.2) \qquad A^\Delta = \{\varphi o p_I \mid \varphi \in \bigotimes_{j \in I} A_j\} \ , \qquad L^\Delta = \{\varphi o p_I \mid \varphi \in \bigotimes_{j \in I} A_j\}$$

Moreover if we put

$$(7.3) \qquad \hat{\bigotimes_{j \in \gamma}} A_j = \text{the } \|\cdot\|\text{-closure of } \bigotimes_{j \in \gamma} A_j$$

$$(7.4) \qquad \overline{\lambda}_{\beta\gamma}(\varphi) = \overline{\lambda}_\gamma(\varphi o p_{\beta\gamma}) \qquad \forall \gamma \in \Gamma \ \forall \beta \subseteq J \ \forall \varphi \in \hat{\bigotimes_{j \in \gamma \cap \beta}} A_j$$

Then it is easily checked that the weak consistency conditions (6.1) are equivalent to:

$$(7.5) \qquad \overline{\lambda}_{\beta\gamma} = \overline{\lambda}_{\alpha\delta} \qquad \forall \gamma, \delta \in \Gamma \quad \forall \beta, \alpha \in 2^J : \beta \cap \gamma = \alpha \cap \delta$$

I.e. if $\mathcal{E}$ is weakly consistent, then we may define λ_β for $\beta \in \Gamma^* = \{\delta \subseteq J \mid \exists \gamma \in \Gamma : \delta \subseteq \gamma\}$ by

$$(7.6) \qquad \overline{\lambda}_\beta = \overline{\lambda}_{\beta\gamma} \qquad \text{if} \quad \gamma \in \Gamma \ \text{ and } \ \gamma \supseteq \beta$$

Thus if $\mathcal{E}$ is weakly consistent, then it is usually no loss of generality to assume that Γ is a <u>hereditary paving</u> (i.e. if $\gamma \in \Gamma$ and $\delta \subseteq \gamma$, then $\delta \in \Gamma$).

Lemma 7.1. Let (T,A) be an algebraic function space and let S be a set. If $\varphi \in B(S) \hat{\otimes} A$ and C is a non-empty subset of S, then the functions:

$$\varphi^*(t) = \sup\{\varphi(s,t) \mid s \in C\} \qquad \forall t \in T$$

$$\varphi_*(t) = \inf\{\varphi(s,t) \mid s \in C\} \qquad \forall t \in T$$

both belong to $\overline{A}$.

Proof. Since $\varphi_* = -(-\varphi)^*$, it suffices to show that $\varphi^* \in \overline{A}$, and since $\varphi \rightsquigarrow \varphi^*$ is a continuous map from $(B(S\times T), \|\cdot\|)$ into $(B(T), \|\cdot\|)$ it is no loss of generality to assume that $\varphi \in B(S) \otimes A$. Hence there exist $f_1,\dots,f_n \in B(S)$ and $\psi_1,\dots,\psi_n \in A$, such that

$$\varphi(s,t) = \sum_{j=1}^{n} f_j(s)\psi_j(t)$$

Let $f = (f_1,\dots,f_n)$ then f maps S into $\mathbb{R}^n$, and $f(S)$ is bounded. Let $K = f(C)$, and let h be the support function of K, i.e.

$$h(x) = \sup\{\sum_{j=1}^{n} x_j y_j \mid (y_1,\dots,y_n) \in K\}$$

for all $x = (x_1,\dots,x_n) \in \mathbb{R}^n$. Since K is bounded it is wellknown that h is finite and continuous (actually h is convex and sub-linear). Moreover we have

$$\varphi^*(t) = h(\psi_1(t),\dots,\psi_n(t))$$

and so $\varphi^* \in \overline{A}$ by Proposition 2.7. □

Lemma 7.2. Let $\mathcal{E} = \{T^J \xrightarrow{p_j} (T_j, A_j, \lambda_j \mid j \in J\}$ be a pure product system, and let $K_j \subseteq T_j$ for all $j \in J$. Let ξ be an infinite cardinal and let $K = \prod_{j \in J} K_j$, then we have

(7.2.1) $\quad K \in \mathcal{K}_\xi(\bigotimes_{j\in J} A_j)$ if $K_j \in \mathcal{K}_\xi(A_j) \quad \forall j \in J$

(7.2.2) $\quad \mathcal{E}^*(\varphi) \leq \sum_{j\in J} (1-\lambda_j^*(K_j)) \qquad \forall \varphi \in \hat{\bigotimes}_{j\in J} A_j : \varphi \leq 1 - 1_K$

(7.2.3) $\quad \mathcal{E}$ is consistent.

Proof. (7.2.1): Let $Q \subseteq \otimes A_j$ with $\operatorname{card}(Q) \leq \xi$. Then there exists algebras $B_j \subseteq A_j$ with $\operatorname{weight}(B_j) \leq \xi$ for all $j \in J$, so that $Q \subseteq \otimes B_j$. By Proposition 2.9 we have that K_j is $\tau(B_j)$-compact for all j, and so K is $\tau(\otimes B_j)$-compact by Tychonov's theorem, since $\tau(\otimes B_j)$ equals the product topology of the $\tau(B_j)$-topologies. Hence $K \in \mathcal{K}(Q)$ and so $K \in \mathcal{K}_\xi(\otimes A_j)$.

(7.2.2): If $K_j = \emptyset$ for some $j \in J$. Then (7.2.2) is trivially satisfied since $\mathcal{E}^*(1) \leq 1$. So suppose that $K_j \neq \emptyset$ for all $j \in J$, and let $\varphi \in \otimes A_j$, such that $\varphi \leq 1 - 1_K$. By definition of the tensor product $\otimes A_j$ there exist a finite set $\gamma \subseteq J$ and $\psi \in \otimes_{j\in\gamma} A_j$, such that $\varphi = \psi \circ p_\gamma$. To simply the notation let us assume that $\gamma = \{1,..,m\}$, and let us define $\psi_1, \psi_2, \ldots, \psi_m$ inductively as follows

$$\psi_1(t_1) = \sup\{\psi(t_1,\ldots,t_m) \mid t_2 \in K_2, \ldots, t_m \in K_m\}$$

$$\psi_j(t_j) = \sup\left\{\psi(t_1,\ldots,t_m) - \sum_{i=1}^{j-1} \psi_i^+(t_i) \;\middle|\; \begin{array}{ll} t_i \in T_i & \forall 1 \leq i < j \\ t_i \in K_i & \forall j < i \leq m \end{array}\right\}$$

Then $\psi_j \in \overline{A}_j$ by Lemma 7.1, and since $\psi \leq 1$ we see that $\psi_j \leq 1$ for all $j = 1,\ldots,m$.

If $(t_1,\ldots,t_m) \in K_1 \times \ldots \times K_m$, then $\psi(t_1,\ldots,t_m) \leq 0$, and so we have that $\psi_1(t_1) \leq 0$ for $t_1 \in K_1$. Similarly if $2 \leq j \leq m$ and $t_i \in T_i$ for $1 \leq i < j$ and $t_i \in K_i$ for $j \leq i \leq m$, then

$$\psi(t_1,\ldots,t_m) - \sum_{i=1}^{j-1} \psi_i^+(t_i)$$

$$\leq [\psi(t_1,\ldots,t_m) - \sum_{i=1}^{j-2} \psi_i^+(t_i)] - \psi_{j-1}(t_{j-1})$$

$$\leq 0$$

Hence $\psi_j(t_j) \leq 0$ for $t_j \in K_j$, and so we find

(i) $$\psi_j^+ \leq 1 - 1_{K_j} \qquad \forall j \in \gamma$$

And by definition of ψ_m we have

(ii) $$\varphi(t) = \psi(p_\gamma(t)) \leq \sum_{j\in\gamma} \psi_j^+(t_j)$$

and so we have

$$\mathcal{E}^*(\varphi) \leq \sum_{j\in\gamma} \overline{\lambda}_j(\psi_j^+) \leq \sum_{j\in\gamma} (1-\lambda_j^*(K_j))$$

Hence (7.2.2) follows.

(7.2.3): Let $\gamma \in 2^{(J)}$ and let $\varphi_j \in A_j$ for $j \in \gamma$, so that $\Sigma \varphi_j \circ p_j \geq 0$. Then

$$\sum_{j\in\gamma} \varphi_j(t_j) \geq 0 \qquad \forall (t_j)_{j\in\gamma} \in T^\gamma$$

Hence if $m_j = m(\varphi_j)$ and $\hat{\varphi}_j = \varphi_j - m_j$, then $\hat{\varphi}_j \in A_j^+$ and $\Sigma m_j \geq 0$. Thus we have

$$\sum_{j\in\gamma} \lambda_j(\varphi_j) = \sum_{j\in\gamma} \lambda_j(\hat{\varphi}_j) + \Sigma m_j \geq 0$$

and so the strong consistency conditions (6.2) holds, and (7.2.3) then follows from Theorem 6.1. □

Theorem 7.3. Let $\{(T_j,A_j) \mid j \in J\}$ be a family of algebraic function spaces and let $\lambda \in \Pr(\underset{j\in J}{\otimes} A_j)$. Put

$$\lambda_\gamma(\varphi) = \lambda(\varphi \circ p_\gamma) \qquad \forall \gamma \subseteq J \quad \forall \varphi \in \underset{j\in\gamma}{\otimes} A_j$$

$$\lambda_j(\varphi) = \lambda(\varphi \circ p_j) \qquad \forall j \in J \quad \forall \varphi \in A_j$$

Let ξ be a given infinite cardinal, and let γ be a given subset of J, then we have

(7.3.1) If λ_j has ξ-compact support (see below) for all $j \in J$, then so has λ

(7.3.2) If γ is countable, if λ_j is ξ-compact $\forall j \in \gamma$, and if $\lambda_{J\setminus\gamma}$ is ξ-compact, then λ is ξ-compact

(7.3.3) If λ_j is perfect for all $j \in J$, then so is λ

(7.3.4) If λ_γ is ξ-perfect and $\lambda_{J\setminus\gamma}$ is ξ-smooth, then λ is ξ-smooth.

Remark. Let (T,A) be an algebraic function space and let $\lambda \in \Pr(A)$, then we say that λ has ξ-compact support, if there exists $K \in \mathcal{K}_\xi(A)$, such that $\lambda^*(K) = 1$. And we put

(7.3.5) $\qquad \Pr_{s\xi}(A) = \{\lambda \in \Pr(A) \mid \lambda$ has ξ-compact support$\}$

Clearly we have that $\Pr_{s\xi}(A) \subseteq \Pr_{c\xi}(A)$.

<u>Proof</u>. Let us consider the pure product system:

$$\mathcal{E} = \{T^J \xrightarrow{p_j} (T_j, A_j, \lambda_j \mid j \in J\}$$

Then $\lambda \in M(\mathcal{E})$, and so we have

(i) $$\mathcal{E}_*(f) \leq \lambda_*(f) \leq \lambda^*(f) \leq \mathcal{E}^*(f) \qquad \forall f : T^J \to \overline{\mathbb{R}}$$

by Proposition 6.2.

(7.3.1): Let $K_j \in K_\xi(A_j)$ so that $\lambda_j^*(K_j) = 1$ for all $j \in J$, and let $K = \prod_{j \in J} K_j$ then

$$\lambda(\varphi) = 1 - \lambda(1-\varphi) \geq 1 - \mathcal{E}^*(1-\varphi)$$

$$\geq 1 - \sum_{j \in J} (1-\lambda_j^*(K_j)) = 1$$

for all $\varphi \in \otimes A_j$ with $\varphi \geq 1_K$ by (7.2.2), and so we have that $\lambda^*(K) = 1$, and $K \in K_\xi(\otimes A_j)$ by (7.2.1). Thus λ has ξ-compact support.

(7.3.2): Let $\varepsilon > 0$ be given. Since γ is countable, there exist $\varepsilon_j > 0$ for $j \in \gamma$, so that $\Sigma_{j \in \gamma} \varepsilon_j < \varepsilon$. Now let $K_j \in K_\xi(A_j)$ be chosen for all $j \in J$, such that $\lambda_j^*(K_j) \geq 1 - \varepsilon_j$, and let

$$C \in K_{J \setminus \gamma}(\bigotimes_{j \in J \setminus \gamma} A_j)$$

be chosen so that $\lambda_{J \setminus \gamma}^*(C) \geq 1 - \varepsilon$. Now put

$$K = C \times \prod_{j \in \gamma} K_j$$

Then $K \in K_\xi(\otimes A_j)$ by (7.2.1) and by (7.2.2) we have

$$\lambda(\varphi) = 1 - \lambda(1-\varphi) \geq 1 - \mathcal{E}^*(1-\varphi)$$

$$\geq 1 - (1-\lambda_{J \setminus \gamma}^*(C)) - \sum_{j \in \gamma} (1-\lambda_j^*(K_j))$$

$$\geq 1 - 2\varepsilon$$

for all $\varphi \in \otimes A_j$ with $\varphi \geq 1_K$. Hence $\lambda^*(K) \geq 1 - 2\varepsilon$ and so λ is ξ-compact.

(7.3.3): Let $A \subseteq \otimes A_j$ be a countably generated algebra. Then there exist a countable set $\beta \subseteq J$ and countably generated algebras $B_j \subseteq A_j$ for $j \in J$, such that

(i) $$A \subseteq \underset{j\in J}{\otimes} B_j \ , \qquad B_j = \{a 1_{T_j} \mid a \in \mathbb{R}\} \ \forall j \in J\setminus\beta$$

Then $T_j \in K(B_j)$ for $j \in J\setminus\beta$, and by Theorem 4.11 and (4.6) there exist $K_j \in K(B_j)$ for $j \in \beta$ such that

$$\sum_{j\in\beta} (1-\lambda_j^*(K_j)) < \varepsilon$$

where $\varepsilon > 0$ is a given number, since β is countable. Now let

$$K = \prod_{j\in\beta} K_j \times \prod_{j\in J\setminus\beta} T_j$$

Then as above we find that $K \in K(\otimes B_j)$ and $\lambda^*(K) \geq 1 - \varepsilon$. Hence λ is perfect by (4.4) and (4.6).

(7.3.4): It is of course no loss of generality to assume that $J = \{1,2\}$, $\gamma = \{1\}$ and $J\setminus\gamma = \{2\}$. Now let $\{\varphi_q \mid q \in Q\}$ be a decreasing net in $A_1 \otimes A_2$ such that $\text{card}(Q) \leq \xi$, $0 \leq \varphi_q \leq 1$, and $\varphi_q \downarrow 0$. Then there is an algebra $B_1 \subseteq A_1$, such that B_1 is generated by a set of cardinality atmost ξ and $\varphi_q \in B_1 \otimes A_2$ for all $q \in Q$. Let $\varepsilon > 0$ be given, then by assumption there exist $K \in K_\xi(B_1)$, such that

(ii) $$\overline{\lambda}_1(\psi) \leq \varepsilon \quad \text{if} \quad \psi \in \overline{B}_1 \quad \text{and} \quad \psi \leq 1 - 1_K$$

Now let us define

$$\hat{\varphi}_q(t_2) = \sup\{\varphi_q(t_1,t_2) \mid t_1 \in K_1\}$$

Then $\hat{\varphi}_q \in \overline{A}_2$ by Lemma 7.1, and $\{\hat{\varphi}_q\}$ decreases. Since the weight of B_1 is atmost ξ we have that $K \in K(B_1)$, and φ_q is $\tau(B_1 \otimes A_2)$-continuous. Hence by Dini's theorem we have that $\hat{\varphi}_q \downarrow 0$ and so

(iii) $\qquad \overline{\lambda}_2(\hat{\varphi}_q) \to 0$

Now let

$$\hat{\psi}_q(t_1) = \sup\{\varphi_q(t_1,t_2) - \hat{\varphi}_q(t_2) \mid t_2 \in T_2\}$$

Then $\hat{\psi}_q \in \overline{B}_1$ by Lemma 7.1, and $\hat{\psi}_q(t_1) \leq 0$ for $t_1 \in K$ by definition of $\hat{\varphi}_q$. Moreover since $\varphi_q \leq 1$ and $\hat{\varphi}_q \geq 0$, we have that $\hat{\psi}_q \leq 1 - 1_K$, and so by (ii) we conclude that

(iv) $\qquad \overline{\lambda}_1(\hat{\psi}_q) \leq \varepsilon \qquad \forall q \in Q$

Now note that by definition of $\hat{\psi}_q$ we have

$$\varphi_q(t_1,t_2) \leq \hat{\psi}_q(t_1) + \hat{\varphi}_q(t_2) \qquad \forall (t_1,t_2) \in T_1 \times T_2$$

and so by (iv) we find

$$\lambda(\varphi_q) \leq \overline{\lambda}_1(\psi_q) + \overline{\lambda}_2(\hat{\varphi}_q) \leq \varepsilon + \lambda_2(\hat{\varphi}_q)$$

Thus letting $q \to \infty$ and $\varepsilon \to 0$, we find from (iii) that $\lim \lambda(\varphi_q) = 0$, and so λ is ε-smooth. □

From (7.3.4) it follows that if $\lambda \in Pr(A_1 \otimes A_2)$ on one marginal of λ is ξ-perfect and the other marginal is ξ-smooth, then λ is ξ-smooth. Our next result shows that ξ-perfectness of one of the marginal is indispensable in general.

Theorem 7.4. Let (T,A) be an algebraic function space, and let ξ be an infinite cardinal. If $\lambda \in Pr(A)$ is not ξ-perfect, then there exist a completely regular Hausdorff space S and probability contents $\mu \in Pr(C(S))$ and $\nu \in Pr(C(S) \otimes A)$ such that

(7.4.1) $\lambda(\varphi) = \nu(\varphi o p_1) \quad \forall \varphi \in A$

(7.4.2) $\mu(\psi) = \nu(\psi o p_2) \quad \forall \psi \in C(S)$

(7.4.3) μ is τ-smooth, and ν is not ξ-smooth

(7.4.4) $\text{weight}(S) \leq \xi$

where $p_1(s,t) = s$ and $p_2(s,t) = t$ for $(s,t) \in S \times T$

Remarks (1): The proof is based of an idea of J. Pachl [17], where a similar result is shown for $\xi = \aleph_o$. In [17, p. 337] J. Pachl gives an example of a subset $T \subseteq [0,1]$ and a functional $\lambda \in Pr_\tau(A) \setminus Pr_\pi(A)$, where $A = C(T)$, such that every $\mu \in Pr(A \otimes A)$ with marginals λ and λ is perfect.

(2): Note that if $\xi = \aleph_o$, then by (7.4.4) we have that S is a separable metric space.

(3): Note that (7.3.4) and Theorem 7.4, show that $\lambda \in Pr(A)$ is ξ-perfect if and only if ν is ξ-smooth whenever $\nu \in Pr(A \otimes B)$ and the first marginal of ν equals μ, the second marginal of ν is ξ-smooth, and (S,B) is an arbitrary algebraic function space (or equivalently (S,B) is an algebraic function space, so that B separates points in S and B is generated by a set of cardinality atmost ξ).

<u>Proof</u>. If $\lambda \notin Pr_\xi(A)$, then the theorem is obvious. So suppose that λ is ξ-smooth but not ξ-perfect. Then there exist an algebra $A_0 \subseteq A$, such that A_0 is generated by a set Q with $card(Q) \leq \xi$, and such that $\lambda | A_0$ is not ξ-compact. Let

$$\theta(t) = (\varphi(t))_{\varphi \in Q}$$

Then θ maps T into $\mathbb{R}^Q$ and $L = cl\, \theta(T)$ is compact in $\mathbb{R}^Q$.

Then θ is continuous from $(T, \tau(A_0))$ into L, and since $weight(A_0) \leq \xi$, we have that

(i) $\qquad \theta^{-1}(B) \in \mathcal{B}(A_0) \subseteq \sigma(A_\xi) \qquad \forall B \in \mathcal{B}(L)$

(ii) $\qquad \theta^{-1}(F) \in \mathcal{F}(A_0) \subseteq \mathcal{F}_\xi(A) \qquad \forall F \in \mathcal{F}(L)$

(iii) $\qquad \theta^{-1}(K) \in \overline{K}(A_0) \qquad \forall K \in K(L): K \subseteq \theta(T)$

Hence $m(B) = \lambda_\xi(\theta^{-1}(B))$ is a τ-smooth Borel probability on $(L, \mathcal{B}(L))$, and since L is compact and Hausdorff we have that m is a Radon probability on L. Since $\lambda | A_0$ is not compact we have by (iii) that

$$m_*(\theta(T)) \leq \sup\{\lambda_\xi(K) \mid K \in K(A_0)\} < 1$$

Let $D \in \mathcal{B}(L)$ be chosen so that

(iv) $\qquad D \subseteq \theta(T)$ and $m(D) = m_*(\theta(T)) < 1$

Since m is Borel regular, there exist $G \subseteq L$ so that

(v) $\qquad G$ is open, $G \supseteq D$ and $m(G) < 1$

Now put $S = D \cup (L \setminus \theta(T))$, since D is m-measurable and $D \cap (L \setminus \theta(T)) = \emptyset$

we have

$$m^*(S) = m(D) + m^*(L\setminus\theta(T))$$

$$= m_*(\theta(T)) + m^*(L\setminus\theta(T))$$

$$= 1$$

Thus if we put

$$m_0(B) = m^*(B) \quad \forall B \in \mathcal{B}(S)$$

then m_0 is a τ-smooth Borel regular probability measure on $(S,\mathcal{B}(S))$ (see Lemma 6.7), and S is a completely Hausdorff space satisfying (7.4.4).

Let $\psi \in C(S)$, then ψ admits a Borel measurable extension $\overline{\psi}$ to L which is unique up to a m-nullset. Hence using a lifting of $L^\infty(m)$, there exist a linear map ρ from $C(S)$ into $B(L)$ satisfying

(vi) $\quad \rho 1_S = 1_L, \quad \rho\psi$ is m-measurable $\quad \forall \psi \in C(S)$

(vii) $\quad \rho\psi = \psi$ m-a.s. on $S \quad \forall \psi \in C(S)$

(viii) $\quad \rho\psi \geq 0$ on L if $\psi \in C(S)$ and $\psi \geq 0$ m_0-a.s.

If $f \in C(S) \otimes A$, then we define

$$(\hat{\rho}f)(s,t) = \rho(f(\cdot,t))(s) \quad \forall (s,t) \in S \times T$$

If $f = \Sigma\psi_j \otimes \varphi_j$ then we have

$$\hat{\varphi}f = \Sigma(\rho\psi_j) \otimes \varphi_j$$

by linearity of ρ, and so $\hat{\rho}$ is a linear positive map from $C(S) \otimes A$ into $B(m) \otimes A$, where $B(m)$ is the set of all bounded m-measurable functions. Now we put

$$\mu(\psi) = \int_S \psi \, dm_0 \quad \forall \psi \in C(S)$$

$$\nu(f) = \int_T (\hat{\rho} f)(\theta(t),t)\lambda_\xi(dt) \quad \forall f \in C(S) \otimes A$$

Then $\mu \in Pr_\tau(C(S))$, since m_0 is τ-smooth, and $\nu \in Pr(C(S) \otimes A)$ by (vi)-(viii). By definition of μ and ν we see that (7.4.1) and (7.4.2) holds.

Now let $f \in C(S) \otimes A$ and $g \in B(m) \otimes A$, such that $f(s,t) \geq g(s,t)$ for all $(s,t) \in S \times T$. Then $\hat{f} = \hat{\rho} f$ belongs to $B(m) \otimes A$ and there exist an m-nullset $N \in \mathcal{B}(L)$ so that

$$\text{(ix)} \qquad \hat{f}(s,t) = f(s,t) \geq g(s,t) \quad \forall (s,t) \in (S \setminus N) \times T$$

Let $h(s) = \inf\{\hat{f}(s,t) - g(s,t) \mid t \in T\}$, then by Lemma 7.1 we have that $h \in B(m)$, and by (ix) we have that $h \geq 0$ on $S \setminus N$. Since $m^*(S \setminus N) = 1$, it follows that $h \geq 0$ m-a.s., and so there exist a m-nullset $N_0 \in \mathcal{B}(L)$ such that

$$\hat{f}(s,t) \geq g(s,t) \quad \forall (s,t) \in (L \setminus N_0) \times T$$

and so we have

$$\hat{f}(\theta(t),t) \geq g(\theta(t),t) \quad \forall t \in T \setminus \theta^{-1}(N_0)$$

and so $\lambda_\xi(\theta^{-1}(N_0)) = m(N_0) = 0$. Thus we find:

$$\text{(x)} \qquad \nu(f) \geq \int_T g(\theta(t),t)\lambda_\xi(dt)$$

whenever $f \in C(S) \otimes A$, $g \in B(m) \otimes A$ and $f \geq g$ on $S \times T$.

Now let

$$F = \{(x,t) \in L \times T \mid x \notin G,\ s = \theta(t)\}$$

where G is the open set from (v). Then F is a closed subset of $L \times T$, when T has its $\tau(A_0)$-topology. And since weight$(C(L)\otimes A_0) \leq \xi$, then by Proposition 2.9 there exist a decreasing net $\{\varphi_q \mid q \in Q\}$ in $C(L) \otimes A_0$ satisfying

(xi) $\qquad \varphi_q \downarrow 1_F \quad$ and $\quad \mathrm{card}(Q) \leq \xi$

Now let ψ be the restriction of φ_q to $S \times T$. Then $\{\psi_q\}$ is a decreasing net in $C(S) \otimes A$, and since

$$(S \setminus G) \cap \theta(T) \subseteq (S \setminus D) \cap \theta(T) = \emptyset$$

we see that $F \cap (S \times T) \supset \emptyset$, and so $\psi_q \downarrow 0$, but by (x) we have

$$\nu(\psi_q) \geq \int_T q(\theta(t),t)\lambda_\xi(dt)$$

$$\geq \int_T 1_F(\theta(t),t)\lambda_\xi(dt)$$

$$= \lambda_\xi(\theta^{-1}(L \setminus G))$$

$$= 1 - m(G)$$

and $1 - m(G) > 0$ by (v). Thus ν is not ξ-smooth. □

<u>Corollary 7.5</u>. <u>Let</u> $\mathcal{E} = \{T^J \xrightarrow{q_\gamma} (T^\gamma, \otimes_{j\in\gamma} A_j, \lambda_\gamma) \mid \gamma \in \Gamma\}$ <u>be a consistent proper product system, and let</u> ξ <u>be a given infinite cardinal, then we have</u>

(7.6.1) <u>If</u> λ_γ <u>has</u> ξ-<u>compact support</u> $\forall\gamma\in\Gamma$, <u>then</u> $\mathcal{E}$ <u>is fully</u> $c\xi$-<u>consistent and there exist</u> $K\in K_\xi(\underset{j\in J}{\otimes} A_j)$ <u>so that</u> $\mu^*(K)=1 \quad \forall\mu\in M(\mathcal{E})$

(7.5.2) <u>If</u> λ_γ <u>is</u> ξ-<u>compact</u> $\forall\gamma\in\Gamma$, <u>and if</u> $J=\cup_1^\infty\gamma_n$ <u>for some sequence</u> $\{\gamma_n \mid n\in\mathbb{N}\}\subseteq\Gamma$, <u>then</u> $\mathcal{E}$ <u>is fully</u> $c\xi$-<u>consistent and</u> $M(\mathcal{E})$ <u>is uniformly</u> ξ-<u>compact.</u>

(7.5.3) <u>If</u> λ_γ <u>is perfect</u> $\forall\gamma\in\Gamma$, <u>then</u> $\mathcal{E}$ <u>is fully</u> π-<u>consistent</u>

(7.5.4) <u>If there exist</u> $\beta,\delta\in\Gamma$, <u>so that</u> $J=\beta\cup\delta$, λ_β <u>is</u> ξ-<u>perfect and</u> λ_δ <u>is</u> ξ-<u>smooth, then</u> $\mathcal{E}$ <u>is fully</u> ξ-<u>consistent.</u>

<u>Proof</u>. Easy consequence of Theorem 7.3, Theorem 4.13 and Lemma 7.1. □

As noted in the introduction, we have that a product system is weakly consistent, if and only if (4.4) holds. In Lemma 7.2 we saw that a pure product system is consistent, in general weak consistency does not imply consistency. The simplest possible counterexample is the following:

$$J=\{1,2,3\}, \qquad \Gamma=\{\{1,2\},\{2,3\},\{1,3\}\}$$

$$T_1=T_2=T_3=\{0,1\}, \qquad A_j=B(T_j)$$

$$\lambda_{12}(f)=\tfrac12 f(0,0)+\tfrac12 f(1,1)$$

$$\lambda_{23}(f)=\tfrac12 f(1,0)+\tfrac12 f(0,1)$$

$$\lambda_{13}(f)=\tfrac12 f(0,0)+\tfrac12 f(1,1)$$

Then the corresponding product system is weakly consistent but not consistent. However we have the following extension of (7.2.3):

Proposition 7.6. Let $\mathcal{E}$ be a weakly consistent product system of the form (4.1) satisfying

(7.6.1) $\quad \forall \pi \in 2^{(\Gamma)} \; \exists \delta \in \Gamma: \gamma \cap \beta \subseteq \delta$ if $\gamma, \beta \in \pi, \quad \gamma \neq \beta$

Then $\mathcal{E}$ is consistent.

Remarks (1): If Γ is a disjoint family of subsets of J then (7.6.1) holds with δ any element of Γ.

(2): Let $k \in J$, then $\Gamma = \{\{k,j\} \mid j \in J\}$ satisfies (7.6.1) with $\delta = \{k\}$

(3): If Γ is filtering upwards, then (7.6.1) holds (if π is a finite subset of Γ, then we take $\delta \in \Gamma$, such that $\delta \geq \gamma$ for all $\gamma \in \pi$).

Proof. By (7.5) and (7.6) it is no loss of generality to assume that Γ is hereditary. Now let $\pi \in 2^{(\Gamma)}$ and let $\varphi_\gamma \in \bigotimes_{j \in \pi} A_j$ for $\gamma \in \pi$, such that

$$\sum_{\gamma \in \pi} \varphi_\gamma \circ p_\gamma \geq 0$$

And let $\delta \in \Gamma$ be chosen according to (7.6.1). Then $\{\gamma \setminus \delta \mid \gamma \in \pi\}$ are mutually disjoint, and so

(i) $$\sum_{\gamma \in \pi} \varphi_\gamma(p_{\gamma\delta}(u), v_\gamma) \geq 0 \quad \forall u \in T^\delta \quad \forall (v_\gamma) \in \prod_{\gamma \in \pi} T^{\gamma \setminus \delta}$$

where we identify T^{γ} with $T^{\gamma\cap\delta} \times T^{\gamma\setminus\delta}$ in the natural way. Let

$$\psi_\gamma(w) = \inf\{\varphi_\gamma(w,v) \mid v \in T^{\gamma\setminus\delta}\} \qquad \forall\, \omega \in T^{\gamma\setminus\delta}$$

Then $\psi_\gamma \in \hat{\otimes}_{j\in\gamma\cap\delta} A_j$ by Lemma 7.1 and by (i) we have

(ii) $$\sum_{\gamma\in\pi} \psi_\gamma \circ p_{\gamma\delta} \geq 0$$

And so we conclude that

$$0 \leq \overline{\lambda}_\delta\Big(\sum_{\gamma\in\pi} \psi_\gamma \circ p_{\gamma\delta}\Big) = \sum_{\gamma\in\pi} \overline{\lambda}_\delta(\psi_\gamma \circ p_{\gamma\delta})$$

$$= \sum_{\gamma\in\pi} \overline{\lambda}_{\gamma\cap\delta}(\psi_\gamma) = \sum_{\gamma\in\pi} \overline{\lambda}_\gamma(\psi_\gamma \circ p_{\delta\gamma})$$

$$\leq \sum_{\gamma\in\pi} \lambda_\gamma(\varphi_\gamma)$$

since $\psi_\gamma \circ p_{\delta\gamma} \leq \varphi_\gamma$ by definition of ψ_γ. Hence we see that $\mathcal{E}$ satisfies the strong consistency conditions and so $\mathcal{E}$ is consistent by Theorem 6.1. □

Definition 7.7. An important way to construct functionals on product spaces goes via multilinear functionals.

Let $\{(T_j, A_j) \mid j \in J\}$ be a family of algebraic function spaces and put

$$A^{\langle J\rangle} = \{(\varphi_j) \in \prod_{j\in J} A_j \mid \exists\, \sigma \in 2^{(J)} : \varphi_j = 1_{T_j} \quad \forall\, j \in J\setminus\sigma\}$$

Then a multilinear functional on $\{(T_j, A_j) \mid j \in J\}$, is a map $\beta : A^{\langle J\rangle} \to \mathbb{R}$ satisfying

(7.7.1) β is linear in each variable separately.

A multilinear functional β is <u>positive</u> if

(7.7.2) $\beta((\varphi_j)) \geq 0$ if $(\varphi_j) \in A^{<J>}$ and $\varphi_j \geq 0$ $\forall j \in J$

And a <u>multi-probability content</u> is a positive multilinear functional β satisfying

(7.7.3) $\beta((1_{T_j})) = 1$

Let $I \subseteq J$ and let β be a multilinear functional, then we define

(7.7.4) $\beta_I(\varphi|\psi) = \beta(<\varphi,\psi>)$ $\forall \varphi \in A^{<I>}$ $\forall \psi \in A^{<J \setminus I>}$

where $<\varphi,\psi>$ denote the element in $A^{<J>}$ defined by

(7.7.5) $<\varphi,\psi> = \begin{cases} \varphi_j & \forall j \in I \\ \psi_j & \forall j \in J \setminus I \end{cases}$

whenever $\varphi = (\varphi_j) \in A^{<I>}$ and $\psi = (\psi_j) \in A^{<J \setminus I>}$. And we write $\beta_i(\varphi|\psi)$ if $I = \{i\}$ for some $i \in I$. And if $j \in J$ then we define <u>the</u> j-<u>th marginal</u> of β by

(7.7.6) $\beta_j(\varphi) = \beta_j(\varphi|(1_{T_i})_{i \neq j})$ $\forall \varphi \in A_j$ $\forall j \in J$

Clearly we have

(7.7.7) $\beta_j \in Pr(A_j)$ if β is a multi-probability content

If $(\varphi_j) \in A^{<J>}$, then $\varphi_j \equiv 1$ for all but finitely many j's, and so we may define the <u>tensor product</u> of (φ_j) as usual:

$$(\bigotimes_{j \in J} \varphi_j)(t) = \prod_{j \in J} \varphi_j(t_j) \quad \forall t = (t_j) \in T^J$$

Note that if $\lambda \in Pr(\bigotimes_{j\in J} A_j)$, then

$$(7.7.8) \qquad \beta((\varphi_j)) = \lambda(\bigotimes_{j\in J} \varphi_j) \qquad \forall\, (\varphi_j) \in A^{<J>}$$

is a multi-probability content on $\{(T_j,A_j) \mid j \in J\}$. And our next theorem shows that all multi-probability contents are of this form.

<u>Theorem 7.8</u>. <u>Let</u> β <u>be a multi-probability content on</u> $\{(T_j,A_j)\}$, <u>then there exist a unique</u> $\lambda \in Pr(\bigotimes_{j\in J} A_j)$ <u>satisfying (7.7.8)</u>.

<u>Proof</u>. Let $i \in J$ and $\psi \in A^{<I>}$ where $I = J\setminus\{i\}$. Then $\psi_j \equiv 1$ for all $j \in I\setminus\sigma$ for some finite set $\sigma \subseteq I$, and since $A_j = A_j^+ - A_j^+$, we see that $\beta_i(\cdot|\psi)$ can be written as an alternating sum of $2^{\mathrm{card}(\sigma)}$ positive linear functionals on A_i. Hence we have

(i) $\qquad \beta_i(\cdot|\psi)$ is linear and $\|\cdot\|$-continuous on A_i.

Now let $\varphi^k = (\varphi_j^k) \in A^{<J>}$ for $k = 1,\dots,m$, we shall then show the following implication:

$$\text{(ii)} \qquad \sum_{k=1}^{m} \bigotimes_{j\in J} \varphi_j^k \geq 0 \Rightarrow \sum_{k=1}^{m} \beta(\varphi^k) \geq 0$$

Let $\sigma = \sigma(\varphi^1,\dots,\varphi^m)$ be the set of all $j \in J$, so that $\varphi_j^k(u) \neq 1$ for some $u \in T_j$ and some $1 \leq k \leq m$. Then σ is finite by definition of $A^{<J>}$, and we shall prove (ii) by induction in $\mathrm{card}(\sigma)$.

If $\mathrm{card}(\sigma) = 0$, then $\varphi_j^k \equiv 1$ for all $j \in J$ and all $1 \leq k \leq m$. Hence by (7.7.2) we have that (ii) holds.

Now suppose that (ii) holds for any positive multilinear functional β and any finite set $\varphi^1,\dots,\varphi^m$ in $A^{<J>}$ with $\mathrm{card}\,\sigma(\varphi^1,\dots\varphi^m) \leq q$ where $q \in \mathbb{N}_0$.

Now let β and $\varphi^1,\dots,\varphi^m \in A^{<J>}$ be given, and suppose that card $\sigma(\varphi^1,\dots,\varphi^m) = q+1$. Let

$$\sigma = \sigma(\varphi^1,\dots,\varphi^m)$$

then $\sigma \neq \emptyset$, so let us choose a point $i \in \sigma$. By (i) we have that $\beta_{ik} = \beta_i(\cdot|\hat{\varphi}^k)$ is $\|\cdot\|$-continuous and linear, where

$$\hat{\varphi}^k = (\varphi_j^k)_{j\in J\setminus\{i\}}$$

So let $\overline{\beta}_{ik}$ be the unique linear $\|\cdot\|$-continuous extension of β_{ik} to $\overline{A}_i$, and choose $c \in \mathbb{R}_+$ so that

$$\text{(iii)} \qquad \sum_{k=1}^{m} |\overline{\beta}_{ik}(\varphi)| \leq c\|\varphi\| \qquad \forall \varphi \in \overline{A}_i$$

If $q = 0$ then $\sigma = \{i\}$ and so $\hat{\varphi}^k = (1_{T_j})_{j\in J\setminus\{i\}}$. Hence

$$\sum_{k=1}^{m} \beta(\varphi^k) = \sum_{k=1}^{m} \beta_i(\varphi_i^k) = \beta_i(\sum_{k=1}^{m} \varphi_i^k) \geq 0$$

since $\beta_i \in \Pr(A_i)$ by (7.7.7), and $\Sigma_k \varphi_i^k \geq 0$ by assumption. Thus (ii) holds if $q = 0$.

So suppose that $q \geq 1$, and let $\theta = (\varphi_i^1,\dots,\varphi_i^m)$, then θ maps T_i into $\mathbb{R}^m$, and $K = \theta(T_i)$ is bounded. Let $\varepsilon > 0$ be given, then there exist continuous functions $f_1,\dots,f_n$ from $\mathbb{R}^m$ into $[0,1]$ satisfying

$$\text{(iv)} \qquad \operatorname{diam}\{x \in K \mid f_v(x) > 0\} \leq \varepsilon/c \qquad \forall v = 1,\dots,n$$

$$\text{(v)} \qquad K \cap \{f_v > 0\} \neq \emptyset \qquad \forall v = 1,\dots,n$$

$$\text{(vi)} \qquad \sum_{v=1}^{n} f_v(x) = 1 \qquad \forall x \in K$$

where we use the encledian metric on $\mathbb{R}^m$. By (v) there exist $y_v \in T_i$, such that $f_v(\theta(y_v)) > 0$. Now let $\eta_v = f_v \circ \theta$ and

$$\xi_k = \sum_{v=1}^{n} \varphi_i^k(y_v) \cdot \eta_v$$

Then $\xi_k \in \overline{A}_i$ by Proposition 2.7, and we have

$$|\xi_k(u) - \varphi_i^k(u)| \leq \sum_{v=1}^{n} |\varphi_i^k(y_v) - \varphi_i^k(u)| \eta_v(u)$$

by (vi). If $\eta_v(u) > 0$ then by (iv) we have that

$$\{\sum_{k=1}^{m} |\varphi_i^k(y_v) - \varphi_i^k(u)|^2\}^{\frac{1}{2}} \leq \varepsilon/c$$

and so by (vi) we find

$$|\xi_k(u) - \varphi_i^k(u)| \leq \frac{\varepsilon}{c} \sum_{v=1}^{n} \eta_v(u) = \varepsilon/c$$

Thus $\|\xi_k - \varphi_i^k\| \leq \varepsilon/c$, and so by (iii) we have

$$\sum_{k=1}^{m} \beta(\varphi^k) = \sum_{k=1}^{m} \beta_{ik}(\varphi_i^k)$$

$$\geq \sum_{k=1}^{m} \overline{\beta}_{ik}(\xi_k) - \sum_{k=1}^{m} \overline{\beta}_{ik}(\xi_k - \varphi_i^k)$$

$$\geq \sum_{k=1}^{m} \overline{\beta}_{ik}(\xi_k) - \varepsilon$$

and since $\overline{\beta}_{ik}$ is linear we find

(vii) $$\sum_{k=1}^{m} \beta(\varphi^k) \geq \sum_{k=1}^{m} \sum_{v=1}^{n} a_{vk}^i \beta_{ik}(\psi_v) - 2\varepsilon$$

where $a_{vk}^i = \varphi_i^k(y_v)$, and $\psi_v \in A_i^+$ is chosen so that $\|\psi_v - \eta_v\|$ is sufficiently small.

Now define $R_v : A^{<J>} \to A^{<J>}$ as follows

$$(R_v\varphi)_j = \begin{cases} \varphi_j & \text{if } j \neq i \\ \psi_v \varphi_i & \text{if } j = i \end{cases}$$

and let $\hat{\beta}_v(\varphi) = \beta(R_v\varphi)$. Then $\hat{\beta}_v$ is a positive multilinear functional on $\{(T_j, A_j) \mid j \in J\}$. Now let $\psi^{vk} = (\psi_j^{vk})$ be given by

$$\psi_j^{vk} = \begin{cases} \varphi_j^k & \text{if } j \neq i,r \\ 1 & \text{if } j = i \\ a_{vk}^i \varphi_r^k & \text{if } j = r \end{cases}$$

where r is an arbitrary but fixed element of $\sigma \setminus \{i\}$ (recall that $q \geq 1$ and so $\mathrm{card}(\sigma) = q + 1 \geq 2$). Then by the assumption in (ii) we have that

$$\sum_{k=1}^{m} \bigotimes_{j \in J} \psi_j^{vk} \geq 0 \qquad \forall\, v = 1,\ldots,n$$

and $\sigma(\psi^{v1},\ldots,\psi^{vm}) \subseteq \sigma \setminus \{i\}$. Hence by induction hypothesis we have that

$$0 \leq \sum_{k=1}^{m} \hat{\beta}_v(\psi^{vk}) = \sum_{k=1}^{m} a_{vk}^i \beta_{ik}(\psi_v)$$

for all $v = 1,\ldots,n$. Thus by (vii) we conclude that

$$\sum_{k=1}^{m} \beta(\varphi^k) \geq -2\varepsilon$$

for all $\varepsilon > 0$, and so the induction step is completed and (ii) holds.

By multilinearity of β we see that (ii) also holds if we reverse the inequalities. Hence

$$\lambda(f) = \sum_{k=1}^{m} \beta(\varphi^k) \quad \text{if} \quad f = \sum_{k=1}^{m} \bigotimes_{j \in J} \varphi_j^k \ , \quad \varphi^k \in A^{<J>}$$

is a welldefined map from $\otimes A_j$ into $\mathbb{R}$. By definition of λ we have that λ is additive, and by multilinearity of β we have that $\lambda(af) = a\lambda(f)$ for all $a \in \mathbb{R}$ and all $f \in \otimes A_j$. Thus λ is linear, and by (ii) we have that λ is positive, and so by (7.7.3) we have that $\lambda \in \Pr(\otimes A_j)$.

Finally the uniqueness of λ is evident from the definition of the tensor product $\underset{j\in J}{\otimes} A_j$. □

Corollary 7.9. *Let β be a multi-probability content on $\{(T_j,A_j) \mid j\in J\}$ with marginals $\beta_j \in \Pr(A_j)$ for $j \in J$. Let $\lambda \in \Pr(\otimes_{j\in J} A_j)$ be the probability content satisfying (7.7.8), and let ξ be a given infinite cardinal. Then we have*

(7.9.1) *If β_j has ξ-compact support for all $j \in J$, then so has λ*

(7.9.2) *If J is countable and β_j is ξ-compact for all $j\in J$, then so is λ*

(7.9.3) *If β_j is perfect $\forall j \in J$, then so is λ*

(7.9.4) *If $J = \{1,2\}$ and β_1 is ξ-perfect, and β_2 is ξ-smooth, then λ is ξ-smooth.*

Proof. Trivial consequence of Theorems 7.3 and 7.8. □

Definition 7.10. Let $\{(T_j,A_j) \mid j \in J\}$ be a family of algebraic function spaces and let $\lambda_j \in \Pr(A_j)$ for all $j \in J$. Then

$$\beta((\varphi_j)) = \prod_{j\in J} \lambda_j(\varphi_j) \qquad \forall\ (\varphi_j) \in A^{<J>}$$

is a welldefined multilinear functional, since $\lambda_j(\varphi_j) = 1$ for all but finitely many j's. So by Theorem 7.8 there exist a unique functional $\lambda \in \Pr(\underset{j\in J}{\otimes} A_j)$ satisfying

(7.10.1) $\lambda(\underset{j\in\gamma}{\otimes}\varphi_j) = \prod_{j\in\gamma}\lambda_j(\varphi_j)$ if $\gamma \in 2^{(J)}$ and $\varphi_j \in A_j \quad \forall j \in \gamma$

The functional λ satisfying (7.10.1) is called the product functional of $\{\lambda_j\}$ and is denoted:

$$\lambda = \underset{j\in J}{\otimes}\lambda_j$$

Product functionals are of course intimately related to product measures, and we shall now study product measures and ξ-smooth Fubini extension of product measures. But first we introduce some notation.

Let $(T_0,\mathcal{B}_0,\mu_0)$ and $(T_1,\mathcal{B}_1,\mu_1)$ be two probability measures and let $\mathcal{B}$ be a σ-algebra on $T_0 \times T_1$. Then a probability measure μ on $(T_0 \times T_1,\mathcal{B})$ is called a complete Fubini product of μ_0 and μ_1 on $\mathcal{B}$ if

(7.10.2) $\mathcal{B} \supseteq \mathcal{B}_1 \otimes \mathcal{B}_2$

(7.10.3) $f(\cdot,t_1) \in B(T_0,\mathcal{B}_0) \quad \forall t_1 \in T_1$

(7.10.4) $f(t_0,\cdot) \in B(T_1,\mathcal{B}_1) \quad \forall t_0 \in T_0$

(7.10.5) $\int f(\cdot,t_1)\mu_1(dt_1) \in B(T_0,\mathcal{B}_0)$

(7.10.6) $\int f(t_0,\cdot)\mu_0(dt_0) \in B(T_1,\mathcal{B}_1)$

(7.10.7) $\int f d\mu = \int d\mu_0 \int f d\mu_1 = \int d\mu_1 \int f d\mu_0$

whenever $f \in B(T_0 \times T_1,\mathcal{B})$. Note (7.10.3) and (7.10.6) simply state that μ_0 is $(\mathcal{B}_1,\mathcal{B})$-admissible (see Definition 4.6), and similarly (7.10.4) and (7.10.5) states that μ_1 is $(\mathcal{B}_0,\mathcal{B})$-admissible, and (7.10.7) then states that:

$$\mu = \mu_0 \vec{\otimes} \mu_1 = \mu_0 \overleftarrow{\otimes} \mu_1 \quad \text{on } \mathcal{B}$$

Now let $\{(T_j, \mathcal{B}_j, \mu_j) \mid j \in J\}$ be a family of probability spaces, and let $T^J = \Pi_{j \in J} T_j$ be the product space. If $\mathcal{B}$ is a σ-algebra on T^J and μ is a probability measure on $(T^J, \mathcal{B})$, then <u>the marginal of</u> μ <u>on</u> T^L is the probability measure μ_L given by

$$\mathcal{B}_L = \{B \subseteq T^L \mid p_L^{-1}(B) \in \mathcal{B}\} \ , \qquad T^L = \prod_{j \in L} T_j$$

$$\mu_L(B) = \mu(p_L^{-1}(B)) \qquad \forall B \in \mathcal{B}_L$$

whenever $L \subseteq J$. We say that μ is a <u>complete Fubini product of</u> $\{\mu_j\}$ <u>on</u> $\mathcal{B}$, if $\mathcal{B}_{\{j\}} = \mathcal{B}$, $\mu_{\{j\}} = \mu_j$ for all $j \in J$, and $\mu_{L \cup M}$ is a complete Fubini product of μ_L and μ_M on $\mathcal{B}_{L \cup M}$, whenever L and M are two disjoint non-empty subsets of J.

Of course the usual product measure on the product σ-algebra is a Fubini product, but we shall search for Fubini product on essentially larger σ-algebras. First we prove a version of Tulcea's theorem:

<u>Theorem 7.11</u>. <u>Let</u> $(T_n, \mathcal{B}_n)$ <u>be a measurable space and let</u> $\mathcal{B}^n$ <u>be an algebra on</u> $T^n = T_1 \times \ldots \times T_n$, <u>such that</u> $\mathcal{B}^n \otimes \mathcal{B}_{n+1} \subseteq \mathcal{B}^{n+1}$ <u>for all</u> $n \geq 1$ <u>and</u> $\mathcal{B}^1 = \mathcal{B}_1$. <u>Let</u>

$$T^{\mathbb{N}} = \prod_{j=1}^{\infty} T_j \ , \quad q_n = \text{the projection: } T^{\mathbb{N}} \to T^n$$

$$\mathcal{B} = \sigma\Big(\bigcup_{n=1}^{\infty} q_n^{-1}(\mathcal{B}^n)\Big)$$

<u>Suppose that</u> ρ_n <u>is a</u> $(\mathcal{B}^n, \mathcal{B}^{n+1})$<u>-admissible Markov kernel on</u> $(T_{n+1}, \mathcal{B}_{n+1}) \big| T^n$ <u>for all</u> $n \geq 1$, <u>and let</u> μ_1 <u>be a probability measure on</u> $(T_1, \mathcal{B}_1)$, <u>then there exist a unique probability measure</u> μ <u>on</u> $(T^{\mathbb{N}}, \mathcal{B})$ <u>satisfying</u>

(7.11.1) $\mathcal{B}^n = \{B \subseteq T^n \mid q_n^{-1}(B) \in \mathcal{B}\}$

(7.11.2) $\mu^1 = \mu_1$

(7.11.3) $\int_{T^{n+1}} f d\mu^{n+1} = \int_{T^n} \mu^n(du) \int_{T_{n+1}} f(u,v)\rho_n(dv|u)$

<u>for all</u> $n \geq 1$ <u>and all</u> $f \in B(T^{n+1}, \mathcal{B}^{n+1})$, <u>where</u> μ^n <u>is the marginal of</u> μ <u>on</u> T^n.

<u>Proof</u> (7.11.1): Let $\mathcal{B}_0^n$ denote the paving on the right hand side of (7.11.1), then evidently we have that $\mathcal{B}_0^n \supseteq \mathcal{B}^n$. Now let $n \geq 1$ and $v = (v_j \mid j \geq n+1)$ be given, such that $v_j \in T_j$ for all $j \geq n+1$, and let

$$h(u) = (u_1, \ldots, u_n, v_{n+1}, v_{n+2}, \ldots)$$

for all $u = (u_1, \ldots, u_n) \in T^n$. Then h maps T^n into $T^{\mathbb{N}}$. Since $\mathcal{B}^n \supseteq \mathcal{B}^k \otimes \mathcal{B}_{k+1} \otimes \ldots \otimes \mathcal{B}_n$ for $k \leq n$ then

$$u \rightsquigarrow (u_1, \ldots, u_k)$$

is measurable: $(T^n, \mathcal{B}^n) \to (T^k, \mathcal{B}^k)$. And since ρ_j is $(\mathcal{B}^j, \mathcal{B}^{j+1})$-admissible for all j we have that

$$u \rightsquigarrow (u_1, \ldots, u_k, v_{n+1}, \ldots, v_k)$$

is measurable: $(T^n, \mathcal{B}^n) \to (T^k, \mathcal{B}^k)$ for $k > n$. Hence $q_k \circ h$ is measurable from $(T^n, \mathcal{B}^n)$ into $(T^k, \mathcal{B}^k)$ for all $k \geq 1$, and so h is measurable from $(T^n, \mathcal{B}^n)$ into $(T^{\mathbb{N}}, \mathcal{B})$. Hence if $B \subseteq \mathcal{B}_0^n$, then

$$B = h^{-1}(q_n^{-1}(B)) \in \mathcal{B}^n$$

and so $\mathcal{B}_0^n \subseteq \mathcal{B}^n$. Thus (7.11.1) holds.

Let us define μ_n on $(T^n, \mathcal{B}^n)$ by induction as follows: $\mu_1 = \mu_1$ and $\mu_{n+1} = \mu_n \vec{\otimes} \rho_n$ on $\mathcal{B}^{n+1}$ (see Definition 4.6), and consider the marginal system:

$$\mathcal{E} = \{T^{\mathbb{N}} \xrightarrow{q_n} (T^n, A^n, \lambda^n) \mid n \in \mathbb{N}\}$$

where $A^n = B(T^n, \mathcal{B}^n)$ and $\lambda^n(f) = \int f d\mu_n$. Then it is easily checked that $\mathcal{E}$ is a projective system, and since q_n is surjective we have that $\mathcal{E}$ is of type $(\mathbb{N}, \sigma)$. Now put

$$\sigma_n(u) = \{u\} \times T_{n+1}$$

$$B(u) = \{v \in T_{n+1} \mid (u,v) \in B\}$$

$$\hat{\rho}_n(B \mid u) = \rho_n(B(u) \mid u)$$

for $B \in \mathcal{B}^{n+1}$ and $u \in T^n$. Then (σ_n) is an atomic maximal sequence for $\mathcal{E}$ (see Definition 6.15) and $(\hat{\rho}_n)$ is a σ-disintegration of $\mathcal{E}$. Moreover by definition of σ_n and $\hat{\rho}_n$ we have that

$$\hat{\rho}_n^*(\sigma_n(u) \mid u) = 1 \quad \forall u \in T^n \quad \forall n \geq 1$$

And so by Theorem 6.16 we have that $\mathcal{E}$ is fully σ-consistent, thus by Theorem 4.1 we conclude that there exist a probability measure μ on $(T^{\mathbb{N}}, \mathcal{B}^{\mathbb{N}})$ satisfying (7.11.2) and (7.11.3).

Let ν be any measure satisfying (7.11.2) and (7.11.3) and let ν^n be the marginal of ν on T^n. Then $\nu^1 = \mu_1$ by (7.11.2) and by induction using (7.11.3) we have that $\nu^n = \mu_n$ for all $n \geq 1$, where μ_n is defined as above. Hence ν and μ coincides on the algebra $\cup_1^\infty q_n^{-1}(\mathcal{B}^n)$ and thus $\nu = \mu$. Hence the uniqueness is proved.

Corollary 7.12. Let F_n be a $(\cup f, \cap\xi)$-stable paving on the set T_n for all $n \geq 1$, where ξ is an infinite cardinal, and put

$$F^n = \text{the } (\cup f, \cap\xi)\text{-closure of } \{\prod_{j=1}^{n} F_j \mid F_j \in F_j \quad \forall 1 \leq j \leq n\}$$

$$B_n = \sigma(F_n) \quad \text{and} \quad B^n = \sigma(F^n)$$

Now let ρ_n be a Markov kernel on $(T_{n+1}, B_{n+1}) \mid T^n$ and let μ_1 be a probability measure on (T_1, B_1) satisfying

(7.12.1) $\rho_n(\cdot \mid u)$ is ξ-supersmooth at F_{n+1} along F_{n+1} $\forall u \in T^n$

(7.12.2) $\rho_n(F \mid \cdot) \in U(T^n, F^n) \quad \forall F \in F_{n+1}$

(7.12.3) μ_1 is ξ-supersmooth at F_1 along F_1.

for all $n \geq 1$. Then $B^n \otimes B_{n+1} \subseteq B^{n+1}$ and ρ_n is (B^n, B^{n+1})-admissible for all $n \geq 1$. Moreover if μ is the probability measure on $(T^{\mathbb{N}}, B)$ satisfying (7.11.2) and (7.11.3) and if

$$F = \text{the } (\cup, \cap\xi)\text{-closure of } \{\prod_{j=1}^{\infty} F_j \mid F_j \in F_j \quad \forall j \in \mathbb{N}\}$$

Then we have

(7.12.4) $F = (\bigcup_{n=1}^{\infty} q_n^{-1}(F^n))_\xi = (\bigcup_{n=1}^{\infty} q_n^{-1}(F^n))_\delta$

(7.12.5) $B = \sigma(F)$

(7.12.6) μ is ξ-supersmooth at F along F.

Proof. Evidently we have that $B^n \otimes B_{n+1} \subseteq B^{n+1}$, and by Theorem 4.7 we have that ρ_n is (B^n, B^{n+1})-admissible for all $n \geq 1$. Moreover if we put

$$\mu_{n+1} = \mu_n \vec{\otimes} \rho_n \quad \text{on} \quad \mathcal{B}^{n+1}$$

Then by induction using Theorem 4.7 we find that μ_n is ξ-supersmooth at $\mathcal{F}^n$ along $\mathcal{F}^n$. Now put

$$\hat{\mathcal{F}}^n = q_n^{-1}(\mathcal{F}^n)$$

and let μ be the measure from Theorem 7.11. Since q_n is surjective we see that μ is ξ-supersmooth at $\hat{\mathcal{F}}^n$ along $\hat{\mathcal{F}}^n$ for all $n \geq 1$, and evidently we have that $\{\hat{\mathcal{F}}^n\}$ is increasing. Let $\hat{\mathcal{F}} = \cup_1^\infty \mathcal{F}^n$, then by Lemma 6.9. (3) we have that $\hat{\mathcal{F}}_\xi = \hat{\mathcal{F}}_\delta \subseteq \mathcal{B}$, and μ is ξ-supersmooth at $\hat{\mathcal{F}}_\xi$ along $\hat{\mathcal{F}}_\xi$.

Since $\hat{\mathcal{F}}$ is $(\cap f, \cup f)$-stable, we have that $\hat{\mathcal{F}}_\xi$ is $(\cap c, \cap\xi, \cup f)$-stable, and so $\hat{\mathcal{F}}_\xi \supseteq \mathcal{F}$, but the converse implication is evident and so (7.12.4) and thus (7.12.5) and (7.12.6) holds. □

<u>Lemma 7.13</u>. <u>Let</u> μ <u>and</u> ν <u>be two probability measures on</u> $(T, \mathcal{B})$ <u>and let</u> $\{\mathcal{F}_q \mid q \in Q\}$ <u>be an increasing net of</u> $(\cap f)$-<u>stable pavings on</u> T. <u>Suppose that</u> ξ <u>is an infinite cardinal such that</u>

(7.13.1) $\hat{\mathcal{F}}_q \subseteq \mathcal{B}$ <u>where</u> $\hat{\mathcal{F}}_q$ = <u>the</u> $(\cup f, \cap \xi)$-<u>closure of</u> $\mathcal{F}_q$

(7.13.2) μ <u>and</u> ν <u>are</u> ξ-<u>supersmooth at</u> $\hat{\mathcal{F}}_q$ <u>along</u> $\hat{\mathcal{F}}_q$

(7.13.3) $\mu(F) = \nu(F) \quad \forall F \in \mathcal{F}_q$

<u>for all</u> $q \in Q$. <u>Let</u> $\mathcal{B}_0 = \sigma(\bigcup_{q \in Q} \hat{\mathcal{F}}_q)$, <u>then we have</u>

(7.13.4) $\mu(B) = \nu(B) \quad \forall B \in \mathcal{B}_0$

Proof. By (7.13.3) and [2, Theorem I.21] we have that μ and ν coincides on $\sigma(F_q)$ for all $q \in Q$. In particular if F^*_q is the $(\cup f, \cap f)$-closure of F_q then μ and ν coincides on F^*_q. Now every set in $\hat{F}_q$ is the intersection of a decreasing net in $\hat{F}^*_q$ of cardinality $\leq \xi$. Hence by (7.13.2) we have that μ and ν coincides on $\hat{F}_q$ for all $q \in Q$. Now let

$$\hat{F} = \bigcup_{q \in Q} \hat{F}_q$$

Since $\{F_q\}$ and thus $\{\hat{F}_q\}$ is filtering upwards and since $\hat{F}_q$ is $(\cap f)$-stable, we have that $\hat{F}$ is $(\cap f)$-stable. And we have just shown that μ and ν coincides on $\hat{F}$, hence by [2, Theorem I.21] we have that μ and ν coincides on $\sigma(\hat{F}) = B_0$, and thus (7.13.4) holds. □

Lemma 7.14. Let (T_0, B_0, μ_0) and (T_1, B_1, μ_1) be two probability spaces, let $B \supseteq B_0 \otimes B_1$ be a σ-algebra on $T_0 \times T_1$, and let μ be a probability measure on $(T_0 \times T_1, B)$. Suppose that $B = \sigma(F)$ where F is a $(\cap f)$-stable paving on $T_0 \times T_1$ satisfying:

For all $F \in F$ there exist an infinite cardinal ξ and $(\cap f)$-stable pavings F_0 and F_1 such that

(7.14.1) $F_j \subseteq B_j$ for $j = 0,1$

(7.14.2) $F \in \sigma(\hat{F}) \subseteq B$

(7.14.3) $\mu(F_0 \times F_1) = \mu_0(F_0)\mu_1(F_1) \quad \forall F_0 \in F_0 \quad \forall F_1 \in F_1$

(7.14.4) μ_j is ξ-supersmooth at $\hat{F}_j$ along $\hat{F}_j$ for $j = 0,1$

(7.14.5) μ is ξ-supersmooth at $\hat{F}$ along $\hat{F}$

where $\hat{F}_j$ is the $(\cup f, \cap \xi)$-closure of $F_j \cup \{\emptyset, T_j\}$ and $\hat{F}$ is the $(\cup f, \cap \xi)$-closure of $\{F_0 \times F_1 | F_0 \in F_0, F_1 \in F_1\}$.

<u>Then</u> μ <u>is a complete Fubini product of</u> μ_0 <u>and</u> μ_1 <u>on</u> $(T_0 \times T_1, \mathcal{B})$.

<u>Proof</u>. (7.10.2) holds by assumption. Now let V be the set of all $f \in B(T_0 \times T_1, \mathcal{B})$ which satisfies (7.10.3)-(7.10.7). Then V is a vector space of bounded functions containing all constant functions, and evidently we have that V is closed under sequential, pointwise, dominated convergence, and so by [2; Theorem I.21] it suffices to show that

(i) $\qquad 1_F \in V \qquad \forall F \in \mathcal{F}$

since $\mathcal{F}$ is $(\cap f)$-stable by assumption.

So let $F \in \mathcal{F}$ and choose $\xi, \mathcal{F}_0$ and $\mathcal{F}_1$ according to (7.14.1)-(7.14.5). Let $\hat{\mathcal{B}}_j = \sigma(\hat{\mathcal{F}}_j)$ for $j = 0,1$ and let $\hat{\mu}_j$ be the restriction of μ_j to $\hat{\mathcal{B}}_j$. Then by (7.14.1), (7.14.4) and Theorem 4.7 we have that $\hat{\mu}_0$ is $(\hat{\mathcal{B}}_1, \hat{\mathcal{B}})$-admissible and $\hat{\mu}_1$ is $(\hat{\mathcal{B}}_0, \hat{\mathcal{B}})$-admissible, where $\hat{\mathcal{B}} = \sigma(\hat{\mathcal{F}})$, and if

$$\nu_1(B) = \int d\hat{\mu}_0 \int 1_B d\hat{\mu}_1 \qquad \forall B \in \hat{\mathcal{B}}$$

$$\nu_2(B) = \int d\hat{\mu}_1 \int 1_B d\hat{\mu}_0 \qquad \forall B \in \hat{\mathcal{B}}$$

then ν_1 and ν_2 are probability measures on $\hat{\mathcal{B}}$, and they are ξ-supersmooth at $\hat{\mathcal{F}}$ along $\hat{\mathcal{F}}$. Moreover we have

$$\nu_1(F_0 \times F_1) = \nu_2(F_0 \times F_1) = \mu(F_0 \times F_1) = \mu_0(F_0)\mu_1(F_1)$$

for all $F_0 \in \mathcal{F}_0$ and all $F_1 \in \mathcal{F}_1$, and so $\mu = \nu_1 = \nu_2$ on $\hat{\mathcal{B}}$ by Lemma 7.13. Now since $F \in \hat{\mathcal{B}}$, we see that $1_F \in V$ and so the lemma is proved. □

Lemma 7.15. Let $(T,\mathcal{B},\mu)$ be a probability space, and let q_j be a measurable map from $(T,\mathcal{B})$ into $(T_j,\mathcal{B}_j)$ for all $j \in J$. Let $\mu_j = q_j\mu$ and let $\mathcal{F}_j \subseteq \mathcal{B}_j$ be an inner proximating paving for μ_j on $\mathcal{B}_j$ $\forall j \in J$. Put

$$\mathcal{F}_0 = \text{the } (\cup f,\cap c)\text{-closure of } \bigcup_{j\in J} q_j^{-1}(\mathcal{F}_j)$$

$$\mathcal{B}_0 = \sigma(\bigcup_{j\in J} q_j^{-1}(\mathcal{B}_j))$$

Then $\mathcal{F}_0$ is an inner approximating paving for μ on $\mathcal{B}_0$.

Now let ξ be an infinite cardinal and let Γ be a paving J, such that $\mathcal{F}_\gamma \subseteq \mathcal{B}$ and μ is ξ-supersmooth at $\mathcal{F}_\gamma$ along $\mathcal{F}_\gamma$ for all $\gamma \in \Gamma$, where

$$\mathcal{F}_\gamma = \text{the } (\cup f,\cap \xi)\text{-closure of } \bigcup_{j\in \gamma} q_j^{-1}(\mathcal{F}_j)$$

$$\mathcal{F}_\Gamma = \text{the } (\cup f,\cap c)\text{-closure of } \bigcup_{\gamma\in \Gamma} \mathcal{F}_\gamma$$

$$\mathcal{B}_\Gamma = \sigma\{\mathcal{F}_\Gamma \cup \bigcup_{\gamma\in\Gamma} \bigcup_{j\in\gamma} q_j^{-1}(\mathcal{B}_j)\}$$

Then $\mathcal{F}_\Gamma$ is an inner approximating paving for μ on $\mathcal{B}_\Gamma$.

Remark. Recall that $\mathcal{F}$ is an inner approximating paving for μ on $\mathcal{G}$, if $\mathcal{G} \subseteq \mathcal{M}(\mu)$ and

$$\mu(G) = \sup\{\mu_*(F) \mid F \in \mathcal{F}, F \subseteq G\} \qquad \forall G \in \mathcal{G}$$

Proof. Let $\mathcal{R}$ be the set of all $B \in \mathcal{B}$, satisfying

(i) $$\mu(B) = \sup\{\mu(F) \mid F \subseteq B, F \in \mathcal{F}_0\}$$

Since F_0 is $(\cup f,\cap c)$ it follows easily that R is $(\cup c,\cap c)$-stable and $q_j^{-1}(B_j) \subseteq R$ for all $j \in J$, since F_j is an inner approximating paving for μ_j on B_j. In particular we have that $\emptyset$ and T belongs to R. Hence

$$R_0 = \{B \mid B \in R \text{ and } T \setminus B \in R\}$$

is a σ-algebra and $R_0 \supseteq q_j^{-1}(B_j)$ for all $j \in J$. Thus we have that $B_0 \subseteq R_0 \subseteq R$ and so F_0 is an inner approximating paving for μ on B_0.

Now define R and R_0 as above but with F_0 replaced by F_Γ. Then as above we have that R is $(\cup c,\cap c)$-stable and R_0 is a σ-algebra such that $B_1 \subseteq R_0$ where

$$B_1 = \sigma\Big(\bigcup_{\gamma\in\Gamma}\ \bigcup_{j\in\gamma} q_j^{-1}(B_j)\Big)$$

Now let F_γ^0 be the $(\cup f,\cap c)$-closure of $\cup_{j\in\gamma} q_j^{-1}(F_j)$. Then $F_\gamma^0 \subseteq B_1 \subseteq R_0$ for all $\gamma \in \Gamma$.

Now let $\gamma \in \Gamma$ and let $F \in F_\gamma$. Since μ is ξ-supersmooth at F along F_γ, then for every $\varepsilon > 0$ there exist $F_0 \in F_\gamma^0$, such that

$$F_0 \supseteq F \quad \text{and} \quad \mu(F_0 \setminus F) < \varepsilon$$

And since $T \setminus F_0 \in R$ we see that $T \setminus F \in R$. Moreover $F \in F_\gamma \subseteq F_\Gamma \subseteq R$, and so $F \in R_0$. Thus $F_\gamma \subseteq R_0$ for all $\gamma \in \Gamma$, and since R_0 is a σ-algebra we see that $B_\Gamma \subseteq R_0 \subseteq R$, and so F_Γ is an inner approximating paving for μ on B_Γ. □

Theorem 7.16. Let (T_j, B_j, μ_j) be a probability space for all $j \in J$, and let μ be a probability measure on the product space (T^J, B^J) where

$$T^J = \prod_{j\in J} T_j \quad \text{and} \quad B^J = \bigotimes_{j\in J} B_j$$

with marginals $\{\mu_j \mid j \in J\}$. Let ξ be an infinite cardinal, such that μ_j is a ξ-compact measure for all $j \in J$. Then μ is a ξ-compact measure.

Proof. By (D.9) and (C.1) there exist a ξ-compact paving $\mathcal{F}_j \subseteq \mathcal{B}_j$, such that $\mathcal{F}_j$ is an inner approcimating paving for μ_j for all $j \in J$, and such that $\mathcal{F}_j$ is $(\cup f, \cap c)$-stable and contains $\emptyset$ and T_j for all $j \in J$. Now let

$$\mathcal{H} = \{ \prod_{j \in J} F_j \mid F_j \in \mathcal{F}_j \}$$

Then it is easily checked that $\mathcal{H}$ is a ξ-compact paving on T^J. Hence by (C.1) we have that

$$\mathcal{F} = \text{the } (\cup f, \cap \xi)\text{-closure of } \mathcal{H}$$

is a ξ-compact paving.

Now let q_j be the projection of T^J onto T_j, then q_j is measurable from $(T^J, \mathcal{B}^J)$ into $(T_j, \mathcal{B}_j)$, and with the notation of Lemma 7.15 we have that $\mathcal{F}_0 \subseteq \mathcal{F}$ and $\mathcal{B}_0 = \mathcal{B}^J$. Thus $\mathcal{F}_0$ is an inner approximating paving for μ on $\mathcal{B}^J$, and $\mathcal{F}_0$ is a ξ-compact paving since $\mathcal{F}_0 \subseteq \mathcal{F}$. Hence μ is a ξ-compact measure. □

Theorem 7.17. Let $\{(T_j, \mathcal{B}_j, \mu_j) \mid j \in J\}$ be a family of probability measures, let ξ be an infinite cardinal, and let $\mathcal{H}_j$ be a $(\cap f)$-stable paving, such that μ_j is ξ-supersmooth at $\mathcal{F}_j$ along $\mathcal{F}_j$ and $\mathcal{B}_j = \sigma(\mathcal{F}_j)$ for all $j \in J$, where

$$\mathcal{F}_j = \text{the } (\cup f, \cap \xi)\text{-closure of } \mathcal{H}_j \cup \{\emptyset, T_j\}$$

Now let $T^J = \prod_{j\in J} T_j$, and put

$$\mathcal{H}_\gamma = \{ \prod_{j\in J} F_j \mid F_j \in \mathcal{F}_j \ \ \forall j \in J,\ F_j = T_j \ \ \forall j \in J \setminus \gamma\}$$

$$\mathcal{F}_\gamma = \text{the } (\cup f, \cap \xi)\text{-closure of } \mathcal{H}_\gamma$$

$$\mathcal{F}_J^0 = \cup\{\mathcal{F}_\gamma \mid \gamma \text{ countable} \subseteq J\}$$

$$\mathcal{B}_J^0 = \sigma(\mathcal{F}_j^0)$$

Then there exist a unique measure μ on $(T^J, \mathcal{B}_J^0)$ satisfying

(7.17.1) μ is ξ-supersmooth at $\mathcal{F}_\gamma$ along $\mathcal{F}_\gamma$ $\forall \gamma$ countable $\subseteq J$

(7.17.2) $\mu(\prod_{j\in J} H_j) = \prod_{j\in\sigma} \mu_j(H_j)$

whenever $\sigma \subseteq J$ is finite, $H_j \in \mathcal{H}_j$ $\forall j \in \sigma$ and $H_j = T_j$ for all $j \in J \setminus \sigma$. Moreover the measure μ satisfies

(7.17.3) μ is a complete Fubini product of $\{\mu_j\}$ on $\mathcal{B}_J^0$

(7.17.4) If $\mathcal{F}_j$ is an inner approximating paving for μ_j on $\mathcal{B}_j$, then $\mathcal{F}_J^0$ is an inner approximating paving for μ on $\mathcal{B}_J^0$.

Remark. If $\xi = \aleph_0$, then $\mathcal{B}_J^0 = \otimes_{j\in J}\mathcal{B}_j$, and μ is just the usual product measure of $\{\mu_j\}$ on $\otimes \mathcal{B}_j$.

Proof. If μ and ν are two measures on $(T^J, \mathcal{B}_J^0)$ satisfying (7.17.1+2), then evidently μ and ν coincides on $\mathcal{H}_\gamma$ for all countable sets $\gamma \subseteq J$. Hence $\mu = \nu$ on $\mathcal{B}_J^0$ by Lemma 7.13, and so the uniqueness is proved.

So let us prove the existence of μ. If J is finite or countable, then the existence follows easily from Corollary 7.12. So suppose that J is uncountable. Let L be a countable subset of J, and put

$$F^L = \text{the } (\cup f, \cap \xi)\text{-closure of } \{\prod_{j\in L} F_j \mid F_j \in F_j \quad \forall j \in L\}$$

Now let $B^L = \sigma(F^L)$, and let μ^L be the unique measure on (T^L, B^L) satisfying (7.17.1+2) with J replaced by L. Then

$$B_J^0 = \cup\{p_L^{-1}(B^L) \mid L \text{ countable} \subseteq J\}$$

and it is easily checked that

$$\mu(B) = \mu^L(B_0) \quad \text{if} \quad B = p_L^{-1}(B_0)$$

is a welldefined probability measure on (T^J, B_J^0) satisfying (7.17.1+2). Thus the existence and uniqueness of μ is established.

Now let μ be the unique measure on (T^J, B_J^0) satisfying (7.17.1+2), then (7.17.4) follows easily from Lemma 7.15, so let us prove that μ is a complete Fubini product of $\{\mu_j\}$ on B_J^0. Hence let

$$B_L = \{B \subseteq T^L \mid p_L^{-1}(B) \in B_J^0\}$$

and let μ_L be the marginal of μ on (T^L, B_L), for all $L \subseteq J$.

Now define $H_{\gamma L}$, $F_{\gamma L}$, F_L^0 and B_L^0 as in theorem but with J replaced by L. Then it is easily checked that p_L is measurable from (T^J, B_J^0) into (T^L, B_L^0) and that $u \curvearrowright (u,v)$ is measurable from (T^L, B_L^0) into (T^J, B_J^0) whenever $v \in T^{J\setminus L}$ is fixed. In particular we have that

(i) $$B_L^0 = B_L = \{B \subseteq T^L \mid p_{LN}^{-1}(B) \in B_N^0\} \quad \forall L \subseteq N \subseteq J$$

Now let L and M be two disjoint non-empty subsets of J and put $N = L \cup M$.

We shall then use Lemma 7.14 to show that μ_N is a complete Fubini product of μ_L and μ_M on $\mathcal{B}_N^0$, where we identify T^N with $T^L \times T^M$ in the natural way. By (i) we have that $\mathcal{B}_N^0 \supseteq \mathcal{B}_L^0 \otimes \mathcal{B}_M^0$, and by definition we have that $\mathcal{B}_N^0 = \sigma(\mathcal{F}_N^0)$. Now let $F \in \mathcal{F}_N^0$, then there exist a countable set $\gamma \subseteq N$, such that $F \in \mathcal{F}_{\gamma N}$. And if $\alpha = \gamma \cap L$ and $\beta = \gamma \cap M$ then it is easily verified that $\mathcal{F}_0 = \mathcal{H}_{\alpha L}$ and $\mathcal{F}_1 = \mathcal{H}_{\beta M}$ satisfies (7.14.1-5). Thus μ_N is a complete Fubini Product of μ_L and μ_M on $\mathcal{B}_N^0 = \mathcal{B}_N$, and so (7.17.3) is proved. And (7.13.4) follows easily from Lemma 7.15. □

Theorem 7.18. Consider the setting describe in Theorem 7.17 and let us put

$$\hat{\mathcal{F}}_J = (\mathcal{F}_J^0)_\xi \ , \qquad \hat{\mathcal{B}}_J = \sigma(\hat{\mathcal{F}}_J)$$

$$S_j = \{u \in T_j \mid u \in F \quad \forall F \in \mathcal{F}_j \ \text{ with } \ \mu_j(F) = 1\}$$

Suppose that $\mathcal{F}_j$ is an inner approximating paving for μ_j on $\mathcal{B}_j$ for all $j \in J$. and that $S_j \neq \emptyset$ for all $j \in J \setminus \alpha$ for some countable set $\alpha \subseteq J$.

Then μ admits a unique σ-additive extension $\hat{\mu}$ to $\hat{\mathcal{B}}_J$, such that $\hat{\mu}$ is ξ-supersmooth at $\hat{\mathcal{F}}_J$ along $\hat{\mathcal{F}}_J$. Moreover the measure $\hat{\mu}$ satisfies

(7.18.1) $\hat{\mu}$ is a complete Fubini product of $\{\mu_j\}$ on $\hat{\mathcal{B}}_J$

(7.18.2) $\hat{\mathcal{F}}_J$ is an inner approximating paving for $\hat{\mu}$ on $\hat{\mathcal{B}}_J$

<u>Proof</u>. We shall apply Theorem 6.10 with $T = T^J$, $B = B_J^0$, $\mu = \mu$, J = all countable subsets of J, and $F_\gamma = F_\gamma$. Then (6.10.1) and (6.10.2) holds by (7.17.1) and $F_\gamma \uparrow F_J^0$.

Now note that F_J^0 is $(\cup f, \cap c)$-stable. So let $D \in F_J^0$ be given. Then there exist a countable set $\gamma \subseteq J$, such that $D \in F_\gamma$ and $S_j \neq \emptyset$ for all $j \in J \setminus \gamma$. So let $a_j \in S_j$ for all $j \in J \setminus \gamma$. If $F \in F_J^0$ then we put $a = (a_j)_{j \in J \setminus \gamma} \in T^{J \setminus \gamma}$ and

$$F_\gamma(a) = \{u \in T^\gamma \mid (u,a) \in F\}$$

$$\rho(F) = F_\gamma(a) \times T^{J \setminus \gamma} = p_\gamma^{-1}(F_\gamma(a))$$

$$T_0 = T^\gamma \times \{a\} = p_{J \setminus \gamma}^{-1}(a)$$

Then ρ is an increasing map from F_J^0 into F_γ satisfying (6.10.3). Let $H \in F_\gamma$ such that $H \cap T_0 = \emptyset$, since $H = p_\gamma^{-1}(H_0)$ for some $H_0 \in T^\gamma$, we see that

$$\emptyset = T_0 \cap H = (T^\gamma \times \{a\}) \cap (H_0 \times T^{J \setminus \gamma}) = H_0 \times \{a\}$$

and so $H_0 = \emptyset$. Thus $H = \emptyset$ and $\mu(H) = 0$, i.e. (6.10.4) holds. Now let $F \in F_J^0$, so that $\mu(D \setminus F) = 0$. Since $D \in F_\gamma$ we have that

$$D = p_\gamma^{-1}(D_0) = D_0 \times T^{J \setminus \gamma}$$

for some $D_0 \in B_\gamma^0$, and so by (7.17.3) we find

(i) $$0 = \mu(D \setminus F) = \int_{D_0} (1 - \mu_{J \setminus \gamma}(F(u)))\, \mu_\gamma(du)$$

where

$$F(u) = \{v \in T^{J \setminus \gamma} \mid (u,v) \in F\} \qquad \forall u \in T^\gamma$$

Now I claim that we have

(ii) $\qquad \mu_{J\setminus\gamma}(F(u)) < 1 \qquad \forall\, u \in T^{\gamma}\setminus F_{\gamma}(a)$

So let $u \in T^{\gamma}\setminus F_{\gamma}(a)$, then $(u,a) \notin F$. If $F(u) = \emptyset$ Then (ii) is clear, so let us assume that $F(u) \neq \emptyset$. By definition of $\mathcal{F}_J^0$, there exist an integer $m \geq 1$ and $F_{kj} \in \mathcal{F}_j$ for $j \in J$ and $1 \leq k \leq m$, auch that

(iii) $\qquad F \subseteq \bigcup_{k=1}^{m} \prod_{j\in J} F_{kj} = H$, and $(u,a) \notin H$

Let $\sigma = \{k \mid 1 \leq k \leq m,\ u_j \in F_{kj}\ \forall\, j \in \gamma\}$, since $F(u) \neq \emptyset$ and $(u,a) \notin H$, then $\sigma \neq \emptyset$ and

(iv) $\qquad F(u) \subseteq \bigcup_{k\in\sigma} \prod_{j\in J\setminus\gamma} F_{kj}, \qquad a \notin \bigcup_{k\in\sigma} \prod_{j\in J\setminus\gamma} F_{kj}$

Since σ is finite there exist a finite set $\alpha \subseteq J\setminus\gamma$, such that

(v) $\qquad \forall\, k \in \sigma\ \exists\, j \in \alpha$ so that $a_j \notin F_{kj}$

Now let $\sigma(j) = \{k \in \sigma \mid a_j \notin F_{kj}\}$ for $j \in J\setminus\gamma$, and let

$$F_j = \begin{cases} \bigcup_{k\in\sigma(j)} F_{kj} & \text{if } j \in J\setminus\gamma \text{ and } \sigma(j) \neq \emptyset \\ \emptyset & \text{if } j \in J\setminus\gamma \text{ and } \sigma(j) = \emptyset \end{cases}$$

Then $F_j \in \mathcal{F}_j$ and $a_j \notin F_j$ for all $j \in J\setminus\gamma$. So if $G_j = T_j\setminus F_j$ then by definition of S_j (recall that $a_j \in S_j\setminus F_j$) we have

(vi) $\qquad \mu_j(G_j) > 0 \qquad \forall\, j \in J\setminus\gamma$

Let $v \in F(u)$, then by (iv) and (v) there exist $k \in \sigma$ and $j \in \alpha$ so that $v_j \in F_{kj}$ and $a_j \notin F_{kj}$. Hence $k \in \sigma(j)$ and so $v_j \in F_j$.

I.e. we have

$$F(u) \subseteq \bigcup_{j\in\alpha} p_j^{-1}(F_j) = T^{J\setminus(\gamma\cup\alpha)} \times (T^\alpha \setminus \prod_{j\in\alpha} G_j)$$

and so by (vi) we find

$$\mu_{J\setminus\gamma}(F(u)) \leq \mu_\alpha(T^\alpha \setminus \prod_{j\in\alpha} G_j)$$

$$= 1 - \prod_{j\in\alpha} \mu_j(G_j)$$

$$< 1$$

since α is finite and μ_α coincides with $\otimes_{j\in\alpha}\mu_j$ on $\otimes_{j\in\alpha}\mathcal{B}_j$. Thus (ii) is proved.

By (i) and (ii) we see that $N_0 = D_0 \setminus F_\gamma(a)$ is a μ_γ-nullset, and so

$$\mu(D) = \mu_\gamma(D_0) \leq \mu_\gamma(F_\gamma(a)) = \mu(\rho(F))$$

Thus (6.10.5) holds, and so by Theorem 6.10 we see μ admits an extension $\hat{\mu}$ to $\hat{\mathcal{B}}_J$ satisfying (7.18.2) and such that $\hat{\mu}$ is ξ-supersmooth at $\hat{F}_J$ along $\hat{F}_J$. And the uniqueness of $\hat{\mu}$ follows easily from Lemma 7.13. So it only remains to show that (7.18.1) holds, but this follows from Lemma 7.14 in the same way as we proved (7.17.3). Thus the theorem is proved. □

<u>Corollary 7.19</u>. <u>Let</u> μ_j <u>be a Borel regular,</u> τ<u>-smooth, Borel probability measure on the topological space</u> T_j <u>for all</u> $j \in J$. <u>Then there exist a unique</u> τ<u>-smooth Borel probability measure</u> μ <u>on</u> $(T^J, \mathcal{B}(T^J))$, <u>where</u> $T^J = \prod_{j\in J} T_j$ <u>with its product topology, satisfying</u>

(7.19.1) $\mu(\prod_{j\in J} F_j) = \prod_{j\in\sigma} \mu_j(F_j)$

<u>whenever</u> $\sigma \subseteq J$ <u>is finite,</u> $F_j \in F(T_j)$ <u>for all</u> $j \in J$, <u>and</u> $F_j = T_j$ <u>for all</u> $j \in J \setminus \sigma$.

<u>Moreover</u> μ <u>is Borel regular,</u> μ <u>is a complete Fubini product of</u> $\{\mu_j\}$ <u>on</u> $B(T^J)$, <u>and if</u> S_j <u>is the support of</u> μ_j <u>then we have</u>

$$(7.19.2) \qquad \mu(B) = \sup\{\mu(F_0) \mid F_0 \in F_0, F_0 \cap \prod_{j \in J} S_j \subseteq B\}$$

<u>for all</u> $B \in B(T^J)$, <u>where</u>

$$F_0 = \{p_\gamma^{-1}(F) \mid \gamma \text{ countable} \subseteq J, \quad F \in F(T^\gamma)\}$$

<u>and</u> $T^\gamma = \prod_{j \in \gamma} T_j$ <u>with its product topology</u> $\forall \gamma \subseteq J$.

<u>Remark</u>. This result was in part proved by P. Ressel [19], when J is countable, and in part proved by I. Ameniy, S. Okada and Y. Okazaki [1], when J is arbitrary and T_j is completely regular for all $j \in J$.

The measure μ will be called the Borel product of $\{\mu_j\}$, and we shall denote it as follows:

$$\mu = \hat{\bigotimes_{j \in J}} \mu_j$$

Clearly we have that $\hat{\otimes} \mu_j$ is an extension of $\otimes \mu_j$.

Note that if

$$(7.19.3) \qquad B(T^\gamma) = \bigotimes_{j \in \gamma} B(T_j) \quad \forall \gamma \text{ countable} \subseteq J$$

$$(7.19.4) \qquad S_j = T_j \text{ for all but countably many } j\text{'s}$$

then by (7.19.2) we have every set in $B(T^J)$ is $\otimes \mu_j$-measurable and $\hat{\otimes} \mu_j$ coincides with the Lebesgue extension of $\otimes \mu_j$ on $B(T^J)$.

Proof. Let $F_j = F(T_j)$ and let ξ be a sufficiently large cardinal, e.g.

$$\xi = \text{card}(2^{T^J})$$

If S_j is defined as in Theorem 7.18, then S_j is the support of μ_j and so $\mu_j(S_j) = 1$ and $S_j \neq \emptyset$. Hence the corollary except (7.19.2) follows from Theorem 7.18.

So let $B \in \mathcal{B}(T^J)$, if $L \subseteq J$ we let μ_L denote the marginal of μ on $(T^L, \mathcal{B}(T^L))$, and we put

$$S^L = \prod_{j \in L} S_j$$

Then it is easily checked that S^L is the support of μ_L and so $\mu_L(S^L) = 1$. Now let $\mu(B) > a$, then by Borel regularity of μ there exist a closed set $F \subseteq B$, such that $\mu(F) > a$. By τ-smoothness of μ there exist a countable set $\gamma \subseteq J$, and a closed set $F_0 \subseteq T^\gamma$, such that

(i) $$F \subseteq p_\gamma^{-1}(F_0) = F_0 \times T^L$$

(ii) $$\mu((F_0 \times T^L) \setminus F) = 0$$

where $L = J \setminus \gamma$. Now let $F_1 = S^\gamma \setminus p_\gamma(S^J \setminus F)$, since the projection p_γ is an open map we have that F_1 is closed, and clearly we have

(iii) $$F_0 \supseteq F_1 \quad \text{and} \quad F_1 \times S^L \subseteq F$$

And I claim that we have

(iv) $$\mu_\gamma(F_0 \setminus F_1) = 0$$

So let $u \in S^{\gamma} \cap (F_0 \setminus F_1)$, then $u \in p_{\gamma}(S^J \setminus F)$, and so there exist $v \in T^L$ with $(u,v) \in S^J \setminus F$. Since F is closed we can find open neighbourhoods U and V of u and v resp. such that $(U \times V) \cap F = \emptyset$. Since $U \cap S^{\gamma} \neq \emptyset$ and $V \cap S^L \neq \emptyset$ we have that

$$\mu_{\gamma}(U) > 0 \quad \text{and} \quad \mu_L(V) > 0$$

since S^{γ} and S^L are the supports of μ_{γ} and μ_L resp. Since $(U \times V) \cap F = \emptyset$, we have that

$$(U \cap F_0) \times V \subseteq (F_0 \times T^L) \setminus F$$

and so by (ii) we find that

$$\mu_{\gamma}(U \cap F_0)\mu_L(V) = 0$$

But $\mu_L(V) > 0$, and so $\mu_{\gamma}(U \cap F_0) = 0$. Hence there exists an open covering $\mathcal{G}$ of $S^{\gamma} \cap (F_0 \setminus F_1)$, such that

(v) $$\mu_{\gamma}(G \cap F_0) = 0 \qquad \forall G \in \mathcal{G}$$

and since μ_{γ} is τ-smooth, there exist $G_0 \in \mathcal{G}_{\sigma}$ such that

$$\mu_{\gamma}(G^* \setminus G_0) = 0 \quad \text{where} \quad G^* = \bigcup_{G \in \mathcal{G}} G$$

Hence by (v) we have that

$$\mu_{\gamma}(F_0 \setminus F_1) = \mu_{\gamma}(S^{\gamma} \cap (F_0 \setminus F_1))$$

$$\leq \mu_{\gamma}(G^* \cap F_0) = \mu_{\gamma}(G_0 \cap F_0) = 0$$

since $\mu_{\gamma}(S^{\gamma}) = 1$ and $G^* \supseteq S^{\gamma} \cap (F_0 \setminus F_1)$. Thus (iv) is proved.

By (ii) and (iv) we have that

$$\mu(p_\gamma^{-1}(F_1)) = \mu_\gamma(F_1) = \mu_\gamma(F_0) = \mu(F) > a$$

and by (iii) we have that

$$p_\gamma^{-1}(F_1) \cap S^L \subseteq F \quad \text{and} \quad p_\gamma^{-1}(F_1) \in \mathcal{F}_0$$

Hence we see that we have $\geq$ in (7.19.1), and the converse inequality follows easily since $\mu(S^J) = 1$. □

Theorem 7.20. Let $\{T_j \mid j \in J\}$ be a family of topological spaces, and let

$$T^L = \prod_{j \in L} T_j \quad \forall L \subseteq J$$

be the product space with its product topology. Let Γ be the set of all countable subsets of J, and let μ_γ be a Borel regular, τ-smooth, Borel probability on T^γ for $\gamma \in \Gamma$, such for all countable set $\gamma \subseteq J$ we have

(7.20.1) $\quad \mu_\gamma(B \times T^{\gamma \setminus \beta}) = \mu_\beta(B) \quad \forall \beta \subseteq \gamma \text{ and } B \in \mathcal{B}(T^\beta)$

Suppose that for all $\varepsilon > 0$ and all $\gamma \in \Gamma$, there exist a lower semicontinuous correspondance $\sigma : T^\gamma \sim\to T^{J \setminus \gamma}$ satisfying

(7.20.2) $\quad \mu_\gamma^*(\sigma^{-1}(T^{J \setminus \gamma})) > 1 - \varepsilon$

(7.20.3) $\quad$ If $\mu_{\gamma \cup \beta}(B \times G) = 0$, where $\beta \in 2^{(J \setminus \gamma)}$, $B \in \mathcal{B}(T^\gamma)$ and G is open in T^α, then $\mu_\gamma(B \cap \sigma_\beta^{-1}(G)) = 0$

where $\sigma_\beta = p_{\beta, J\setminus\gamma} \circ \sigma$ for all $\beta \subseteq J\setminus\gamma$.

Then there exist a unique Borel regular, τ-smooth Borel probability μ on T^J, such that μ_γ is the marginal of μ on $(T^\gamma, \mathcal{B}(T^\gamma))$ for all $\gamma \in \Gamma$.

Remarks (1): Recall that if σ is an ordinary point map, then (7.20.2) holds trivially and σ is lower semicontinuous, if and only if σ is continuous.

(2): Suppose that $\gamma, \beta \in \Gamma$, $\gamma \cap \beta = \emptyset$ and

$$\mu_{\gamma \cup \beta}(B_0 \times B_1) = \int_{B_0} \rho_\beta^\gamma(B_1 | u) \mu_\gamma(du)$$

for all $B_0 \in \mathcal{B}(T^\gamma)$ and all $B_1 \in \mathcal{B}(T^\beta)$, where ρ_β^γ is a μ_γ-measurable Markov kernel on $(T^\beta, \mathcal{B}(T^\beta)) | T$. Then it is easily checked that (7.20.3) holds if

$$(7.20.4) \qquad \mu_\gamma^*(u \in T^\gamma | \sigma_\beta(u) \subseteq S_\beta^\gamma(u)) = 1$$

where $S_\beta^\gamma(u)$ is the support of $\rho_\beta^\gamma(\cdot | u)$.

Proof. Let $F_\gamma = p_\gamma^{-1}(F(T^\gamma))$ and $\mathcal{B}_\gamma = p_\gamma^{-1}(\mathcal{B}(T^\gamma))$ for $\gamma \in \Gamma$. Then $\mathcal{B}_\gamma = \sigma(F_\gamma)$, and if

$$F = \bigcup_{\gamma \in \Gamma} F_\gamma, \qquad \mathcal{B}_0 = \bigcup_{\gamma \in \Gamma} \mathcal{B}_\gamma$$

then $\mathcal{B}_0 = \sigma(F)$, since Γ is $(\cup c)$-stable. For the same reason, then by (7.20.1) there exist a unique probability measure μ_0 on $\mathcal{B}_0$, such that μ_γ is the marginal of μ_0 on $(T^\gamma, \mathcal{B}^\gamma)$ for all $\gamma \in \Gamma$.

We shall now apply Theorem 6.10 to $(T^J, \mathcal{B}_0, \mu_0)$ with $J = 2^{(J)}$ and $\mathcal{F}_\gamma = F_\gamma$ for all γ finite $\subseteq J$. Then (6.10.1) is evident and (6.10.2) follows from τ-smoothness of μ_γ, since p_γ is surjective. Moreover if

$$\mathcal{F}_0 = \bigcup_{\gamma \in 2^{(J)}} \mathcal{F}_\gamma$$

then $F = (\mathcal{F}_0)_\delta$ by (6.10.10), and by Borel regularity of μ_γ, we see that

(i) $\quad F$ is an inner approximating paving for μ_0 on $\mathcal{B}_0$

So let $\varepsilon > 0$ and $D \in (\mathcal{F}_0)_\delta$ be given. Then $D \in F$, and so there exist a countable set $\gamma \subseteq J$, such that $D \in F_\gamma$. Let us then choose a lower semicontinuous correspondance σ from T^γ into T^L, where $L = J \setminus \gamma$, satisfying (7.20.2) and (7.20.3). Let $F \in \mathcal{F}_0$ and put

$$\hat{\sigma}(u) = \{u\} \times \sigma(u), \quad \hat{\rho}(F) = \{u \in T^\gamma \mid \hat{\sigma}(u) \subseteq F\}$$

$$\rho(F) = p_\gamma^{-1}(\hat{\rho}(F)), T_0 = \{t \in T^J \mid p_L(t) \in \sigma(p_\gamma(t))\}$$

Then $\hat{\sigma}: T^\gamma \to T^J$ is lower semicontinuous, and so $\hat{\rho}(F)$ is closed in T^γ. Thus ρ is an increasing map from $\mathcal{F}_0$ into $\mathcal{F}_\gamma$. Now let $t \in T_0 \cap \rho(F)$ and let $u = p_\gamma(t)$, and $v = p_L(T)$. Then $u \in \hat{\rho}(F)$ and $v \in \sigma(u)$, hence

$$t = (u,v) \in \hat{\sigma}(u) \subseteq F$$

and so (6.10.3) holds. Now let $H \in \mathcal{F}_\gamma$ and suppose that $H \cap T_0 = \emptyset$, then $H = p_\gamma^{-1}(H_0)$ for some closed set H_0 in T^γ, and since

$$(H_0 \times T^L) \cap T_0 = \emptyset$$

we see that $\sigma(u) = \emptyset$ for all $u \in H_0$. Hence H_0 is contained in $T^{\gamma} \setminus \sigma^{-1}(T^L)$, and so by (7.20.2) we have

$$\mu(H) = \mu_{\gamma}(H_0) \leq 1 - \mu_{\gamma}^*(\sigma^{-1}(T^L)) < \varepsilon$$

Thus (6.10.4) holds. Now suppose that $F \in F_0$ so that $\mu(D \setminus F) = 0$, then there exist a finite set β such that $F \in F_{\gamma \cup \beta}$. Let $D_0 \in F(T^{\gamma})$ and $F_0 \in F(T^{\gamma \cup \beta})$ be chosen so that

$$D = D_0 \times T^L = (D_0 \times T^{\beta}) \times T^{L \setminus \beta}, \quad F = F_0 \times T^{L \setminus \beta}$$

Then we have

(ii) $$\mu(D \setminus F) = \mu_{\gamma \cup \beta}((D_0 \times T^{\beta}) \setminus F_0) = 0$$

Let $H_0 = \hat{\rho}(F)$, and let $u_0 \in D_0 \setminus H_0$, then there exist $v_0 \in \sigma_{\beta}(u_0)$, so that $(u_0, v_0) \notin F_0$. Since F_0 is closed we can find open neighhourhoods G_0 of u_0 and G_1 of v_0, so that $(G_0 \times G_1) \cap F_0 = 0$. Hence

$$(G_0 \cap D_0) \times G_1 = (G_0 \times G_1) \cap (D_0 \times T^{\beta}) \subseteq (D_0 \times T^{\beta}) \setminus F_0$$

and so by (ii) we have that

$$\mu_{\gamma \cup \beta}((G_0 \cap D_0) \times G_1) = 0$$

Hence by (7.20.3) we have

$$\mu_{\gamma}(G_0 \cap D_0 \cap \sigma_{\beta}^{-1}(G_1)) = 0$$

Now note that $G = G_0 \cap \sigma_{\beta}^{-1}(G_1)$ is an open neighbourhood of u_0, since σ_{β} is lower semicontinuous. I.e. there exist an open covering $\mathcal{G}$ of $D_0 \setminus H_0$ such that

$$\mu_\gamma(G \cap D_0) = 0 \qquad \forall G \in \mathcal{G}$$

Hence by τ-smoothness of μ_γ we conclude that $\mu_\gamma(D_0 \setminus H_0) = 0$, and so (6.10.5) holds.

But then by (i) and Theorem 6.10, we see that μ_0 admits a unique τ-smooth, Borel regular extension. □

Theorem 7.21. Let $\{(T_j, A_j) \mid j \in J\}$ be a family of algebraic function spaces, and let $\lambda_j \in \mathrm{Pr}(A_j)$ for all $j \in J$. Let ξ be an infinite cardinal and put

$$\lambda = \bigotimes_{j\in J} \lambda_j, \qquad A = \bigotimes_{j\in J} A_j$$

Then we have

(7.21.1) $\quad \lambda \in \mathrm{Pr}_\sigma(A) \iff \lambda_j \in \mathrm{Pr}_\sigma(A_j) \quad \forall j \in J$

(7.21.2) $\quad \lambda \in \mathrm{Pr}_\tau(A) \iff \lambda_j \in \mathrm{Pr}_\tau(A_j) \quad \forall j \in J$

(7.21.3) $\quad \lambda \in \mathrm{Pr}_\pi(A) \iff \lambda_j \in \mathrm{Pr}_\pi(A_j) \quad \forall j \in J$

(7.21.4) $\quad \lambda \in \mathrm{Pr}_{s\xi}(A) \iff \lambda_j \in \mathrm{Pr}_{s\xi}(A_j) \quad \forall j \in J$

Now let S_j be the support of λ_j, i.e. let

$$S_j = \{t \in T_j \mid \varphi(t) \le 0 \quad \forall \varphi \in A \ \text{so that} \ \overline{\lambda}(\varphi^+) = 0\}$$

If $S_j \neq \emptyset$ for all but countably j's, then

(7.21.5) $\quad \lambda \in \mathrm{Pr}_\xi(A) \iff \lambda_j \in \mathrm{Pr}_\xi(A_j) \quad \forall j \in J$

Finally we have

(7.21.6) $\lambda \in Pr_{c\xi}(A)$ <u>if and only if</u> $\exists \gamma$ <u>countable</u> $\subseteq J$ <u>so that</u> $\lambda_j \in Pr_{c\xi}(A_j)$ <u>for all</u> $j \in \gamma$, <u>and</u> $\lambda_j \in P_{s\xi}(A_j)$ <u>for all</u> $j \in J \setminus \gamma$

<u>Remarks</u> (1): Recall that $Pr_{s\xi}(A)$ is the set of all probability contents on A with ξ-compact support.

(2): I do not know wether (7.21.5) holds in general without the assumption of non-emptyness of sufficiently many supports. Note that by (7.21.1) and (7.21.2) this is so if $\xi = \aleph_0$ or if $\xi \geq \text{weight}(A_j)$ for all $j \in J$.

(3): By Theorem 7.16 we have that the product of ξ-compact probability measures is ξ-compact, and by Theorem 4.10 we have that every ξ-compact probability content has a ξ-compact representing measure. Hence by (7.21.6) we see that there exist ξ-compact probabilit measures which are <u>not</u> induced by a ξ-compact probability content, see also Theorem 4.12.

<u>Proof</u> (7.21.1): Suppose that $\lambda_j \in Pr_\sigma(A_j)$ for all $j \in J$, then by Theorem 7.17 we have that $\lambda \in Pr_\sigma(A)$, and the converse follows from Theorem 4.13.

(7.21.2): Folloes from Corollary 7.19 and Theorem 4.13 exactly as above.

(7.21.3): Follows from Theorem 7.3 and Theorem 4.13.

(7.21.4): By Proposition 2.10 we have that $p_j(K) \in K_\xi(A_j)$ whenever $j \in J$ and $K \in K_\xi(A)$. Hence if $\lambda \in Pr_{s\xi}(A)$ then $\lambda_j \in Pr_{s\xi}(A_j)$ for all $j \in J$, and the converse follows from Theorem 7.3.

(7.21.5): I claim that the set S_j defined here equals the set S_j defined in Theorem 7.18, i.e. that

$$S_j = \{u \in T_j \mid u \in F \ \ \forall F \in F_\xi(A_j) \ \text{ with } \ \lambda_\xi(F) = 1\}$$

So suppose that $u_0 \in S_j$ and let $F \in F_\xi(A_j)$ such that $\lambda_\xi(F) = 1$. By Proposition 2.9 there exist a family $\{\varphi_q \mid q \in Q\} \subseteq A_j$ so that

$$F = \bigcap_{q \in Q} \{\varphi_q \leq 0\}$$

Hence $\varphi_q(u) \leq 0$ for all $u \in F$ and all $q \in Q$ and so $\overline{\lambda}_j(\varphi_q^+) = 0$. Hence $\varphi_q(u_0) \leq 0$ for all $q \in Q$ by definition of S_j and so $u_0 \in F$. The converse inclusion is easy. Thus (7.21.5) follows from Theorem 7.18 and Theorem 4.13.

(7.21.6): The "if"-part follows easily from Theorem 7.3. So suppose that $\lambda \in Pr_{c\xi}(A)$, then $\lambda_j \in Pr_{c\xi}(A_j)$ by Theorem 4.13. Let $K \in K_\xi(A)$ be chosen such that

(i) $$\lambda^*(K) \geq \tfrac{1}{2}$$

and put $C_j = p_j(K)$ and $C = \Pi_{j \in J} C_j$. Then $C_j \in K_\xi(A_j)$ by Proposition 2.10 and $K \subseteq C$. Now define

$$\gamma_n = \{j \in J \mid \lambda_j^*(C_j) \leq \frac{n}{n+1}\} \qquad \forall n \geq 1$$

Let $\varphi_j \in A_j$ for $j \in \gamma_n$ be chosen so that

(ii) $$\varphi_j \geq 1_{C_j} \quad \text{and} \quad \lambda_j(\varphi_j) \leq 1 - 2^{-n}$$

If β is a finite subset of γ_n then

$$\bigotimes_{j \in \beta} \varphi_j \geq 1_K$$

and so

$$\tfrac{1}{2} \leq \lambda\big(\bigotimes_{j\in\beta} \varphi_j\big) = \prod_{j\in\beta} \lambda_j(\varphi_j) \leq \left(\frac{n}{n+1}\right)^k$$

where $k = \operatorname{card}(\beta)$. Hence

$$k = \operatorname{card}(\beta) \leq \frac{\log 2}{\log(n+1) - \log n}$$

for all finite sets $\beta \subseteq \gamma_n$. Thus γ_n is finite and so

$$\gamma = \{j \in J \mid \lambda_j^*(C_j) < 1\}$$

is countable. Thus λ_j has ξ-compact support for all $j \in J \setminus \gamma$. □

Appendices

In the appendices below I shall fix the notation from probability theory, measure theory and topology that I have used in this exposition, and I shall state some of the basic results from these areas that has been used in the previous sections.

A. Arithmetic on $\overline{\mathbb{R}}$. We let $\mathbb{R} =]-\infty,\infty[$ denote the real line, and $\overline{\mathbb{R}} = [-\infty,\infty]$ denotes the extended real line. We put $\mathbb{R}_+ = [0,\infty[$ and $\overline{\mathbb{R}}_+ = [0,\infty]$. We shall use the common rules for addition and multiplication on $\overline{\mathbb{R}}$, together with the convention

$$0\cdot(\pm\infty) = 0$$

This leaves $\infty-\infty$ undefined and we shall now introduce 3 conventions for defining $\infty-\infty$:

$$\infty-\infty = 0, \qquad \infty \dot{-} \infty = +\infty, \qquad \infty \underset{\cdot}{-} \infty = -\infty$$

If $\{a_j \mid j \in J\} \subseteq \overline{\mathbb{R}}$ and $a_1, a_2 \in \overline{\mathbb{R}}$ we define

$$\Sigma\, a_j = \Sigma\, a_j^+ - \Sigma\, a_j^- , \quad a_1 + a_2 = \sum_1^2 a_j , \quad a_1 - a_2 = a_1 + (-a_2)$$

$$\Sigma^*a_j = \Sigma\, a_j^+ \dot{-} \Sigma\, a_j^- , \quad a_1 \dot{+} a_2 = \sum_1^{2}{}^{*} a_j , \quad a_1 \dot{-} a_2 = a_1 \dot{+} (-a_2)$$

$$\Sigma_*a_j = \Sigma\, a_j^+ \underset{\cdot}{-} \Sigma\, a_j^- , \quad a_1 \underset{\cdot}{+} a_2 = \sum_{*1}^{2} a_j , \quad a_1 \underset{\cdot}{-} a_2 = a_1 \underset{\cdot}{+} (-a_2)$$

Then we have

(A.1) $\dot{+}$, $\underset{\cdot}{+}$ and $+$ are commutative, and $\dot{+}$ and $\underset{\cdot}{+}$ are associative on $\overline{\mathbb{R}}$

(A.2) $a \underset{\bullet}{+} (b \overset{\bullet}{+} c) \le (a \underset{\bullet}{+} b) \overset{\bullet}{+} c, \quad a + b \le (a+c) \overset{\bullet}{+} (b-c)$

(A.3) $\Sigma_* a_j \le \Sigma a_j \le \Sigma\, a_j$ with equality if either $\Sigma_* a_j > -\infty$ or $\Sigma^* a_j < \infty$

(A.4) $b\Sigma_* a_j = \Sigma_* b a_j, \quad b\Sigma^* a_j = \Sigma^* b a_j$ if $b \in \overline{\mathbb{R}}_+$

(A.5) $b\Sigma_* a_j = \Sigma^* b a_j, \quad b\Sigma^* a_j = \Sigma_* b a_j$ if $b \in \overline{\mathbb{R}}_-$

(A.6) $b\Sigma a_j = \Sigma b a_j$ if $b \in \overline{\mathbb{R}}$

(A.7) $\Sigma_* a_j \underset{\bullet}{+} \Sigma_* b_j \le \Sigma_* (a_j \underset{\bullet}{+} b_j) \le \Sigma_* (a_j \overset{\bullet}{+} b_j) \le \Sigma_* a_j \overset{\bullet}{+} \Sigma^* b_j$

(A.8) $\Sigma_* a_j \underset{\bullet}{+} \Sigma^* b_j \le \Sigma^* (a_j \underset{\bullet}{+} b_j) \le \Sigma^* (a_j \overset{\bullet}{+} b_j) \le \Sigma^* a_j \overset{\bullet}{+} \Sigma^* b_j$

(A.9) $\sum_{j=1}^{n}{}^{*}$ is upper semicontinuous on $\mathbb{R}^n$

(A.10) If $\Sigma^* a_j < \infty$ then $\Sigma^*_{j \in J}$ is upper semicontinuous on $\{x \in \overline{\mathbb{R}}^J \mid x_j \le a_j \ \forall j \in J\}$

(a.11) $\inf_i a_i \underset{\bullet}{+} \inf_j b_j \le \inf_{i,j} (a_i \underset{\bullet}{+} b_j)$

(A.12) $\inf_{i,j} (a_i \overset{\bullet}{+} b_j) = \inf_i a_i \overset{\bullet}{+} \inf_j b_j$

(A.13) $\sup_i a_i \underset{\bullet}{+} \sup_j b_j = \sup_{i,j} (a_i \underset{\bullet}{+} b_j)$

(A.14) $\sup_{i,j} (a_i \overset{\bullet}{+} b_j) \le \sup_i a_i \overset{\bullet}{+} \sup_j b_j$

Also note that Σ^* and Σ_* are the upper and lower integrals with respect to <u>the counting measure</u> i.e. the measure $\#$ on $(J, 2^J)$ defined by

$$\# A = \begin{cases} n & \text{if } A \text{ has } n \text{ elements, } n \in \mathbb{N}_0 \\ \infty & \text{if } A \text{ is infinite} \end{cases}$$

where $\mathbb{N} = \{1,2,\dots\}$ is the set of positive integers and $\mathbb{N}_0 = \{0,1,\dots\}$ is the set of non-negative integers. □

B. Ordered sets. A preordered set $(X,\leq)$ is a set X with a reflexive (i.e. $x \leq x$) and transitive (i.e. $x \leq y,\ y \leq z \Rightarrow x \leq z$) relation $\leq$. An ordering is an antisymmetric (i.e. $x \leq y,\ y \leq x \Rightarrow x = y$) preordering. A preordering is called linear, if for all $x,y \in X$ we have that either $x \leq y$ or $y \leq x$. A preordered set $(X,\leq)$ is said to be upwards directed or filtering upwards, if for all $x,y \in X$ there exist $z \in X$ such that $z \geq x$ and $z \geq y$. Downwards directed or downwards filtering sets are defined similarly.

Let $(X,\leq)$ be a preordered set, and let $S \subseteq X$, then we say that S is cofinal in X, if for all $x \in X$ there exist $y \in S$ such that $x \leq y$. And we say that S is final in X if there exist $x \in X$, such that $y \leq x$ for all $y \in S$. Increasing and decreasing maps between two preordered space are defined in the obvious way. If f is a map from M into the preordered space $(X,\leq)$, then f is said to be cofinal resp. final if $f(M)$ is so, and if $(M,\leq)$ is preordered we say that f is exhausting if for all $x_0 \in X$ there exist $a \in M$, such that $f(v) \geq x_0$ for all $v \geq a$. If $(X,\leq)$ admits a countable cofinal subset, we say that $(X,\leq)$ is countably cofinal.

Let $(X,\leq)$ be an upwards directed set, then we define two cardinals $\mathrm{cof}(X)$ and $\mathrm{fin}(X)$ as follows

$$\mathrm{cof}(X) = \min\{\mathrm{card}(S) \mid S \text{ cofinal in } X\}$$
$$\mathrm{fin}(X) = \min\{\mathrm{card}(S) \mid S \text{ is not final in } X\}$$

with the convention: $\min \emptyset = 1$, then it is easily checked that

(B.1) $\operatorname{fin}(X) \leq \operatorname{cof}(X)$ with equality if either X is countably cofinal or if $\leq$ is linear

(B.2) $\operatorname{fin}(S) = \operatorname{fin}(X)$ and $\operatorname{cof}(S) = \operatorname{cof}(X)$, if S is cofinal in X

(B.3) If $\operatorname{fin}(X) = \operatorname{cof}(X) = \xi$ then there exist an increasing, cofinal injection $f:\xi \to X$

where we in (B.3) consider ξ as an ordinal with its usual ordering: $\xi = \{\alpha \mid \alpha < \xi\}$. An upwards directed set $(X,\leq)$ is said to be the finitely founded if $\{x \mid x \leq a\}$ is finite for all $a \in X$. If X is finitely founded then we have

(B.4) If $(Y,\leq)$ is an upwards directed set with $\operatorname{card}(Y) \leq \operatorname{card}(X)$, then there exist an increasing, cofinal function $f:X \to Y$.

Let M be a set, then a net on M is a map $x \leadsto v_x$ from some upwards directed set $(X,\leq)$ into M. A net on M is usually denoted $\{v_x \mid x \in X\}$. If $\{v_x \mid x \in X\}$ is a net on M, then a subnet of $\{v_x\}$ is a net $\{w_y \mid y \in Y\}$ of the form $w_y = v_{f(y)}$ where $(Y,\leq)$ is an upwards directed set and f is an exhausting map:$Y \to X$. If $\{v_x \mid x \in X\}$ is a net in a topological space M, then we define $\lim_x v_x$ as usual, see e.g. [5], and if $M = \overline{\mathbb{R}}$, then we define

$$\limsup_x v_x = \inf_{y \in X}\{\sup_{x \geq y} v_x\}$$

$$\liminf_x v_x = \sup_{y \in x}\{\inf_{x \geq y} v_x\} \qquad \square$$

C. Stable pavings and function spaces. If S and T are set, then S^T denotes the set of all functions from T into S, 2^T denotes the set of all subsets of T, and $2^{(T)}$ denote the set of all non-empty finite subsets of T. We shall identify 2^T with $\{0,1\}^T \subseteq \overline{\mathbb{R}}^T$ in the usual way: $F \sim 1_F$ for $F \subseteq T$, where 1_F is the indicator function of F, i.e.

$$1_F(t) = \begin{cases} 1 & \text{if } t \in F \\ 0 & \text{if } t \in T \setminus F \end{cases}$$

Thus if $H \subseteq \overline{\mathbb{R}}^T$, μ is a map: $H \to M$, and $F \subseteq T$, then $F \in H$ means that $1_F \in H$, and $\mu(F) := \mu(1_F)$.

We shall use the usual notation, see e.g. [2], [18] or [25], for stability of subsets of $\overline{\mathbb{R}}^T$ under the operations:

$$\begin{array}{l} \wedge f,\ \vee f,\ \Sigma^* f,\ \Sigma_* f,\ \cdot f \\ \wedge c,\ \vee c,\ \Sigma^* c,\ \Sigma_* c,\ \uparrow c,\ \downarrow c \\ \wedge \xi,\ \vee \xi,\ \Sigma^* \xi, \Sigma_* \xi,\ \uparrow \xi,\ \downarrow \xi \\ \wedge a,\ \vee a,\ \Sigma^* c,\ \Sigma_* a,\ \uparrow a,\ \downarrow a \end{array}$$

where f: = finite, c: = countable, a: = arbitrary and $\xi \geq 1$ is a cardinal number. E.g. $H \subseteq \overline{\mathbb{R}}^T$ is $(\uparrow\xi)$-stable if $\sup f_q \in H$ whenever $\{f_q | q \in Q\}$ is an upwards filtering subset of H of cardinality $\leq \xi$, and H is $(\cdot f)$-stable if $f_1 \cdot f_2 \in H$ whenever $f_1, f_2 \in H$.

More generally if φ is a map from $D \subseteq \overline{\mathbb{R}}^Q$ into $\overline{\mathbb{R}}$, then we say that $H \subseteq \overline{\mathbb{R}}^T$ is φ-stable, if $\varphi \circ f \in H$, whenever $f(t) = (f_q(t))$ is a map from T into D such that $f_q \in H$ for all $q \in Q$. If $(\alpha, \beta, \ldots)$ is a list of operations, then we say that H is $(\alpha, \beta, \ldots)$-stable if H is α-stable, β-stable etc. And the $(\alpha, \beta, \ldots)$-closure of a set $F \subseteq \overline{\mathbb{R}}^T$, is the intersection of all $(\alpha, \beta, \ldots)$-stable subset of $\overline{\mathbb{R}}^T$ containing F.

We use a similar notation for stability of pavings under the operations

$$\cap f,\ \cup f,\ \Sigma f,\ c,\ \setminus,\ -$$
$$\cap c,\ \cup c,\ \Sigma c,\ \uparrow c,\ \downarrow c$$
$$\cap \xi,\ \cup \xi,\ \Sigma \xi,\ \uparrow \xi,\ \downarrow \xi$$
$$\cap a,\ \cup a,\ \Sigma a,\ \uparrow \xi,\ \downarrow \xi$$

where $\Sigma :=$ disjoint union, $c :=$ complement, $\setminus :=$ difference, $- :=$ proper difference, and $\xi \geq 1$ is a cardinal number.

If $F \subseteq \overline{\mathbb{R}}^T$, then we put

$$F^{\xi} = \text{the } (\vee\xi)\text{-closure of } F$$
$$F_{\xi} = \text{the } (\wedge\xi)\text{-closure of } F$$
$$F_{\sigma} = \text{the } (\vee c)\text{-closure of } F$$
$$F_{\delta} = \text{the } (\wedge c)\text{-closure of } F$$
$$\sigma(F) = \text{the } \sigma\text{-algebra generated by } F$$

i.e. $\sigma(F)$ is the smallest σ-algebra on T making all functions in F measurable. And similarly if $\mathcal{F}$ is a paving on T.

In particular if $(T_j, \mathcal{B}_j)$ is a <u>measurable space</u> (i.e. T_j is a set and $\mathcal{B}_j$ is a σ-algebra on T_j) for all $j \in J$, then <u>the product σ-algebra</u>:

$$\bigotimes_{j\in J} \mathcal{B}_j = \sigma\{p_j^{-1}(B) \mid j \in J, B \in \mathcal{B}_j\}$$

is the smallest σ-algebra on $T = \Pi_{j\in J} T_j$ making all the projections $p_j : T \to T_j$ measurable.

If $\{K_j \mid j \in J\}$ is a family of subsets of T, we say that $\{K_j\}$ has <u>the finite intersection property</u>, if $\cap_{j\in\gamma} K_j \neq \emptyset$ for all $\gamma \in 2^{(J)}$.

Let ξ be an infinite cardinal, then we say that K is ξ-compact (resp. ξ-monocompact), if $\cap_{j\in J} K_j \neq \emptyset$, whenever $\{K_j \mid j \in J\} \subseteq K$, such that $\text{card}(J) \leq \xi$ and $\{K_j\}$ has the finite intersection property (resp. $\text{card}(J) \leq \xi$, $K_j \neq \emptyset$ $\forall j \in J$ and $\{K_j\}$ is filtering downwards). It is wellknown that we have

(C.1) If K is ξ-compact, then so it the $(\cup f, \cap\xi)$-closure of $K \cup \{\emptyset, T\}$

see e.g. [2; Theorem III.4], where (C,1) is proved for $\xi = C$, and observe that the proof functions for all infinite cardinals ξ. If $\xi = C$ we write semicompact in place of ξ-compact, and if $\xi \geq \text{card}(K)$ we write compact in place of ξ-compact.

Let H be a subset of $\overline{\mathbb{R}}^T$, then we define the equivalence relation on T:

$$t' \equiv t'' \pmod H \iff f(t') = f(t'') \quad \forall f \in H$$

The equivalence class containing t is called the H-atom containing t. If all H-atoms are singleton we say that H separates points in T, i.e.

(C.2) H separates point in T, if and only if $\forall t' \neq t''$ there exist $f \in H$ so that $f(t') \neq f(t'')$.

A set $S \subseteq T$ is said to be H-saturated, if S is a union of H-atoms. The reader easily verifies that the following 3 statements are equivalent

(C.3) S is H-saturated

(C.4) $s \in S,\ t \in T,\ f(s) = f(t)\ \forall f \in H \Rightarrow t \in S$

(C.5) $\exists S_0 \subseteq \overline{\mathbb{R}}^H$ so that $S = p_H^{-1}(S_0)$

where p_H is the map: $T \to \overline{\mathbb{R}}^H$ given by

$$p_H(t) = (h(t))_{h \in H}$$

clearly we have

(C.6) The set of all H-saturated subsets of H is φ-stable for all $\varphi : D \to \{0,1\}$, with $D \subseteq \overline{\mathbb{R}}^Q$

Clearly we have similar results for pavings. □

D. Measures. Let $(T, \mathcal{B}, \mu)$ be a positive measure space, then μ^* and μ_* denote the outer and inner μ-measure, and $M(\mu)$ denotes the set of all Lebesgue μ-measurable sets:

$$M(\mu) = \{M \subseteq T \mid \exists B_0, B_1 \in \mathcal{B}: B_0 \subseteq M \subseteq B_1,\ \mu(B_1 \setminus B_0) = 0\}$$

And we put

$$L^1(\mu) = \{f \in \overline{\mathbb{R}}^T \mid f \text{ is } \mu\text{-integrable}\}$$
$$\overline{L}(\mu) = \{f \in \overline{\mathbb{R}}^T \mid f \text{ is } \mu\text{-measurable}\}$$
$$L(\mu) = \{f \in \overline{L}(\mu) \mid \int f^+ d\mu < \infty \text{ or } \int f^- d\mu < \infty\}$$
$$\int f d\mu = \int f^+ d\mu - \int f^- d\mu \quad \forall f \in L(\mu)$$
$$\int^* h d\mu = \inf\{\int f d\mu \mid f \in L(\mu), f \geq h\} \quad \forall h \in \overline{\mathbb{R}}^T$$
$$\int_* h d\mu = \sup\{\int f d\mu \mid f \in L(\mu), f \leq h\} \quad \forall h \in \overline{\mathbb{R}}^T$$

If μ is finitely founded, i.e. if μ has no infinite atoms, then we may replace $L(\mu)$ by $L^1(\mu)$ in the definition of the upper integral: $\int^* h d\mu$, and the lower integral $\int_* h d\mu$. It is wellknown that we have

(D.1) $\int_* f d\mu \underset{\cdot}{+} \int_* g d\mu \le \int_* (f \underset{\cdot}{+} g) d\mu \le \int_* (f \dot{+} g) d\mu \le \int_* f d\mu \dot{+} \int^* g d\mu$

(D.2) $\int_* f d\mu \underset{\cdot}{+} \int^* g d\mu \le \int^* (f \underset{\cdot}{+} g) d\mu \le \int^* (f \dot{+} g) d\mu \le \int^* f d\mu \dot{+} \int^* g d\mu$

(d.3) $\int^* (af) d\mu = a \int^* f d\mu, \int_* (af) d\mu = a \int_* f d\mu \quad \forall a \in \mathbb{R}_+$

(D.4) $\int^* (af) d\mu = a \int_* f d\mu, \int_* (af) d\mu = a \int^* f d\mu \quad \forall a \in \mathbb{R}_-$

(D.5) $\int^* f d\mu = \lim_{n\to\infty} \int^* f_n d\mu \quad \text{if} \quad f_n \uparrow f, \ \int^* f_1 d\mu > -\infty$

(D.6) $\int_* f d\mu = \lim_{n\to\infty} \int_* f_n d\mu \quad \text{if} \quad f_n \downarrow f, \ \int_* f_1 d\mu < \infty$

(D.7) $\int^* (\liminf_{n\to\infty} f_n) d\mu \le \liminf_{n\to\infty} \int^* f_n d\mu \quad \text{if} \quad \int^* (\inf_n f_n) d\mu > -\infty$

(D.8) $\int_* (\limsup_{n\to\infty} f_n) d\mu \ge \limsup_{n\to\infty} \int_* f_n d\mu \quad \text{if} \quad \int_* (\sup_n f_n) d\mu < \infty$

If $0 < p < \infty$, then we put

$$L^p(\mu) = \{f \in \overline{L}(\mu) \mid \int |f|^p d\mu < \infty\}$$
$$L^\infty(\mu) = \{f \in L^0(\mu) \mid \exists\, a \in \mathbb{R}_+ : \ |f| \le a \quad \mu\text{-a.e.}\}$$
$$L^0(\mu) = \{f \in \overline{L}(\mu) \mid \ |f| < \infty \quad \mu\text{-a.e.}\}$$

and we define the $\|\cdot\|_p$ - "norm" by

$$\|f\|_0 = \int_T (\operatorname{Arctg}|f|) d\mu$$

$$\|f\|_p = \int_T |f|^p d\mu \quad \text{if} \quad 0 < p \le 1$$

$$\|f\|_p = \{\int_T |f|^p d\mu\}^{1/p} \quad \text{if} \quad 1 \le p < \infty$$

$$\|f\|_\infty = \inf\{a \in \mathbb{R}_+ \mid \ |f| \le a \quad \mu\text{- a.e.}\}$$

for $f \in L^p(\mu)$ and $0 \le p \le \infty$

A paving $\mathcal{F}$ on T is said to be an <u>inner approximating paving</u> for μ on $\mathcal{B}_0$, if $\mathcal{B}_0 \subseteq M(\mu)$ and

$$\mu(B) = \inf\{\mu_*(F) \mid F \in \mathcal{F}, F \subseteq B\} \qquad \forall B \in \mathcal{B}_0$$

It is wellknown (and easily checked) that we have

(D.9) If $\mathcal{F}$ is an inner approximating paving for μ on $\mathcal{B}$, then so is $\mathcal{F}_\delta \cap \mathcal{B}$

Let ξ be an infinite cardinal, then we say that μ is ξ-<u>compact</u>, ξ-<u>monocompact</u>, ξ-<u>semicompact</u> or <u>compact</u>, if μ admits an inner approximating paving on $\mathcal{B}$ of the corresponding type. And μ is said to be <u>perfect</u> if the restriction of μ to every countably generated sub σ-algebra of $\mathcal{B}$ is semicompact, see e.g. [12], [14], [15], [16], [17], [21] and [24].

Let f be a μ-measurable map from $(T,\mathcal{B},\mu)$ into the measurable space $(S,\mathcal{A})$. Then f denote the <u>image measure</u> on $(S,\mathcal{A})$:

$$(f\mu)(A) = \mu(f^{-1}(A)) \qquad \forall A \in \mathcal{A}.$$

And if g is a map from S into T, then $g^{-1}\mu$ denotes <u>the coinmage measure</u> on $(S,g^{-1}(\mathcal{B}))$:

$$(g^{-1}\mu)(A) = \mu^*(g(A)) \qquad \forall A \in g^{-1}(\mathcal{B})$$

It is wellknown that $f\mu$ and $g^{-1}\mu$ are measures and that we have

(D.10) $(f\mu)_* \leq \mu_* \circ f^{-1} \leq \mu^* \circ f^{-1} \leq (f\mu)^*$

(D.11) $(g^{-1}\mu)^*(B) = \mu^*(g(B)) \qquad \forall B \subseteq T$

(D.12) $\mu(B) = (f^{-1}(f\mu))(B) \qquad \forall B \in f^{-1}(\mathcal{A})$

(D.13) $\mu = g(g^{-1}\mu) \iff \mu_*(T \setminus g(S)) = 0 \qquad \square$

E. Topology. As far as topological spaces are concerned we shall use the terminology and notation of [5]. Let ξ be a cardinal, then we say that a topological space is ξ-Lindelöf, if every open cover of T has a subcover of cardinality $\leq \xi$. If $\xi = \aleph_0$ then we just say that T is Lindelöf. If every subset of T is ξ-Lindelöf or Lindelöf, then we say that T is hereditarily ξ-Lindelöf or hereditarily Lindelöf. The weight of T is defined by

$$\text{weight}(T) = \min\{\text{card}(\mathcal{G}) \mid \mathcal{G} \text{ is an open base for } T\}$$

Let T be a topological space and let $K \subseteq T$. If ξ is an infinite cardinal we say that K is ξ-compact (resp. relatively ξ-compact in T) if every open cover $\mathcal{G}$ of K (resp. of T) with $\text{card}(\mathcal{G}) \leq \xi$ has a finite subset which covers K. It is easily checked that K is ξ-compact (resp. relatively ξ-compact in T), if and only if every net $\{t_\gamma \mid \gamma \in \Gamma\}$ in K with $\text{card}(\Gamma) \leq \xi$ has a limit joint in K (resp. in T). If $\xi = \aleph_0$ we say that K is countably compact resp. relatively countably compact. If every sequence in K has a subsequence which convergence to a point in K (resp. to a point in T) we say that K is sequentially compact (resp. relatively sequentially compact in T).

Let T be a topological space, then $\mathcal{G}(T)$, $\mathcal{F}(T)$, $\mathcal{K}(T)$ and $\overline{\mathcal{K}}(T)$ denotes the set of all open, closed, compact and closed compact subsets of T resp. And Lsc(T), Usc(T), $C(T)$ and C(T) denotes the set of all $\overline{\mathbb{R}}$-valued lower semicontinuous, $\overline{\mathbb{R}}$-valued upper semicontinuous, $\mathbb{R}$-valued continuous and bounded $\mathbb{R}$-valued continuous functions on T resp. And we define the Borel σ-algebra $\mathcal{B}(T)$, and the Baire σ-algebra $\mathcal{B}a(T)$ as follows

$$\mathcal{B}(T) = \sigma(\mathcal{G}(T)) = \sigma(\mathcal{F}(T)) = \sigma(\text{Lsc}(T)) = \sigma(\text{Usc}(T))$$

$$\mathcal{B}a(T) = \sigma(C(T)) = \sigma(\text{C}(T))$$

If T_0 is a subset of T, then we have

(E.1) $\mathcal{B}(T_0) = \{B \cap T_0 \mid B \in \mathcal{B}(T)\}$

(E.2) $\mathcal{B}a(T_0) \supseteq \{B \cap T_0 \mid B \in \mathcal{B}a(T)\}$

with equality in (E.2) in either of the following 4 cases (1): T is perfectly normal. (2): T is completely regular and T_0 is Lindelöf. (3): T is normal and $T_0 \in \mathcal{F}_\sigma(T)$. (4): There exist Baire measurable maps $\varphi_n : T \to T_0$ for $n \in \mathbb{N}$, such that $\varphi_n(t) \to t \quad \forall t \in T_0$.

If S and T are topological spaces, then we have

(E.3) $\mathcal{B}(S \times T) \supseteq \mathcal{B}(S) \otimes \mathcal{B}(T)$

(E.4) $\mathcal{B}a(S \times T) \supseteq \mathcal{B}a(S) \otimes \mathcal{B}a(T)$

We have equality in (E.3) if S has countable weight. And we have equality in (E.4) if $S \times T$ is Lindelöf or if there exist measurable maps φ_n from $(S \times T, \mathcal{B}a(S) \otimes \mathcal{B}a(T))$ into $(S \times T, \mathcal{B}a(S \times T))$ for $n \in \mathbb{N}$, such that $\varphi_n(s,t) \to (s,t)$ for all $(s,t) \in S \times T$.

Finally we have

(E.5) $\mathcal{B}a(T) \subseteq \mathcal{B}(T)$

with equality in (E.5) if T is perfectly normal. □

F. Function spaces. Let T be a set, and let $f \in \overline{\mathbb{R}}^T$, then we define

$$\|f\| = \sup_{t \in T} |f(t)|$$

$$M(f) = \sup_{t \in T} f(t) \ , \qquad m(f) = \inf_{t \in T} f(t)$$

And if $S \subseteq T$, then we define $\|f\|_S$, $M_S(f)$ and $m_S(f)$ similarly. We put

$$B(T) = \{f \in \overline{\mathbb{R}}^T \mid \|f\| < \infty\}$$
$$B^*(T) = \{f \in \overline{\mathbb{R}}^T \mid M(f) < \infty\}$$
$$B_*(T) = \{f \in \overline{\mathbb{R}}^T \mid m(f) > -\infty\}$$
$$B(T,\mathcal{B}) = \{f \in B(T) \mid f \text{ is } \mathcal{B}\text{-measurable}\}$$

whenever $\mathcal{B}$ is a σ-algebra on T.

If $H \subseteq \overline{\mathbb{R}}^T$, then $\tau(H)$ denote <u>the weakest topology on</u> T <u>making all functions in</u> H <u>continuous</u>, and we put

$$F(H) = F(T,\tau(H)) \ , \quad G(H) = G(T,\tau(H))$$
$$K(H) = K(T,\tau(H)), \quad \overline{K}(H) = \overline{K}(T,\tau(H))$$
$$\mathcal{B}(H) = \mathcal{B}(T,\tau(H)), \quad \mathcal{B}a(H) = \mathcal{B}a(T,\tau(H))$$

Now let ξ be cardinal, then we define

$$F_\xi(H) = \cup\{F(Q) \mid Q \subseteq H, \ \text{card}(Q) \leq \xi\}$$
$$G_\xi(H) = \cup\{G(Q) \mid Q \subseteq H, \ \text{card}(Q) \leq \xi\}$$
$$K_\xi(H) = \cap\{K(Q) \mid Q \subseteq H, \ \text{card}(Q) \leq \xi\}$$
$$\overline{K}_\xi(G) = K_\xi(H) \cap F(H), \ \overline{\overline{K}}_\xi(H) = K_\xi(H) \cap F_\xi(H)$$

If $\xi = \aleph_0$ we write $F_0(H)$, $G_0(H)$, $K_0(H)$, $\overline{K}_0(H)$ and $\overline{\overline{K}}_0(H)$. It is easily checked that we have

(F.1) $F_\xi(H)$ and $\overline{\overline{K}}_\xi(H)$ are $(\cup f, \cap\xi)$-stable

(F.2) $K_\xi(H)$ and $\overline{K}_\xi(H)$ are $(\cup f)$-stable

(F.3) $\overline{\overline{K}}_\xi(H)$ is ξ-compact

(F.4) $F \cap K \in K_\xi(H) \quad \forall F \in F_\xi(H) \ \ \forall K \in K_\xi(H)$

(F.5) $\text{cl}(K) \in \overline{K}_\xi(H) \quad \forall K \in K_\xi(H)$

where cl(K) in (F.5) denotes the closure of K in the topology $\tau(H)$. Moreover if

$$p_H(t) = (h(t))_{h\in H}: T \to \overline{\mathbb{R}}^H$$

then we have

(F.6) $\quad \mathcal{F}(H) = \{p_H^{-1}(F) \mid F \in \mathcal{F}(\overline{\mathbb{R}}^H)\}$

(F.7) $\quad \mathcal{K}(H) = \{K \subseteq T \mid p_H(K) \in \mathcal{K}(\overline{\mathbb{R}}^H)\}$

(F.8) $\quad \overline{\mathcal{K}}(H) = \{p_H^{-1}(K) \mid K \in \mathcal{K}(\overline{\mathbb{R}}^H), \ K \subseteq p_H(T)\}$

Note that $\|\cdot\|$ is a seminorm on $\overline{\mathbb{R}}^T$, and so it induces a metric topology on $\overline{\mathbb{R}}^T$. If $H \subseteq \overline{\mathbb{R}}^T$ we put

$$\overline{H} = \text{the } \|\cdot\|\text{-closure of } H \text{ in } \overline{\mathbb{R}}^T$$

If $\mathcal{L}$ denote one of the symbols $\mathcal{F}_\xi$, $\mathcal{G}_\xi$, $\mathcal{K}_\xi$, $\overline{\mathcal{K}}_\xi$ or $\overline{\mathcal{K}}_\xi$, where ξ is an infinite cardinal, then we have

(F,9) $\quad \mathcal{L}(H) = \mathcal{L}(\overline{H})$ and $\tau(H) = \tau(\overline{H})$ □

G. Baire and Borel measures. Let T be a topological space, then a Baire (resp. Borel) measure on T is a measure defined on the Baire (resp. Borel) σ-algebra. We put

$$\mathcal{F}_0(T) = \mathcal{F}_0(C(T)) \ , \quad \mathcal{G}_0(T) = \mathcal{G}_0(C(T))$$

sets in $\mathcal{F}_0(T)$ resp. $\mathcal{G}_0(T)$ are called functionally closed resp. functionally open, see [5]. Then we have (see Proposition 2.9):

(G.1) $\quad Ba(T) = (\mathcal{F}_0(T)) = \sigma(\mathcal{G}_0(T))$

Let μ be a Borel (resp. Baire) measure on T, then we say that μ is τ-smooth if

$$\mu(G) = \sup_{i \in I} \mu(G_i)$$

whenever G and $\{G_i\}$ are open (resp. functionally open) and $\{G_i\}$ filters upwards to G. And we say that μ is Borel (resp. Baire) regular, if $\mathcal{F}(T)$ (resp. $\mathcal{F}_0(T)$) is an inner approximating paving for μ on $\mathcal{B}(T)$ (resp. $\mathcal{B}a(T)$). It is wellknown that we have

(G.2) Every finite Baire measure is Baire regular

(G.3) A finite Baire measure is τ-smooth, if and only if $\lim_{\gamma} \int \varphi_\gamma d\mu = 0 \quad \forall \{\varphi_\gamma\} \subseteq C(T): \varphi_\gamma \downarrow 0$

(G.3) If T is Lindelöf, then every finite Baire on T is τ-smooth

(G.4) If T is paracompact and weight$(T) < RM$, then every finite Baire measure on T is τ-smooth

where RM is the smallest real measurable cardinal, see. e.g. [23]. Recall that it is consistent with set theory to assume that RM do not exist, in which case every finite Baire on a paracompact space is τ-smooth. Similarly we have

(G.5) If T is hereditarily Lindelöf, then every finite Borel measure on T is τ-smooth.

(G.6) If T is regular, then every τ-smooth Borel measure on T is Borel regular.

(G.7) If T is hereditarily paracompact and weight$(T) < RM$, then every finite Borel measure on T is τ-smooth.

By Proposition 3.14 we have that a finite Borel measure μ on T is τ-smooth, if and only if

(G.8) $$\lim_{\gamma} \int_T g_\gamma d\mu = \int_T g d\mu$$

whenever $\{g_\gamma\} \subseteq Lsc^+(T)$ and $g_\gamma \uparrow g$.

A <u>Radon measure</u> on T is a Borel measure μ on T, such that $\overline{K}(T)$ is an inner approximating paving for μ on $B(T)$, and μ is finite on $\overline{K}(T)$. Clearly we have

(G.9) Every Radon measure is regular and τ-smooth.

(G.10) Every Radon measure is compact and perfect. □

<u>H. Correspondances</u>. A <u>correspondance</u> from S into T is a map θ from S into 2^T. If θ is a correspondance from S into T, then we write $\theta: S \rightsquigarrow T$, and we define

$$\theta(A) = \bigcup_{s \in A} \theta(s) \qquad \forall A \subseteq S$$

$$\theta^{-1}(t) = \{s \in S \mid t \in \theta(s)\} \qquad \forall t \in T$$

$$\theta^{-1}(B) = \bigcup_{t \in B} \theta^{-1}(t) = \{s \in S \mid \theta(s) \cap B \neq \emptyset\} \qquad \forall B \subseteq T$$

$$Gr(\theta) = \{(s,t) \in S \times T \mid t \in \theta(s)\}$$

Then θ^{-1} is a correspondance: $T \rightsquigarrow S$, and we have

(H,1) $$\theta(\bigcup_i A_i) = \bigcup_i \theta(A_i) \ , \qquad \theta(\bigcap_i A_i) \subseteq \bigcap_i \theta(A_i)$$

(H.2) $$\theta = (\theta^{-1})^{-1}$$

Let S and T be topological spaces, and let θ be a correspondance: $S \rightsquigarrow T$. Then we say that θ is <u>lower</u> (resp. <u>upper</u>) <u>continuous</u> if $\theta^{-1}(G)$ is open for all $G \in \mathcal{G}(T)$ (resp. if $\theta^{-1}(F)$ is closed for all $F \in \mathcal{F}(T)$). And θ is <u>lower</u> (<u>upper</u>) <u>continuous on</u> S_0, where $S_0 \subseteq S$, if θ restricted to S_0 is lower (upper) continuous. (NB: <u>upper</u> and <u>lower semicontinuous</u> is also commonly used)

If θ is upper continuous on S_0, and if $\theta(s)$ is compact for all $s \in S_0$, then we have

(H.3) $\quad \theta(K) \in \mathcal{K}(T) \quad \forall K \in \mathcal{K}(S)$ so that $K \subseteq S_0$

Let θ be a correspondance from S into T, and let $F \subseteq \overline{\mathbb{R}}^S$ and $H \subseteq \overline{\mathbb{R}}^T$, such that

(H.4) $\quad \forall h \in H \ \exists f \in F: f(s) = h(t) \quad \forall (s,t) \in Gr(\theta)$

Then it is easily checked that we have

(H.5) $\quad \theta$ is upper and lower continuous on $\theta^{-1}(T)$

(H.6) $\quad \theta(s)$ is compact $\forall s \in S$

(H.7) $\quad \theta(K) \in \mathcal{K}_\xi(H) \quad \forall K \in \mathcal{K}_\xi(F)$ so that $K \subseteq \theta^{-1}(T)$

If S and T are equipped with the $\tau(F)$-topology resp. the $\tau(H)$-topology.

R E F E R E N C E S

[1] I. Ameniya, S. Okada and Y. Okazaki, Pre-Radon measures on topological spaces, Kodai Math. J. 1 (1978), p. 101-132.

[2] C. Dellacherie and P.-A. Meyer, Probability and Potentials, North Holland, Amsterdam 1978.

[3] R.M. Dudley, Probability and metrics, Mat. Inst., Aarhus Univ., Lecture Notes Series No. 45. 1976.

[4] N. Dunford and J.T. Schwartz, Linear operators I, Interscience Publishers Inc. 1958, New York.

[5] R. Engelking, General Topology, PWN, Warszawa, 1977.

[6] J. Hoffmann-Jørgensen, How to make a divergent sequence convergent by Martin's exiom, Matematisk Institut, Aarhus Universitet, Preprint Series 1977/78, No. 21.

[7] J. Hoffmann-Jørgensen, Existence of conditional probabilities, Math. Scand. 28 (1971), p. 257-265.

[8] J. Hoffmann-Jørgensen, A general "in between theorem", Math. Scand., 50 (1982), p. 55-65.

[9] J. Hoffmann-Jørgensen, Weak compactness and tightness of subsets of M(X), Math. Scand. 31 (1972), p. 127-150.

[10] H.G. Kellerer, Duality theorems for marginal problems, Preprint, Dept. of Math., Univ. of Munich (1984).

[11] G. Köthe, Topological vector spaces I, Springer Verlag 1969, GMW 159.

[12] E. Marczewski, On compact measures, Fund. Math. 40 (1953).

[13] I. Mitoma, S. Okada and Y. Okazaki, Cylindrical σ-algebra and cylindrical measure, Osaka J. Math. 14 (1977), 635-647.

[14] K. Musial, Projective limits of perfect measures, Fund. Math. 110 (1980), p. 163-189.

[15] K. Musial, Inheritness of compactness and perfectness of measures by thick subsets, Proc. Conf. on Measure Theory 1975, Springer Verlag 1976, LNS 541, p. 31-42.

[16] J.K. Pachl, Disintegration and compact measures, Math. Scand. 43 (1978), p. 157-168.

[17] J.K. Pachl, Two classes of measures, Coll. Math. 52 (1979), p. 331-340.

[18] D. Pollard and F. Topsøe, A unified approach to Riesz type representation theorems, Stud. Math. 54 (1975).

[19] P. Ressel, Some continuity and measurability results on spaces of measures, Math. Scand. 40 (1977), p. 69-78.

[20] C.A. Rogers et al., Analytic sets, Academic Press, London 1980.

[21] C. Ryll-Nardzewski, On quasi-compact measures, Fund. Math. 40 (1953), p. 125-130.

[22] V. Strassen, The existence of measure with given marginals, Ann. Math. Stat. 36 (1965), p. 423-439.

[23] M. Talagrand, Pettis integral and measury theory, Mem. Amer. Math. Soc. 1984, vol. 51 N° 307.

[24] F. Topsøe, Approximating pavings and constructions of measures, Coll. Math. 52 (1974), p. 377-385.

[25] F. Topsøe, Topology and measure, Springer Verlag 1979, LNS 133.

LIST OF SYMBOLS

1. Spaces of functions and pavings

2. Topological spaces

3. Ordered sets, $\overline{\mathbb{R}}$, and correspondances

4. Stability operations

5. Functionals

6. Measures

7. Probability contents

8. Outer probability contents

9. Marginal systems

Subject Index

INVARIANT SUBSPACES OF SHIFTS IN QUATERNIONIC HILBERT SPACE

Salih Suljagić

A left vector space $\mathcal{H}$ over the noncommutative field $\mathbb{Q}$ of all quaternions complete in respect to a scalar product $(\ ,\ ) : X\times X \to \mathbb{Q}$ is said to be <u>quaternionic Hilbert space</u>. Suppose $1, i, j, k$ is a basis in four dimensional real vector space $\mathbb{Q}$ such that $1q=q=q1$, $\forall q \in \mathbb{Q}$, $i^2=j^2=k^2=-1$, $ij=k$, $jk=i$, $ki=j$. $\mathbb{Q}$ can be considered as a noncommutative field over the set of all complex numbers $\mathbb{C}$. So, $\mathcal{H}$ can be considered as a complex vector space H. The part of $(\ ,\ )$ spanded by 1 and i, in notation $\langle\ ,\ \rangle$, is scalar product in H, and H is complete in respect to it. So, H is complex Hilbert space and it is said to be <u>symplectic image of $\mathcal{H}$</u>. Let $\varkappa : H \to H$ be the operator defined by $\varkappa x = kx$. It is antiunitary and $\varkappa^2 = -I$, where I denotes the identity. A subspace H_0 in H is symplectic image of a subspace $\mathcal{H}_0$ in $\mathcal{H}$ if and only if $\varkappa H_0 = H_0$. Let $\mathcal{A} : \mathcal{H} \to \mathcal{H}$ be a linear bounded operator. Then $\mathcal{A}$ can be considered as an operator $A : H \to H$, $Ax = \mathcal{A}x$. A is linear, bounded, the norm of A is equal to that of $\mathcal{A}$. The operator A is said to be <u>symplectic image of $\mathcal{A}$</u> ([6],[7],[3]).

<u>A subspace</u> $\mathcal{L}$ is said to be <u>wandering</u> for an isomerty $\mathcal{W}$ in $\mathcal{H}$ if $\mathcal{W}^n\mathcal{L} \perp \mathcal{W}^m\mathcal{L}$ for $n \neq m$; $n,m \in \mathbb{N}_0$ (the set of all nonnegative integers). If $\mathcal{U} : \mathcal{H} \to \mathcal{H}$ is unitary and $\mathcal{L}$ is wandering subspace for $\mathcal{U}$, then it follows $\mathcal{U}^n\mathcal{L} \perp \mathcal{U}^m\mathcal{L}$ for $n \neq m$; $n,m \in \mathbb{Z}$ (the set of all integers). An isometry $\mathcal{S}$ in $\mathcal{H}$ having a wandering subspace $\mathcal{L}$ such that $\mathcal{H} = \bigoplus_{n=0}^{\infty} \mathcal{S}^n\mathcal{L}$ is said to be <u>unilateral shift</u>. A unitary operator $\mathcal{U} : \mathcal{H} \to \mathcal{H}$ having a wandering subspace $\mathcal{L}$ such that $\mathcal{H} = \bigoplus_{n\in\mathbb{Z}} \mathcal{U}^n\mathcal{L}$ is said to be bilateral shift. The dimension of $\mathcal{L}$ is said to be <u>multiplicity</u> of shift. Let $\mathcal{A} : \mathcal{H} \to \mathcal{H}$ be a linear bounded operator and $\mathcal{M}$ an invariant subspace for $\mathcal{A}$. $\mathcal{M}$ is said to reduce $\mathcal{A}$ if it is invariant for $\mathcal{A}^*$ too. $\mathcal{A}$ is said to be irreducible if there exists no nontrivial subspace in $\mathcal{H}$ which reduces $\mathcal{A}$.

The existence of nontrivial invariant subspace of uni- and bilateral shifts is not in question. Thus, only the characterization of them as it has been done for complex Hilbert space in [1] is of interests. Further, having in mind that Spectral Theorem for normal operators [6] and Wold's decomposition for isometries [3] hold true in quaternionic Hilbert space, it follows that only those invariant subspaces, restrictions of shift on which are irreducible, are to be considered.

Lemma. Let $\mathcal{U} : \mathcal{H} \to \mathcal{H}$ be a bilateral shift, $\mathcal{M}$ invariant subspace of $\mathcal{U}$, M symplectic image of $\mathcal{M}$, and U symplectic image of $\mathcal{U}$. Then M is invariant for U and

(i) if $\mathcal{M}$ reduces $\mathcal{U}$, then M reduces U,

(ii) if $\mathcal{U}|\mathcal{M}$ is irreducible, then $U|M$ is irreducible.

Proof: Let $\mathcal{M}$ be an invariant subspace of $\mathcal{U}$, and M, U symplectic image of $\mathcal{M}, \mathcal{U}$ respectively. Obviously M is invariant for U. Suppose that $\mathcal{M}$ contains a subspace $\mathcal{M}_0$ wich reduces $\mathcal{U}$. Then the symplectic image M_0 of $\mathcal{M}_0$ is invariant for U and U^*. Suppose $\mathcal{U}|\mathcal{M}$ is irreducible and $U|M$ is not. Then there exists a subspace $M_0 \neq \{0\}$, $M_0 \subset M$, which reduces U. Since $\varkappa U = U\varkappa$, the subspace M_0 reduces U too. Hence the smallest closed subspace $\varkappa M_0 \vee M_0$ spanned by M_0 and $\varkappa M_0$ reduces U. From $\varkappa(M_0 \vee \varkappa M_0) = M_0 \vee \varkappa M_0$ it follows that $\varkappa M_0 \vee M_0$ is symplectic image of a subspace $\mathcal{M}_0$ of $\mathcal{M}$. Symplectic image of $\mathcal{M}_0$ reduces U, thus $\mathcal{M}_0$ reduces $\mathcal{U}$. Obviously $\mathcal{M}_0 \neq \{0\}$. Contradiction.

Theorem 1: Let $\mathcal{U} : \mathcal{H} \to \mathcal{H}$ be a bilateral shift. Let $\mathcal{M}$ be an invariant subspace of $\mathcal{U}$ such that $\mathcal{U}|\mathcal{M}$ is irreducible. Then there exists a wandering subspace $\mathcal{N}$ for $\mathcal{U}$ such that $\mathcal{M} = \bigoplus_{n+0}^{\infty} \mathcal{U}^n \mathcal{N}$.

Proof: Let H, U, M be the symplectic image of $\mathcal{H}, \mathcal{U}, \mathcal{M}$ respectively. Then $U: H \to H$ is a bilateral shift and M is an invariant subspace of U such that $U|M$ is irreducible ([3]). It follows ([1]) that there exists a subspace N in H wandering for U such that $M = \bigoplus_{n=0}^{\infty} U^n N$. Let us prove that N is symplectic image of a subspace $\mathcal{N}$ in $\mathcal{H}$ which is wandering for $\mathcal{U}$ and such that $\mathcal{M} = \bigoplus_{n=0}^{\infty} \mathcal{U}^n \mathcal{N}$. From $\varkappa M = M$ it follows $M = \bigoplus_{n=0}^{\infty} U^n N = \bigoplus_{n=0}^{\infty} U^n \varkappa N$. Hence $\varkappa N = M \ominus UM = N$. Thus N is symplectic image of a subspace $\mathcal{N}$ in $\mathcal{H}$. For arbitrary $x, y \in \mathcal{N}$, and $n \in \mathbb{N}$

$$(\mathcal{U}^n x, y) = \langle U^n x, y \rangle + \langle U^n x, ky \rangle k = 0.$$

Hence $\mathcal{N}$ is wandering for $\mathcal{U}$ and

$$\text{sympl. image}\Big(\bigoplus_{n=0}^{\infty} \mathcal{U}^n \mathcal{N}\Big) = \bigoplus_{n=0}^{\infty} \text{sympl. image}\,(\mathcal{U}^n \mathcal{N}) = \bigoplus_{n=0}^{\infty} U^n N = M.$$

Thus $\mathcal{M} = \bigoplus_{n=0}^{\infty} \mathcal{U}^n \mathcal{N}$.

Corollary 1: Let $\mathcal{S} : \mathcal{H} \to \mathcal{H}$ be a unilateral shift and $\mathcal{M}$ an invariant subspace of $\mathcal{S}$. Then there exists a wandering subspace $\mathcal{N}$ for $\mathcal{S}$ such that $\mathcal{M} = \bigoplus_{n=0}^{\infty} \mathcal{S}^n \mathcal{N}$.

Corollary 2: Let $\mathcal{S} : \mathcal{H} \to \mathcal{H}$ be a unilateral shift. Let $\mathcal{H}_0$ be the wandering subspace for $\mathcal{S}$ such that $\mathcal{H} = \bigoplus_{n=0}^{\infty} \mathcal{S}^n \mathcal{H}_0$, and let $\mathcal{M}$ be an invariant subspace which reduces $\mathcal{S}$. Then there exists a subspace $\mathcal{N} \subset \mathcal{H}_0$ such that $\mathcal{M} = \bigoplus_{n=0}^{\infty} \mathcal{S}^n \mathcal{N}$.

Thus we have geometric characterization of invariant subspaces of shifts. Now we give a functional-analytic approach. Let $\mathcal{K}$ be a separable quaternionic Hilbert space. Denote by $L^2_{\mathbb{Q}}(\mathcal{K})$ the set of all classes of equivalence a.e. equal, Lebesgue measurable, square integrable $\mathcal{K}$-valued functions defined on $[0,2\pi]$. $L^2_{\mathbb{Q}}(\mathcal{K})$ is separable quaternionic Hilbert in respect to the scalar product

$$(f,g) = \frac{1}{2\pi} \int_0^{2\pi} (f(t),g(t))_{\mathcal{K}}\, dt.$$

Let $(e_n)_{n\in I}$ be an orthonormal basis in $\mathcal{K}$. Put $E_{mn}(t) = e^{imt}e_n$, $m \in \mathbb{Z}$, $n \in I$. $(E_{mn})_{m\in\mathbb{Z}, n\in I}$ is an orthonormal basis in $L^2_{\mathbb{Q}}(\mathcal{K})$, and let $H^2_{\mathbb{Q}}(\mathcal{K})$ denote the subspace in $L^2_{\mathbb{Q}}(\mathcal{K})$ generated by $(E_{mn})_{m \geq 0, n\in I}$. Let $\mathcal{U} : \mathcal{H} \to \mathcal{H}$ be a bilateral shift, $\mathcal{H}_0$ a wandering subspace for $\mathcal{U}$ such that $\mathcal{H} = \bigoplus_{n\in\mathbb{Z}} \mathcal{U}^n \mathcal{H}_0$ and $(e_n)_{n\in I}$ an orthonormal basis in $\mathcal{H}_0$. Put

$$(\Phi\, \mathcal{U}^m e_n)(t) = e^{imt} e_n = E_{mn}(t), \qquad m\in\mathbb{Z},\ n\in I.$$

Then Φ can be extended to the isometry from $\mathcal{H}$ onto $L^2_{\mathbb{Q}}(\mathcal{H}_0)$. Let $\mathcal{T}_E : L^2_{\mathbb{Q}}(\mathcal{H}_0) \to L^2_{\mathbb{Q}}(\mathcal{H}_0)$ be the linear bounded operator defined by the formula

$$\mathcal{T}_E\, E_{mn} = E_{m+1\, n}, \qquad m\in\mathbb{Z}, \quad n\in I.$$

It follows

$$\Phi\, \mathcal{U} = \mathcal{T}_E\, \Phi .$$

Hence $\mathcal{T}_E$ is a representation of $\mathcal{U}$ on $L^2_{\mathbb{Q}}(\mathcal{H}_0)$. Let Θ_n, $n\in\mathbb{Z}$ be a linear bounded operator on the subspace $\Phi\mathcal{H}_0$ of all constants in $L^2_{\mathbb{Q}}(\mathcal{H}_0)$, such that

$$\Theta_n E_{op} = \sum_{m\in I} \vartheta^{(p)}_{nm} E_{om}, \qquad n\in\mathbb{Z}, \quad p\in I$$

and suppose that $\sup_p \sum_{n\in\mathbb{Z}} \|\Theta_n E_{op}\|^2 < \infty$.

Denote by $\mathcal{T}_\Theta$ the linear bounded operator in $L^2_{\mathbb{Q}}(\mathcal{H}_o)$ defined by the formula:

$$\mathcal{T}_\Theta E_{rp} = \sum_{n\in\mathbb{Z}} \sum_{m\in I} \vartheta^{(p)}_{nm} E_{r+n\,m}, \quad r\in\mathbb{Z}, \quad p\in I.$$

$\mathcal{T}_\Theta$ commutes with $\mathcal{T}_E$. Hence $\mathcal{T}_\Theta H^2_{\mathbb{Q}}(\mathcal{H}_o)$ is an invariant subspace of $\mathcal{T}_E$ such that $\mathcal{T}_E \mid \mathcal{T}_\Theta H^2_{\mathbb{Q}}(\mathcal{H}_o)$ is irreducible.

Theorem 2. Let $\mathcal{U}: \mathcal{H} \to \mathcal{H}$ be a bilateral shift, $\mathcal{H}_o$ a wandering subspace for $\mathcal{U}$ such that $\mathcal{H} = \bigoplus_{n\in\mathbb{Z}} \mathcal{U}^n \mathcal{H}_o$, and $\mathcal{M}$ an invariant subspace of $\mathcal{U}$ such that $\mathcal{U}|\mathcal{M}$ is irreducible. Then there exists a $\mathcal{T}_\Theta$ such taht $\Phi\mathcal{M} = \mathcal{T}_\Theta H^2_{\mathbb{Q}}(\mathcal{H}_o)$.

Proof: $\quad \Phi\mathcal{M} = \Phi \bigoplus_{n=0}^{\infty} \mathcal{U}^n \mathcal{N} = \bigoplus_{n=0}^{\infty} \mathcal{T}_E^n \Phi\mathcal{N}$,

where $\Phi\mathcal{N} \subset L^2_{\mathbb{Q}}(\mathcal{H}_o)$. Since $\dim \Phi\mathcal{N} = \dim \mathcal{N} \leq \dim \mathcal{H}_o$([3]), there exists a partial isometry $\mathcal{A}: \Phi\mathcal{H}_o \to L^2_{\mathbb{Q}}(\mathcal{H}_o)$ such that

$$\mathcal{A}\Phi\mathcal{H}_o = \Phi\mathcal{N}.$$

$(E_{om})_{m\in I}$ is an orthonormal basis in $\Phi\mathcal{H}_o$. Put

$$\mathcal{A} E_{om} = \sum_{p\in\mathbb{Z}} \sum_{r\in I} \vartheta^{(m)}_{pr} E_{pr} = \sum_{p\in\mathbb{Z}} \mathcal{T}_E^p \,\Theta_p E_{om},$$

where $\Theta_p : \Phi\mathcal{H}_o \longrightarrow \Phi\mathcal{H}_o$, $p\in\mathbb{Z}$, is defined by formula

$$\Theta_p E_{om} = \sum_{r\in I} \vartheta^{(m)}_{pr} E_{or}.$$

Then

$$\mathcal{A} = \sum_{p\in\mathbb{Z}} \mathcal{T}_E^p \Theta_p.$$

Extend $\mathcal{A}$ on the set of all linear combinations of $(E_{mn})_{m\in\mathbb{Z}, n\in I}$, putting, by definition,

$$\Theta_p E_{mn} = \mathcal{T}_E^m \,\Theta_p E_{on}.$$

Thus

$$\mathcal{A} E_{mn} = \sum_{p\in\mathbb{Z}} \mathcal{T}_E^p \Theta_p E_{mn} = \sum_{p\in\mathbb{Z}} \mathcal{T}_E^{m+p} \,\Theta_p E_{on}$$

$$= \mathcal{T}_E^m \sum_{p\in\mathbb{Z}} \mathcal{T}_E^p \Theta_p E_{on} = \mathcal{T}_E^m \mathcal{A} E_{on}.$$

By continuity exted $\mathcal{A}$ to the partial isometry $\mathcal{T}_\Theta : L^2_{\mathbb{Q}}(\mathcal{H}_o) \to L^2_{\mathbb{Q}}(\mathcal{H}_o)$. Further, from

$$\mathcal{T}_E^n \Phi \mathcal{N} \perp \mathcal{T}_E^m \Phi \mathcal{N}, \quad n,m \in \mathbb{Z}, \quad n \neq m$$

it follows

$$\mathcal{T}_\Theta \mathcal{T}_E^n \Phi \mathcal{H}_0 \perp \mathcal{T}_\Theta \mathcal{T}_E^m \Phi \mathcal{H}_0, \quad n,m \in \mathbb{Z}, \quad n \neq m.$$

Thus

$$\Phi \mathcal{M} = \bigoplus_{n=0}^{\infty} \mathcal{T}_E^n \mathcal{T}_\Theta \Phi \mathcal{H}_0 = \bigoplus_{n=0}^{\infty} \mathcal{T}_\Theta \mathcal{T}_E^n \Phi \mathcal{H}_0$$

$$\mathcal{T}_\Theta \bigoplus_{n=0}^{\infty} \mathcal{T}_E^n \Phi \mathcal{H}_0 = \mathcal{T}_\Theta H_{\mathbb{Q}}^2(\mathcal{H}_0).$$

Take the same notations as in Theorem 2. $\mathcal{U} | \bigoplus_0^\infty \mathcal{H}_0 = \mathcal{S}$ is a unilateral shift. Define an isometry $\Phi^+ : \bigoplus_0^\infty \mathcal{U}^n \mathcal{H}_0 \to H_{\mathbb{Q}}^2(\mathcal{H}_0)$ putting $(\Phi^+ \mathcal{U}^m e_n)(t) = e^{imt} e_n = E_{mn}(t)$, $m \in \mathbb{N}_0$, $n \in I$.
Let $\mathcal{T}_E^+ : H_{\mathbb{Q}}^2(\mathcal{H}_0) \to H_{\mathbb{Q}}^2(\mathcal{H}_0)$ be the linear bounded operator such that

$$\mathcal{T}_E^+ E_{mn} = E_{m+1\, n}, \quad m \in \mathbb{N}_0, \quad n \in I.$$

It follows $\mathcal{T}_E^+ \Phi^+ = \Phi^+ \mathcal{S}$. Put $\Theta_n : \Phi^+ \mathcal{H}_0 \to \Phi^+ \mathcal{H}_0$ to be the linear bounded operator such that

$$\Theta_n E_{op} = \sum_{m \in I} \vartheta_{nm}^{(p)} E_{om}, \quad n \in \mathbb{N}_0, \quad p \in I,$$

and $\sup_p \sum_{n \in \mathbb{N}_0} \| \Theta_n E_{op} \|^2 < \infty$. Denote by $\mathcal{T}_\Theta^+$ the linear bounded operator in $H_{\mathbb{Q}}^2(\mathcal{H}_0)$ such that

$$\mathcal{T}_\Theta^+ E_{rp} = \sum_{n \in \mathbb{N}_0} \sum_{m \in I} \vartheta_{nm}^{(p)} E_{r+n\, m}$$

$\mathcal{T}_\Theta^+$ commutes with $\mathcal{T}_E^+$. Hence $\mathcal{T}_\Theta^+ H_{\mathbb{Q}}^2(\mathcal{H}_0)$ is an invariant subspace of $\mathcal{S}$. If in the proof of the Theorem 2. put

$$\mathcal{T}_E^+, \ \Phi^+, \ H_{\mathbb{Q}}^2(\mathcal{H}_0), \ \mathbb{N}_0$$

instead of $\mathcal{T}_E$, Φ, $L_{\mathbb{Q}}^2(\mathcal{H}_0)$, $\mathbb{Z}$, then it follows:

<u>Theorem 3</u>. Let $\mathcal{S} : \mathcal{H} \to \mathcal{H}$ be a unilateral shift, $\mathcal{H}_0$ the wandering subspace for $\mathcal{S}$ such that $\mathcal{H} = \bigoplus_{n=0}^{\infty} \mathcal{S}^n \mathcal{H}_0$, and $\mathcal{M}$ an invariant subspace of $\mathcal{S}$. Then there exists a $\mathcal{T}_\Theta^+$ such that $\Phi^+ \mathcal{M} = \mathcal{T}_\Theta^+ H_{\mathbb{Q}}^2(\mathcal{H}_0)$.

In case of shifts have multiplicity 1 it can be said more. $L_{\mathbb{Q}}^2(\mathcal{H}_0)$ and $H_{\mathbb{Q}}^2(\mathcal{H}_0)$ are now $L_{\mathbb{Q}}^2$ and $H_{\mathbb{Q}}^2$ where the meaning of these symbols is clear. $\mathcal{T}_{\theta}$, $\mathcal{T}_{\theta}^{+}$ is replaced with $\mathcal{T}_f$ on $L^2_{\mathbb{Q}}$, $H^2_{\mathbb{Q}}$ where $f \in L^{\infty}_{\mathbb{Q}}$, $H^{\infty}_{\mathbb{Q}}$ resp. and

$$\mathcal{T}_f \sum_n u_n e^{int} = \sum_n u_n f(t)\, e^{int}.$$

The consequences of Thms 2. and 3. are

Theorem 4. ([2]). Let X be an invariant subspace for $\mathcal{T}_E$ which does not reduce $\mathcal{T}_E$. Then there exists an $f \in L^{\infty}_{\mathbb{Q}}$ such that $X = \mathcal{T}_f H^2_{\mathbb{Q}}$. Moreover f can be chosen so that $\mathcal{T}_f$ is an isometry.

Theorem 5. ([2]). Let X be an invariant subspace for $\mathcal{T}_E^{+}$. Then there exists an $f \in H^{\infty}_{\mathbb{Q}}$ such that $X = \mathcal{T}_f H^2_{\mathbb{Q}}$. Moreover, if $X \neq \{0\}$ then f can be chosen so that $\mathcal{T}_f$ is an isometry.

One can define inner and outer factors of an $f \in H^{\infty}_{\mathbb{Q}}$. So, for shifts of miltiplicity 1 there is a satisfactory analogy with the Beurling theorem. For details see [2].

REFERENCES

[1] P.R.Halmos, Shifts on Hilbert spaces, J. reine angew. Math. 208 (1961) 102-112.

[2] S.Suljagić, Quaternionic Beurling's theorem, Glasnik Mat. 15 (35) (1980) 327-339.

[3] S.Suljagić, Contribution to functional calculus in quaternionic Hilbert spaces, Ph.D. thesis (in Croatian), Zagreb 1979.

[4] S.Suljagić, Invariant subspaces of shifts in separable quaternionic Hilbert space, Glasnik Mat. (to appear).

[5] B.Sz.-Nagy and C. C.Foiaş, Harmonic analysis of operators in Hilbert space, (in Russian), Mir, Moscow, 1970.

[6] O.Teichmüller, Operatoren im Wachsschen Raum, J. reine angew. Math. 174 (1936) 73-124.

[7] K.Viswanath, Normal operators on quaternionic Hilbert space, Trans.Amer.Math.Soc. 162 (1971) 337-350.

ENERGY IN MARKOV PROCESSES

Z.R.Pop-Stojanović

Introduction

The goal of this presentation is to illustrate the role which the concept of energy plays in the Potential Theory of General Markov Processes. The motivation for development here comes from the classical Potential Theory. In the classical Potential Theory spectacular success was achieved by H.Cartan [4] when he showed how the concepts such as balayage, equilibrium potentials, etc., could be considered as special cases of projections in a Hilbert space. This also opened a new way for treating these difficult concepts. The main tool employed here was the concept of energy. A.Beurling and J.Deny used this tool to develop their "kernel free" potential theory [1],[6].

In classical theory the symmetry of the kernels plays the decisive role. To illustrate the point let us consider the potential of a measure $\mu \geq 0$, namely

$$U(y) = \int |x-y|^{-1}\, d\mu(x) .$$

This potential may be infinite but it has certain regularity properties. For instance, it is continuous from below and it is superharmonic. In the case when U is the potential of a measure with a twice continuously differentiable density h of compact support, one has:

$$\int |\text{grad } U(x)|^2 dx = -\int U(x)\Delta U(x)dx = 4\pi \int U(x)h(x)dx = 4\pi\int\int |x-y|^{-1} h(x)h(y)dx\, dy ,$$

where integration by parts is used. This suggests that it is natural to consider the quadratic form I of C.F.Gauss [12]:

(A) $$I(\rho) = \int\int |x-y|^{-1}\, \rho(x)\rho(y)d\Gamma(x)d\Gamma(y)$$

where ρ is a charge on a surface Γ with density $\rho(y) \geq 0$, which can be extended to measure μ which gives:

(B) $$(\mu,\mu) = \int U(x)d\mu(x) = \int\int |x-y|^{-1}d\mu(x)d\mu(y).$$

In the case when the measure μ is positive and the last integral exists, (B) represents the energy of μ. Note here that in both (A) and (B) the corresponding kernel is symmetric, which will be a starting assumption for the whole theory subsequently developed.

However, there are a few papers dealing with non-symmetric potential densities [13],[21],[24]. The main part of this paper will deal with characterizations of convergence in energy in probabilistic potential theory of general non-symmetric Markov processes and which K. Murali Rao and the author of this paper developed earlier. (See [24], [25],[27]).

Green functions

Let Ω denote a domain in R^d, $d \geq 3$, and Δ the Laplacian operator:

$$\Delta = \sum_{i=1}^{d} \frac{\partial^2}{\partial x_i^2} .$$

The following is known: [2]

There is a unique non-negative symmetric function $G(\cdot,\cdot)$ on $\Omega \times \Omega$ which is continuous off the diagonal and identically infinite on it such that

(1) $$G\Delta\varphi = \int G(\cdot,y)\Delta\varphi(y)dy = -\varphi$$

for each C^2 function φ with compact support in Ω.

(2) For each bounded measurable function f, Gf vanishes at "most points" of $\partial\Omega$. Here, we will not dwell on the meaning of "most points" of $\partial\Omega$.

The function G introduced here is called the Dirichlet Green function for Ω (relative to the Laplacian oeprator). Up to a constant multiple G has the following physical interpretation: The Newtonian potential at $z \in R^d$ due to a unit charge placed at $x \in \Omega$, is

$$|x-y|^{-d+2} .$$

If the boundary $\partial\Omega$ of Ω is grounded the potential at y due to the unit charge at x is $G(x,y)$.

<u>Examples</u>. The Green function for R^d is up to a constant factor equal to

$$| x-y |^{-d+2} .$$

For the ball with center at 0 and radius r, it is (up to a constant factor)

$$G(x,y) = |x-y|^{-d+2} - r^{d-2}|y|^{-d+2}|x-y^*|^{-d+2}$$

where y^* is the inverse of y relative to $\partial B(0,r)$, i.e., $y^* = r^2|y|^{-2}y$.

The Green function for the half-space $(x_1,x_2,\ldots,x_d: x_d > 0)$ is

$$|x-y|^{-d+2} - |x-y^*|^{-d+2} ,$$

where

$$y^* = (y_1,\ldots,-y_d) \quad \text{if} \quad y = (y_1,\ldots,y_d) .$$

See [29].

G is excessive (or superharmonic) in the following sense: for all $x,y \in \Omega$, Ω domain in R^d, and for all r such that the ball $B(y,r)$ with center y and radius r is completely contained in Ω, one has:

(3) $(1/|B|) \int_B G(x,z)dz \leq G(x,y)$, with $|B|$ = volume of B,

and the limit of the above integral is $G(x,y)$ as $r \to 0$.

By using (3) one can deduce that G is strictly positive in Ω. Indeed, if $G(x_0,y_0) = 0$ for some x_0,y_0, one sees using (3) and the lower semi-continuity of G that $G(x_0,y) = 0$. After applying this argument to $G(y,x_0)$, one gets $G(x,y) = 0$, which contradicts (1).

Another useful consequence of (3) is the following fact:

Let K be a compact subset of Ω and $\delta = \text{dist}(K,\partial\Omega)$. Let φ be a smooth function on R^1, which decreases on $[0,\delta]$, $\varphi(\delta) = 0$ and such that

(4) $$-\omega_d \int_0^\delta r^d \varphi'(r)dr = 1 ,$$

where ω_d = area of the unit sphere in R^d. Then one has:

(5) $$\int G(x,y-z)\varphi(|z|)dz \leq G(x,y)$$

for all $x \in \Omega$ and $y \in K$.

Indeed fixing $x \in \Omega$ and $y \in K$, the left side in (4) is close to, (for a suitable subdivision $0 < r_1 < r_2 < \ldots < r_n = \delta$),

(6) $$\sum_i \varphi(r_i)[\phi(r_{i+1}) - \phi(r_i)] ,$$

where

$$\phi(r) = \int_{|z| \leq r} G(x,y-z)dz .$$

After necessary rearrangement we see that the sum in (6) is close to

$$-\int_0^\delta \phi(s)\varphi'(s)ds .$$

By taking into account that $-\varphi'(s) \geq 0$, $\phi(s) \leq s^d \omega_d G(x,y)$ and (4), we arrive at (5).

Potentials

For any positive measure μ the function $G\mu$ defined by

(7) $$G\mu(x) = \int G(x,y)\mu(dy)$$

is called the <u>potential of</u> μ.

The strict positivity of G implies that μ must be a Radon measure unless $G\mu$ is identically infinite. Condition (3) further implies that unless identically infinite, $G\mu$ is locally integrable. We only consider measures μ whose potentials are not identically infinite.

We say that a measure μ or its potential $G\mu$ has finite energy if

(8) $$||\mu||_e^2 = \int G\mu \, d\mu < \infty .$$

$||\mu||_e$ is called the <u>energy of</u> μ .

For any two positive measures μ,ν , the quantity

(9) $$(\mu,\nu)_e = \int G\mu\, d\nu = \int G\nu\, d\mu$$

is called their mutual energy.

We have the following important (only a sketch of the proof is given here)

THEOREM 1. For all positive measures μ,ν

(10) $$2(\mu,\nu)_e \leq \|\mu\|_e^2 + \|\nu\|_e^2 .$$

Proof. Monotone convergence theorem permits us to a assume that μ,ν have compact support K in Ω. Let $\delta = \mathrm{dist}(K,\partial\Omega)$. For each n let $\rho_n \geq 0$ be smooth on R^1, decreasing for $r \geq 0$, $\rho_n(\delta/n) = 0$ and

$$-\omega_d \int_0^{\delta/n} \rho_n'(r)dr = 1 .$$

Write also $\rho_n(z) = \rho_n(|z|)$, $z \in R^d$. Let $f_n=\mu*\rho_n$, $g_m=\nu*\rho_m$. Then, f_n, g_m are smooth function on R^d. From the observation made before $u_n=Gf_n \leq G\mu$, $v_m = Gg_m \leq G\nu$. Also, u_n, v_m are C^∞ in Ω .

Now suppose $\|\mu\|_e < \infty$. Then

$$\| f_n \|_e^2 = \int u_n f_n dx \leq \int G\mu f_n dx = \int Gf_n d\mu \leq \int G\mu\, d\mu = \|\mu\|_e^2 .$$

Similarly for the energy of $g_m(x)dx$. Also integrating by parts one gets:

$$\|f_n\|_e^2 = \int u_n f_n dx = -\int u_n \Delta u_n dx = \int |\mathrm{grad}\, u_n|^2 dx$$

$$(f_n,g_m)_e = \int u_n g_m = -\int u_n \Delta u_m = \int(\mathrm{grad}\, u_n, \mathrm{grad}\, u_m)dx .$$

Therefore one has for all m,n:

$$2(u_n,g_m)_e \leq (u_n,f_n)_e+(u_m,g_m)_e \leq \|\mu\|_e^2 + \|\nu\|_e^2 .$$

By letting $n \to \infty$, one obtains:

$$2\int G\mu g_m \leq \|\mu\|_e^2 + \|\nu\|_e^2 \quad \text{, i.e.}$$

$$2\int Gg_m d\mu \leq \|\mu\|_e^2 + \|\nu\|_e^2 .$$

By taking $m \to \infty$ one gets the final conclusion.

DEFINITION. Let μ be a difference of positive measures of finite energy. We define

$$\|\mu\|_e^2 = \int G\mu \, d\mu$$

and call $\|\mu\|_e$ the energy of the signed measure μ . It follows from the Theorem 1. that the energy of a signed measure is a well-defined real number. Another consequence of Theorem 1. is that $\|\mu\|_e \geq 0$ and $\|\mu\|_e = 0$ implies $\mu = 0$ (the last follows from the proof of Theorem 1).

Theorem 1. shows that $(\mu,\nu)_e$ is an inner product on the space of signed measures of finite energy. This space is not complete in energy norm. [However, the space of positive measures of finite energy is complete [16] and this fact has proved to be useful in potential theory]. It is preferable to give the energy norm to $G\mu$., i.e.,

$$\|G\mu\|_e = \|\mu\|_e = \int G\mu d\mu \quad .$$

Then, the completion of the pre-Hilbert space mentioned above is exactly Sôbolev space W_0^1 or the space of BLD functions: W_0^1 is the completion in gradient norm of C^∞-functions with compact support in Ω .

Brownian Motion

Let W denote the space of all continuous R^d-valued functions on $[0,\infty)$. Let $\mathcal{B}$ denote the σ-field generated by the coordinate maps. Then, there exists a probability measure P on $\mathcal{B}$ [2] with the following properties: If $X_t(w) = w(t)$ for $w \in W$, then

1) $P[X_0 = 0] = 1$.
2) $X = (X_t)$ has independent increments: for $0 \leq t_1 < \ldots < t_n$, the random variables $X_{t_1}, X_{t_2} - X_{t_1}, \ldots, X_{t_n} - X_{t_{n-1}}$ are independent.
3) $E[\exp(i\alpha X_t)] = \exp(-(|\alpha|^2 t)/2)$, where E denotes expectation or integral relative to measure P .

Now consider the map $w \to x+w$ of W into W with $x \in R^d$. We denote the image measure by P^x and the corresponding expectation operator by E^x. We call (W,P^x) the Brownian Motion process. Further, for each $t \geq 0$ we denote by θ_t the shift operator $W \to W$:

$$(\theta_t w)(s) = w(t+s) .$$

A fundamental concept which is in a "natural" way connected with development here is that of stopping or Markov time. In order to define this concept let (F_t) denotes the family of σ-fields generated by the random variables X_s for $s \leq t$. A function

$$T: W \to [0,\infty]$$

is called a Markov time if the set $(T < t)$ is F_t-measurable for all $t \geq 0$, i.e.,

$$(T < t) \in F_t,\ t \geq 0.$$

Then, we denote by F_T ("random events prior to T") the σ-field of sets A such that:

$$A \cap (T < t) \in F_t,\ t \geq 0.$$

Now we can formulate the "Strong Markov property" of Brownian Motion process:

For each measurable $f \geq 0$ on W one has:

$$E^x[f(\theta_T)I_{T<\infty}] = E^x[E^{X_T}[f]I_{T<\infty}].$$

The Brownian Motion process is associated with the Laplace operator Δ in the following way. For each bounded continuous function f on R^d the function u defined by

$$u(t,x) = E^x[f(X_t)]$$

satisfies:

$$\frac{\partial u}{\partial t} = (\Delta/2)u \ ,\ u(0,x) = f(x).$$

Given a bounded subdomain Ω of R^d. We define the Brownian motion "killed upon exit from Ω" as follows: Let ζ be the first exit time from Ω:

$$\zeta(w) = \inf[t > 0: X_t \notin \Omega],$$

and define:

$$X_t^{\zeta}(w) = w(t) \ , \text{ if } \ t < \zeta \ ,$$
$$\delta \ , \text{ if } \ t \geq \zeta \ .$$

Here, the state δ denotes the so-called "cemetery". With this definition $X^{\zeta} = (X_t^{\zeta})$ is a new process possessing also the strong Markov property as explained earlier. Its "state space", i.e. its range is $\Omega \cup \delta$.

Potential theory associated with the killed Brownian Motion

Let G be the Green function of Ω as described in §1. The relationship between G and the Brownian Motion in Ω is given by the following equation:

$$\text{(11)} \qquad Gf(x) = E^x[\int_0^{\zeta} f(X_s)ds] \ , \ x \in \Omega,$$

where f is any non-negative measurable function on Ω.

In order to write a probabilistic "formula" for $G\mu$ we need to introduce the notion of an additive functional. To motivate this write

$$A_t = \int_0^t f(X_s)ds$$

and note that

$$A_{t+s}(w) = A_s(w)+A_t(\theta_s w) \ .$$

This leads to the following definition of an additive functional: A family $A = (A_t)$ of measurable functions is an additive functional if:

1) For every $t > 0$, A_t is F_t-measurable, $A_0=0$.
2) For every s,t, one has:

$$A_{t+s} = A_t+A_s(\theta_t) \ .$$

We say that additive functional A is continuous, increasing, etc., if A is continuous in t, increasing in t, etc. The following result is valid.

THEOREM. Let μ be a positive measure of finite energy. Assume that $G\mu$ is finite everywhere. Then, there is an unique non-negative additive functional such that

$$G\mu(x) = E^x[A_\infty] ,$$

where $A_\infty = \lim_{t\to\infty} A_t$.

The proof of this theorem in much greater generality one can find in [2].

Let us now look for a "probabilistic version" of energy. In order to simplify notations for the rest of this section we shall denote by $X = (X_t)$ the Brownian Motion killed upon exit from Ω. We can write (11) as

$$(12) \qquad s = Gf = \int_0^\infty P_t f\, dt$$

where

$$P_t f(x) = E^x[f(X_t)] .$$

Family of operators (P_t) is a semi-group; property $P_{t+s}=P_tP_s$, for any t,s, follows as a simple consequence of Markov property for X . Denote by $(\cdot,\cdot)$ the inner product in $L^2(\Omega)$ with respect to Lebesgue measure. After realizing that

$$(s-P_ts)/t = (1/t)\int_0^t P_u f\, du ,$$

one gets the following limit relation:

$$(13) \qquad \lim_{t\to 0} [(1/t)(s,s-P_ts)] = (s,f) = \|f\|_e^2 .$$

Thus, the left-hand side of (13) can be used to define the energy of s .

Now by using symmetry of P_t it is not difficult to see that for each $f \in L^2$

$$(14) \qquad (1/t)(f,f-P_tf)$$

is a decreasing function of t. Using this fact it is possible to show that the space H_0^1 is simply the space of functions in $L^2(\Omega)$ for which the limit in (14) as $t \to 0$ is finite.

M.Fukushima [11] defines the limit

$$(15) \qquad \lim_{t\to 0} (1/2t)E^m [(s(X_t)-s(X_0))^2]$$

as the square of the energy of s, where $s \in L^2$, m is the Lebesgue measure and E^m means $\int E^x[\ ]dm(x)$. He shows that these two definitions agree. From a probabilistic point of view the definition in (15) makes more sense. Note here that $s(X_t)-s(X_0)$ is an additive functional. Further, M.Fukushima shows this additive functional is the unique sum of a Martingale additive functional $M=(M_t)$ and an additive functional $A=(A_t)$ of zero energy:

$$s(X_t)-s(X_0) = M_t+A_t ,$$

where additive functionals M and A satisfy the following properties: $A_0=M_0=0$ and $E^x[M_t] = 0$, $\lim_{t\to 0} (1/t)E^m[A_t^2] = 0$.

Levy Processes

Several authors have worked on generalizations of previously introduced concepts of energy: M.Fukushima [11], M.Silverstein [34], J. Bliedner [3], Berg-Forst [10], to name a few. Here, however, we will concentrate our attention on Levy processes.

Let (F_t) be a convolution semi-group of probability measures on R^d. By this we mean:

$$F_t * F_s = F_{t+s} , \quad 0 \leq t,s < \infty .$$

Under some mild conditions there is a stochastic process $X=(X_t)$ which is right-continuous in t with $X_0=0$, with stationary independent increments (and, in particular a Markov process), and such that X_t has distribution F_t. The Fourier transform of F_t has the form:

(1) $$\hat{F}_t = \int \exp(i(\alpha,x))dF_t(x) = \exp(-t\phi(\alpha)) ,$$

where ϕ satisfies:

(2) $$\phi(\alpha) = i(a,\alpha)+Q(\alpha)+ \int[1-e^{i(\alpha,y)}+ \frac{i(\alpha,y)}{1+|y|^2}]\nu(dy)$$

where a is a vector, $Q \geq 0$ a quadratic form and ν a measure satisfying

$$\int (1+|y|^2)^{-1}\nu(dy)< \infty .$$

This is the famous Levy-Khinchine formula. Here, ϕ is called the exponent of the process. The dependence of X on ϕ has been extensively studied.

Now define a family (P_t) for $f \geq 0$, f measurable, by:

(3) $$P_t f(x) = \int f(x+y) F_t(dy) .$$

Then, family (P_t) is a semi-group of operators $P_t P_s = P_{t+s}$, for all $s,t \geq 0$. The above definition makes sense if for example, $f \in L^p$, $p \geq 1$ and P_t is a contraction semi-group on L^p for each $p \geq 1$. Its resolvent U^λ is defined by:

(4) $$U^\lambda f = \int_0^\infty e^{-\lambda t} P_t f \, dt , \; \lambda > 0 .$$

We call a non-negative function f, λ-excessive if

(5) $$\sup_{t>0} e^{-\lambda t} P_t f = f .$$

We will assume that all excessive functions are lower semi-continuous. This happens if and only if U^λ has density u^λ. In this case we can select u^λ such that:

$$y \to u^\lambda(-y)$$

is excessive. (See J.Hawkes [15]). We will assume that this holds for the rest of this paper. With the notion of stopping time introduced earlier, we now define:

(6) $$P_T^\lambda f(x) = E^x[\exp(-\lambda T) f(X_T)] .$$

DEFINITION. An λ-excessive function s is called a class (D) potential if:

$$P_{T_n}^\lambda s(x)$$

decreases to zero for almost all x as $n \to \infty$; here

$$T_n = \inf[t > 0: s(X_t) > n].$$

It is not difficult to show the following Proposition. (See for example, [31]).

PROPOSITION 1. Every class (D) potential is a sum of bounded class (D) potentials.

The following theorem is proved in much greater generality in [2].

THEOREM 2. To every finite class (D) potential s corresponds an unique additive functional $A = (A_t)$ such that:

(7) $$s = E^{\bullet}[A_\infty]$$

where A and X have no common discontinuities (in t).

Combining Proposition 1 and Theorem 2 , we get:

THEOREM 3. Theorem 2 remains valid even if s is not assumed to be finitely valued.

Now we can introduce the notion of energy. Let s be a class (D) potential whose additive functional is A. It is easy to see that

$$p = E^{\bullet}[A_\infty^2]$$

is itself a class (D) potential, provided it is finite almost everywhere . We define $\|s\|_e$ = energy of s by

(8) $$\|s\|_e^2 = \lim_{t\to\infty} (1/t) \int_0^t (p - P_u p)\, du \,.$$

It can be shown that $(1/t)(p - P_t p, 1)$ is an increasing function of t so that the limit in (8) certainly exists.

Details of the above considerations can be found in [27].

It is a known fact [2] that every class (D) potential s can be written as

$$s(x) = \int u(y-x)\mu(dy).$$

Moreover, if s has finite energy then

$$p = E^{\bullet}[A_\infty^2]$$

has the representation

$$p(x) = \int u(y-x)\nu(dx) \,.$$

Then the energy of s is:

(9) $$\|s\|_e^2 = \nu(1) \,.$$

(Here, ν denotes the so-called Revuz measure [33]).

The motivation for the previous comes from the following particular case. Suppose s has finite energy and s is of the form

$$s = Uf$$

for some $f \geq 0$. The additive functional of s is then $\int_0^t f(X_s)ds$.

This implies that

$$P(x) = E^x[(\int_0^\infty f(X_s)ds)^2] .$$

This one can rewrite as:

$$p = 2E^{\cdot}[\int_0^\infty f(X_t)dt \int_t^\infty f(X_s)ds] = 2E^{\cdot}[\int_0^\infty f(X_t)s(X_t)dt]$$
$$= 2\int u(y-x)f(y)s(y)dy .$$

Thus, the energy of s in this particular case is:

$$(10) \qquad \|s\|_e^2 = 2\int s(y)f(y)dy .$$

Unfortunately, this formula <u>does not hold in general</u> when $s=U\mu$. Indeed it holds only if the famous <u>Hypothesis</u> (<u>H</u>) of G.A.Hunt is satisfied.

It is more convenient to express the energy of s in terms of the exponent ϕ. We do this for 1-potentials. However, first we present the following Proposition where we use the fact that λU^λ is a contraction in L^2.

PROPOSITION 2. Let $f \in L^2$. Then for all $\beta > \alpha > 0$,

$$(11) \qquad (U^\alpha f, U^\beta f) \geq (\alpha/\beta) \|U^\alpha f\|^2 .$$

In particular,

$$(12) \qquad (U^\alpha f, f) \geq \alpha \|U^\alpha f\|^2 .$$

<u>Proof</u>. Using the resolvent equation

$$U^\alpha f = U^\beta f + (\beta-\alpha)U^\alpha U^\beta f ,$$

one gets,

$$(U^\alpha f, U^\beta f) = \|U^\alpha f\|^2 - (U^\alpha f, (\beta-\alpha)U^\alpha U^\beta f)$$
$$\geq \|U^\alpha f\|^2 - \|U^\alpha f\| (\beta-\alpha) \|U^\beta U^\alpha f\|$$
$$\geq \|U^\alpha f\|^2 - ((\beta-\alpha)/\beta) \|U^\alpha f\|^2 = (\alpha/\beta) \|U^\alpha f\|^2 .$$

This proves (11). To get (12), multiply both sides of (11) by β and let $\beta \uparrow \infty$. Q.E.D.

Relation (12) says in particular, that for $\alpha > 0$, if the α-energy of $U^\alpha f$ is finite, then $U^\alpha f$ is in L^2. Now, let us express the α-energy in terms of the exponent ψ. Suppose first that $f \in L^2$. Using Plancherel transformation it follows that:

$$(13) \qquad \|U^\alpha f\|^2_{e,\alpha} = 2(U^\alpha f, f) = 2(\hat{U}^\alpha f, \hat{f}) = (1/(\alpha+\psi), |\hat{f}|^2)$$
$$= 2\int(\alpha + \mathrm{Re}(\psi))|\hat{U}^\alpha f|^2 .$$

This result holds whether or not $f \in L^2$. To extend (13) to arbitrary α-potentials of finite energy we need a weak convergence result for whose proof we refer the reader to Theorem 1.5 of [27].

LEMMA 1. Let (s_n) be a sequence of α-excessive functions which increases to s as $n \to \infty$. Assume that s is a potential of finite energy. Then, (S_n) converges to s <u>weakly in energy</u> as $n \to +\infty$.

LEMMA 2. Let s be a 1-potential of finite energy. Then

$$(14) \qquad \|s\|_1^2 = \int (1+\mathrm{Re}\ (\psi))|\hat{s}|^2 .$$

(Here, $\|\cdot\|_1 = \|\cdot\|_{e,1}$) .

<u>Proof</u>. Clearly, it is sufficient to prove (14) assuming that $s = U^\lambda \mu$, with μ a finite measure. Let us first show that s is square integrable. Let $s_n = U^\lambda f_n$ with $f_n \in L^2$ and assume that (s_n) increases to s as $n \to +\infty$. Then $f_n(x)dx$ converges weakly to μ. Now,

$$(15) \qquad \|s\|_1^2 = 2\int s_n f_n = 2\int \mathrm{Re}(1/(1+\psi))|\hat{f}_n|^2\, dx .$$

By the previous Lemma, (s_n) converges weakly in energy to s. By taking convex combinations we may assume that (s_n) converges to s <u>strongly in energy norm</u> as $n \to +\infty$. Now, by applying Fatou's lemma in (15), we get:

$$(16) \quad \|s\|_1^2 \geq 2\int((1+\mathrm{Re}(\psi))/|1+\psi|^2)|\hat{\mu}|^2 = 2\int(1+\mathrm{Re}(\psi))|\hat{s}|^2.$$

This shows that s is necessarily in L^2. By taking for f_n's the following special choice

$$f_n = nU^{n+1}\mu ,$$

one gets from (15):

$$\|s\|_1^2 = 2\int (\mathrm{Re}(1/(1+\psi))n^2/(|n+1+\psi|^2))|\hat{\mu}|^2$$

$$= 2\int ((1+\mathrm{Re}(\psi))n^2/(|n+1+\psi|^2))|\hat{s}|^2$$

Here, because of (16) one may use the dominated convergence theorem. Q.E.D.

However, we can characterize measures of finite energy without using Fourier transforms. Here is the way:

PROPOSITION 3. Let μ be a probability measure and let $\tilde{\mu}$ be its reflection, i.e., $\tilde{\mu}(E) = \mu(-E)$. Then, μ has finite 1-energy if and only if the 1-potential of $\mu*\tilde{\mu}$ is bounded. (Here, $*$ denotes the convolution product of measures).

<u>Proof</u>. Let $\tilde{u}^1(x) = u^1(-x)$. Here, u^1 denotes the density of 1-potential U^1. Then, the Fourier transform of $(u^1+\tilde{u}^1)*\mu*\tilde{\mu}$ is

$$2((1+\mathrm{Re}(\psi))/|1+\psi|^2)|\hat{\mu}|^2 \geq 0 .$$

Thus, by Corollary to Theorem 3 p. 482, in W.Feller [9] ,μ has finite energy if and only if $(u^1+\tilde{u}^1)*\mu*\tilde{\mu}$ is bounded. Since $\tilde{u}^1*\mu*\mu(-x) =$ $= u^1*\mu*\tilde{\mu}(x)$, we see that μ has finite energy if and only if the 1-potential of $\mu*\tilde{\mu}$ is bounded. Q.E.D.

<u>Capacity</u>

Let us return to the Brownian Motion process given in a bounded domain in R^d and let K be a compact subset of this domain. The <u>capacity</u> (Newtonian) of K is defined as maximum mass on K whose potential is less or equal to 1 everywhere. More precisely,

(17) $$C(K) = \sup_{\mu} \{\mu(1)\}$$

where the supremum is taken over all measures μ supported by K and satisfying $G\mu \leq 1$, where $G\mu$ is the potential of μ. It is known that the supremum is attained for an unique measure μ called the <u>capacitary</u> measure for K. An equivalent definition of capacity is given by:

$$(18) \qquad 1/C(K) = \inf_{\mu} \| G \|_e^2$$

where the infimum extends over all probability measures μ on K.

A compact set of capacity zero is also called <u>polar</u> set. This is so beacuse every compact polar set is the set of infinities (poles) of an excessive function. Thus, polar sets are very "small". For example, the Cantor ternary set in $[0,1]$ is not a polar set for two-dimensional Brownian Motion process.

Probabilistically speaking a polar set is "never hit". Conversely, a set that is "never hit" is a polar set.

In general, the capacity of Levy processes can be defined as in (17). Here, (18) also makes sense. However, it is not clear at all that these two definitions agree. For sets of zero capacity these two definitions agree: namely, a compact set has zero capacity if and only if every measure on it has infinite energy.

For $\lambda > 0$, λ-capacities are defined in an analogous way. Now we present the following Lemma whose proof is omitted.

LEMMA 3. Let s be the λ-potential of a measure μ with compact support K. Then

$$(19) \qquad \| s \|_\lambda^2 \geq (\mu(1))^2 / C^\lambda(K) .$$

In particular, the λ-potential of every measure on a compact set has infinite energy if and only if the set is polar.

By combining this Lemma with Lemma 2, we get the following application.

COROLLARY. (H.Kesten,[22]). Let X be a Levy process on the real line whose exponent is ϕ. Then, points are non-polar sets of X if

and only if:
$$\int_0^\infty \mathrm{Re}(1/(1+\psi(z)))dz < \infty .$$

For general Levy processes capacity is usually defined by (17). In the classical case (i.e., non-probabilistic), there is only one measure μ at which the infimum in (18) is attained. This measure is called capacitary distribution. It is interesting to realize that the following Proposition holds in much greater generality:

PROPOSITION 4. Let K be a compact set. There is an unique probability measure μ on K such that $s = U^1\mu$ has smallest 1-energy.

Proof. Let
$$\alpha = \inf\{\|U^1\mu\|_1^2 : \mu(1) = 1,\ \mu \text{ on } K\}.$$

Choose a sequence of measures (μ_n) such that $\mu_n \to \mu$ weakly, and $\|U^1\mu_n\|_1 \to \alpha$, as $n \to +\infty$. Let $s_n = U^1\mu_n$, and $r = U^1g \leq U^1\mu$. Let A_n be the additive functional of s_n and B that of U^1g. Then,
$$E^\bullet[A_{n,\infty}B_\infty] = U^1[r\mu_n] + U^1[s_n g] .$$
Now by letting $n \uparrow \infty$, and using lower semi-continuity of the first term and Fatou lemma on the second term of the right-hand side in the last equality, we get:
$$\liminf_{n\to\infty} E^\bullet[A_{n,\infty}B_\infty] \geq U^1[r\mu] + U^1[sg] = E^\bullet[A_\infty B_\infty].$$
In other words,
$$\liminf_{n\to\infty} (s_n, U^1g)_1 \geq (s, U^1g)_1 .$$
(Here, $(\cdot,\cdot)_1 = (\cdot,\cdot)_{e,1}$).

Since the set $\{U^1g\}$ is dense in energy norm (see Theorem 1.4, [27]), we get that $\|s\|_1^2 = \alpha$. Now if two measures μ, ν have this property so does $(\mu+\nu)/2$. But then,
$$\|U\mu - U\nu\|_1 = 0$$
implying $U\mu = U\nu$. This completes the proof.

We do not know the connection of the distribution given by this Proposition to the capacitary distribution. For example, the potential of the capacitary distribution is bounded. However, we do not know if the same statement holds for the distribution given by this Proposition.

Hypothesis (H) of G.A.Hunt

Let us return to the classical case. Let (s_n) be a decreasing sequence of excessive functions converging to s. Function s can fail to be excessive only by failing to be lower semi-continuous. This can be seen directly from the definition of an excessive function which we gave earlier. The so-called lower semi-continuous regularization $\overline{s}$ of s, (which always exists), is excessive. It was a question of great interest to determine the "size" of the set where $\overline{s} - s > 0$. This set can be written as the union:

$$\bigcup_n (s-\overline{s} \geq 1/n).$$

It was found quickly that each of the sets

$$(s-\overline{s} \geq 1/n)$$

was "thin". Let us illustrate the concept of a "thin" set in terms of the Brownian Motion process. A "thin" set is a set which in order to visit, the Brownian traveller will need positive amount of time, regardless where he starts his journey. A simple argument then tells us that it is possible to visit a thin set at most countably often. A later development, using the continuity principle of Evans-Vasilescu showed that a thin set is indeed a polar set, i.e., the Brownian Motion process never visits such a set.

The previous question is also connected with the Dirichlet problem in the following way. As it is well-known, the Dirichlet problem is to find a harmonic function in a given domain with given "boundary data". Several solutions were given: the Perron method using super and subharmonic function, the Wiener method using expanding subdomains, the Stochastic solution, etc. All these solutions agreed when the domain had smooth boundary. For arbitrary domains those points of the boundary where the solution agreed with the boundary data, were called regular. It was also known that the set of irregular points of a boundary was a countable union of thin sets, or, a semi-polar set. In terms of the Brownian Motion process, a semi-polar set can only be visited a countable number of times. Thus, the distinction between a semi-polar and polar set is that the former is visited only a countable number while the later never at all. For the Brownian Motion process semi-polar sets are polar. This fact leads to a "strong" potential theory in the classical case.

In series of papers [17],[18],[19], G.A.Hunt set a foundation for Potential Theory of General Markov Processes. In order to get a strong analogue of the classical potential theory, G.A.Hunt postulated that semi-polar sets were polar. This postulate is known today as Hunt's Hypothesis (H). It appears to be a difficult problem to decide what is the class of Levy processes satisfying this hypothesis.

One important consequence of Hypothesis (H) is that all additive functionals are continuous. In fact, this is equivalent to Hypothesis (H). (See the Chapter 6 in [2]). M.Kanda [21], K.Murali Rao [30], have shown that all stable Levy processes satisfy hypothesis (H).Also, [14] gives a large class of Levy processes satisfying hypothesis (H).

Semi-polar sets are "small" sets but it seems difficult to say what "small" is. In the classical potential theory a compact cet K is thin if and only if for every measure μ on K, $U\mu$ is infinite μ -a.e. (Here, U denotes a potential operator defined earlier). We give below a simple analogue of this fact in the probabilistic potential theory.

A process X is called transient if it permanently leaves every compact set in finite time.

PROPOSITION 5. Let K be a compact set and assume that K is thin relative to X, where X is a transient process. Let μ be a probability measure on K . Then, $s=U\mu$ is discontinuous at μ-almost all points.

Proof. Let $L = \{x: s$ is continuous at $x\}$. Suppose $\mu(L) > 0$. We may assume L is compact set. Let $s_L = U(\mu|L)$. Since

$$s_L + U(\mu|K-L) = s ,$$

s is continuous on L and each summand is lower semi-continuous, we must have s_L continuous at each point of L. Since $\mu(L) > 0$, $L \subset K$. Let (D_n) be a sequence of open sets which decreases to L. Then, the hitting times T_n of D_n , $n=1,2,\ldots,$ increase to the hitting time T_L of L. Now

$$P_{D_n} s_L = s_L \quad \text{and} \quad X_{T_L} \in L .$$

Now, by continuity one gets:

$$E^{\cdot}[s_L(X_{T_L})] = s_L .$$

Strong Markov property of the process X can be used repeatedly to conclude that

$$E^{\cdot}[s_L(X_{T_{n,L}})] = s_L ,$$

where $T_{n,L}$ is the n-th hitting time of L. We can repeat this procedure by using the transfinite induction. Using thinness we can thus assert the existence of a sequence (R_n) of stopping times such that $R_n \uparrow \infty$, as $n \to \infty$, $X_{R_n} \in K$ and $E^{\cdot}[s_L(X_{R_n})] = s_L$. However, this contradicts the assumption that X is a transient process. Q.E.D.

ACKNOWLEDGMENT. This presentation is based on the manuscript of the forthcoming study on energy in Markov process which Murali Rao and the author are preparing and which will appear in the near future. The author wishes to express his gratitude to Murali Rao for his help concerning this work.

REFERENCES:

[1] Beurling, A., and Deny, J., Espaces de Dirichlet I. Le cas elementaire, Acta Math. 99,3-4 (1958).

[2] Blumenthal, R.M., and Getoor, R.K., Markov processes and their potential theory, Academic Press, New York (1968).

[3] Bliedtner,J., Functional spaces and their exceptional sets, Seminar on Potential Theory II Lecture Notes in Math., 226, Springer-Verlag, Berlin-Heidelberg-New York, (1971).

[4] Cartan, H., Sur les fondaments de la theorie du potentiel, Bull. Soc.Math.France, 69,71-96 (1941).

[5] Cartan, H., Theorie du potentiel newtonian: energie, capacite, suites de potentiels, Bull.Soc.Math.France, 73, 74-106 (1945).

[6] Deny,J., Sur les infinis d'un potentiel, C.R.Acad.Sci.Paris, 224,524-525 (1947).

[7] Deny, J., Les potentiels d'energie finie, Acta Math., 82, 107-183 (1950).

[8] Doob,J.L., Classical Potential Theory and its Probabilistic Counterpart, Springer-Verlag, New York, Berlin, Heidelberg, Tokyo, (1984).

[9] Feller,W., An Introduction to Probability Theory and Its Applications, Vol.II, Second Edition, Wiley, (1970).

[10] Forst, G., A characterization of non-symmetric translation invariant Dirichlet forms, Theorie du potentiel et analyse harmonique, Lecture Notes in Math., 404, 113-125, Springer Verlag (1974).

[11] Fukushima,M., Dirichlet Forms and Markov Processes, North-Holland/ Kodansha (1980).

[12] Gauss,C.F., Allgemeine Lehrsatze in Beziehung auf die im verkehztem Verhaltnisse des Quadrats der Entfernung wizkende Anziehungs - und Abstossangskraffe, Leipzig (1840).

[13] Glover,J., Energy and Maximum Principle for nonsymmetric Hunt Processes, Probability Theory and its Applications, XXVI,4, 757-768, Moscow, (1981).

[14] Glover,J., and Rao,Murali, Hunt's Hypothesis (H) and Getoor's Conjecture, Annals of Probability, (to appear).

[15] Hawkes,J., Potential Theory of Levy processes, Proc. of London Math.Soc., 38, 335-352, (1979).

[16] Helms, L.L., Introduction to Potential Theory, Wiley (1969).

[17] Hunt,G.A., Markoff Processes and Potentials, Illinois Journal of Mathematics, Vol. I, 44-93, (1957).

[18] Hunt, G.A., Markoff Processes and Potentials, Illinois Journal of Mathematics, vol.I, 316-369, (1957).

[19] Hunt,G.A., Markoff Processes and Potentials, Illinois Journal of Mathematics, vol.II, 151-213, (1958).

[20] Ito,K., and McKean,H.P., Jr. Diffusion Processes and their simple Paths, Springer-Verlag, New York, Berlin, Heidelberg, (1965).

[21] Kanda,M., Two theorems on capacity for Markov processes with independent increments, Z.W. und V.Gebiete, 35,159-166 (1976).

[22] Kesten,H., Hitting probabilities of single points for processes with stationary independent increments, Memoirs of the AMS , 93, (1969).

[23] Meyer, P.A., Probability and Potentials, Blaisdell, (1966).

[24] Pop-Stojanović, Z.R.,Rao,Murali, Some results on Energy, Seminar on Stohastic Processes 1981, 135-150, Boston, Birkhauser (1981).

[25] Pop-Stojanović, Z.R.,Rao, Murali, Remarks on Energy, Seminar on Stohastic Processes 1982, 229-237, Boston, Birkhauser (1983).

[26] Pop-Stojanović, Z.R., Rao, Murali, Further Results on Energy, Seminar on Stohastic Processes 1983, 143-150, Boston, Birkhauser, 61984).

[27] Pop-Stojanović,Z.R.,Rao, Murali, Convergence in Energy, Z.W. und V.Gebiete, 69, 593-608,(1985).

[28] Port,S., and Stone,C., Brownian Motion and Classical Potential Theory, Academic Press, New York (1978).

[29] Rao, Murali, Brownian Motion and Classical Potential Theory, Aarhus University, Lecture Notes Series, No.47 (1977).

[30] Rao, Murali, On a result of M.Kanda, Z.W. und V.Gebiete, 41, 35-37 (1977).

[31] Rao, Murali, A note on Revuz measure, Seminaire de probabilites XIV, 1978/79, Lecture Notes in Mathematics 784, 1980, 418-436.

[32] Rao, Murali, Representation of excessive functions, Math. Scand. 51, 367-381,(1982).

[33] Revuz,D., Measures associees a functionnelles additives de Markov I, Trans.Am.Math.Soc., 148, 501-531,(1970).

[34] Silverstein,M.L., The sector condition implies that semipolar sets are quasi-polar, Z.W.verw.Gebiete, 41, 13-33,(1977).

[35] Weil,M., Quasi-processus et energie, Seminaire de probabilites V Lecture Notes 191, 347-361, Berlin-Heidelberg-New York, Springer, (1971).

[36] Wiener,N., The Dirichlet Problem, J.Math.Phys. 3, 127-146,(1924).

ON THE ALMOST CONVERGENCE

D.Butković, H.Kraljević, N.Sarapa

In this paper we prove some new results on almost convergent sequences. These theorems are connected with some results of B.E. Rhoades, who applied them to Markov chains; therefore, as an application, we obtain generalizations of some results on the almost convergence of the Markov chain transition probabilities.

1. Generalities on matrix limiting methods

We denote by S the set of all sequences $x=(x_n)_{n\in\mathbf{N}}$ of complex numbers. Furthermore, let C be the set of all convergent sequences $x \in S$. Then S is a complex vector space and C is a subspace. The mapping $\lim: C \to \mathbf{C}$, $\lim(x) = \lim_{n\to\infty} x_n$, is a linear form on C.

Let $A = [a_{nk}]_{n,k\in\mathbf{N}}$ be a complex infinite matrix. Denote by $S(A)$ the set of all $x \in S$, such that Ax is a well-defined sequence, i.e. such that the series

$$(Ax)_n = \sum_{k=1}^{\infty} a_{nk}x_k$$

converges for every $n \in \mathbf{N}$. $S(A)$ is a subspace of S. Now, set

$$C(A) = \{x \in S(A): Ax \in C\} ;$$

the subspace $C(A)$ is called the convergence domain of the matrix A. Define the linear form A-lim on $C(A)$ by $A\text{-}\lim = \lim\circ A$, i.e.

$$A\text{-}\lim(x) = A\text{-}\lim_{n\to\infty} x_n = \lim (Ax) = \lim_{n\to\infty} \sum_{k=1}^{\infty} a_{nk}x_k .$$

The elements of $C(A)$ are called A-convergent sequences.

The matrix A is called conservative (or convergence preserving) if $C \subseteq C(A)$. It is called regular (or limit preserving), if it is coservative and if in addition $\lim = A\text{-}\lim|C$, i.e.

$$A\text{-}\lim_{n\to\infty} x_n = \lim_{n\to\infty} x_n , \quad \forall x \in C.$$

The following two theorems characterize conservative and regular matrices. For the proofs see [6]; very nice functional-analytic proofs can be found in [15].

1.1. THEOREM (Kojima-Schur). The matrix A is conservative if and only if the following three conditions are satisfied:

(i) $$\sup_{n\in\mathbf{N}} \sum_{k=1}^{\infty} |a_{nk}| < +\infty .$$

(ii) For every $k \in \mathbf{N}$ there exists $a_k = \lim_{k\to\infty} a_{nk}$.

(iii) There exists $a = \lim_{n\to\infty} \sum_{k=1}^{\infty} a_{nk}$.

In this case for every $x \in C$

$$A\text{-}\lim(x) = a\cdot\lim(x) + \sum_{k=1} a_k\cdot(x_k - \lim(x)). \quad \blacksquare$$

1.2. THEOREM (Toeplitz-Silverman). The matrix A is regular if and only if the following three conditions are satisfied:

(i) $$\sup_{n\in\mathbf{N}} \sum_{k=1}^{\infty} |a_{nk}| < +\infty .$$

(ii) For every $k \in \mathbf{N}$ $\lim_{n\to\infty} a_{nk} = 0$.

(iii) $\lim_{n\to\infty} \sum_{k=1}^{\infty} a_{nk} = 1$. $\blacksquare$

1.3. REMARK. Let B be the space of all bounded sequences. We note that $B \subsetneq S(A)$ for every conservative matrix A; it is an obvious consequence of the condition (i) in 1.1.

2. Almost convergence and strong regularity

The best known example of a regular matrix is the so called Cesàro matrix:

$$C=[a_{nk}]=\begin{bmatrix} 1 & 0 & 0 & \cdots \\ \frac{1}{2} & \frac{1}{2} & 0 & \cdots \\ \frac{1}{3} & \frac{1}{3} & \frac{1}{3} & 0 \cdots \\ \cdots & & & \end{bmatrix} ; \quad a_{nk} = \begin{cases} \frac{1}{n} & 1 \le k \le n, \\ 0 & k > n . \end{cases}$$

Then

$$C\text{-}\lim_{n\to\infty} x_n = \lim_{n\to\infty} \frac{1}{n}(x_1+\ldots+x_n);$$

this convergence is called the convergence in the Cesàro sense. For

$r \in \mathbf{N}$ we define the shifted Cesàro matrix $C^{(r)}$ by

$$C^{(r)} = [a_{nk}] = \begin{bmatrix} \overbrace{0 \ldots 0}^{r} & 1 & 0 & 0 & \ldots \\ 0 \ldots 0 & \frac{1}{2} & \frac{1}{2} & 0 & \ldots \\ 0 \ldots 0 & \frac{1}{3} & \frac{1}{3} & \frac{1}{3} & 0 \ldots \\ \ldots & \ldots & \ldots & \ldots & \ldots \end{bmatrix};$$

$$a_{nk} = \begin{cases} \frac{1}{n} & r < k \leq r+n, \\ 0 & k \leq r \text{ or } k > r+n \end{cases}.$$

The $C^{(r)}$-convergence of a sequence $x=(x_1,x_2,\ldots)$ is exactly the C-convergence of the shifted sequence $(x_{r+1},x_{r+2},\ldots)$.

2.1. LEMMA. For every $r \in \mathbf{N}$ $\mathcal{C}(C) = \mathcal{C}(C^{(r)})$. Furthermore, for every $x \in \mathcal{C}(C)$, $C^{(r)}$-lim (x) = C-lim (x).

Proof. It follows immediately from the equalities

$$\frac{1}{n}(x_{r+1}+\ldots+x_{r+n}) = \frac{n+r}{n}\cdot\frac{1}{n+r}(x_1+\ldots+x_{r+n}) - \frac{1}{n}(x_1+\ldots+x_r);$$

$$\frac{1}{n+r}(x_1+\ldots+x_{r+n}) = \frac{1}{n+r}(x_1+\ldots+x_r) + \frac{n}{n+r}\cdot\frac{1}{n}(x_{r+1}+\ldots+x_{r+n}) . \blacksquare$$

Therefore, for every $x \in \mathcal{C}(C)$ and for a = C-lim(x) we have

$$a = \lim_{\to\infty} \frac{1}{n}(x_{r+1}+\ldots+x_{r+n}), \quad r \in \mathbf{N} \cup \{0\}. \tag{1}$$

If the convergence in (1) is uniform with respect to r, the sequence $x = (x_n)$ is called <u>almost convergent.</u> We denote by AC the space of all almost convergent sequences.

2.2. LEMMA. $AC \subseteq \mathcal{B}$.

Proof. Let $x \in AC$. Set a = C-lim(x) and $y_{n,r} = (C^{(r)}x)_n = \frac{1}{n}(x_{r+1}+\ldots+x_{r+n})$. Then

$$\forall \varepsilon > 0, \exists m \in \mathbf{N}, \forall r \in \mathbf{N}, n \geq m \Rightarrow |y_{n,r}-a| < \varepsilon .$$

Especially, there exists $m \in \mathbf{N}$ such that

$$|y_{n,r}-a| < 1 \ , \quad r,n \in \mathbf{N} \ , \ n \geq m.$$

Then

$$(2) \qquad |y_{n,r}| < 1+|a| \ , \quad r,n \in \mathbf{N} \ , \ n \geq m.$$

Now, for every $r \in \mathbf{N}$ we have:

$$(m+1)(y_{m+1,r-1}-y_{m,r}) = \sum_{i=r}^{r+m} x_i - \frac{m+1}{m} \sum_{i=r+1}^{r+m} x_i =$$

$$= x_r + (1-\frac{m+1}{m}) \sum_{i=r+1}^{r+m} x_i = x_r - y_{m,r} .$$

Thus, using (2) we obtain that for every $r \in \mathbf{N}$

$$|x_r| \leq |x_r-y_{m,r}|+|y_{m,r}| = (m+1)|y_{m+1,r-1} - y_{m,r}|+|y_{m,r}| <$$

$$< 2(m+1)(1+|a|) + 1 +|a| = (2m+3)(1+|a|). \ \blacksquare$$

The C-limit (or $C^{(r)}$-limit) of $x \in AC$ is usually denoted by $\mathrm{Lim}(x) = \mathrm{Lim}_{n\to\infty} x_n$. Therefore, $\mathrm{Lim} = C\text{-}\lim|AC$.

2.3. LEMMA. $C \subseteq AC$.

Proof. Let $x \in C$ and $a = \lim(x)$. Let $\varepsilon > 0$ be arbitrary. Choose $M > 0$ so that $|x_n-a| \leq M \ \forall n$. Let $m \in \mathbf{N}$ be such that

$$n > m \Rightarrow | x_n-a| < \frac{\varepsilon}{2} .$$

Choose $p \in \mathbf{N}$ so that $p \geq m$ and $\frac{m}{p} M < \frac{\varepsilon}{2}$. Then we have for every $r \in \mathbf{N}$ and every $n > p$:

$$| \frac{1}{n} (x_{r+1}+\ldots+x_{r+n})-a| \leq \frac{1}{n} \sum_{j=1}^{n} |x_{r+j}-a| =$$

$$= \frac{1}{n} \sum_{j=1}^{m} |x_{r+j}-a| + \frac{1}{n} \sum_{j=m+1}^{n} |x_{r+j}-a| <$$

$$< \frac{p}{n} \cdot \frac{m}{p} M + \frac{n-m}{n} \cdot \frac{\varepsilon}{2} < \frac{p}{n} \cdot \frac{\varepsilon}{2} + \frac{n-m}{n} \cdot \frac{\varepsilon}{2} \leq \varepsilon . \ \blacksquare$$

2.4. REMARK. It can be easily shown by examples that AC is a proper subspace of $\mathcal{B} \cap C(C)$ and that C is a proper subspace of AC.

A matrix A is called <u>strongly regular</u> if it is regular and if $AC \subseteq C(A)$.

2.5. THEOREM. Let $A = [a_{nk}]$ be a regular matrix. The following four properties are mutually equivalent:

(i) A is strongly regular.

(ii) $\lim_{n\to\infty} \sum_{k=1}^{\infty} |a_{n,k} - a_{n,k+1}| = 0$.

(iii) $\lim_{m\to\infty} \sum_{k=m}^{\infty} |a_{n,k}-a_{n,k+1}| = 0$ uniformly in $n \in \mathbf{N}$.

(iv) For every $x \in \mathcal{B}$ $\lim_{n\to\infty} \sum_{k=1}^{\infty} (a_{n,k}-a_{n,k+1})x_k = 0$.

In this case $A\text{-}\lim|AC = \mathrm{Lim}$.

Proof. (i) $\Rightarrow$ (ii). Suppose that (ii) is not satisfied. Then there exist an $\varepsilon > 0$ and an infinite subset J of $\mathbf{N}$ such that

$$\sum_{k=1}^{\infty} |a_{n,k}-a_{n,k+1}| \geq 4\varepsilon , \quad n \in J .$$

Then for every $n \in J$ either

$$\sum_{k=1}^{\infty} |a_{n,2k}-a_{n,2k+1}| \geq 2\varepsilon \tag{3}$$

or

$$\sum_{k=1}^{\infty} |a_{n,2k-1}-a_{n,2k}| \geq 2\varepsilon \tag{4}$$

holds true. We will suppose that there is an infinite subset $I \subseteq J$ such that (3) is satisfied for every $n \in I$ (the other case is treated similarly).

Now, we construct inductively two strictly increasing sequences (n_j) in I and (p_j) in $2\mathbf{N}$ in such a way that the following conditions are satisfied:

$$\sum_{k < p_j} |a_{n_j,k}| < \frac{\varepsilon}{4} , \ j \in \mathbf{N} ; \tag{5}$$

$$\sum_{k \geq p_{j+1}} |a_{n_j,k}| < \frac{\varepsilon}{4} , \ j \in \mathbf{N} ; \tag{6}$$

$$\sum_{k=0}^{\frac{p_{j+1}-p_j}{2}-1} |a_{n_j,p_j+2k} - a_{n_j,p_j+2k+1}| > \varepsilon , \ j \in \mathbf{N} . \tag{7}$$

Let us explain the procedure of choosing p_j and n_j. First take $p_1=2$. Since A is regular, we have $\lim_n a_{n,1} = 0$, thus we can choose $n_1 \in I$ so that (5) is satisfied for $j=1$. Now, because of the regularity of A and because of (3), we can choose $p_2 > p_1$, $p_2 \in 2\mathbf{N}$, so that (6) and (7) are satisfied for $j=1$. Suppose now that $i \geq 2$ and that $p_1,\dots,p_i$ and $n_1,\dots,n_{i-1}$ were chosen so that (5), (6) and (7) are satisfied for $1 \leq j \leq i-1$. Choose $n_i \in I$ so that $n_i > n_{i-1}$

and that

$$|a_{n_i,k}| < \frac{\varepsilon}{4(p_i-1)} \qquad \text{for} \quad 1 \leq k < p_i \ .$$

Then (5) is satisfied for $j=i$. Since

$$\sum_{k=1}^{(p_i-2)/2} |a_{n_i,2k}-a_{n_i,2k+1}| \leq 2\sum_{k=1}^{p_i-2} |a_{n_i,k}| < \frac{\varepsilon}{2} \ ,$$

using (3) for $n=n_i$ and the regularity of A we see that we can choose $p_{i+1} \in 2\mathbf{N}$ so that $p_{i+1} > p_i$ and that (6) and (7) are satisfied for $j=i$.

Now we define the sequence (x_n) in $\mathbf{C}$ as follows:

$$x_{p_j+2k} = \begin{cases} 0 & \text{if} \quad a_{n_j,p_j+2k}=a_{n_j,p_j+2k+1} \\ (-1)^j \dfrac{|a_{n_j,p_j+2k}-a_{n_j,p_j+2k+1}|}{a_{n_j,p_j+2k}-a_{n_j,p_j+2k+1}} & \text{otherwise} \end{cases}$$

$$x_{p_j+2k+1} = -x_{p_j+2k} \ , \ j \geq 1, \ 0 \leq k \leq \frac{p_{j+1}-p_j-2}{2} \ . \quad \text{Then}$$

(8) $$a_{n_j,p_j+2k}x_{p_j+2k}+a_{n_j,p_j+2k+1}x_{p_j+2k+1} = (-1)^j|a_{n_j,p_j+2k}-a_{n_j,p_j+2k+1}|$$

Let $y = Ax$. Then by (8) we have

$$y_{n_j} = (-1)^j \sum_{k=0}^{(p_{j+1}-p_j-2)/2} |a_{n_j,p_j+2k}-a_{n_j,p_j+2k+1}| + \sum_{k<p_j} a_{n_j,k}x_k + \sum_{k\geq p_{j+1}} a_{n_j,k}x_k$$

Hence, using (5), (6), (7) and $|x_k| \leq 1$ we find that for every $j \in \mathbf{N}$

$$|y_{n_j}-y_{n_{j+1}}| \geq \sum_{k=0}^{(p_{j+1}-p_j-2)/2} |a_{n_j,p_j+2k}-a_{n_j,p_j+2k+1}| +$$

$$+ \sum_{k=0}^{(p_{j+2}-p_{j+1}-2)/2} |a_{n_{j+1},p_{j+1}+2k}-a_{n_{j+1},p_{j+1}+2k+1}| - \sum_{k<p_j} |a_{n_j,k}| -$$

$$- \sum_{k\geq p_{j+1}} |a_{n_j,k}| - \sum_{k<p_{j+1}} |a_{n_{j+1},k}| - \sum_{k\geq p_{j+2}} |a_{n_{j+1},k}| >$$

$$> \varepsilon + \varepsilon - \frac{\varepsilon}{4} - \frac{\varepsilon}{4} - \frac{\varepsilon}{4} - \frac{\varepsilon}{4} = \varepsilon \ .$$

Thus $y = (y_n)$ does not converge, i.e. $x \notin C(A)$. On the other hand, since for every $n \in \mathbf{N}$ $|x_n| \leq 1$ and $x_{n+1}=-x_n$ or $x_{n+2}=-x_{n+1}$, it is easy to see that

$$\left| \frac{1}{n} \ (x_{r+1}+\ldots+x_{r+n})\right| \leq \frac{2}{n} \ , \quad r \geq 0 \ , \ n \geq 1 \ .$$

Thus, $x \in AC$ (and $\mathrm{Lim}(x) = 0$). This shows that A is not strongly regular.

(ii) $\Rightarrow$ (iii). Suppose that (ii) holds true and set $b_{n,k} = |a_{n,k} - a_{n,k+1}|$. Let $\varepsilon > 0$. By (ii) there exists $n_0 \in \mathbf{N}$ such that

(9) $$\sum_{k=1}^{\infty} b_{n,k} < \varepsilon \quad \text{for} \quad n \geq n_0 .$$

Furthermore, since $\sum_{k=1}^{\infty} b_{n,k} \leq 2 \sum_{k=1}^{\infty} |a_{n,k}| < +\infty$, we can find $k_0 \in \mathbf{N}$ such that

(10) $$\sum_{k>k_0} b_{n,k} < \varepsilon \quad \text{for} \quad 1 \leq n < n_0 .$$

(9) and (10) show that (iii) holds true.

(iii) $\Rightarrow$ (ii). Suppose that (iii) holds true and let $\varepsilon > 0$. Then there exists $k_0 \in \mathbf{N}$ such that

$$\sum_{k>k_0} b_{n,k} < \frac{\varepsilon}{2} , \quad n \in \mathbf{N} .$$

Now, choose $n_0 \in \mathbf{N}$ so that

$$\sum_{k=1}^{k_0} |a_{n,k}| < \frac{\varepsilon}{4} , \quad n \geq n_0 ;$$

this is possible because A was supposed to be regular. Then we have for $n \geq n_0$:

$$\sum_{k=1}^{\infty} b_{n,k} \leq 2 \sum_{k=1}^{k_0} |a_{n,k}| + \sum_{k>k_0} b_{n,k} < 2 \cdot \frac{\varepsilon}{4} + \frac{\varepsilon}{2} = \varepsilon .$$

(ii) $\Rightarrow$ (i). Assume that (ii) is fulfilled. Let $x \in AC$ and let $a = \mathrm{Lim}(x)$. Set

$$A_n = \sum_{k=1}^{\infty} a_{nk} , \qquad y = Ax .$$

Since $x \in AC$, by 2.2. x is bounded, hence there is an $N > 0$ such that

(11) $$|x_k| \leq N , \qquad k \in \mathbf{N}.$$

Furthermore, by the regularity of A we can choose $M > 0$ such that

(12) $$\sum_{k=1}^{\infty} |a_{n,k}| \leq M , \quad n \in \mathbf{N} .$$

Let $\varepsilon > 0$ be arbitrary. Since $x \in AC$ and $a = \mathrm{Lim}(x)$, there exists $m \in \mathbf{N}$ such that

(13) $$|b_r| < \frac{\varepsilon}{4M}, \quad b_r = \frac{1}{m}(x_{r+1}+\ldots+x_{r+m})-a, \quad r=0,1,2,\ldots$$

Choose $n_0 \in \mathbf{N}$ so that

(14) $$\sum_{k=1}^{\infty} |a_{n,k}-a_{n,k+1}| < \frac{\varepsilon}{2N(m-1)}, \quad n \geq n_0,$$

(15) $$|a_{n,k}| < \frac{\varepsilon}{6N(m-1)}, \quad n \geq n_0, \quad k < m,$$

and

(16) $$|A_n a - a| < \frac{\varepsilon}{4}, \quad n \geq n_0.$$

This is possible because of (ii) and because of the regularity of A (note that $\lim_n A_n = 1$).

Now, we have

$$a = \frac{1}{m}(x_k+\ldots+x_{k+m-1})-b_{k-1}, \quad k \in \mathbf{N},$$

hence

$$y_n - A_n a = \sum_{k=1}^{\infty} a_{n,k}x_k + \sum_{k=1}^{\infty} a_{n,k}b_{k-1} - \frac{1}{m}\sum_{k=1}^{\infty}\sum_{j=0}^{m-1} a_{n,k}x_{k+j} =$$

$$= \sum_{k=1}^{\infty} a_{n,k}x_k + \sum_{k=1}^{\infty} a_{n,k}b_{k-1} - \frac{1}{m}\sum_{j=0}^{m-1}\sum_{k=j+1}^{\infty} a_{n,k-j}x_k.$$

Thus

(17) $$y_n - A_n a = \sum_{k=1}^{m-1} a_{n,k}x_k - \frac{1}{m}\sum_{j=0}^{m-1}\sum_{k=j+1}^{m-1} a_{n,k-j}x_k + \sum_{k=1}^{\infty} a_{n,k}b_{k-1} +$$

$$+ \sum_{k=m}^{\infty} x_k \{a_{n,k} - \frac{1}{m}\sum_{j=0}^{m-1} a_{n,k-j}\}.$$

Now, suppose that $n \geq n_0$. By (11) and (15) we get

(18) $$|\sum_{k=1}^{m-1} a_{n,k}x_k| \leq N\cdot\sum_{k=1}^{m-1} |a_{n,k}| < N(m-1)\cdot\frac{\varepsilon}{6N(m-1)} = \frac{\varepsilon}{6}.$$

Furthermore, since the number of summands in the second member of the right hand side of (17) is $m(m-1)/2$, we obtain

(19) $$|\frac{1}{m}\sum_{j=0}^{m-1}\sum_{k=j+1}^{m-1} a_{n,k-j}x_k| < \frac{N}{m}\cdot\frac{m(m-1)}{2}\cdot\frac{\varepsilon}{6N(m-1)} = \frac{\varepsilon}{12}.$$

Using (12) and (13) we find

$$(20)\qquad \left|\sum_{k=1}^{\infty} a_{n,k}b_{k-1}\right| < M\cdot\frac{\varepsilon}{4M} = \frac{\varepsilon}{4}\ .$$

Finally, let us estimate the apsolute value of the last member in the right hand side of (17).

$$\left|\sum_{k=m}^{\infty} x_k \{a_{n,k}-\frac{1}{m}\sum_{j=0}^{m-1} a_{n,k-j}\}\right| \le \frac{N}{m}\sum_{j=0}^{m-1}\sum_{k=m}^{\infty}|a_{n,k}-a_{n,k-j}| \le$$

$$\le \frac{N}{m}\sum_{j=0}^{m-1}\sum_{k=m}^{\infty}\sum_{p=1}^{j}|a_{n,k-p+1}-a_{n,k-p}| =$$

$$= \frac{N}{m}\sum_{p=1}^{m-1}(m-p)\sum_{k=m}^{\infty}|a_{n,k-p+1}-a_{n,k-p}| =$$

$$= \frac{N}{m}\sum_{p=1}^{m-1}(m-p)\sum_{q=m-p}^{\infty}|a_{n,q+1}-a_{n,q}| =$$

$$= \frac{N}{m}\sum_{q=1}^{\infty}|a_{n,q+1}-a_{n,q}|\sum_{p=\max\{1,m-q\}}^{m-1}(m-p) \le$$

$$\le \frac{N}{m}\cdot\frac{m(m-1)}{2}\cdot\sum_{q=1}^{\infty}|a_{n,q+1}-a_{n,q}|\ .$$

Thus, by (14)

$$(21)\qquad \left|\sum_{k=m}^{\infty} x_k \{a_{n,k}-\frac{1}{m}\sum_{j=0}^{m-1} a_{n,k-j}\}\right| \le \frac{N}{m}\cdot\frac{m(m-1)}{2}\cdot\frac{\varepsilon}{2N(m-1)} = \frac{\varepsilon}{4}\ .$$

Now, from (16), (17), (18), (19), (20) and (21) we obtain

$$n \ge n_0 \Rightarrow |y_n-a| < \frac{\varepsilon}{6}+\frac{\varepsilon}{12}+\frac{\varepsilon}{4}+\frac{\varepsilon}{4}+\frac{\varepsilon}{4} = \varepsilon\ .$$

Therefore, the sequence y converges. This means that $x \in C(A)$. So, we have proved that $AC \subseteq C(A)$, i.e. A is strongly regular. Furthermore, we have obtained that $A\text{-}\lim(x) = \lim(y) = a = \mathrm{Lim}(x)$.

(ii) => (iv) is trivial.

(iv) => (ii). Suppose that (ii) is not satisfied. Then there exists an $\varepsilon > 0$ and an infintie subset J of $\mathbf{N}$ such that

$$\sum_{k=1}^{\infty}|a_{n,k}-a_{n,k+1}| \ge 2\varepsilon\ ,\quad n \in J$$

Similarly, as in the proof of (i) => (ii) we can construct inductively two strictly increasing sequences (n_j) in J and (p_j) in $\mathbf{N}$ in such a way that the following conditions are satisfied:

(22) $$p_1=1\ ,\quad \sum_{k\le p_j} |a_{n_j,k}| < \frac{\varepsilon}{4}\ ,\qquad j\in\mathbf{N}\ ;$$

(23) $$\sum_{k\ge p_{j+1}} |a_{n_j,k}| < \frac{\varepsilon}{4}\ ,\qquad j\in\mathbf{N}\ ;$$

(24) $$\sum_{k=p_j}^{p_{j+1}-1} |a_{n_j,k}-a_{n_j,k+1}| > \varepsilon\ ,\qquad j\in\mathbf{N}\ .$$

Now, we define the sequence $x=(x_k)$ in C as follows: if $p_j \le k < p_{j+1}$ we set

$$x_k = \begin{cases} 0 & \text{if } a_{n_j,k} = a_{n_j,k+1}\ ,\\ \dfrac{|a_{n_j,k}-a_{n_j,k+1}|}{a_{n_j,k}-a_{n_j,k+1}} & \text{if } a_{n_j,k} \ne a_{n_j,k+1}\ .\end{cases}$$

Then $x \in \mathcal{B}(|x_k| \le 1)$ and for any $j\in\mathbf{N}$

$$(a_{n_j,k}-a_{n_j,k+1})x_k = |a_{n_j,k}-a_{n_j,k+1}|,\quad p_j \le k < p_{j+1}\ .$$

Thus

$$\Big|\sum_{k=1}^{\infty} (a_{n_j,k}-a_{n_j,k+1})x_k\Big| \ge \Big|\sum_{k=p_j}^{p_{j+1}-1} |a_{n_j,k}-a_{n_j,k+1}|\Big| -$$

$$- \Big|\sum_{k<p_j} (a_{n_j,k}-a_{n_j,k+1})x_k\Big| - \Big|\sum_{k\ge p_{j+1}} (a_{n_j,k}-a_{n_j,k+1})x_k\Big| \ge$$

$$\ge \sum_{k=p_j}^{p_{j+1}-1} |a_{n_j,k}-a_{n_j,k+1}| - 2\sum_{k\le p_j} |a_{n_j,k}| - 2\sum_{k\ge p_{j+1}} |a_{n_j,k}| >$$

$$> 2\varepsilon - 2\cdot\frac{\varepsilon}{4} - 2\cdot\frac{\varepsilon}{4} = \varepsilon\ .$$

This shows that (iv) is not satisfied. ■

2.6. REMARK. The equivalence (i) $\Leftrightarrow$ (ii) was proved by G.G. Lorentz in [13]. The equivalence (ii) $\Leftrightarrow$ (iv) is due to R.G. Cooke [6], and (ii) $\Leftrightarrow$ (iii) was proved by D. Butković [3]; the condition (iii) was first considered by L.W. Cohen [5].

Denote by SR the set of all strongly regular matrices. By definition we have that $AC \subseteq C(A)$ for every $A \in SR$. The following theorem shows that in fact

(25) $$AC = \bigcap_{A\in SR} C(A)\ .$$

The result is even better: it is enough to take the intersection over a much smaller class of matrices.

A matrix $A = [a_{n,k}]$ is called <u>generalized Cesàro matrix</u> if it is obtained from the Cesàro matrix C by shifting rows; i.e. for every $n \in \mathbf{N}$ there exists $p(n) \in \mathbf{N}$ such that

$$(26) \qquad a_{nk} = \begin{cases} 0 & \text{if } 1 \leq k < p(n) \text{ or } k \geq p(n)+n, \\ \frac{1}{n} & \text{if } p(n) \leq k < p(n)+n. \end{cases}$$

Let G denote the set of all generalized Cesàro matrices. Then obviously $G \subseteq SR$.

2.7. THEOREM. $AC = \bigcap_{A \in G} C(A)$.

Proof. Suppose on the contrary that there exists a sequence x such that $x \in C(A)$ for every $A \in G$ but that $x \notin AC$.

Suppose that $A\text{-lim}(x) \neq B\text{-lim}(x)$ for some $A, B \in G$, $A=[a_{n,k}]$, $B=[b_{n,k}]$. Define $D=[d_{n,k}]$ by

$$d_{n,k} = \begin{cases} a_{n,k} & \text{if } n \text{ is odd}, \\ b_{n,k} & \text{if } n \text{ is even}. \end{cases}$$

Then $D \in G$, but $x \notin C(D)$, contrary to the assumption. Thus A-limits of x coincide for all $A \in G$. Denote this number by a.

Now, since x is not almost convergent, there exist $\varepsilon > 0$, an infinite subset J of $\mathbf{N}$ and a function $p: J \to \mathbf{N}$ such that

$$(27) \qquad \left| \frac{1}{n} (x_{p(n)} + x_{p(n)+1} + \ldots + x_{p(n)+n-1}) - a \right| \geq \varepsilon, \quad n \in J.$$

Extend the function p to $\mathbf{N}$ by $p(n) = 1$, $n \in \mathbf{N} \setminus J$, and define $A = [a_{n,k}] \in G$ by (26). Then (27) shows that $a \neq A\text{-lim}(x)$, a contradiction. ■

2.8. REMARK. The above proof is adapted from the proof of a more general theorem of G.M.Petersen [16]. The Petersen's results were rediscovered by H.T.Bell [1]. B.E.Rhoades, evidently, unaware of these results, obtained in [17] a weaker result:

$$(28) \qquad AC = \bigcap_{A \in H} C(A),$$

where H is the set of the so called "hump-matrices": $A=[a_{n,k}]$ is a hump-matrix if A is regular, if

$$\lim_{n\to\infty} \sup_k |a_{n,k}| = 0 ,$$

and if for every $n \in \mathbf{N}$ there exists $p(n) \in \mathbf{N}$ such that

$$\begin{array}{lll} a_{n,k} \leq a_{n,k+1} & \text{if} & 1 \leq k < p(n) \\ a_{n,k} \geq a_{n,k+1} & \text{if} & k \geq p(n) . \end{array}$$

Then $G \subset H \subset SR$ and so Rhoades' result also implies (25).

2.9. REMARK. We note that it was shown by Lorentz [14] that there does not exist any countable subset $N \subset SR$ such that

$$AC = \bigcap_{A \in N} C(A).$$

3. A theorem on almost convergent sequences

B.E.Rhoades has obtained a result on the almost convergence of the Markov chain transition probabilities [17] ; it was proved using a theorem that can be reformulated as follows: Let $(x_{(m-1)d+j})_{m\in\mathbf{N}}$ be convergent sequences for $j=1,\ldots,d$; then the sequence $(x_n)_{n\in\mathbf{N}}$ is almost convergent and

$$\text{(29)} \qquad \operatorname{Lim}_{n\to\infty} x_n = \frac{1}{d} \sum_{j=1}^{d} \lim_{m\to\infty} x_{md+j} .$$

It is tempting to check a more general statement: supposing that the sequences $(x_{(m-1)d+j})_{m\in\mathbf{N}}$ are only almost convergent, is it true that $(x_n)_{n\in\mathbf{N}}$ is almost convergent and does the formula (29) still holds true after replacing lim by Lim in the right-hand side? In fact this will be an easy consequence of a more general theorem which we are going now to explain and prove.

Let $\omega\colon \mathbf{N} \to \mathbf{N}$ be a strictly increasing function. For $x \in S$ denote by x^ω the sequence defined as follows

$$x_n^\omega = \begin{cases} x_k & \text{if} \quad \omega(k) = n , \\ 0 & \text{if} \quad n \neq \omega(k) \text{ for every } k \in \mathbf{N}. \end{cases}$$

We say that the sequence x^ω is obtained from the sequence x by <u>dilution</u> (with the dilution function ω).

Let ω be as above. Define the nondecreasing function $\lambda_\omega\colon \mathbf{N} \to \mathbf{N}$ by

$$\lambda_\omega(n) = \max \{k \in \mathbf{N}\colon \omega(k) \leq n\}.$$

Note that the mapping $\omega \mapsto \lambda_\omega$ is a bijection from the set $\mathcal{D}$ of all strictly increasing functions $\omega: \mathbf{N} \to \mathbf{N}$ onto the set $\mathcal{E}$ of all non-decreasing functions $\lambda: \mathbf{N} \to \mathbf{N}$ such that $\lambda(n+1) - \lambda(n) \leq 1 \;\forall n$ and such that $\lim_n \lambda(n) = +\infty$.

We say that ω is a <u>dilution function with density</u> if there exists

$$\rho(\omega) = \lim_{n\to\infty} \frac{1}{n} \lambda_\omega(n) .$$

The number $\rho(\omega) \in [0,1]$ is called the <u>density of the dilution</u> ω. For every $r \in \mathbf{N}$ we have

$$\frac{1}{n} (\lambda_\omega(n+r)-\lambda_\omega(r)) = \frac{n+r}{n} \cdot \frac{1}{n+r} \lambda_\omega(n+r) - \frac{1}{n}\lambda_\omega(r)$$

and this obviously converges to $\rho(\omega)$ as n tends to infinity. If this convergence is uniform with respect to $r \in \mathbf{N}$, we say that ω is a <u>dilution function with uniform density</u>.

Let $\mathcal{F}$ be the set of all sequences $x = (x_n)$ such that $x_n \in \{0,1\}$ for every n and such that $x_n = 1$ for infinitely many $n \in \mathbf{N}$. We have an obvious bijection $\varphi: \mathcal{D} \to \mathcal{F}$; $x = \varphi(\omega)$ is defined by

$$x_n = \begin{cases} 1 & \text{if } n = \omega(k) \text{ for some } k \in \mathbf{N} , \\ 0 & \text{otherwise} . \end{cases}$$

In this case

$$\lambda_\omega(n) = x_1 + \ldots + x_n .$$

Therefore, ω is a dilution function with density if and only if $\varphi(\omega) \in \mathcal{C}(C)$ (and in this case $\rho(\omega) = C\text{-lim}(\varphi(\omega))$); furthermore, ω is a dilution function with uniform density if and only if $\varphi(\omega) \in \mathcal{AC}$.

3.1. THEOREM. (i) Let $x \in \mathcal{C}(C)$ and let ω be a dilution function with density $\rho(\omega)$. Then $x^\omega \in \mathcal{C}(C)$ and $C\text{-lim}(x^\omega) = \rho(\omega) \cdot C\text{-lim}(x)$.

(ii) Suppose that $x \in \mathcal{AC}$ and that ω has uniform density. Then $x^\omega \in \mathcal{AC}$ (and of course $\mathrm{Lim}(x^\omega) = \rho(\omega) \cdot \mathrm{Lim}(x)$).

Proof. (i) Set

$$a_n = \frac{1}{n} (x_1 + \ldots + x_n)$$
$$b_n = \frac{1}{n} (x_1^\omega + \ldots + x_n^\omega) .$$

Then obviously

$$b_n = \frac{1}{n} \lambda_\omega(n) a_{\lambda_\omega(n)}$$

and since $\lambda_\omega(n) \to \infty$ as n tends to infinity, we obtain the assertion immediately.

(ii) Set

$$y_{n,r} = \frac{1}{n} (x_{r+1} + \ldots + x_{r+n}) ,$$

$$z_{n,r} = \frac{1}{n} (x^\omega_{r+1} + \ldots + x^\omega_{r+n}) .$$

Then it is not difficult to see that

$$z_{n,r} = \frac{1}{n} (\lambda_\omega(r+n) - \lambda_\omega(r)) \, y_{\lambda_\omega(r+n)-\lambda_\omega(r), \lambda_\omega(r)} .$$

Again the assertion follows immediately. ■

Let $\Omega = (\omega_1, \ldots, \omega_d)$ be a d-tuple of dilution functions, such that **N** is the disjoint union of the ranges of $\omega_1, \ldots, \omega_d$. Let $x_1, \ldots, x_d \in S$. Define the sequence x by

$$x = x_1^{\omega_1} + \ldots + x_d^{\omega_d} .$$

We say that x is the <u>superposition of the sequences</u> $x_1, \ldots, x_d$ with the distribution Ω. We have an immediate consequence of 3.1:

3.2. COROLLARY. Let x be the superposition of the sequences $x_1, \ldots, x_d$ with the distribution $\Omega = (\omega_1, \ldots, \omega_d)$.

(i) If $x_1, \ldots, x_d \in C(C)$ and if the functions $\omega_1, \ldots, \omega_d$ have densities, then $x \in C(C)$ and

$$\text{C-lim}(x) = \rho(\omega_1) \cdot \text{C-lim}(x_1) + \ldots + \rho(\omega_d) \cdot \text{C-lim}(x_d).$$

(ii) If $x_1, \ldots, x_d \in AC$ and if the functions $\omega_1, \ldots, \omega_d$ have uniform densities, then $x \in AC$ and

$$\text{Lim}\,(x) = \rho(\omega_1) \cdot \text{Lim}(x_1) + \ldots + \rho(\omega_d) \cdot \text{Lim}\,(x_d) . \quad ■$$

Especially, we obtain a generalization of the Rhoades' theorem:

3.3. COROLLARY. Let x be a sequence such that the sequences $(x_{(m-1)d+j})_{m \in \mathbf{N}}$ are almost convergent $(j=1, \ldots, d)$. Then x is almost convergent and

$$\text{Lim}(x) = \frac{1}{d} \sum_{j=1}^{d} \lim_{m \to \infty} x_{md+j} .$$

Proof. Let $x_1,\dots,x_d$ be the almost convergent sequences defined by

$$x_{j,m} = x_{(m-1)d+j} \qquad j=1,\dots,d,\ m \in \mathbf{N} .$$

Then x is the superposition of $x_1,\dots,x_d$ with distribution $\Omega = (\omega_1,\dots,\omega_d)$, where

$$\omega_j(m) = (m-1)d+j , \quad j=1,\dots,d,\ m \in \mathbf{N}.$$

Then

$$\lambda_{\omega_j}(n) = [\frac{n+d-j}{d}] \quad j=1,\dots,d , \quad n \in \mathbf{N}$$

($[a]$ denoting the greatest integer k such that $k \leq a$), and so

$$\rho(\omega_j) = \lim_{n\to\infty} \frac{1}{n} \lambda_{\omega_j}(n) = \frac{1}{d} , \qquad j = 1,\dots,d .$$

Furthermore,

$$0 \leq \frac{1}{n} (\lambda_{\omega_j}(n+r)-\lambda_{\omega_j}(r)) = \frac{1}{n} [\frac{n+r+d-j}{d}] - \frac{1}{n} [\frac{r+d-j}{d}] \leq$$

$$\leq \frac{n+r+d-j}{nd} - \frac{r-1}{nd} = \frac{n+d-j+1}{nd}$$

and we see that $\omega_1,\dots,\omega_d$ have uniform densities. Now, the assertion follows from 3.2. ■

We note that the Rhoades' theorem was proved in a more complicated way using (28) , i.e. by checking that $x \in C(A)$ for every $A \in H$.

4. A formula for Markov chains

Let P be a transition matrix for a discrete time stationary Markov chain with countable state space, and let $P^n = [p_{ij}^{(n)}]$ be the n-th step transition matrix. An old result of Kolmogorov [12] states that for every $i,j \in \mathbf{N}$ there exists

$$\pi_{ij} = \lim_{n\to\infty} \frac{1}{n} \sum_{k=1}^{n} p_{ij}^{(k)},$$

which means that there exists the Cesàro limit

$$\Pi(P) = \lim_{n\to\infty} \frac{1}{n} \sum_{k=1}^{n} P^k = C\text{-}\lim_{n\to\infty} P^n .$$

Of course, we have also

$$\Pi(P) = \lim_{n\to\infty} \frac{1}{n} \sum_{k=0}^{n-1} P^k , \quad P^0 = I .$$

In [17] B.E.Rhoades proved that

$$\Pi(P) = A\text{-}\lim_{n\to\infty} P^n \tag{30}$$

for every strongly regular matrix A ; this means that (P^n) is almost convergent. Rhoades obtained this result by adapting the Kolmogoroff's proof to the matrices $A \in H$ and using (28). On the other hand, for $A \in G$ there is a very simple proof of (30), which also gives the almost convergence of (P^n) according to 2.7. It can be obtained by modifying the first proof of Yosida and Kakutani in [18]. If $A \in G$ is given by (26) we have to put

$$q_{ij}^{(n)} = \frac{1}{n} \sum_{m=p(n)}^{p(n)+n-1} p_{ij}^{(m)}$$

and to check that

$$\left| \sum_{k=1}^{\infty} p_{ik} q_{kj}^{(n)} - q_{ij}^{(n)} \right| = \frac{1}{n} \left| p_{ij}^{(p(n)+n)} - p_{ij}^{(p(n))} \right| \le \frac{2}{n} \quad ;$$

by the diagonal method, we have to find an increasing sequence $(n_k)_{k\in\mathbf{N}}$ in $\mathbf{N}$ such that $\pi_{ij} = \lim_k q_{ij}^{(n_k)}$ exists, and then conclude that $\pi_{ij} = A\text{-}\lim_{n\to\infty} p_{ij}^{(n)}$.

For a matrix sequence (P_n) 3.3. gives

$$\operatorname{Lim}_n P_n = \frac{1}{d} \sum_{j=1}^{d} \operatorname{Lim}_m P_{md+j} \tag{31}$$

in case that the sequences $(P_{(m-1)d+j})_{m\in\mathbf{N}}$ are all almost convergent. From this we obtain

4.1. THEOREM. For the Markov transition matrix P and every $d \in \mathbf{N}$

$$\operatorname{Lim}_n P^n = \frac{1}{d} \sum_{j=0}^{d-1} \operatorname{Lim}_m P^{md+j} \, . \tag{32}$$

Proof. We have the existence of $\operatorname{Lim}_n P^n$ and $\operatorname{Lim}_m P^{md}$ which imply the existence of $A\text{-}\lim_m P^{md}$ for every $A \in SR$. The apsolute convergence of $\sum_{k=1}^{\infty} p_{ik}^{(j)}$ gives that

$$P^j \cdot A\text{-}\lim_m P^{md} = A\text{-}\lim_m P^{md+j} \, ,$$

which in turn gives the existence of $\operatorname{Lim}_m P^{md+j}$, $j=0,\dots,d-1$. The proof now completes by (31).

Alternatively, knowing that $(P^{(m-1)d+j})_{m\in\mathbf{N}}$ and $(P^n)_{n\in\mathbf{N}}$ are almost convergent $(j=0,1,\dots,d-1)$, we can obtain the above formula if we prove that

(33) $$\operatorname{C-lim}_n P^n = \frac{1}{d} \sum_{j=0}^{d-1} \operatorname{C-lim}_n P^{md+j}$$

because of $\text{C-lim}|AC = \text{Lim}$. And (33) can be obtained by going to the limit in the identity

(34) $$P^{d-1} \cdot \frac{1}{nd} \sum_{j=1}^{nd} P^j = \frac{1}{d} (I+P+\ldots+P^{d-1}) \cdot \frac{1}{n} \sum_{j=1}^{n} P^{jd} .$$

The same argument as above allows going with P^k inside the limit in the right-hand side of (34). The proof concludes by the fact that $P^{d-1}\Pi(P) = \Pi(P)$ (see [18]). ■

From 4.1. we see clearly how taking an appropriate d we can calculate the elements of $\operatorname{Lim}_n P^n$ via simple limits (as in [4]).

Let j be a nonrecurrent state. Then $\lim_n p_{ij}^{(n)} = 0$ and so $\pi_{ij} = 0$. This agrees with the value of π_{ij} obtained from the right-hand side of (32), since $\lim_n p_{ij}^{(n)} = 0$ implies that $\lim_m p_{ij}^{(md+k)} = 0$ for every k.

Let j be a recurrent state. Then it belongs to some essential class with the period d_j and the mean recurrence time m_{jj}. But then $\lim_m p_{ij}^{(md_j+k)}$ exists. In [4] this limit is denoted by

$$f_{ij}^{*}(k) \cdot \frac{d_j}{m_{jj}}$$

and it turns out that $f_{ij}^{*}(k)$ is the probability of finding the process that starts at i in the state j at the step $n \equiv k \pmod{d_j}$. The right-hand side of (32) gives

(35) $$\pi_{ij} = \frac{1}{d_j} \sum_{k=0}^{d_j-1} f_{ij}^{*}(k) \frac{d_j}{m_{jj}} = \frac{f_{ij}^{*}}{m_{jj}} ,$$

where f_{ij}^{*} is the probability of finding the process at j if it starts at i. If we put $m_{jj} = \infty$ for a nonrecurrent state (as in [4]) we can write $\pi_{ij} = f_{ij}^{*}/m_{jj}$, regardless the recurrence of the state j

Therefore, we can look at (32) as at a formula for reducing Lim to lim. Instead of $\sum_{q=0}^{d-1} P^q$ we can take a sum with q from any set of representatives of (mod d)-classes: $\operatorname{Lim}_m P^{md}$ absorbes powers of P^m . In case of a chain transient in the sense of [11], i.e. of a chain with recurrent classes consisting of single elements ("absorbing states"), $\lim_n P^n$ exists and (32) reduces to a trivial identity.

Another form of (32), showing the relationship between the generalized limits (Lim) of the iterates of transition probabilities of the process and of the process observed at every d^{th} step, is the following:

$$\Pi(P) = \frac{1}{d}(I+P+\ldots+P^{d-1})\Pi(P^d) . \tag{36}$$

This can be regarded as a generalization of the formula obtained by Bharoucha-Reid, who considered chains with the property $P^{d+1} = P$ and $d \in \mathbf{N}$ is the smallest possible. Then P^d is an idempotent matrix and so $\Pi(P^d) = P^d$. Thus, we obtain the result of [2], i.e.

$$\Pi(P) = \frac{1}{d}(P^d+P+\ldots+P^{d-1}) = \frac{1}{d}(P+P^2+\ldots+P^d).$$

In [2] chains with $P^d=P$ are said to be periodic and the smallest such d is the period of the chain. Because the usual periodicity is a class property, it is natural to compare chains periodic in one sense or another, which consist of one essential class only. In the finite state space, periodicity in the sense of [2] yields periodicity in the usual sense with the same period d. This follows from the Frobenius theory and the fact that $P^{d+1}=P$ gives precisely d distinct roots of unity as eigenvalues of P (see e.g.[7]). (36) then gives that the (i,j)-th entry of $\mathrm{Lim}\, P^n$ is $p_{ij}^{(q(i,j))}/d$ with $q(i,j) \in \{1,\ldots,d\}$ such that $p_{ij}^{(q(i,j))n} \neq 0$. In particular, $p_{ii}^{(d)}/d$ is the reciprocal of the mean recurrence time for i and the original chain $(X_n, n \in \mathbf{N})$; it is just d times smaller than (i,i)-th entry in P^d, which is the reciprocal of the mean recurrence time for i in the chain $(X_{dn})_{n\in\mathbf{N}}$. An example of such a chain is given by

$$P = \begin{bmatrix} 0 & 0 & \frac{1}{2} & \frac{1}{2} \\ 0 & 0 & \frac{1}{2} & \frac{1}{2} \\ \frac{1}{2} & \frac{1}{2} & 0 & 0 \\ \frac{1}{2} & \frac{1}{2} & 0 & 0 \end{bmatrix} .$$

5. Some remarks on chains and strong regularity

5.1. REMARK. The existence of A-limits of the iterates of transient probabilities for strongly regular methods implies that a number of theorems involving Cesàro means can be generalized to theorems involving sums with matrix elements as coefficients. As an example we state a result analogous to Theorem 3 in [4], p. 38. It can be obtained by the same proof, keeping in mind that the interchange of the order of summation in (37) bellow is justified by a discrete version of the Fubini's theorem.

THEOREM. Let $P^k = [p_{ij}^{(k)}]$ be the k-th power of a Markov chain transition matrix, let $\Pi(P) = [\pi_{ij}]$ and let $A = [a_{nk}] \in SR$. Then the series

$$(37) \qquad \sum_{j=1}^{\infty} \left(\sum_{k=1}^{\infty} a_{nk} p_{ij}^{(k)} \right)$$

in j converges uniformly with respect to $n \in \mathbf{N}$ if and only if $\sum_j \pi_{ij} = 1$.

5.2. **REMARK.** An interesting interpretation of the condition (ii) in 2.5. is possible in cases when the matrix $A = [a_{nk}]$ is for a fixed $i \in \mathbf{N}$ given by

$$a_{nk} = p_{ik}^{(k)}, \quad n,k \in \mathbf{N} .$$

here $[p_{ik}^{(n)}] = P^n$ and P is a Markov chain transition matrix; we consider transient or null-recurrent chain that starts at i (see[10]). A necessary and sufficient condition for a lower triangular matrix A to come from a Markov chain in this way is given in [10] (Th. 4.2,p.19). If the characteristic function of a set $E \subseteq \mathbf{N}$ is described by a sequence $x = (x_j)$ ($x_j=1$ if $j \in E$, $x_j=0$ otherwise), the condition (ii) in 2.5. implies that

$$\lim_{n\to\infty} \sum_{k=1}^{\infty} p_{ik}^{(n)} x_k$$

(the probability to find the process in E after a long time) does not depend on shifting the set E.

For most of the usual limiting methods "almost all" sequences of 0's and 1's are evaluated to 1/2 (the so called Borel property of a limiting method, see [8] and [9]), but this property appearently does not imply strong regularity (see [14], p. 255). Of course, if a method includes a method with the Borel property, it has the Borel property itself.

We list some usual methods that can be associated with Markov chains.

5.3. **EXAMPLE.** The Euler methods E_t ($0 < t < 1$). The matrix $E_t = [e_{nk}(t)]$ is given by

$$e_{nk}(t) = \begin{cases} \binom{n-1}{k-1} t^{k-1}(1-t)^{n-k} & \text{if } 1 \le k \le n , \\ 0 & \text{if } k > n . \end{cases}$$

All these methods can be associated to chains (see [10], p.21); all Euler methods have the Borel property [8], Th. 3.2, p. 562).

5.4. EXAMPLE. Balanced methods (simple Riesz means). Let $b_0, b_1, b_2, \ldots$ be a sequence of non-negative real numbers with $b_0=1$ and $\Sigma b_k = +\infty$. Set

$$s_n = b_0+b_1+\ldots+b_n \ , \quad a_{nk} = \begin{cases} b_{k-1}/s_{n-1} & \text{if } 1 \le k \le n , \\ 0 & \text{if } k > n . \end{cases}$$

If the sequence (b_k) is non-increasing, the matrix $A = [a_{nk}]$ is regular and can be derived from chains ([10],p.22). These methods include the ordinary Cesàro method with the matrix C (ibid.,p.24). Hence, these methods have the Borel property (see also [9], 2.2, p.229) and are strongly regular. The ordinary Cesàro method is a balanced method with $b_k = 1 \ \forall k$.

5.5. EXAMPLE. Nörlund methods. With the same b_k and s_n as in 5.4. these methods are given by the matrices

$$a_{nk} = \begin{cases} b_{n-k}/s_{n-1} & \text{if } 1 \le k \le n , \\ 0 & \text{if } k > n . \end{cases}$$

If the sequence (b_k) is non-increasing, the corresponding method includes the Cesàro methods and thus they are strongly regular with the Borel property ([9],p. 233). The question what Nörlund methods may be associated to Markov chains was solved only in particular cases (see [10], p.25). An example of a class of Nörlund methods coming from chains are the Cesàro methods (C,α), $\alpha > 0$, given by the matrices $C^{\alpha} = [c^{\alpha}_{nk}]$:

$$c^{\alpha}_{nk} = \begin{cases} \dfrac{\alpha(n-1)!}{(\alpha+n-1)(\alpha+n-2)\ldots(\alpha+n-k)(n-k)!} & \text{if } 1 \le k \le n , \\ 0 & \text{if } k > n . \end{cases}$$

In this case the sequence $b_0, b_1, b_2, \ldots$ given by:

$$b_0=1, \ b_k = \frac{\alpha(\alpha+1)\ldots(\alpha+k-1)}{k!} \ , \ k \in \mathbf{N} .$$

(The ordinary Cesàro method is just $(C,1)$, i.e. $C^1 = C$). All (C,α), $\alpha > 0$, have the Borel property (see [8], Th. 1.7,p.557).

References:

[1] H.T.Bell, Order summability and almost convergence, Proc.Amer. Math.Soc. 38 (1973), 548-552.

[2] A.T.Bharoucha-Reid, Ergodic projections for semi-groups of periodic operators, Studia Math. 17 (1958), 189-197.

[3] D.Butković, On the summability of convolution sequences of measures, Glasnik Mat. 13 (1978), 69-74; Correction, ibid. 18 (1983), 391-392.

[4] K.L.Chung, Markov chains with stationary transition probabilities, 2nd ed., Springer-Verlag, New York, 1967.

[5] L.W.Cohen, On the mean ergodic theorem, Ann.of Math. 41 (1940), 505-509.

[6] R.C.Cooke, Infinite matrices and sequence sapces, Macmillan, London, 1950.

[7] F.R.Gantmacher, The theory of matrices, vol.II, Chelsea, New York, 1959.

[8] J.D.Hill, Summability of sequences of 0's and 1's, Ann.of Math. 46 (1945), 556-562.

[9] J.D.Hill, Remarks on the Borel property, Pacific J.Math. 4 (1954), 227-242.

[10] J.G.Kemeny and J.L.Snell, Markov chains and summability methods, Z.Wahrscheinlichkeitstheorie verw. Geb. 18 (1971), 17-33.

[11] J.G.Kemeny, J.L.Snell and A.W.Knapp, Denumerable Markov chains, Van Nostrand, Princeton, 1966.

[12] A.Kolmogoroff, Anfangsgrunde der Theorie der Markoffschen Ketten mit unendlich vielen möglichen Zustanden, Mat.Sbornik (Recueil Math.) 1 (1936), 607-610.

[13] G.G.Lorentz, A contribution to the theory of divergent sequences, Acta Math. 80 (1948), 167-190.

[14] G.G.Lorentz, Direct theorems on methods of summability II, Canad. J.Math. 3 (1951), 236-256.

[15] I.J.Maddox, Elements of functional analysis, Cambridge Univ.Press, Cambridge, 1970.

[16] G.M.Petersen, Almost convergence and uniformly distributed sequences, Quart.J.Math.Oxford 7 (1956), 188-191.

[17] B.E.Rhoades, Some applications of strong regularity to Markov chains and fixed point theorems, Approximation Theory III, Acad.Press 1980, 736-740.

[18] K.Yosida and S.Kakutani, Markoff process with an enumerable infinite number of possible states, Japan J.Math. 16 (1940), 47-55.

p-NUCLEAR OPERATORS AND CYLINDRICAL MEASURES ON TENSOR PRODUCTS OF BANACH SPACES

Neven Elezović

1. Introduction.

We consider some relations between various Radonification theorems for cylindrical measures defined on tensor products of Banach spaces.

An operator is called p-Radonifying, if it maps every cylindrical measure of type p (defined on a Banach space E or on the tensor product of two Banach spaces) into a Radon probability of order p. It is well known (cf. [9]) that, for $1 < p < \infty$, $w : E \to G$ is p-summing if and only if it is p-Radonifying. Similarly, $w_1 \otimes w_2 : E \otimes F \to G \hat{\otimes}_\alpha H$ is $\tilde{p}$-summing if and only if it is p-Radonifying, for $1 < p < \infty$ and if the norm α satisfies certain natural assumptions (see Theorem 4.5.).

Every p-nuclear operator $w : E \to G$ is p-summing and also p-Radonifying, even in the case $0 < p \leqslant 1$. In the present paper we consider p-Radonifying operators of nuclear type (but not of the form $w_1 \otimes w_2$), from $E \otimes F$ into a Banach space which is not necessarily the completion of some normed tensor product $G \otimes_\alpha H$. In the main Theorem 4.6., it will be shown that $(s', r, p')^\sim$-nuclear operators $T : E \otimes F \to G$ are p-Radonifying, if it holds $p > 1$ and $\frac{1}{p} + \frac{1}{\delta(s,p)} \leqslant \frac{1}{r}$, with $\delta(s,p)$ defined as in (4.2).

In Section 2 we recall the definition of $\bar{r}_p$-nuclear operators. These operators appear in the problem of Radonification of F-cylindrical probabilities defined on $E \otimes F$. We establish a connection between $(p,r,s)^\sim$- nuclear operators and operators of the form $W \circ (1 \otimes w)$, where W is $\bar{r}_p$-nuclear and w is p-nuclear operator. In Section 3 we define the image of a cylindrical measure on $E \otimes F$ by linear mappings which are not necessarily of the form $w_1 \otimes w_2$ (but have a factorization throught such operators). The main result, anounced before, is proved in Section 4. In the last section we give some examples of $(s', r, p')^\sim$-nuclear and p-Radonifying operators between ℓ_q-spaces.

2. Some classes of nuclear operators

Throughout this paper E, F, G, H will denote real Banach spaces with duals E', F', G', H'. All operators will be linear. $L(E,G)$ denotes the space of all continuous operators from E into G. For $w \in L(F,H)$, $\tilde{w} : E \otimes F \to E \otimes H$ denotes the operator $1_E \otimes w$, defined by $\tilde{w}(x \otimes y) := x \otimes wy$, $x \in E$, $y \in F$. $\langle x,x' \rangle$ denotes the canonical pairing and $[u,x'] \in F$ the action of an element $u \in E \otimes F$ on vectors in E'.

Let $\{x_j\}$ be a sequence in E. By $N_p(\{x_j\}_j)$, or shortly by $N_p(x_j)$ we denote the number, finite or not

$$N_p(x_j) := \begin{cases} (\sum \|x_j\|^p)^{1/p} & , \quad 0 < p < \infty \\ \sup_j \|x_j\| & , \quad p = \infty \end{cases}$$

and by

$$M_p(x_j) := \sup\{ N_p(\langle x_j, x' \rangle) : \|x'\| \leq 1 \} \quad , \quad 0 < p \leq \infty$$

For $\{v_j\} \subset L(F,G)$ we denote

$$S_p(v_j) := \sup\{ N_p(v_j y) : \|y\| \leq 1 \}$$

Let us recall briefly several classes of nuclear operators which shall be used later. In the sequell, p will satisfy $1 \leq p \leq \infty$, and p' is defined by $\frac{1}{p} + \frac{1}{p'} = 1$.

Operators $w : E \to G$, $W : E \otimes F \to G$, $T : E \otimes F \to G$ which have a representation of the form

$$w = \sum_{j=1}^{\infty} x_j' \otimes z_j \quad , \quad x_j' \in E', \; z_j \in G \tag{2.1}$$

$$W = \sum_{j=1}^{\infty} x_j' \otimes v_j \quad , \quad x_j' \in E', \; v_j \in L(F,G) \tag{2.2}$$

$$T = \sum_{j=1}^{\infty} x_j' \otimes y_j' \otimes z_j \quad , \quad x_j' \in E', \; y_j' \in F', \; z_j \in G \tag{2.3}$$

such that it holds

$$g_p(w) := \inf\{ N_p(x_j') \, M_{p'}(z_j) \} < \infty \tag{2.4}$$

$$\bar{r}_p(W) := \inf\{ N_p(x_j') \, S_{p'}(v_j') \} < \infty \tag{2.5}$$

$$\tilde{n}_p(T) := \inf\{ M_\infty(x_j') \, N_p(y_j') \, M_{p'}(z_j) \} < \infty \tag{2.6}$$

are called *p-nuclear*, $\bar{r}_p$*-nuclear* and $\tilde{p}$*-nuclear*, respectively (the infimum in (2.4)-(2.6) is taken over all representations of the form (2.1) -(2.3)).

If W is $\bar{r}_p$-nuclear, and $u \in E \otimes F$, then $W(u)$ is well defined by

$$W(u) = \sum_{j=1}^{\infty} v_j([u, x_j'])$$

(cf. [1, Theorem 2]), similarly for $\tilde{p}$-nuclear operators.

An important example is due in [1], Theorem 4 : $w \otimes 1_F : E \otimes F \to G \hat{\otimes}_{d_p} F$ is $\bar{r}_p$-nuclear, if w is p-nuclear (see [8] for the definition of the norm d_p, and all others tensor norms used here). Also, $\tilde{p}$-nuclear operators are $\bar{r}_p$-nuclear: denote $\bar{y}_j' := \|y_j'\|^{-1} \cdot y_j'$ and define $v_j : F \to G$ by $v_j y := \langle y, \bar{y}_j' \rangle z_j$. Hence, a $\tilde{p}$-nuclear operator T has a representation $T = \sum \|y_j'\| x_j \otimes v_j$ and $\bar{r}_p$-nuclearity follows easily.

PROPOSITION 2.1. *If* $w : F \to H$ *is p-nuclear and* $W : E \otimes H \to G$ *is* $\bar{r}_p$*-nuclear, then* $W \circ \tilde{w} : E \otimes F \to G$ *is* $\tilde{p}$*-nuclear and*

$$\tilde{n}_p(W \circ \tilde{w}) \leq \bar{r}_p(W)\, g_p(w) \tag{2.7}$$

Proof. Take a representation $w = \sum y_j' \otimes h_j$, $W = \sum x_k' \otimes v_k$. Then, for $x \otimes y \in E \otimes F$ it holds

$$(W \circ \tilde{w})(x \otimes y) = W(x \otimes wy) = \sum_k \langle x, x_k' \rangle\, v_k(wy)$$

$$= \sum_k \langle x, x_k' \rangle\, v_k\Big(\sum_j \langle y, y_j' \rangle\, h_j \Big)$$

$$= \sum_{k,j} \langle x, x_k' \rangle \langle y, y_j' \rangle\, v_k h_j$$

(since this series converges in G). Denote $\bar{x}_k' := \|x_k'\|^{-1} \cdot x_k'$. Hence, $W \circ \tilde{w}$ has a representation

$$W \circ \tilde{w} = \sum_{k,j=1}^{\infty} \bar{x}_k' \otimes \|x_k'\|\, y_j' \otimes v_k h_j$$

for which it holds

$$M_\infty(\{\bar{x}_k'\}_{k,j}) = \sup_{k,j} \|\bar{x}_k'\| = 1$$

$$N_p(\{\|x_k'\|\, y_j\}_{k,j}) = N_p(x_k')\, N_p(y_j') < \infty$$

$$\begin{aligned} M_{p'}(\{v_k h_j\}_{k,j}) &= \sup_{\|z'\|\leq 1} \{ \sum_{k,j} |\langle v_k h_j , z' \rangle|^{p'} \}^{1/p'} \\ &= \sup_{\|z'\|\leq 1} \{ \sum_{k,j} |\langle h_j , v_k' z' \rangle|^{p'} \}^{1/p'} \\ &\leq \sup_{\|z'\|\leq 1} \{ \sum_{k} \| v_k' z' \|^{p'} \cdot \sup_{\|h'\|\leq 1} (\sum_j |\langle h_j , h' \rangle|^{p'}) \}^{1/p'} \\ &= S_{p'}(v_k') \, M_{p'}(h_j) < \infty \end{aligned}$$

Thus, $W \circ \tilde{W}$ is $\tilde{p}$-nuclear, and (2.7) follows.

Q.E.D.

In Section 4 we need slightly general operators than $\tilde{p}$-nuclear. Let $1 \leq p,s \leq \infty$, $0 < r \leq \infty$ be such that it holds $\frac{1}{p} + \frac{1}{r} + \frac{1}{s} \geq 1$. An operator $T : E \otimes F \to G$ is said to be *(p,r,s)*$^{\sim}$*-nuclear*, if it has a representation of the form

$$T = \sum_{j=1}^{\infty} x_j' \otimes y_j' \otimes z_j$$

for which it holds

$$\tilde{n}_{p,r,s}(T) := \inf \{ M_p(x_j') \, N_r(y_j') \, M_s(z_j) \} < \infty$$

(see [2] for additional properties). Hence, $\tilde{p}$-nuclear means $(\infty,p,p')^{\sim}$ -nuclear.

By $\{e_j'\}$, $\{f_j'\}$, $\{g_j\}$ we denote the canonical bases in different ℓ_q-spaces. All of them are defined by the sequence $\{\delta_{ij}\}_{i \in \mathbb{N}}$, different notation helps us to identify a basis with a space. By [2], Theorem 5, $T : E \otimes F \to G$ is $(p,r,s)^{\sim}$-nuclear if and only if it has a factorisation of the form

$$\begin{array}{ccc} E \otimes F & \xrightarrow{\quad T \quad} & G \\ \downarrow{\scriptstyle a_1 \otimes a_2} & & \uparrow{\scriptstyle b} \\ \ell_p \otimes \ell_\infty & \xrightarrow[\quad D \quad]{} & \ell_{s'} \end{array} \qquad (2.8)$$

where a_1 , a_2 , b are continuous, and D is defined by

$$D(u) = \sum_{j=1}^{\infty} d_j \langle u , e_j' \otimes f_j' \rangle g_j \quad , u \in \ell_p \otimes \ell_\infty \tag{2.9}$$

for an element $\{d_j\} \in \ell_r$. Moreover, D is itself $(p,r,s)^{\sim}$-nuclear.

A partial converse of Proposition 2.1. will be more interesting to us:

PROPOSITION 2.2. *Let* $1 \leq p < \infty$. *Each* $(\infty,\frac{p}{2},p')^{\sim}$*-nuclear operator* $T : E \otimes F \to G$ *can be factorized in the form* $W \circ \tilde{w}$, *where* $w : F \to \ell_p$ *is p-nuclear and* $W : E \otimes \ell_p \to G$ *is* $\bar{r}_p$*-nuclear. Further*

$$\bar{r}_p(W)\; g_p(w) \leq \tilde{n}_{\infty,p/2,p'}(T) \tag{2.10}$$

Proof. Let $T = \sum x_j' \otimes y_j' \otimes z_j$ be a representation of T , for which it holds $\tilde{n}_{\infty,p/2,p'}(T) < M_\infty(x_j')\, N_{p/2}(y_j')\, M_{p'}(z_j) + \eta$, $\eta > 0$ being given in advance. Consider the following factorization

$$E \otimes F \xrightarrow{\;1_E \otimes w\;} E \otimes \ell_p \xrightarrow{\;W\;} G$$

where w and W are defined by

$$w = \sum_{j=1}^{\infty} \|y_j'\|^{-1/2} \cdot y_j' \otimes e_j$$

$$W = \sum_{j=1}^{\infty} \|y_j'\|^{1/2} \, x_j' \otimes f_j' \otimes z_j$$

The operator w is p-nuclear:

$$N_p(\|y_j'\|^{-1/2} \cdot y_j') = (\sum \|y_j'\|^{p/2})^{1/p} = (N_{p/2}(y_j))^{1/2}$$

$$M_{p'}(e_j) = \sup \{ N_{p'}(\langle e_j, x' \rangle) : \|x'\| \leq 1 \ , \ x' \in \ell_{p'} \} = 1$$

Define $v_j : F \to G$ by $v_j y := \langle y, f_j' \rangle z_j$. Then

$$N_p(\|y_j'\|^{1/2} x_j') \leq M_\infty(x_j')(N_{p/2}(y_j'))^{1/2}$$

$$S_{p'}(v_j') = \sup_{\|z'\| \leq 1} N_{p'}(\langle z_j, z' \rangle f_j') = M_{p'}(z_j)$$

which shows that W is $\bar{r}_p$-nuclear. By definitions of the norms and inequalities above, we have

$$\bar{r}_p(W)\; g_p(w) \;\leqslant\; N_{p/2}(y_j')\; M_\infty(x_j')\; M_{p'}(z_j)$$

and (2.10) follows.

Q.E.D.

3. Cylindrical measures on $E \otimes F$ and its images.

We recall a definition and some properties of cylindrical measures defined on tensor products of Banach spaces (cf. [5] for details). By $FC(E)$ is denoted the family of all closed subspaces of E of the finite codimension, and by $\pi_N : E \to E/N$, $\pi_{N_2N_1} : E/N_1 \to E/N_2$ $(N_1 \subset N_2)$ the canonical projections. It is obvious that the following diagram commutes:

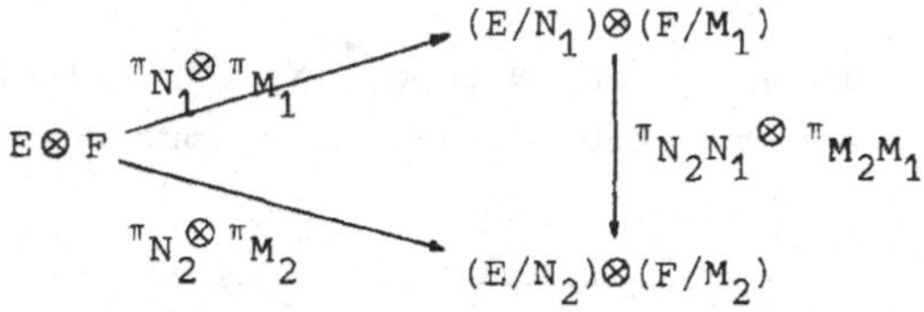

A *cylindrical measure* λ on $E \otimes F$ is a projective system $\{\lambda_{N\otimes M}, \pi_N \otimes \pi_M, N \in FC(E), M \in FC(F)\}$ of Radon probabilities on the finite dimensional spaces $(E/N)\otimes(F/M)$. We say that λ is of *type p*, $p > 0$, if

$$\|\lambda\|_p^* := \sup_{\substack{\|x'\| \leqslant 1 \\ \|y'\| \leqslant 1}} \left\{ \int_{E \otimes F} |\langle u, x' \otimes y' \rangle|^p \, d\lambda(u) \right\}^{1/p} < \infty$$

$M_p^c(E \otimes F)$ denotes the space of all cylindrical measures of type p.

Let α be a norm on $E \otimes F$ (not necessarily a tensor norm) which satisfies the following:

$$|\langle u, x' \otimes y' \rangle| \;\leqslant\; \alpha(u)\, \|x'\|\, \|y'\| \tag{3.1}$$

for all $u \in E \otimes F$, and

$$\psi : E \hat{\otimes}_\alpha F \to L(E', F) \quad \text{is one to one} \tag{3.2}$$

where ψ is the canonical embedding, which exists since (3.1) holds.

If μ is a Radon probability on $E \hat{\otimes}_\alpha F$, then there exists a unique cylindrical measure λ on $E \otimes F$, determined by μ. λ is defined by $\lambda_{N \otimes M} := (\pi_N \hat{\otimes} \pi_M)(\lambda)$, cf. [5]. Conversely, if α satisfies (3.1) and (3.2) and if a cylindrical measure λ on $E \otimes F$ defines a Radon probability μ on $E \hat{\otimes}_\alpha F$, then μ is unique. Of course, a cylindrical measure need not define any probability on $E \hat{\otimes}_\alpha F$. If λ has a (unique) extension to a Radon probability on $E \hat{\otimes}_\alpha F$, we simply say that λ is a Radon probability on $E \hat{\otimes}_\alpha F$.

A Radon probability μ on G is of *order p*, $p > 0$, if

$$\| \mu \|_p := \{ \int_G \| z \|^p \, d\mu(z) \}^{1/p} < \infty$$

An operator $T : E \otimes F \to G$, for which the image $T(\lambda)$ is a Radon probability of order p, for all $\lambda \in M_p^c(E \otimes F)$ is called *p-Radonifying*.

The first difficulty is to define the image $T(\lambda)$. Let us recall a definition of this image for operators of the form $w_1 \otimes w_2 : E \otimes F \to G \otimes H$.

Take $X \in FC(G)$. Then $N := w_1^{-1}(X) \in FC(E)$ and all the operators in the following commutative diagram are continuous:

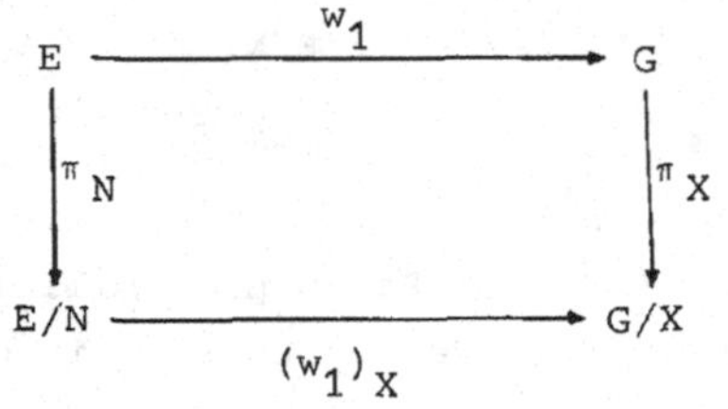

Denote similarly $M := w_2^{-1}(Y) \in FC(F)$ for $Y \in FC(H)$. Then a Radon probability $\lambda_{N \otimes M}$ on the space $(E/N) \otimes (F/M)$ is well defined. Thus, we can define a cylindrical measure $(w_1 \otimes w_2)(\lambda)$ on $G \otimes H$ by

$$(w_1 \otimes w_2)(\lambda)_{X \otimes Y} := ((w_1)_X \otimes (w_2)_Y)(\lambda_{N \otimes M})$$

In several occasions we need the image of a cylindrical measure which is not obtained by "elementary" operators of the form $w_1 \otimes w_2$. Let $T : E \otimes F \to G$ be an operator. The image $T(\lambda)$ of a cylindrical measure λ on $E \otimes F$ cannot be defined in a usual way (to be a cylindrical measure on G). Namely, for $X \in FC(G)$, the space $(E \otimes F)/T^{-1}(X)$ is not of the form $(E/N) \otimes (F/M)$ and $\lambda_{T^{-1}(X)}$ is not defined. However we can define $T(\lambda)$ for a large class of operators, provided this is a Radon probability on G :

The image $T(\lambda)$ is, by definition, a Radon probability $\mu = b(\nu)$ on G , if there exists a factorization of T of the form

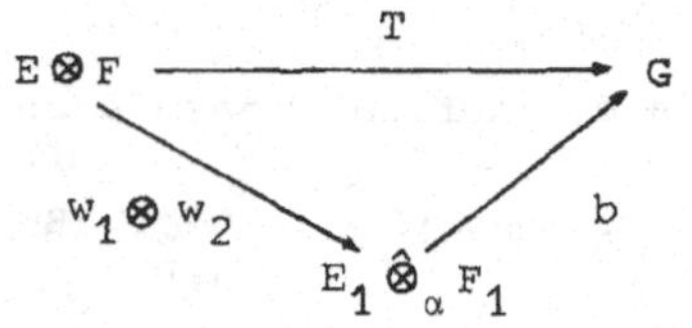

for some Banach spaces E_1 , F_1 and a norm α which satisfies (3.1), (3.2), where w_1 , w_2 , b are continuous and $\nu := (w_1 \otimes w_2)(\lambda)$ is a Radon probability on $E_1 \hat{\otimes}_\alpha F_1$.

4. p-Radonifying operators of nuclear type.

F-cylindrical probability ν on $E \otimes F$ is a projective system $\{ \nu_N , \pi_N \otimes 1_F , N \in FC(E) \}$ of Radon probabilities on $(E/N) \otimes F$ (cf. [7] for details). Each F-cylindrical probability defines a cylindrical measure on $E \otimes F$. Conversely, if a cylindrical measure $\lambda_N := \{ \lambda_{N \otimes M} , \pi_N \otimes \pi_M , M \in FC(F) \}$ is a probability for all $N \in FC(E)$, then a cylindrical measure $\lambda := (\lambda_{N \otimes M})$ defines uniquely a F-cylindrical probability on $E \otimes F$.

F-cylindrical probability ν is said to be of *type (p,F)*, $p > 0$, if

$$\| \nu \|^*_{p,F} := \sup_{\|x'\| \leq 1} \{ \int_{E \otimes F} \| [u,x'] \|^p \, d\nu(u) \}^{1/p} < \infty$$

$M^c_{p,F}(E \otimes F)$ denotes the space of all F-cylindrical probabilities on $E \otimes F$ of type (p,F).

The Radonification problem can be solved in two steps. In the first step, we observe operators $w : F \to H$ such that $(1_E \otimes w)(\lambda) \in M^c_{p,H}(E \otimes H)$, for all $\lambda \in M^c_p(E \otimes F)$. In the second, we must find operators $W : E \otimes H \to G$ such that $W(\nu)$ is a Radon probability on G of order p, for all $\nu \in M^c_{p,H}(E \otimes H)$.

The first step can be easily solved using classical Schwartz´s theorems. Here we give one result which uses p-nuclear operators:

PROPOSITION 4.1. *Let* $\lambda \in M_p^c(E\otimes F)$ *and* $w : F \to H$ *be p-nuclear,* $1 \leq p < \infty$ *Then* $\tilde{w}(\lambda) \in M^c_{p,H}(E\otimes H)$ *and*

$$\|\tilde{w}(\lambda)\|^*_{p,H} \leq g_p(w)\|\lambda\|_p$$

Proof. Let $\lambda_{x'}$ be a cylindrical measure on F defined by

$$\lambda_{x'}\{y\in F : (\langle y,y'_j\rangle)_{1\leq j\leq n} \in B\}$$

$$:= \lambda\{u\in E\otimes F : (\langle u, x'\otimes y'_j\rangle)_{1\leq j\leq n} \in B\}$$

where B is a Borel set in $\mathbb{R}^n$. This measure coincides with the image $\tilde{\lambda}_{x'} := (x'\otimes 1_F)(\lambda)$ on $\mathbb{R}\otimes F$ by the isomorphism $F \simeq \mathbb{R}\otimes F$. Since we have $(x'\otimes 1_H)(1_E\otimes w) = (1_{\mathbb{R}}\otimes w)(x'\otimes 1_F) = x'\otimes w$, it holds

$$[\tilde{w}(\lambda)]_{x'} = (x'\otimes 1_H)(1_E\otimes w)(\lambda)$$

$$= (1_{\mathbb{R}}\otimes w)(x'\otimes 1_F)(\lambda)$$

$$= (1_{\mathbb{R}}\otimes w)(\tilde{\lambda}_{x'}) = w(\lambda_{x'})$$

Further,

$$\|\lambda_{x'}\|^*_p = \sup_{\|y'\|\leq 1}\left\{\int_F |\langle y,y'\rangle|^p \, d\lambda_{x'}(y)\right\}^{1/p}$$

$$= \sup_{\|y'\|\leq 1}\left\{\int_{E\otimes F} |\langle u, x'\otimes y'\rangle|^p \, d\lambda(u)\right\}^{1/p}$$

$$\leq \|\lambda\|^*_p \|x'\|$$

and this shows that $\lambda_{x'}$ is of type p, for all x' in E'. By Schwartz' result ([9, Prop. 2.6.]), $w(\lambda_{x'})$ is a Radon probability on F of order p, and

$$\|w(\lambda_{x'})\|_p \leq g_p(w)\|\lambda_{x'}\|^*_p$$

Thus, $(x'\otimes w)(\lambda)$ is a Radon probability, hence, a forteriori, $(\pi_N\otimes w)(\lambda)$ is a Radon probability on $(E/N)\otimes H$, for all $N \in FC(E)$. Since it holds $[(1_E\otimes w)(\lambda)]_N = (\pi_N\otimes w)(\lambda)$, it follows that $\tilde{w}(\lambda)$ is a H-cylindrical probability on $E\otimes H$ of type

$$\|\tilde{w}(\lambda)\|^*_{p,H} = \sup_{\|x'\|\leq 1} \|\tilde{w}(\lambda)_{x'}\|_p$$

$$= \sup_{\|x'\|\leq 1} \|w(\lambda_{x'})\|_p \leq g_p(w)\,\|\lambda\|^*_p$$

Q.E.D.

The second step can be solved as in [4] and [7] :

THEOREM 4.2. ([4]) *Let* H *be reflexive and* $1 \leq p < \infty$. *If* $W : E \otimes H \to G$ *is* $\bar{r}_p$*-nuclear and* ν *H-cylindrical probability on* $E \otimes H$ *of type* (p,H), *then* $W(\nu)$ *is a Radon probability on* G *of order p and*

$$\|W(\nu)\|_p \leq \bar{r}_p(W)\,\|\nu\|^*_{p,H}$$

COROLLARY 4.3. *Let* H *be reflexive,* $w : F \to H$ *p-nuclear and* $W : E \otimes H \to G$ $\bar{r}_p$*-nuclear,* $1 \leq p < \infty$. *Then* $W \circ \tilde{w}$ *is p-Radonifying.*

Proof. Follows from Proposition 4.1. and Theorem 4.2.

Q.E.D.

COROLLARY 4.4. *An* $(\infty, \frac{p}{2}, p')^{\sim}$*-nuclear operator* $T : E \otimes F \to G$ *is p-Radonifying, for* $1 < p < \infty$.

Proof. By Proposition 2.3., T has the factorization

$$E \otimes F \xrightarrow{\;1_E \otimes w\;} E \otimes \ell_p \xrightarrow{\;W\;} G$$

where w is p-nuclear and W is $\bar{r}_p$-nuclear. Proposition follows from Corollary 4.4., since ℓ_p is reflexive.

Q.E.D.

An operator $T : E \otimes F \to G$ is said to be $\tilde{p}$*-summing* if it exists $C \geq 0$ such that for all $u_1, \ldots, u_n \in E \otimes F$ it holds

$$\{\sum \|T(u_j)\|^p\}^{1/p} \leq C \sup_{\substack{\|x'\|\leq 1 \\ \|y'\|\leq 1}} \{\sum |\langle u_j, x' \otimes y'\rangle|^p\}^{1/p} \qquad (4.1)$$

The infimum of all constants C in (4.1) is denoted by $\tilde{\pi}_p(T)$.

$\tilde{p}$-summing operators are crucial in the Radonification problem:

THEOREM 4.5. ([5]) *Let* $1 < p < \infty$ *and* α *be a norm on* $G \otimes H$. $W := w_1 \otimes w$: $E \otimes F \to G \hat{\otimes}_\alpha H$ *is* $\tilde{p}$*-summing if and only if it is p-Radonifying, and for* $\lambda \in M_p^c(E \otimes F)$ *it holds*

$$\|W(\lambda)\|_p \leqslant \tilde{\pi}_p(W) \|\lambda\|_p^*$$

The main example of p-Radonifying operators of such form is the tensor product $w_1 \otimes w_2$ of two p-summing operators. This operator is $\tilde{p}$-summing whenever the norm α satisfies $\alpha \leqslant /d_p$ or $\alpha \leqslant g_p\backslash$ (cf. [3, Theorem 3.]).

We shall now prove the main result, which is a strenghtening of Corollary 4.4. Denote

$$\delta(s,p) := \begin{cases} p & , \text{ if } 1 \leqslant p < s \leqslant 2 \text{ or } s < p \\ s & , \text{ if } 2 \leqslant p < s \\ (\frac{1}{p} + \frac{1}{s} - \frac{1}{2})^{-1} & , \text{ if } p < 2 < s \end{cases} \tag{4.2}$$

THEOREM 4.6. *Let* $1 < p < \infty$, $T : E \otimes F \to G$ *be* $(s', r, p')^\sim$*-nuclear and*

$$\frac{1}{p} + \frac{1}{\delta(s,p)} \leqslant \frac{1}{r}$$

Then T *is p-Radonifying.*

<u>Proof</u>. Define q by

$$\frac{1}{q} := \frac{1}{r} - \frac{1}{p} \tag{4.4}$$

The sequence $\{d_{rj}\} \in \ell_r$, which defines the operator $D : \ell_{s'} \otimes \ell_\infty \to \ell_r$ in (2.9) can be factorized in a way $d_{rj} = d_{pj} \cdot d_{qj}$, $j \in \mathbb{N}$, where $\{d_{qj}\} \in \ell_q$ and $\{d_{pj}\} \in \ell_p$. Hence, D can be factorized into the product $i \circ (d_q \otimes d_p)$, where d_q (resp. d_p) is the diagonal operator of the multiplication by the element $\{d_{qj}\}$ (resp. $\{d_{pj}\}$), and $i : \ell_p \otimes \ell_p \to \ell_p$ is defined by

$$i(u) := \sum_{j=1}^{\infty} \langle u, e'_j \otimes f'_j \rangle \, g_j$$

This operator is continuous, from $\ell_p \otimes_{g_p\backslash} \ell_p$ into ℓ_p :

$$\|i(u)\| = N_p(\langle u, e'_j \otimes f'_j \rangle) =$$

$$= N_p(\langle \hat{u} e_j', f_j' \rangle)$$

$$\leq N_p(\hat{u} e_j')$$

$$\leq \pi_p(\hat{u})\, M_p(e_j')$$

$$= g_{p\backslash}(u)$$

since

$$M_p(e_j') = \sup \{ N_p(\langle x, e_j' \rangle) : \|x\| \leq 1 \ , \ x \in \ell_p \} = 1$$

and since it holds $\pi_p(\hat{u}) = g_{p\backslash}(u)$ (cf. [8, Théorème 3.4.]), where $\hat{u} : (\ell_p)' \to \ell_p$ is defined by $\hat{u}\, e' := [u, e']$.

Thus, from factorization (2.8) we obtain the new one:

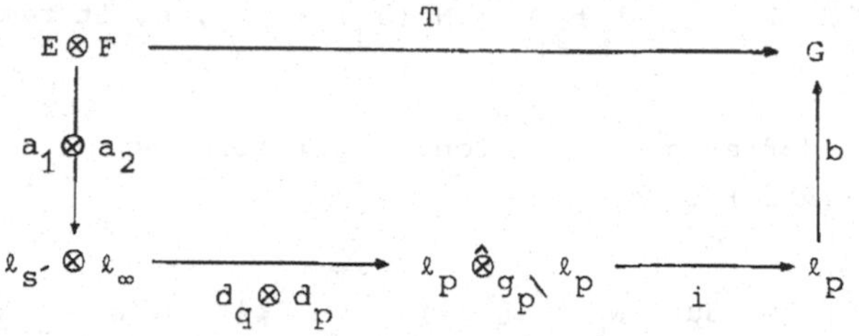

By Garling's result, ([6, Theorem 9]), $d_p : \ell_\infty \to \ell_p$ is p-summing. The same theorem implies that $d : \ell_{s'} \to \ell_p$ is p-summing if and only if $d \in \ell_{\delta(s,p)}$. Since (4.3) and (4.4) imply $q \leq \delta(s,p)$, it follows that $d_q : \ell_{s'} \to \ell_p$ is p-summing. Thus, by [3], Theorem 3, $d_q \otimes d_p : \ell_{s'} \otimes \ell_\infty \to \ell_p \hat{\otimes}_{g_{p\backslash}} \ell_p$ is $\tilde{p}$-summing.

Now, we can apply Theorem 4.5.: $d_q \otimes d_p$ is p-Radonifying. Since $\nu := (a_1 \otimes a_2)(\lambda)$ is a cylindrical measure on $\ell_{s'} \otimes \ell_\infty$ of type p (cf. [5, Proposition 5.1.]), it follows that $(d_q \otimes d_p)(\nu)$ is a Radon probability on $\ell_p \hat{\otimes}_{g_{p\backslash}} \ell_p$ of order p, and the theorem follows.

Q.E.D.

5. Examples: p-Radonifying operators between tensor products of ℓ_q-spaces.

We give some examples of $(s',r,p')^{\sim}$-nuclear operators, which can be useful in studying of random linear operators between sequence spaces.

EXAMPLE 5.1. Suppose $s=1$, $p \geq 2r > 1$, $1 \leq t_1,t_2,q_1,q_2 \leq \infty$, and $\frac{1}{q_1} + \frac{1}{q_2} \leq 1 + \frac{1}{p}$. Then $T : \ell_{t_1} \otimes \ell_{t_2} \to \ell_{q_1} \hat{\otimes}_\varepsilon \ell_{q_2}$, defined by

$$T = \sum_{j=1}^{\infty} d_j \, e_j' \otimes f_j' \otimes (e_j \otimes f_j) \quad , \quad \{d_j\} \in \ell_r \tag{5.1}$$

is p-Radonifying. Here, ε is the least reasonable norm.

First of all we show that T is $(\infty,r,p')^{\sim}$-nuclear. Namely, it holds $M_\infty(e_j') = 1$, $N_r(d_j f_j') = N_r(d_j) < \infty$, so it remains to prove $M_{p'}(e_j \otimes f_j) < \infty$.

Let us define one-dimensional operators $v_j : \ell_{t_2} \to \ell_{q_1} \hat{\otimes}_\varepsilon \ell_{q_2}$ by $v_j := f_j' \otimes (e_j \otimes f_j)$. Then

$$\begin{aligned} S_{p'}(v_j') &= \sup \{ N_{p'}(v_j' u') : \|u'\| \leq 1 , u' \in (\ell_{q_1} \hat{\otimes}_\varepsilon \ell_{q_2})' \} \\ &= \sup \{ N_{p'}(\langle e_j \otimes f_j, u' \rangle) : \|u'\| \leq 1 , u' \in (\ell_{q_1} \hat{\otimes}_\varepsilon \ell_{q_2})' \} \\ &= M_{p'}(e_j \otimes f_j) \end{aligned}$$

On the other hand, [1], Proposition 1.4. gives:

$$\begin{aligned} S_{p'}(v_j') &= \sup \{ \varepsilon(\textstyle\sum v_j y_j) : y_j \in \ell_{t_2} , N_p(y_j) \leq 1 \} \\ &= \sup_{\substack{\|x'\| \leq 1 \\ \|y'\| \leq 1}} \{ |\textstyle\sum \langle y_j, f_j' \rangle \langle e_j, x' \rangle \langle f_j, y' \rangle | : N_p(y_j) \leq 1 \} \\ &\leq \sup_{N_p(y_j) \leq 1} \sup_{\substack{\|x'\| \leq 1 \\ \|y'\| \leq 1}} \{ N_p(\langle y_j, f_j' \rangle) \times \\ &\qquad \times N_{q_1'}(\langle e_j, x' \rangle) \, N_{q_2'}(\langle f_j, y' \rangle) \} \end{aligned}$$

$$= \sup_{N_p(y_j)\leq 1} \{ N_p(<y_j,f'_j>) \; M_{q'_1}(e_j) \; M_{q'_2}(f_j) \} \leq 1$$

By Theorem 4.6., T is p-Radonifying, whenever $p \geq 2r$, since it holds $\delta(1,p) = p$. Observe that T is not of the form $w_1 \otimes w_2$.

Further, the norm ε cannot be always substituted by some stronger p-norm. Take e.q. $q_1 = q_2 = 2$, then T is not $(\infty,r,p')^{\sim}$-nuclear for $r > 1$, if the norm ε on $\ell_2 \otimes \ell_2$ is replaced by any of the norms d_q, $/d_q$, g_q, $g_q\backslash$ $(1 < q < \infty)$, since all of them are equivalent to the norm g_2 on $\ell_2 \otimes \ell_2$, and on $\ell_2 \hat{\otimes}_{g_2} \ell_2$ it holds

$$g_2(\sum v_j y_j) = g_2(\sum <y_j,f'_j> e_j \otimes f_j) = N_2(<y_j,f'_j>)$$

so that

$$S_{p'}(v'_j) \sim \sup \{ N_2(<y_j,f'_j>) : N_p(y_j) \leq 1 \} = \infty$$

EXAMPLE 5.2. Suppose $2 \leq p < s$, $\frac{1}{p} + \frac{1}{s} \leq \frac{1}{r}$, $\frac{1}{q_1} + \frac{1}{q_2} \leq 1 + \frac{1}{p}$. Then $T : \ell_{s'} \otimes \ell_{s'} \to \ell_{q_1} \hat{\otimes}_{\varepsilon} \ell_{q_2}$ defined by (5.1) is $(s',r,p')^{\sim}$-nuclear and p-Radonifying.

In this case we have $\delta(s,p) = s$ and $M_{s'}(e'_j) = 1$.

EXAMPLE 5.3. Let $T : \ell_{s'} \otimes \ell_{s'} \to \ell_p \hat{\otimes}_{d_p} \ell_p$ be defined by

$$T = \sum_{j,k=1}^{\infty} d_k \, s_{kj} \, e'_j \otimes f'_k \otimes (e_j \otimes f_k)$$

where $N_{s'}(d_k) < \infty$, $N_r(\{s_{jk}\}_{j,k}) < \infty$. Then T is $(s',r,p')^{\sim}$-nuclear. Namely, it holds

$$M_{s'}(d_k e'_j) = \sup_{\substack{\|x\| \leq 1 \\ x \in \ell_{s'}}} \{ \sum_j \sum_k |d_k|^{s'} |<x,e'_j>|^{s'} \}^{1/s'}$$

$$= N_{s'}(d_k) \; M_{s'}(e'_j)$$

$$N_r(\{s_{jk} f'_k\}_{j,k}) = \{ \sum_j \sum_k \|s_{jk} f'_k\|^r \}^{1/r} = N_r(s_{jk}) < \infty$$

and for $e_j \otimes f_k \in \ell_p \otimes_{d_p} \ell_p$, $1 < p < \infty$, it holds

$$M_{p'}(\{e_j \otimes f_k\}_{j,k}) = \sup_{\substack{\|u'\| \leq 1 \\ u' \in (\ell_p \otimes_{d_p} \ell_p)'}} \{ \sum_j \sum_k |\langle e_j \otimes f_k, u' \rangle|^{p'} \}^{1/p'}$$

$$= \sup_{\substack{\pi_{p'}(v) \leq 1 \\ v \in \Pi_{p'}(\ell_p, \ell_{p'})}} \{ \sum_j \sum_k |\langle v e_j, f_k \rangle|^{p'} \}^{1/p'}$$

$$= \sup_{\substack{\pi_{p'}(v) \leq 1 \\ v \in \Pi_{p'}(\ell_p, \ell_{p'})}} \{ \sum_j \| v e_j \|^{p'} \}^{1/p'}$$

$$= M_{p'}(e_j) = 1$$

since $(E \otimes_{d_p} F)'$ is isometrically isomorphic to the space $\Pi_{p'}(E, F')$ of all p'-summing operators, cf. [8] , Theorem 3.2.

REFERENCES

[1] N.Elezović: On some classes of nuclear operators defined on tensor products of Banach spaces, to appear in Glasnik Mat.

[2] N.Elezović: $(p,r,s)^\sim$-nuclear and $(p,r,s)^\sim$-compact operators defined on tensor products of Banach spaces, preprint

[3] N.Elezović: $\tilde{p}$-summing operators defined on tensor products of Banach spaces, preprint

[4] N.Elezović: Radonification theorem for F-cylindrical probabilities, to appear in Hokkaido Math. J.

[5] N.Elezović: Radonification problem for cylindrical measures on tensor products of Banach spaces, Publ RIMS Kyoto Univ, 22, 2 (1986)

[6] D.J.H.Garling: Diagonal mappings between sequence spaces, Studia Math. 51 (1974) 129-138

[7] B.Maurey: Rappels sur les operateurs sommants et radonifiants, Sem. Maurey-Schwartz 1973/74, Exposé I+II, 9+10 p.

[8] P.Saphar: Produits tensoriels déspaces de Banach et classes d'applications lineaires, Studia Math. 38 (1970) 71-100

[9] L.Schwartz: Applications p-sommantes et p-radonifiantes, Sem. Maurey-Schwartz 1972/73, Exposé III, 12 p.

Elektrotehnički fakultet

Unska 3, P.B. 170

41001 Zagreb, Yugoslavia

1090: Differential Geometry of Submanifolds. Proceedings, 4. Edited by K. Kenmotsu. VI, 132 pages. 1984.

1091: Multifunctions and Integrands. Proceedings, 1983. Edited 3. Salinetti. V, 234 pages. 1984.

1092: Complete Intersections. Seminar, 1983. Edited by S. Greco R. Strano. VII, 299 pages. 1984.

1093: A. Prestel, Lectures on Formally Real Fields. XI, 125 pages. 4.

1094: Analyse Complexe. Proceedings, 1983. Edité par E. Amar, 3ay et Nguyen Thanh Van. IX, 184 pages. 1984.

1095: Stochastic Analysis and Applications. Proceedings, 1983. ted by A. Truman and D. Williams. V, 199 pages. 1984.

1096: Théorie du Potentiel. Proceedings, 1983. Edité par G. kobodzki et D. Pinchon. IX, 601 pages. 1984.

1097: R.M. Dudley, H. Kunita, F. Ledrappier, École d'Éte de babilités de Saint-Flour XII – 1982. Edité par P.L. Hennequin. X, 3 pages. 1984.

1098: Groups – Korea 1983. Proceedings. Edited by A.C. Kim B.H. Neumann. VII, 183 pages. 1984.

1099: C.M. Ringel, Tame Algebras and Integral Quadratic Forms. , 376 pages. 1984.

1100: V. Ivrii, Precise Spectral Asymptotics for Elliptic Operators ing in Fiberings over Manifolds with Boundary. V, 237 pages. 4.

1101: V. Cossart, J. Giraud, U. Orbanz, Resolution of Surface gularities. Seminar. VII, 132 pages. 1984.

1102: A. Verona, Stratified Mappings – Structure and Triangu-lity. IX, 160 pages. 1984.

1103: Models and Sets. Proceedings, Logic Colloquium, 1983, t I. Edited by G.H. Müller and M.M. Richter. VIII, 484 pages. 1984.

1104: Computation and Proof Theory. Proceedings, Logic loquium, 1983, Part II. Edited by M.M. Richter, E. Börger, W. Ober-elp, B. Schinzel and W. Thomas. VIII, 475 pages. 1984.

1105: Rational Approximation and Interpolation. Proceedings, 3. Edited by P.R. Graves-Morris, E.B. Saff and R.S. Varga. 528 pages. 1984.

1106: C.T. Chong, Techniques of Admissible Recursion Theory. 214 pages. 1984.

1107: Nonlinear Analysis and Optimization. Proceedings, 1982. ted by C. Vinti. V, 224 pages. 1984.

1108: Global Analysis – Studies and Applications I. Edited by .G. Borisovich and Yu.E. Gliklikh. V, 301 pages. 1984.

1109: Stochastic Aspects of Classical and Quantum Systems. oceedings, 1983. Edited by S. Albeverio, P. Combe and M. Sirugue-llin. IX, 227 pages. 1985.

1110: R. Jajte, Strong Limit Theorems in Non-Commutative obability. VI, 152 pages. 1985.

1111: Arbeitstagung Bonn 1984. Proceedings. Edited by F. Hirze-uch, J. Schwermer and S. Suter. V, 481 pages. 1985.

1112: Products of Conjugacy Classes in Groups. Edited by Arad and M. Herzog. V, 244 pages. 1985.

1113: P. Antosik, C. Swartz, Matrix Methods in Analysis. IV, pages. 1985.

1114: Zahlentheoretische Analysis. Seminar. Herausgegeben n E. Hlawka. V, 157 Seiten. 1985.

1115: J. Moulin Ollagnier, Ergodic Theory and Statistical Mechan-s. VI, 147 pages. 1985.

1116: S. Stolz, Hochzusammenhängende Mannigfaltigkeiten d ihre Ränder. XXIII, 134 Seiten. 1985.

Vol. 1117: D.J. Aldous, J.A. Ibragimov, J. Jacod, Ecole d'Été de Probabilités de Saint-Flour XIII – 1983. Édité par P.L. Hennequin. IX, 409 pages. 1985.

Vol. 1118: Grossissements de filtrations: exemples et applications. Seminaire, 1982/83. Edité par Th. Jeulin et M. Yor. V, 315 pages. 1985.

Vol. 1119: Recent Mathematical Methods in Dynamic Programming. Proceedings, 1984. Edited by I. Capuzzo Dolcetta, W.H. Fleming and T. Zolezzi. VI, 202 pages. 1985.

Vol. 1120: K. Jarosz, Perturbations of Banach Algebras. V, 118 pages. 1985.

Vol. 1121: Singularities and Constructive Methods for Their Treatment. Proceedings, 1983. Edited by P. Grisvard, W. Wendland and J.R. Whiteman. IX, 346 pages. 1985.

Vol. 1122: Number Theory. Proceedings, 1984. Edited by K. Alladi. VII, 217 pages. 1985.

Vol. 1123: Séminaire de Probabilités XIX 1983/84. Proceedings. Edité par J. Azéma et M. Yor. IV, 504 pages. 1985.

Vol. 1124: Algebraic Geometry, Sitges (Barcelona) 1983. Proceed-ings. Edited by E. Casas-Alvero, G.E. Welters and S. Xambó-Descamps. XI, 416 pages. 1985.

Vol. 1125: Dynamical Systems and Bifurcations. Proceedings, 1984. Edited by B.L.J. Braaksma, H.W. Broer and F. Takens. V, 129 pages. 1985.

Vol. 1126: Algebraic and Geometric Topology. Proceedings, 1983. Edited by A. Ranicki, N. Levitt and F. Quinn. V, 423 pages. 1985.

Vol. 1127: Numerical Methods in Fluid Dynamics. Seminar. Edited by F. Brezzi, VII, 333 pages. 1985.

Vol. 1128: J. Elschner, Singular Ordinary Differential Operators and Pseudodifferential Equations. 200 pages. 1985.

Vol. 1129: Numerical Analysis, Lancaster 1984. Proceedings. Edited by P.R. Turner. XIV, 179 pages. 1985.

Vol. 1130: Methods in Mathematical Logic. Proceedings, 1983. Edited by C.A. Di Prisco. VII, 407 pages. 1985.

Vol. 1131: K. Sundaresan, S. Swaminathan, Geometry and Nonlinear Analysis in Banach Spaces. III, 116 pages. 1985.

Vol. 1132: Operator Algebras and their Connections with Topology and Ergodic Theory. Proceedings, 1983. Edited by H. Araki, C.C. Moore, Ş. Strătilă and C. Voiculescu. VI, 594 pages. 1985.

Vol. 1133: K.C. Kiwiel, Methods of Descent for Nondifferentiable Opti-mization. VI, 362 pages. 1985.

Vol. 1134: G.P. Galdi, S. Rionero, Weighted Energy Methods in Fluid Dynamics and Elasticity. VII, 126 pages. 1985.

Vol. 1135: Number Theory, New York 1983–84. Seminar. Edited by D.V. Chudnovsky, G.V. Chudnovsky, H. Cohn and M.B. Nathanson. V, 283 pages. 1985.

Vol. 1136: Quantum Probability and Applications II. Proceedings, 1984. Edited by L. Accardi and W. von Waldenfels. VI, 534 pages. 1985.

Vol. 1137: Xiao G., Surfaces fibrées en courbes de genre deux. IX, 103 pages. 1985.

Vol. 1138: A. Ocneanu, Actions of Discrete Amenable Groups on von Neumann Algebras. V, 115 pages. 1985.

Vol. 1139: Differential Geometric Methods in Mathematical Physics. Proceedings, 1983. Edited by H. D. Doebner and J. D. Hennig. VI, 337 pages. 1985.

Vol. 1140: S. Donkin, Rational Representations of Algebraic Groups. VII, 254 pages. 1985.

Vol. 1141: Recursion Theory Week. Proceedings, 1984. Edited by H.-D. Ebbinghaus, G.H. Müller and G.E. Sacks. IX, 418 pages. 1985.

Vol. 1142: Orders and their Applications. Proceedings, 1984. Edited by I. Reiner and K. W. Roggenkamp. X, 306 pages. 1985.

Vol. 1143: A. Krieg, Modular Forms on Half-Spaces of Quaternions. XIII, 203 pages. 1985.

Vol. 1144: Knot Theory and Manifolds. Proceedings, 1983. Edited by D. Rolfsen. V, 163 pages. 1985.

Vol. 1145: G. Winkler, Choquet Order and Simplices. VI, 143 pages. 1985.

Vol. 1146: Séminaire d'Algèbre Paul Dubreil et Marie-Paule Malliavin. Proceedings, 1983–1984. Edité par M.-P. Malliavin. IV, 420 pages. 1985.

Vol. 1147: M. Wschebor, Surfaces Aléatoires. VII, 111 pages. 1985.

Vol. 1148: Mark A. Kon, Probability Distributions in Quantum Statistical Mechanics. V, 121 pages. 1985.

Vol. 1149: Universal Algebra and Lattice Theory. Proceedings, 1984. Edited by S. D. Comer. VI, 282 pages. 1985.

Vol. 1150: B. Kawohl, Rearrangements and Convexity of Level Sets in PDE. V, 136 pages. 1985.

Vol 1151: Ordinary and Partial Differential Equations. Proceedings, 1984. Edited by B.D. Sleeman and R.J. Jarvis. XIV, 357 pages. 1985.

Vol. 1152: H. Widom, Asymptotic Expansions for Pseudodifferential Operators on Bounded Domains. V, 150 pages. 1985.

Vol. 1153: Probability in Banach Spaces V. Proceedings, 1984. Edited by A. Beck, R. Dudley, M. Hahn, J. Kuelbs and M. Marcus. VI, 457 pages. 1985.

Vol. 1154: D.S. Naidu, A.K. Rao, Singular Pertubation Analysis of Discrete Control Systems. IX, 195 pages. 1985.

Vol. 1155: Stability Problems for Stochastic Models. Proceedings, 1984. Edited by V.V. Kalashnikov and V.M. Zolotarev. VI, 447 pages. 1985.

Vol. 1156: Global Differential Geometry and Global Analysis 1984. Proceedings, 1984. Edited by D. Ferus, R.B. Gardner, S. Helgason and U. Simon. V, 339 pages. 1985.

Vol. 1157: H. Levine, Classifying Immersions into $\mathbb{R}^4$ over Stable Maps of 3-Manifolds into $\mathbb{R}^2$. V, 163 pages. 1985.

Vol. 1158: Stochastic Processes – Mathematics and Physics. Proceedings, 1984. Edited by S. Albeverio, Ph. Blanchard and L. Streit. VI, 230 pages. 1986.

Vol. 1159: Schrödinger Operators, Como 1984. Seminar. Edited by S. Graffi. VIII, 272 pages. 1986.

Vol. 1160: J.-C. van der Meer, The Hamiltonian Hopf Bifurcation. VI, 115 pages. 1985.

Vol. 1161: Harmonic Mappings and Minimal Immersions, Montecatini 1984. Seminar. Edited by E. Giusti. VII, 285 pages. 1985.

Vol. 1162: S.J.L. van Eijndhoven, J. de Graaf, Trajectory Spaces, Generalized Functions and Unbounded Operators. IV, 272 pages. 1985.

Vol. 1163: Iteration Theory and its Functional Equations. Proceedings, 1984. Edited by R. Liedl, L. Reich and Gy. Targonski. VIII, 231 pages. 1985.

Vol. 1164: M. Meschiari, J.H. Rawnsley, S. Salamon, Geometry Seminar "Luigi Bianchi" II – 1984. Edited by E. Vesentini. VI, 224 pages. 1985.

Vol. 1165: Seminar on Deformations. Proceedings, 1982/84. Edited by J. Ławrynowicz. IX, 331 pages. 1985.

Vol. 1166: Banach Spaces. Proceedings, 1984. Edited by N. Kalton and E. Saab. VI, 199 pages. 1985.

Vol. 1167: Geometry and Topology. Proceedings, 1983–84. Edited by J. Alexander and J. Harer. VI, 292 pages. 1985.

Vol. 1168: S.S. Agaian, Hadamard Matrices and their Applications. III, 227 pages. 1985.

Vol. 1169: W.A. Light, E.W. Cheney, Approximation Theory in Tensor Product Spaces. VII, 157 pages. 1985.

Vol. 1170: B.S. Thomson, Real Functions. VII, 229 pages. 1985.

Vol. 1171: Polynômes Orthogonaux et Applications. Proceedings, 1984. Edité par C. Brezinski, A. Draux, A.P. Magnus, P. Maroni et A. Ronveaux. XXXVII, 584 pages. 1985.

Vol. 1172: Algebraic Topology, Göttingen 1984. Proceedings. Edited by L. Smith. VI, 209 pages. 1985.

Vol. 1173: H. Delfs, M. Knebusch, Locally Semialgebraic Spaces. X 329 pages. 1985.

Vol. 1174: Categories in Continuum Physics, Buffalo 1982. Semin Edited by F.W. Lawvere and S.H. Schanuel. V, 126 pages. 1986.

Vol. 1175: K. Mathiak, Valuations of Skew Fields and Project Hjelmslev Spaces. VII, 116pages. 1986.

Vol. 1176: R.R. Bruner, J.P. May, J.E. McClure, M. Steinberg H_∞ Ring Spectra and their Applications. VII, 388 pages. 1986.

Vol. 1177: Representation Theory I. Finite Dimensional Algebr Proceedings, 1984. Edited by V. Dlab, P. Gabriel and G. Michler. 340 pages. 1986.

Vol. 1178: Representation Theory II. Groups and Orders. Proc dings, 1984. Edited by V. Dlab, P. Gabriel and G. Michler. XV, 3 pages. 1986.

Vol. 1179: Shi J.-Y. The Kazhdan-Lusztig Cells in Certain Affine We Groups. X, 307 pages. 1986.

Vol. 1180: R. Carmona, H. Kesten, J.B. Walsh, École d'Été Probabilités de Saint-Flour XIV – 1984. Édité par P.L. Hennequin. 438 pages. 1986.

Vol. 1181: Buildings and the Geometry of Diagrams, Como 198 Seminar. Edited by L. Rosati. VII, 277 pages. 1986.

Vol. 1182: S. Shelah, Around Classification Theory of Models. VII, 2 pages. 1986.

Vol. 1183: Algebra, Algebraic Topology and their Interactions. Proc dings, 1983. Edited by J.-E. Roos. XI, 396 pages. 1986.

Vol. 1184: W. Arendt, A. Grabosch, G. Greiner, U. Groh, H.P. Lo U. Moustakas, R. Nagel, F. Neubrander, U. Schlotterbeck, O parameter Semigroups of Positive Operators. Edited by R. Nag X, 460 pages. 1986.

Vol. 1185: Group Theory, Beijing 1984. Proceedings. Edited by Tu H.F. V, 403 pages. 1986.

Vol. 1186: Lyapunov Exponents. Proceedings, 1984. Edited by Arnold and V. Wihstutz. VI, 374 pages. 1986.

Vol. 1187: Y. Diers, Categories of Boolean Sheaves of Sim Algebras. VI, 168 pages. 1986.

Vol. 1188: Fonctions de Plusieurs Variables Complexes V. Sémina 1979–85. Edité par François Norguet. VI, 306 pages. 1986.

Vol. 1189: J. Lukeš, J. Malý, L. Zajíček, Fine Topology Methods in R Analysis and Potential Theory. X, 472 pages. 1986.

Vol. 1190: Optimization and Related Fields. Proceedings, 19 Edited by R. Conti, E. De Giorgi and F. Giannessi. VIII, 419 pag 1986.

Vol. 1191: A.R. Its, V.Yu. Novokshenov, The Isomonodromic Def mation Method in the Theory of Painlevé Equations. IV, 313 pag 1986.

Vol. 1192: Equadiff 6. Proceedings, 1985. Edited by J. Vosmansky a M. Zlámal. XXIII, 404 pages. 1986.

Vol. 1193: Geometrical and Statistical Aspects of Probability Banach Spaces. Proceedings, 1985. Edited by X. Femique, Heinkel, M.B. Marcus and P.A. Meyer. IV, 128 pages. 1986.

Vol. 1194: Complex Analysis and Algebraic Geometry. Proceedin 1985. Edited by H. Grauert. VI, 235 pages. 1986.

Vol. 1195: J.M. Barbosa, A.G. Colares, Minimal Surfaces in $\mathbb{R}^3$. X, pages. 1986.

Vol. 1196: E. Casas-Alvero, S. Xambó-Descamps, The Enumera Theory of Conics after Halphen. IX, 130 pages. 1986.

Vol. 1197: Ring Theory. Proceedings, 1985. Edited by F.M.J. Oystaeyen. V, 231 pages. 1986.

Vol. 1198: Séminaire d'Analyse, P. Lelong – P. Dolbeault – H. Sko Seminar 1983/84. X, 260 pages. 1986.

Vol. 1199: Analytic Theory of Continued Fractions II. Proceedin 1985. Edited by W.J. Thron. VI, 299 pages. 1986.

Vol. 1200: V.D. Milman, G. Schechtman, Asymptotic Theory of Fi Dimensional Normed Spaces. With an Appendix by M. Gromov. V 156 pages. 1986.